THE HISTORY OF LARGE FEDERAL DAMS: PLANNING, DESIGN, AND CONSTRUCTION IN THE ERA OF BIG DAMS

David P. Billington
Donald C. Jackson
Martin V. Melosi

U.S. Department of the Interior
Bureau of Reclamation
Denver Colorado
2005

INTRODUCTION

The history of federal involvement in dam construction goes back at least to the 1820s, when the U.S. Army Corps of Engineers built wing dams to improve navigation on the Ohio River. The work expanded after the Civil War, when Congress authorized the Corps to build storage dams on the upper Mississippi River and regulatory dams to aid navigation on the Ohio River. In 1902, when Congress established the Bureau of Reclamation (then called the "Reclamation Service"), the role of the federal government increased dramatically. Subsequently, large Bureau of Reclamation dams dotted the Western landscape.

Together, Reclamation and the Corps have built the vast majority of major federal dams in the United States. These dams serve a wide variety of purposes. Historically, Bureau of Reclamation dams primarily served water storage and delivery requirements, while U.S. Army Corps of Engineers dams supported navigation and flood control. For both agencies, hydropower production has become an important secondary function.

This history explores the story of federal contributions to dam planning, design, and construction by carefully selecting those dams and river systems that seem particularly critical to the story. Written by three distinguished historians, the history will interest engineers, historians, cultural resource planners, water resource planners and others interested in the challenges facing dam builders. At the same time, the history also addresses some of the negative environmental consequences of dam-building, a series of problems that today both Reclamation and the U.S. Army Corps of Engineers seek to resolve.

While Reclamation and the U.S. Army Corps of Engineers funded this history, we gratefully acknowledge the work of the National Park Service, which managed the project. It may be possible that some federal dams warrant inclusion in the National Historic Landmarks program, which the National Park Service administers. The appendices to this book include material that will enable cultural resource managers to make informed decisions about the historic value of particular dams.

We hope that you find this study as useful and informative as we do.

John W. Keys III
Commissioner
Bureau of Reclamation

Carl A. Strock
Lieutenant General, US Army
Chief of Engineers

PREFACE

The concept for this study emerged in discussions between the undersigned in the early 1990s. The U.S. Army Corps of Engineers and Reclamation agreed to fund the project with the costs equally divided. The Tennessee Valley Authority decided not to participate.

Eventually, the Park Service awarded a contract to the Public Works Historical Society of the American Public Works Association. Dr. Howard Rosen, a distinguished engineering historian, then head of the Society, and later at the School of Engineering at the University of Wisconsin, worked with us to assemble the team of historians necessary to ensure a quality product. David P. Billington, Gordon Y. S. Wu Professor of Engineering at Princeton University (and a distinguished engineering historian) became the principal investigator. Joining him were Professor Donald C. Jackson of Lafayette College, Professor Martin V. Melosi, Distinguished University Professor of History at the University of Houston; and Ann Emmons of History Research Associates, in Missoula developed the material relating to evaluation guidelines and dam nominations for the National Historic Landmarks program. Acting as a peer reviewer was Donald J. Pisani, who holds the Merrick Chair of Western American History at the University of Oklahoma.

Through a period of years, we met with Robie Lange, Don Pisani, and the team of authors to review drafts and discuss progress. The objectives of the study were twofold: a history of federal dam development, concentrating on key projects and river systems, and the drafting of documents to assist cultural resource managers and others interested in nominating dams to the National Historic Landmarks program (see appendices). The history is organized into chapters which sometimes include both the Corps and Reclamation, but each chapter is devoted mostly to the work of one or the other of the agencies. While the book is very much a collaborative effort, in general each author had the primary responsibility for specific chapters: Martin V. Melosi for chapters 1 and 9; David P. Billington for chapters 5, 6, and 8; and Donald Jackson for chapters 4 and 7. Professors Billington and Jackson shared responsibility for chapters 2 and 3.

We acknowledge the support that the leadership in each agency has given this project over the years. We also wish to thank the countless number of Corps and Reclamation rangers, cultural resource managers, dam operators, and others who have contributed their time and invaluable knowledge to the study.

Many reviewers at Reclamation, the Corps, and the National Park Service reviewed this manuscript as it was edited for publication, and their comments have been most helpful. Most editorial work was done by Brit Storey at Reclamation, with the particular assistance of Andrew Gahan and David Muñoz

Last, it is worth noting that the manuscript for this book was completed before the World Commission on Dams studies on Grand Coulee were available for review.

Brit Allan Storey, Ph.D.	Martin A. Reuss, Ph.D.
Senior Historian	Senior Historian
Bureau of Reclamation	Water Resources
	U.S. Army Corps of Engineers

ACKNOWLEDGMENTS

This work benefitted greatly from the continual assistance of Martin Reuss, Senior Historian of the U.S. Army Corps of Engineers and Brit Storey, Senior Historian of the Bureau of Reclamation. The provided us with both documents and advice, which were essential to the project. We also appreciate the kind assistance from the personnel of the U.S. Army Corps of Engineers and the Bureau of Reclamation at various dam sites. Robie Lange of the National Historic Landmarks Survey, National Park Service, provided essential guidance on the organization of the writing and was a fine tour leader on several dam study trips.

J. Wayman Williams helped us to organize the work and produced various versions of the manuscript. He skillfully brought together the work of the three authors. Donald Pisani of the University of Oklahoma served as a consultant at a crucial period during which time he did much needed research, and most importantly helped us reorganize early drafts. He also read a complete draft and made valuable suggestions. Ann Emmons of Historical Research Associates, Inc., led a parallel effort to prepare guidelines for nominating dams as national historic landmarks and completed five nominations supported by much of the material presented in this manuscript. David P. Billington, Jr., spent many hours in the National Archives ferreting out materials that enriched the manuscript greatly.

This complex project was initially organized by Howard Rosen, Ph.D., then with the American Public Works Association (APWA), which administered the project. William J. Bertera, Dale Crandall, and Dennis H. Ross, P.E. in turn ably administered the contract for that association. Connie Hartline, of APWA, provided final editing of the manual to ensure consistency of layout and format.

Finally, the authors would like to thank the hundreds of individuals whom we have contacted over the course of this project that have provided commentary and insight into the history and importance of the various dams built by the federal government. Without their cooperation and assistance, this project could have been even more monumental than some of the dams described herein.

David P. Billington, Princeton University
Donald C. Jackson, Lafayette College
Martin V. Melosi, University of Houston

TABLE OF CONTENTS

INTRODUCTION iii

PREFACE v

ACKNOWLEDGMENTS vii

TABLE OF CONTENTS ix

LIST OF FIGURES xvii

CHAPTER 1: "IMPROVING" RIVERS IN AMERICA: FROM THE
REVOLUTION TO THE PROGRESSIVE ERA - RIVERS IN
EARLY AMERICA 1
    Rivers as Resource: The American Watershed System 1
    The Rise of an Industrializing Nation 2
    The Origins of Federal Water Resource Policy 3
WATER LAW AND THE USE OF RIVERS 5
    Mills and Dams in the Early Industrial Era 5
    Water Law in the West 7
    The Western Setting 8
    The California Doctrine: 1851-1886 8
THE U.S. ARMY CORPS OF ENGINEERS IN THE
NINETEENTH CENTURY 11
    The Corps and the French Engineering Tradition 11
    Commerce, Navigation and the "Steamboat Case" 13
    French Tradition versus Frontier Techniques 14
    Navigation and the Beginning of River Dams: 1824-1865 15
    Postwar Navigation and the Ohio River: 1866-1885 17
    The Upper Mississippi and the Headwater Dams:1866-1899 18
WATER IN THE WEST: ORIGINS OF THE RECLAMATION
SERVICE 21
    The West Before the Nineteenth Century 21
    Water and Mormon Migration 22
    California Water Development 22
    The Exploits of John Wesley Powell 23
    The Sentimental and Practical during the 1890s 25
    The Chittenden Survey of 1897 26
    Newell, Roosevelt and the Move to Reclamation 27

PROGRESSIVISM AND WATER RESOURCES 31
Progressivism, Conservation, and Efficiency 31
Multipurpose Stirrings 31
The Hydroelectric Challenge 36

CHAPTER 2: THEORIES AND COMPETING VISIONS FOR CONCRETE DAMS 49
PRE-TWENTIETH CENTURY THEORIES: THE MASSIVE TRADITION 49
Massive or Structural 49
Gravity Dam Design Theory 50
European Origins of the "Profile of Equal Resistance" 52
The Middle Third 54
A Limit to Gravity Design Innovation 54
PRE-TWENTIETH CENTURY THEORIES - THE STRUCTURAL TRADITION 58
Gravity Dams vs. Structural Dams 58
Arch Dam Theory: European Origins and the Cylinder Formula 59
American Arch Dams 60
Constant Angle Designs 61
American and European Dam Design Practice 61
THE STRUCTURAL TRADITION AND THE RATIONAL DESIGN OF CONCRETE STRUCTURES 62
The Beginning of Rational Design 62
Form and Mass in Structure 63
Arch and Cantilever Behavior in Dams 64
The Constant-Angle Arch Dam 68
Noetzli and the Curved Dams 70
The Trial Load Method 73
The Stevenson Creek Test Dam 75
Buttress Dams 79
Costs in Form and Mass 81
Concrete Form and Masonry Mass 82

CHAPTER 3: EARLY MULTIPURPOSE DAMS: ROOSEVELT AND THE RECLAMATION SERVICE, WILSON AND THE U.S. ARMY CORPSOF ENGINEERS 89
FROM SINGLE TO MULTIPURPOSE DAMS 89
Roosevelt and Wilson Dams 89
YEARS OF TURMOIL: FROM RECLAMATION SERVICE TO BUREAU 92
Newell, Davis and Wisner 92
The Salt River Project 97
Roosevelt Dam 98

  Crises and Rebirth 107
  Elwood Mead and Reclamation 107
 YEARS OF INDECISION: FROM NAVIGATION TO POWER 110
  Floods and Politics 110
  Muscle Shoals: The Battle for Control of Hydropower 111
  Hugh Cooper, McCall Ferry, and Keokuk 114
  From Muscle Shoals to Wilson Dam 117
  From the Tennessee River to House Document No. 308 119
  The 308 Reports 121

CHAPTER 4: THE BOULDER CANYON PROJECT, WATER
 DEVELOPMENT IN THE COLORADO RIVER BASIN, AND
 HOOVER DAM 129
 THE COLORADO RIVER: IRRIGATION AND FLOOD 129
  The River 129
  The Bureau of Reclamation and the West 131
  Early Developments and the Imperial Valley 135
 THE FEDERAL INITIATIVES 142
  The Fall/Davis Report 142
  The Design of Hoover Dam 145
  Selection of the Black Canyon Site 152
  Congressman Phil Swing 154
  The Colorado River Compact 157
 CALIFORNIA AND POWER 160
  Municipal Demand: Los Angeles and the Metropolitan
   Water District of Southern California 160
  The Politics of Hydroelectric Power and Approval of
   the Boulder Canyon Project 165
 HOOVER DAM 171
  Construction of the Dam 171
  The Legacy of Hoover Dam 179
 GLEN CANYON DAM 182
  The Upper Basin 182
  Design of Glen Canyon Dam and Beyond 184

CHAPTER 5: BONNEVILLE AND GRAND COULEE DAMS AND
 THE COLUMBIA RIVER CONTROL PLAN 191
 EXPLORING A CHANGING RIVER 191
 BONNEVILLE DAM: EMPLOYMENT VERSUS MARKETS 194
  From Survey to Construction 194
  From Construction to Dedication 196
  From Dedication to Debate 203

BONNEVILLE AND GRAND COULEE: CONFLICT OF POWER ... 203
   The Columbia 308 Report...203
   The Grande Coulee Stimulus to Bonneville..................................208
GRAND COULEE: COLUMBIAN COLOSSUS 210
   The Design Questions 210
   Raising the Dam 212
   Completing the Dam 218
THE BONNEVILLE POWER ADMINISTRATION AND THE
CONTROL PLAN..222
   Power and the Restudy of the River Basin...................................222
   The Flood and the Control Plan..223
   Power and Salmon 226

CHAPTER 6: EARTH DAMS ON THE MISSOURI RIVER:
FORT PECK AND GARRISON DAMS AND
THE PICK-SLOAN PLAN 235
THE BASIN PLAN..235
   From Navigation to Power..235
EARTHFILL DAMS AND FORT PECK 239
   The Birth of Soil Mechanics...239
   Construction of Hydraulic-Fill Dams...241
   Fort Peck: Skilled Generalship and Ultra Modern Mechanization ... 244
   The Spillway and the Castle 252
   Slide and Recovery 256
   Completion and Controversy 265
THE PICK-SLOAN PLAN...269
   The Pick Plan 269
   Congressional Conflict and the Creeping Commerce Clause 272
   The Sloan Report 273
   Pick-Sloan and the Valley Authority..275
GARRISON DAM 278
   The Garrison Design Controversies..278
   Garrison Dam Design and Construction 1948-1950 280
   Garrison Dam Completion 1951-1955 283
   Spillway Designs...287
PICK'S PLAN AND THE MISSOURI BASIN 288
   From Garrison to Big Bend 288

CHAPTER 7: THE CENTRAL VALLEY PROJECT: SHASTA
AND FRIANT DAMS 301
WATER ISSUES IN CALIFORNIA..301
   Agriculture and Topography of the Central Valley......................301
   "Move the Rain": The Central Valley Project 304

    Miller & Lux and Early Agriculture 306
    The Board of Commissioners' Survey 308
    Miller & Lux and Riparian Rights 310
    The Federal Hiatus: 1873-1934 312
    The Early Reclamation Service in California 314
    The Orland Project 315
    The Iron Canyon Project and the Sacramento Valley 316
    Rice and Drought 319
STATE PLANS FOR CALIFORNIA 320
    The "Marshall Plan" and the Evolution of a State Water Plan 320
    The Early Studies of the State Engineer 322
    The "Water Surge" and Riparian Rights 323
    The 1927 State Engineer's Reports and Kennett Dam 324
    The Federal/State Interaction 326
    A "Federal Reclamation Project" 329
SHASTA AND FRIANT DAMS 330
    Design of Shasta and Friant Dams 330
    Shasta Dam: Planning and Bidding 333
    The Construction Process 335
    Construction Schedule 341
    Friant Dam Construction 342
    Aftermath 344

CHAPTER 8: DAMS FOR NAVIGATION AND FLOOD:
TYGART AND MAINSTEM DAMS ON THE OHIO, UPPER
MISSISSIPPI, AND TENNESSEE RIVERS 353
    The Ohio River Floods and Tygart Dam 353
    Navigation from New Orleans to Monongahela 355
    Floods, Droughts, and Power 356
    Cove Creek and Tygart 361
TYGART DAM 363
    The Planning and Design of Tygart: 1933-1934 363
    Design During Construction at Tygart: 1935-1936 366
    Tygart Completion and Operation: 1937-39 370
    Flood Control and Pittsburgh 371
SLACKWATER NAVIGATION SYSTEMS 372
    Modernization of the Ohio 372
    The Upper Mississippi Basin 373
    The Design of Dam Gates 374
    Shipping, Fish and Wildlife 376

CHAPTER 9: THE ENVIRONMENTAL IMPACT OF THE
BIG DAM ERA 383
 Conservation and Controversies 383
 The Economic and Social Impacts of Dams........................... 386
 Environmental Threats: Flooding and Silting........................ 387
 Dam Failures and Dam Safety 389
 A Fish Story: Conflicts over Resource Development............ 392
 The Echo Park Controversy 395
 Dams and the Modern Environmental Movement 399
 Flood Control and Non-Structural Alternatives..................... 403
 Water Quality and Other Environmental Impacts 404
 Decaying Dams: The Impending Crisis in Dam Safety 406
 Two Fish Stories: Pacific Northwest Salmon
  the Tellico Snail Darter 407
 Environmentalism Comes of Age: Rampart Dam and
  the Grand Canyon Dams.. 409
 Conclusion... 411

APPENDIX A: GUIDELINES FOR APPLYING THE
NATIONAL HISTORIC LANDMARKS CRITERIA
TO STORAGE DAMS 421
INTRODUCTION 421
GUIDELINES 421
 Criterion 1 422
 Criterion 2 426
 Criterion 3 427
 Criterion 4 428
Design Signifiance.. 430
Construction Signifiance 434
 Criterion 5 435
 Criterion 6 436
National Historic Landmark Exclusion.. 437
Integrity... 438
Boundary Justification/Identification of Contributing and
 Noncontributing Resources.. 439

APPENDIX B: RESEARCH METHODOLOGY AND LIST OF
POTENTIAL NATIONAL HISTORIC LANDMARK
MULTIPLE-PURPOSE DAMS 445
I. Consultation with Project Historians 445
II. Secondary Source Review 445
III. Comparative Analysis.. 447
Historic Context.. 448
Corps 449
Reclamation 449

Multiple Purpose River Development 451
Engineering Considerations - Prepared by the Institute for
    the History of Technology and Industrial Archaeology, 1994 453
Consideration of Hydraulic Designs 457
List of Potential National Historic Landmark Dams 458
    Anderson Ranch Dam 458
    Arrowrock Dam 459
    Bartlett Dam 459
    Buffalo Bill Dam 459
    Central Valley Project 460
    Shasta Dam 460
    Friant Dam 460
    Elephant Butte Dam 461
    Fort Peck Dam 461
    Garrison Dam 462
    Gatun Dam 462
    Gibson Dam 462
    Glen Canyon Dam 463
    Grand Coulee Dam 463
    Owyhee Dam 463
    Pathfinder Dam 464
    Pine Flat Dam 464
    Sardis Dam 464
    Troy Dam #1 465
Comprehensive Coordinated River-Basin Flood-Control Systems
    (U.S. Army Corps of Engineers) 465
    Ohio River Basin 466
Allegheny and Monongahela River System 466
    Tygart Dam 467
    Youghiogheny Dam 467
    North Atlantic Division 468
    Connecticut River Basin 468
    Claremont 469
    Union Village 469
    Surry Mountain Reservoir 469
    Birch Hill 469
    Knightville Reservoir 469
    Tully Reservoir 469
Los Angeles County Flood Control and Water Conservation
    District, California 469
    Red River, Texas 470
    Denison Dam 470

BIBLIOGRAPHY 473

INDEX 529

List of Figures

Figure 2-1:	Cross sections of Almans' and Alicante Dams	51
Figure 2-2:	F. Emile DeLocre design for Furen's Dam, 1858	52
Figure 2-3:	DeLocre design of Furen's Dam	53
Figure 2-4:	Comparison profiles in Wegmann's *Design and Construction*	55
Figure 2-5:	DeLocre profile and Zola Dam	60
Figure 2-6:	Upstream face of Pathfinder Dam	66
Figure 2-7:	Pathfinder Dam construction	67
Figure 2-8:	Pathfinder Dam analysis	68
Figure 2-9:	Buffalo Bill Dam	69
Figure 2-10:	Calculating loading on Pathfinder Dam	71
Figure 2-11:	Trial load method of dam analysis, Dam No.1	73
Figure 2-12:	Trial load method of dam analysis, Dam No.2	75
Figure 2-13:	John (Jack) L. Savage	77
Figure 3-1:	Frederick H. Newell	93
Figure 3-2:	Arthur Powell Davis	94
Figure 3-3:	Theodore Roosevelt Dam	95
Figure 3-4:	Cable towers at Roosevelt Dam	96
Figure 3-5:	Roosevelt Dam construction	97
Figure 3-6:	Elwood Mead	108
Figure 3-7:	Wooden formwork, Wilson Dam	112
Figure 3-8:	Wilson Dam at Muscle Shoals	112
Figure 3-9:	Wilson Dam	113
Figure 4-1:	Colorado River Basin map	130
Figure 4-2:	Colorado River below Hoover Dam and southern California	134
Figure 4-3:	Grapefruit orchard, Imperial Valley	136
Figure 4-4:	Strawberry field, Imperial Valley	137
Figure 4-5:	The Black Canyon damsite	145
Figure 4-6:	Preliminary design for Black Canyon dam	146
Figure 4-7:	Weymouth design for Hoover Dam	147
Figure 4-8:	Savage plan for Hoover Dam	150
Figure 4-9:	Dignitaries at Black Canyon damsite	151
Figure 4-10:	Hoover (Boulder) Dam promotional pamphlet cover	166
Figure 4-11:	Hoover (Boulder) Dam promotional flier	167
Figure 4-12:	Talking Points," Hoover (Boulder) Dam promotion	168
Figure 4-13:	Alternate routes for Colorado River aqueduct	170

Figure 4-14:	Profiles of preliminary Colorado River Aqueduct routes	172
Figure 4-15:	View of Hoover Dam from downstream	173
Figure 4-16:	Construction of Hoover Dam	176
Figure 4-17:	Hoover Dam and Lake Mead	177
Figure 4-18:	Glen Canyon Dam construction	183
Figure 4-19:	Glen Canyon Dam	185
Figure 5-1:	Cross sections of early Corps concrete dams	192
Figure 5-2:	Bonneville Dam cross section	193
Figure 5-3:	Coffer dams at Bonneville Dam	195
Figure 5-4:	Bonneville Dam under construction	197
Figure 5-5:	Main spillway at Bonneville Dam	198
Figure 5-6:	The Bonneville spillway dam	199
Figure 5-7:	Bonneville Dam locks and powerhouse	204
Figure 5-8:	Bonneville Dam powerhouse	201
Figure 5-9:	Generators #1 and #2, Bonneville powerhouse	204
Figure 5-10:	Fishladder, Bonneville Dam	205
Figure 5-11:	Bradford Island and Bonneville Dam plan	206
Figure 5-12:	Coffer dams at Bonneville Dam	207
Figure 5-13:	Cross-river coffer dam at Bonneville Dam	209
Figure 5-14:	Spillway section of Grand Coulee Dam	215
Figure 5-15:	Outlet tubes for Grand Coulee Dam spillway	216
Figure 5-16:	Grand Coulee Dam and Lake Roosevelt	220
Figure 5-17:	Reservoirs in Main Control Plan, Columbia River Basin	222
Figure 5-18:	Power network map for Washington state	225
Figure 5-19:	Dams on the Columbia River system	225
Figure 5-20:	Grand Coulee Dam	227
Figure 5-21:	Aerial view of Grand Coulee Dam and Banks Lake	229
Figure 5-22:	West coffer dam at Grand Coulee	229
Figure 5-23:	Excavation at Grand Coulee behind west coffer dam	230
Figure 6-1:	Missouri River Basin map	236
Figure 6-2:	General Lytle Brown	238
Figure 6-3:	Plan of Fort Peck Dam	242
Figure 6-4:	Spillway gate structure at Fort Peck Dam	245
Figure 6-5:	Fort Peck Dam and spillway	246
Figure 6-6:	Hydraulic fill area at Fort Peck Dam	247
Figure 6-7:	Hydraulic fill area at Fort Peck Dam	248
Figure 6-8:	East section of Fort Peck Dam before slide	248
Figure 6-9:	Spreading hydraulic fill at Fort Peck Dam	249
Figure 6-10:	Dredge *Jefferson* at Fort Peck Dam	250
Figure 6-11:	Upstream face of Fort Peck Dam	251

Figure 6-12:	Core pool near the top of Fort Peck Dam	251
Figure 6-13:	Concrete placement at Fort Peck Dam	254
Figure 6-14:	Fort Peck Dam spillway gate structure	254
Figure 6-15:	President Roosevelt at Fort Peck Dam	255
Figure 6-16:	Slide on the upstream embankment, Fort Peck Dam	257
Figure 6-17:	Fort Peck Dam slide	258
Figure 6-18:	Cross section of hydraulic fill, Fort Peck Dam	260
Figure 6-19:	Plan for Garrison Dam	261
Figure 6-20:	Section through Garrison Dam	262
Figure 6-21:	General Julian L. Schley	264
Figure 6-22:	General Lewis A. Pick	270
Figure 6-23:	(William) Glenn Sloan	270
Figure 6-24:	General Eugene Reybold	271
Figure 6-25:	Colonel Lews A. Pick and Glenn Sloan at Yellowtail Dam	277
Figure 6-26:	Spillway gate at Garrison Dam	284
Figure 6-27:	Water intake at Garrison Dam	285
Figure 6-28:	General William E. Potter	285
Figure 6-29:	Riprap on Garrison Dam	286
Figure 6-30:	Switchyard at Garrison Dam	286
Figure 6-31:	Profile of mainstem reservoirs, Missouri River	289
Figure 6-32:	Missouri River mainstem storage	289
Figure 6-33:	Missouri River mainstem hydroelectric generation	289
Figure 6-34:	Commercial tonnage on the Missouri River	290
Figure 6-35:	Missouri River annual flow	290
Figure 6-36:	Missouri River flow at Ft. Randall	291
Figure 7-1:	Central Valley Project	302
Figure 7-2:	Kings River irrigation diversion	304
Figure 7-3:	Irrigation canal, San Joaquin Valley	306
Figure 7-4:	Plan view of Shasta Dam	327
Figure 7-5:	Cross section Shasta Dam	332
Figure 7-6:	Commissioner John C. Page at Shasta Dam	334
Figure 7-7:	Construction at Shasta Dam	336
Figure 7-8:	Night view of construction, Shasta Dam	336
Figure 7-9:	Plan view of Shasta Dam	337
Figure 7-10:	Construction at Shasta Dam	338
Figure 7-11:	Keyways and galleries in Shasta Dam	339
Figure 7-12:	Penstocks and powerhouse, Shasta Dam	340
Figure 7-13:	Shasta Dam spillway	340
Figure 7-14:	Aerial view, Shasta Dam	341
Figure 7-15:	Friant Dam under construction	343

Figure 7-16:	Friant Dam and Millerton Lake	345
Figure 7-17:	Drying almonds in the Central Valley	348
Figure 8-1:	Tygart Dam construction	354
Figure 8-2:	Trestle for placing concrete at Tygart Dam	355
Figure 8-3:	Baffles and stilling dam, Tygart Dam	357
Figure 8-4:	Tygart Dam construction	358
Figure 8-5:	Tygart Dam	359
Figure 8-6:	Tygart Dam	360
Figure 8-7:	Tygart Dam	361
Figure 8-8:	Cutoff wall construction at Kinzua Dam	372
Figure 8-9:	Kinzua Dam	373
Figure 8-10:	Mississippi River Dam #15	375
Figure 8-11:	Roller Gates, Mississippi River Dam #15	376
Figure 8-12:	Tainter gates, Mississippi River Dam #17	377
Figure 8-13:	Mississippi River Dam #24	377
Figure 8-14:	Construction at Mississippi River Dam #24	378

CHAPTER 1

"IMPROVING" RIVERS IN AMERICA: FROM THE REVOLUTION TO THE PROGRESSIVE ERA - RIVERS IN EARLY AMERICA

Rivers as Resource: The American Watershed System

Fresh water is a precious resource, and water from rivers, streams, and lakes has often been regarded as an economic commodity in the United States as in much of the world. Water is essential not only for human consumption and for a variety of domestic purposes, but for fire protection, military defense, transporting people and goods, irrigating farmlands, manufacturing, and generating power. The great rivers and their tributaries in the United States are the primary source of the water bounty and are major symbols of American regionalism, ultimately binding together disparate areas into a powerful whole.

The American watershed system is an awesome force. The Mississippi Basin alone drains more than 40 percent of the country's land from the Appalachian Mountains in the East to the Rockies in the West. To the North, the St. Lawrence River drains the Great Lakes. In the Southwest, the Colorado traverses seven states and Mexico on its route to the Gulf of California, and the Rio Grande forms part of the nation's southern boundary. Along the Pacific Coast, the Columbia gathers water from the Rocky Mountains and the Cascades, and the Sacramento and San Joaquin Rivers collect water from the Sierra Nevada, linking inland valleys to the Pacific Ocean. The geological and human history of the United States is linked inextricably to its rivers.[1]

American rivers were symbols of a burgeoning nation in the eighteenth and nineteenth centuries. They inspired romantic renderings at the hands of artists, and in some cases—as with painters of the Hudson River School in the 1820s—they were depicted as detailed landscape features with physical and even human qualities.[2] But at times they were regarded as untapped or under-utilized resources, raw material waiting to be harnessed, managed, and exploited for human benefit. In the neoclassical tradition of the eighteenth and early nineteenth centuries, "The 'proper' channel for a river is not necessarily the one it has carved for itself: By means of canals and locks it can be guided by men along a straight and level line, thereby improving upon natural design." Rivers, therefore, were most attractive "when they yielded to humanity's needs, whether as mechanisms of transportation or as sites for nascent towns."[3] For aesthetic and for practical reasons, wild rivers served little purpose, historian Theodore Steinberg noted:

> As the [nineteenth] century progressed, a consensus emerged on the need to exploit and manipulate water for economic gain. A stunning cultural transformation was taking place, a shift in people's very perception of nature. By the latter part of the nineteenth century, it was commonly assumed, even expected, that water should be tapped, controlled, and dominated in the name of progress–a view clearly reflected in the law.[4]

Steamboats, canals, and dams became the technologies of choice to accomplish those goals.

To appreciate the importance of dams in the process of "harnessing" American rivers, it is necessary to establish a context for understanding water resource management in the United States, including relevant economic, political, institutional, and legal issues in the nineteenth and early twentieth centuries. Large federal dams would be designed and built from about 1930 to 1965 to further the objectives of economic growth in a modern industrializing nation beginning to blossom in the early nineteenth century.

The Rise of an Industrializing Nation

The impulse to "improve" waterways was stimulated by the profound changes transforming the young nation. Beginning as early as 1820, the Industrial Revolution ushered in a period of unprecedented economic development for the United States. Manufacturing began to challenge agriculture as the nation's leading economic enterprise. While agriculture was responsible for the largest single share of production income before the Civil War, the growth and importance of manufacturing, especially in the East and along the Great Lakes, rose rapidly during the decades that followed the war. In 1859, there were 140,000 industrial establishments in the United States—many of them hand or neighborhood industries. Forty years later, there were 207,000 industrial plants, excluding hand and neighborhood industries.[5]

The economic transformation of the nation paralleled the rise of cities. The first federal census, in 1790, showed that city dwellers represented less than 4 percent of the nation's population. Urban growth stagnated until 1820, but by the end of the decade the urban population had almost doubled.[6] While only 7 out of every 100 Americans lived in cities or towns at that time, the urban population grew by 552 percent (from 1.1 million to 6.2 million) between 1830 and 1860, which was the fastest rate of urbanization the nation had ever experienced.[7]

Industrialization also inspired the mechanization of agriculture and stimulated demand for a variety of products that helped to build a national market economy. Irrigation ultimately became a tool for expanding the agricultural market in the West to supply a variety of goods for growing urban centers at home and abroad. As early as the 1770s, an emerging capitalist economic system was evident in the Northeast, the Mid-Atlantic region, the South, and the back country. A booming transatlantic market for grain and other agricultural products, a rising number of American capitalist entrepreneurs, surplus labor available to work for wages, and state and national governments encouraging and promoting economic growth underlay the emergence of a market revolution along the American rural frontiers.[8]

The promise of economic growth had long attracted the interest of government. In the manufacturing belt of the East coast and the Great Lakes, the states and the federal government had been active agents in stimulating commerce and industrialization. Competition between the states beyond the Appalachians for access to ports on the Atlantic had been intense. Rivalries between the states for a variety of public works projects focused on economic opportunities to be won and lost.

The Origins of Federal Water Resource Policy

With respect to water resource issues, rivalries between the states suggest a partial answer for an increased federal government role. However, no comprehensive water resource policy ever emerged in the nineteenth or twentieth centuries. Federal navigation policy, flood control policy, and irrigation policy were conceived and administered separately over the years, and water issues even today remain a combination of local, state, and national interest.[9] Supporters of national initiatives for water and navigation projects chronically vied with advocates of states' rights, who opposed outright subsidies for waterway construction. Steering a middle course, an emerging "water bureaucracy"– including the U.S. Army Corps of Engineers, the Bureau of Reclamation, and the Tennessee Valley Authority—often urged government planning without directly challenging state control of water projects.[10]

There is merit in Richard N. L. Andrews' observation that federal responsibility for water resource management "evolved almost unintentionally" from the convergence of nineteenth-century public-land and internal-improvements policies.[11] Disposal of public lands set several precedents about how the federal government would deal with the nation's resources. At one time or another, more than 78 percent of the nation's 2.3 billion acres was owned by the federal government. There was no uniform method of land distribution during early colonial days. Since much of the frontier remained within the boundaries of the states after the American Revolution, state legislatures often developed the first land schemes to deal with estates confiscated from Loyalists. Land speculation on federal lands initially focused on the Ohio River region, that area wedged between the new nation and the vast frontier. After the Louisiana Purchase, new land law that lowered the minimum purchase to a quarter section (160 acres), made western migration attractive to easterners and European immigrants. Between 1850 and 1900 the number of farms in American territory increased from 1.4 million to 5.7 million. Indian land rights, however, were often ignored or manipulated in providing settlers with land. In essence, much of the productive land in the West had already been claimed before the famous 1862 Homestead Act, and after its initial disposal, former public land increasingly became a speculative commodity.[12]

The disposal of public lands was not merely an end in itself. From the first land ordinances in the eighteenth through the nineteenth century, the federal government intended to generate revenue and to stimulate economic development by a rapid transfer of public lands to private individuals. This was not accomplished without fierce debate, characterized most graphically by Thomas Jefferson's image of a nation of self-sufficient yeomen farmers and Alexander Hamilton's promotion of manufacturing, inland navigation, and the development of new economic markets.

In dealing with the states, the federal government could offer public lands in exchange for their support on development projects or other policies. Public lands also were used to provide capital for private businesses, such as the railroads. The first land laws in the 1780s and 1790s (including the Northwest Ordinance of 1787), however, were primarily directed toward using land to raise revenue, to retire the public debt, and to create a market in western lands.[13]

While land subsidies for public works projects were not provided for in federal law, many land grants were made to subsidize road building, river improvements, and railroad construction. For public lands to have value, they needed to be accessible to facilitate settlement and for the transportation of raw materials and crops to the East and to Europe. The federal government funded "internal improvements" through general revenues, the sale of public lands, and land grants.[14] But as John Lauritz Larson perceptively observed, "The campaign for internal improvements, so universally appealing in the abstract, proved incredibly controversial at all levels of government as soon as workmen struck their spades into the earth."[15]

Prior to 1789, private investors provided internal improvements. At the constitutional convention, Benjamin Franklin was the primary advocate for federal sponsorship for internal improvements, but he could not carry the day. The Constitution ultimately reserved that responsibility for the states. However, with poor economic conditions in many states, Congress began appropriating funds for specific improvements beginning in 1802. In 1808, Secretary of Treasury Albert Gallatin submitted his report recommending federal aid for a system of roads and canals that would link the Atlantic seaports with the nation's interior.[16]

Artificial canals became the foremost technology in the early nineteenth century to connect the riverine system to the sea. The virtue of such canals was to 'free' rivers from their natural courses and to direct them into channels that would serve the economic ends of the nation.[17] East coast rivers were only navigable up to the fall line, a barrier at the foothills of the Appalachians. In the late eighteenth century, several short canals and the 27-mile Middlesex Canal in Massachusetts had been constructed, but by 1816 only about 100 miles of canals existed in the United States. These manmade waterways proved to be demanding engineering feats and financial liabilities, and it became difficult to find investors for new projects.

The construction of the Erie Canal, linking Albany and Buffalo by means of an artificial waterway 364-miles in length, set off a canal boom in the United States that ultimately attracted federal dollars to future projects. The New York legislature authorized the construction of the Erie Canal in 1817 without a promise of federal support, and the canal was completed in 1825. By 1840, various states had invested approximately $125 million in 3,200 miles of canals. Between 1815 and 1860, the total public and private expenditures for canal construction was about $195 million. While the federal government had refused to help New York State build its canal, and states were the primary financial contributors in the early canal era, the federal government ultimately provided financial support through land grants and subscribed more than $3 million in canal company stock. Expensive enlargement programs, the Panic of 1837, and competition from railroads brought the canal boom to an end by the 1840s.[18]

WATER LAW AND THE USE OF RIVERS

Mills and Dams in the Early Industrial Era

Complicating the creation of a national water resource program was the fact that fresh water, unlike land, was common as opposed to private property. Navigable waterways, for example, could not be treated like the public lands, that is, could not be disposed of to generate revenue or to promote economic development. They were open to common use and thus required special treatment. Water usage also was subject to unique practices imbedded in the law.[19]

Water, among other things, was an important source of energy before and during the early stages of the Industrial Revolution, and was, thus, the focus of voluminous litigation over water rights. The bulk of litigation arose from disputes over the use of streams for waterpower.[20] Mills and dams raised for the first time legal questions over the relationship between property law and private development, when "antidevelopmental doctrines of the common law first clashed with the spirit of economic improvement." Evolving water rights law had a greater impact on the effort to adapt private law doctrines to the promotion of economic growth than any other branch of law.[21]

The water mill inevitably came into conflict with other stream uses. Aside from the waterwheel, the dam was the most essential element of a mill. Pre-industrial dams were low, crude structures designed to increase water fall by raising the stream level. The dam created a storage reservoir, or millpond, which not only obstructed navigation and log floats but also the seasonal movement of fish.[22]

Water mills challenged prevailing water rights law and practices such as riparian rights, commonly recognized in the eastern United States in the eighteenth and nineteenth centuries. This English common-law doctrine granted ownership of a water privilege with the land bordering the two banks of the stream. The landowner did not own the stream, but only the rights of water usage. Even usage was subject to rights and claims of other users, including navigation interests, owners of riparian farmlands above and below a specific water privilege, lumber and other commercial interests, upstream communities, and mill owners themselves.[23]

Before the nineteenth century, common law doctrines were generally based on the natural flow of water, and jurists rarely looked with favor on the use of water to irrigate or to run machinery. Possessing a narrow view of the productive capacity of water, they generally placed strict limits on its appropriation.[24] With the onset of the Industrial Revolution, the increasing number of conflicting claims and shades of interpretation of privilege challenged the water rights of riparian owners.[25]

Since navigation rights had priority on streams sufficiently large to carry regular traffic, the parts of the law referring to that activity were the least controversial. As power needs increased, government officials began to favor mill owners—especially in New England—over other riparians. This also was true for capitalists who wanted to divert water from natural sources to build canals.[26]

The most typical water rights controversy pitted downstream riparian landowners against upstream owners whose dams obstructed the natural flow of water for mills or irrigation. Other cases pitted upstream mill owners against downstream mill owners or landowners flooded by the dam. Some courts virtually refused to recognize any right to interfere with the flow of water to a mill.[27]

"Reasonable use," or a balancing test, was the most important challenge to the common law doctrine of riparian water rights. Although the concept did not find general acceptance until around 1825, some early decisions set the stage. By the Civil War, most courts accepted a balancing test in which "reasonable use" of a stream depended on the extent of detriment to riparian landowners downstream.[28]

In determining "reasonable use," it was common to take into account what constituted a proportionate share of the water. In *Cary* v. *Daniels* (1844), however, Massachusetts Chief Justice Lemuel Shaw tended to weaken the standard of proportionality by giving priority to the proprietor who first erected his dam, thus placing greater emphasis on maximizing economic development at the expense of equal distribution of the water privilege.[29] Not until the nineteenth century was a theory of priority used offensively to maintain a right to obstruct the flow.[30] What brought on the change was the building of large dams, which widened the possibilities for injury by causing potential damage to mill owners both upstream and downstream from the dam.

The two doctrines–reasonable use and prior appropriation–were becoming less and less interchangeable, at least as they operated within the context of economic development in the emerging industrial age. Thus a tension between the two–which had moved beyond the natural rights doctrine characteristic of pre-industrial societies–found its way into the courts. By mid-century, almost all courts rejected prior appropriation because it so obviously interfered with competition. Riparian rights, modified by "reasonable use," prevailed in the East in dealing with economic development. In addition, the advent of the steam engine and the railroad made concessions to mill dams and canals temporary.[31]

Water Law in the West

While waterpower development and canal building framed much of the water law in the East, in the West mining activity and agriculture helped shape the law.[32] The traditional interpretation stresses that water rights in the nineteenth-century West, as opposed to the East, have been closely associated with the prior appropriation doctrine.[33] When Anglo American settlers arrived in the West, neither land nor water rights issues had been clearly resolved. Until the Civil War, the federal government controlled the public domain. Legislation enacted by Congress in the 1860s and 1870s, however, recognized the rights of settlers to utilize water on the public lands for a variety of purposes. Thus, the prior appropriation doctrine in the West owes a great deal to local circumstance.[34]

Donald Pisani, however, has persuasively argued that "Water law evolved slowly in both California and the West, constructed piece by piece, like a quilt, rather than from whole cloth." The courts and legislatures, he added, "rarely looked beyond immediate economic needs" in determining water rights.[35]

The Western Setting

In humid eastern America, water is an essential resource. But control over water resources does not define the central character of that society. In contrast, water is dramatically scarce in the arid West and that "precious liquid" occupies a pivotal position in regional development and in the larger political economy. Much of the West's historical character arises from a pervasive lack of rainfall.[36] It has become clear that water resources development is a key factor in regional growth.[37] Moreover, in the history of western water use, the work of the federal government–in particular the U.S. Reclamation Service after 1902 and the U.S. Army Corps of Engineers after 1933–has had enormous influence in transforming the environment and fostering economic development.[38]

Precipitation in the West is not evenly distributed over the landscape, and while billions of gallons of water might be dumped on the desert in the period of a few days or weeks, such storms can be spaced years apart. With much surface water originating either as seasonal snowmelt or infrequent torrential rainstorms, the ability to support widespread agriculture—as well as mining, municipal growth, and hydroelectric power development—has by necessity become dependent upon artificial means of controlling water. Leaving aside groundwater that can be lifted to the surface by either windmills or electric pumps, irrigated agriculture depends upon water diverted from rivers, transported in canals, and then distributed over fields to sustain crops. The engineering techniques and the political instruments devised to foster irrigation in the West later comprised the basis for water resources development throughout the nation. Water in the West, of course, served many needs other than agriculture, including mining and urban development.

The California Doctrine: 1851-1886

During the California gold rush, the right to a claim went to the first person working it. Not surprisingly, this "first in time, first in right" principle (or prior appropriation doctrine) could also apply to water—a commodity essential to mining. A miner did not acquire property in the running water itself, but only its use if he continued to work the claim. But this prior appropriation doctrine coexisted with riparian rights in the 1840s and early 1850s, since many miners did not want streams diverted from their natural courses. The California State Legislature, eager to promote mining, supported prior appropriation for the gold country in 1851, the state court accepted it in 1855, followed by its congressional endorsement for public lands generally in 1866.[39] The federal action endorsed prior appropriation not only for mining, but also for agricultural, manufacturing, and other uses, and it further acknowledged the states' power to regulate water rights. The prior appropriation doctrine promoted economic development, but gave no preference to communities over individuals. Eventually every western state endorsed some form of the doctrine, and nine states adopted it as its sole water law.[40]

In practice, prior appropriation worked well enough when water was abundant, but when scarce it created confusion. An appropriator could sue to defend his rights, and the courts reviewed the records to determine a prior claim. But the amount of water available was not always known. A title established in one case protected an appropriator from one claimant only. Although the states gradually evolved more orderly approaches, the system remained confused.[41]

Although California set a precedent in the application of prior appropriation, riparianism also gained legal recognition early in the state's history. In 1850, the first legislature adopted as its basic system English common law, subsequently modified by state courts in response to statutory and case law. "For nearly three decades the state dealt with the problem of two contradictory legal systems by reaffirming the legitimacy of both and seeking to soften their differences...." However, when irrigation appeared necessary for some forms of agriculture, the courts demonstrated flexibility, "taking a cue from eastern states, which had begun modifying their riparian law tradition in favor of some appropriation practices...."[42]

Drought in the 1860s and 1870s, and especially increased irrigation, threatened to challenge the uneasy *status quo*. The development of refrigerated railroad cars, for example, meant that high-profit fruit and vegetable crops produced through irrigation could be shipped to distant markets.[43]

While the California courts ruled in favor of some irrigation under riparianism by the 1870s, accommodation had not been made for an irrigation boom. During the 1880s, the area of irrigated land in the arid West increased four-or five-fold. The clash of the water doctrines reached an acme in 1886. In *Lux* v. *Haggin*, the California Supreme Court affirmed a dual system of water rights, the so-called "California Doctrine."[44]

The court held that riparianism was law in California, applicable in all private lands and public lands that became privately owned. An appropriator could have a superior claim if he used the water before a riparian user had acquired the property. Timing was crucial.[45] As unpopular as the decision was within the public at large—since large landholders would be affected much less than small farmers—the California Doctrine eventually was adopted along the Pacific Coast (Washington and Oregon) and in the Great Plains (Nebraska, Oklahoma, Texas, Kansas, North Dakota, and South Dakota).[46]

In the 1880s, Colorado invalidated riparian rights to surface water and began enforcing appropriative rights under state authority. Prior appropriation became the sole water right and came to dominate much of the Rocky Mountain region. Seven other states (Utah, Wyoming, Arizona, New Mexico, and Idaho) soon accepted the "Colorado Doctrine," with Montana and Alaska following in the early twentieth century.[47]

A third approach developed in Wyoming, emphasizing a different type of enforcement. The state constitution gave the state title to all water. Officials could reject water claims and overturn existing appropriations not believed to be in "the public interest." In essence, the so-called "Wyoming Doctrine" gave greater protection to appropriators than under the Colorado system. Besides Wyoming, Nebraska, Oklahoma, North Dakota, and South Dakota claimed full control over their water.[48]

Despite the flurry of activity that led to the three major water doctrines, water rights—let alone water policy—were not completely rationalized, nor were conflicts ended among economic interest groups. Battles over irrigation, farming and livestock raising, mining, and the demands of urban growth kept the water issue center stage.

Inevitably, the federal government would be active in the controversies–welcomed by some, not welcomed by others. The commitment in the early twentieth century to the construction of federal dams in every major watershed occurred in the wake of contested uses of water underway for years. That water law favored the states only complicated the ability of federal dams to provide stored water to a variety of consumers. However, under the property clause (Article 4, Section 3) of the U.S. Constitution, the federal government had legal authority to accept, manage, and dispose of public domain lands, and this provided the basis for subsequent laws and regulations pertaining to public lands and other resources. With regard to water resource policy, the federal government presumably holds "reserve rights" to enormous amounts of annual water flows in the West, since it was the earliest formal owner of the public lands. However, the federal government has never fully asserted these rights and the U.S. Supreme Court has never formally recognized them.[49]

Prior appropriation exacted heavy social and environmental costs in the West. Water was an economic commodity, although private gain resulting from the use of water did not translate into revenue for the states. Instead, several large corporations and monopolies benefitted, and many farmers adopted wasteful irrigation practices. Prior appropriation led to a rapid economic development that "exacerbated the boom-and-bust mentality endemic to the mining industry, encouraging speculation and maximum production." Moreover, it failed to preserve water quality as did riparian rights, and it allowed vast environmental destruction.[50]

Environmental policy was in the developmental stages in the late nineteenth century.[51] The emergence of resource conservationism, as opposed to nature preservation, emerged out of concern about the depletion of natural resources, which could stall further economic development.[52] Resource exploitation was central to the actions of a rapidly industrializing society; laissez-faire capitalism was more regaled than condemned for stoking the fires of economic growth.[53] Particularly in the West, where the forests, rivers, and mineral wealth were directly linked to economic opportunity, conservationism was largely dismissed in the nineteenth century.[54] But even practical concerns, such as the marshaling of such a scarce resource as water, generated intense conflict. A more widely held interest was how to tap yet-to-be exploited water sources.

THE U.S. ARMY CORPS OF ENGINEERS IN THE NINETEENTH CENTURY

The Corps and the French Engineering Tradition

In a March 16, 1802, congressional act, the U.S. Army Corps of Engineers was separated from the Corps of Artillerists and Engineers and stationed at West Point, New York. This act not only marked the reestablishment of a separate U.S. Army Corps of Engineers (first created in 1779), but also founded the U.S. Military Academy. The Military Academy remained under the charge of the Corps until 1866.[55]

The American engineering profession in the nineteenth century was being shaped by two European traditions. One emphasized the civilian "builder-mechanic" model of the British; the other, the military, formally trained engineer in the French (or Continental European) style.[56] Of the two European engineering traditions, the French was the older—linked to the rise of a powerful monarchy in the sixteenth and seventeenth centuries. French engineering successes included a high point in canal engineering with the 149 mile long Languedoc Canal opened in 1681.

For a variety of largely non-military tasks deemed essential to the national interest, the royal French government established the *Corps des pont et chaussees* (Corps of bridges and roads) in 1716, and founded the *Ecole des Ponts et Chaussees* in 1747. In 1794, it founded the *Ecole polytechnique*, which quickly became the international leader in technical education. Government sponsored education, furthermore, was linked to government employment.[57]

The U.S. Army Corps of Engineers became the chief American standard-bearer for French engineering. The key to the Corps' preference for French engineering rested not only in its connection to the military, but to the role of the military within the state.

The influence of French engineering in the United States actually began during the American Revolution. The reputation of the French engineers, plus their country's sympathies for the rebelling colonies, resulted in a period of seven years when French military engineers organized and trained the American army's engineering corps.[58] It is not surprising that when the U.S. Army Corps of Engineers was reestablished in 1802, it embraced the French engineering tradition and sought to implant it at West Point.[59]

The appointment of Claudius Crozet as professor of civil engineering in 1816 signaled a strong commitment to the French model. Educated at the *Ecole polytechnique* and the artillery program at Metz, Crozet had been an artillery officer under Napoleon. A bridge builder, he also served as an engineer in Holland and Germany, studying the sluices and navigation jetties. In 1817, he applied geometry to canal design in the first scientific course on construction taught in the United States.[60]

In that same year, superintendent Major Sylvanus Thayer—a great admirer of Napoleonic engineers—introduced methods of instruction using the *Ecole polytechnique* as his model. He also insisted on the importance of studying the French language, which he viewed as the "sole repository of military science." In 1837, Captain Dennis Hart Mahan completed the first American textbook based on French engineering practice.[61] The influence of the French at West Point did not simply shape the engineering style and dictate the engineering methods, but it also imbued the Corps with an interest in scientific design, natural philosophy, applied mathematics, and a commitment to large, state-supported projects.[62]

But even those engineers who entered the Corps did not focus exclusively on military projects. They helped map the West, constructed coastal fortifications and lighthouses, built jetties and piers for harbors, and mapped navigation channels.[63]

Commerce, Navigation, and the "Steamboat Case"

The Corps' water projects in the early nineteenth century focused primarily on navigation. With the economic climate of the nation improving after the War of 1812, the steamboat came of age. In the West, the steamboat was vital to commerce and travel. Only 17 steamboats operated on western rivers in 1817, but there were no less than 727 by 1855.[64]

While states' rights advocates typically objected to direct federal subsidies for waterway construction, they were less likely to block indirect types of federal aid, such as scientific surveys.[65] In 1820, Congress appropriated $5,000 for a navigation survey of the Ohio and Mississippi rivers from Louisville to New Orleans. In the next few years, the Corps also made surveys of harbors, coastal areas, and lead mines on the upper Mississippi. It also built jetties and breakwaters along the Massachusetts coast and at Presque Isle in Lake Erie. But the monetary value of all federal river and harbor projects between 1802 and 1823 was a meager $85,500.[66] Army engineers had demonstrated their ability to deal with a variety of civilian projects. Nevertheless, direct federal aid to waterways fared little better than other forms of internal improvements in the early nineteenth century.

The 1824 Supreme Court ruling in *Gibbons* v. *Ogden* (the "Steamboat Case") changed that, and also initiated the Corps' regular participation in civil works and led to its role in maintaining the nation's inland waterways.[67] The case, producing the Supreme Court's first interpretation of the commerce power of the federal government, originated in 1807, when Robert R. Livingston and Robert Fulton acquired a steamboat monopoly from the New York legislature. Subsequently, they also petitioned other states and territorial legislatures for similar monopolies in the hope of developing a national network of steamboat lines. Only Orleans Territory accepted their petition and awarded them a monopoly on the lower Mississippi.

Competitors, aware of the potential of steamboat navigation, challenged Livingston and Fulton arguing that the commerce power of the federal government was exclusive and superseded state laws. Legal challenges followed, and in response the monopoly attempted to undercut its rivals by selling them franchises or buying their boats. Former New Jersey governor Aaron Ogden had tried to defy the monopoly, but ultimately purchased a license from its assignees in 1815.[68] He entered business that year with Thomas Gibbons from Georgia, but the partnership collapsed three years later when Gibbons ran an unlicensed steamboat on Ogden's route. The former partners ended up in the New York Court of Errors, which granted a permanent injunction against Gibbons in 1820.

Gibbons appealed to the U.S. Supreme Court, arguing as he did in New York that the monopoly conflicted with federal law. After several delays, the Court began discussing the meaning of the commerce clause in 1824, which by that time had become an issue of wider interest. Congress was debating a bill to provide a federal survey of road and canal routes. Southerners, in particular, were growing increasingly sensitive to what the resolution of these issues would mean to them as sectional disputes, especially over slavery, were heating up.

Chief Justice John Marshall could not ignore the political ramifications of *Gibbons*, and thus in the unanimous decision he avoided stating flatly that the federal government had exclusive power over commerce. Marshall articulated a broad construction of the commerce clause, but he also tried to accommodate state regulation of local problems and state demands for the principle of free trade. The New York monopoly was struck down, however, based on the argument that national law took precedence over state law in case of conflict.[69]

While *Gibbons* did not settle the issue of the extent of federal power over commerce, it did provide an expansive interpretation of commerce. Marshall stated that "Commerce, undoubtedly, is traffic, but it is something more; it is intercourse. It describes the commercial intercourse between nations, and parts of nations, in all its branches."[70] This included river navigation, giving impetus to further federal river and harbor improvements, and thus providing an opportunity for the Corps to play a central role in planning and construction along commercial routes.[71] Although the Corps began to assume responsibility for flood control in the 1880s, river and harbor work comprised a large part of its mission in the nineteenth century.[72]

French Tradition versus Frontier Techniques

Shortly after the Supreme Court rendered its judgment in *Gibbons*, President Monroe signed the General Survey Act on April 30, 1824, which gave him the authority to employ engineers and Corps officers to survey "routes of such roads and canals as he may deem of national importance in a commercial or military point of view, or necessary for the transportation of the public mail." In gaining this role in civil works, including the planning and politics of internal improvements, the Corps essentially became "the engineering department of the federal government."[73] One month later, Monroe signed an additional bill for improving navigation over sand bars in the Ohio River and for removing snags from the Ohio and Mississippi. The $75,000 appropriation was the first that Congress had issued for work in inland navigable waters.[74]

In 1826, Congress passed the Rivers and Harbors Act authorizing surveys and construction for more than 20 water projects on the Atlantic and Gulf coasts and on the Great Lakes. Combining both planning and construction, it is regarded as the first law of its kind, and it eventually became the model of enabling legislation for the Corps' navigation improvement program and later for flood control. The act significantly expanded the work of the Corps in waterways engineering.[75]

The expanding federal program on rivers and harbors was shrouded in controversy between 1824 and the beginning of the Civil War. The Corps could not escape the controversy in these volatile years. It was caught between the forces contesting the internal improvements issue, especially as the primary agent for executing federal rivers and harbors projects. It also continued to be locked in a contest over the application of "the polytechnic orientation that promoted theory and standardization at the expense of frontier technique."[76]

Between 1824 and 1831, the Corps attempted to develop a comprehensive, national system of internal improvements through its Board of Engineers for Internal Improvements. It consisted of the French Army Engineer Brigadier General Simon Bernard (who served under Napoleon), Colonel Joseph G. Totten of the U.S. Army Corps of Engineers, and civil engineer John L. Sullivan.[77] The plan called for three main projects: (1) canals between the Chesapeake and the Ohio and between the Ohio and Lake Erie and improvements to navigation on the Ohio and Mississippi Rivers; (2) a series of canals connecting the bays to the north of Washington, D.C.; and (3) a road from Washington to New Orleans. By 1827, thirty-five examinations and surveys were conducted, but nothing more. By 1830, local political considerations became more influential than the overall plan in defining priorities, and it soon became the practice of Congress to adopt laws with this in mind.[78]

Navigation and the Beginning of River Dams: 1824–1865

For navigation improvements on the Ohio and Mississippi Rivers in the early nineteenth century, the Corps focused on snag removal and channel deepening. Thousands of snags—possibly more than 50,000—threatened transportation daily, and accounted for the majority of steamboat losses before 1826 and, along with other isolated obstructions, were responsible for three-fifths of all steamboat accidents until 1849.

Under the 1827 Rivers and Harbors Act, Congress made the first in a series of annual appropriations (through 1838) for the removal of obstructions, reflecting a clearer understanding—after one failed contract—that snag removal had to be ongoing. A year earlier, Henry M. Shreve had been appointed Superintendent of Western River Improvements and was given responsibility for snag removal. He built the first steam-power snagboat, the *Heliopolis*, launched in 1829. Shreve's operation was so successful that no boats were lost on the Ohio River due to snags in 1832, and the drop in insurance rates on steamboat cargoes between 1827 and 1835 reflected the vast improvements in clearing the channels.[79]

The other major improvement in river navigation in the 1820s was to deepen channels across sand and gravel bars.[80] Major Stephen H. Long conducted an early experiment with wing dams (or "spur dikes") on the Ohio River near Henderson Island, Kentucky, about 100 miles below Louisville.[81] The structure (two rows of 1,400 piles filled with brush) extended from the bank at a 45-degree angle. It narrowed the width of the channel, thus increasing the velocity of the current and deepening the channel itself. The wing dam was the primary method of deepening channels on the Ohio and several of its tributaries until the late nineteenth century. Long's project at Henderson Island, and a similar use of a wing dam carried out under the direction of Shreve at the Grand Chain near the mouth of the Ohio, led to a congressional appropriation in 1831 for additional dikes on the Ohio River. Some bars were dealt with effectively, but no system of wing dams was in place before the Civil War.[82]

Snag and boulder removal in some relatively minor tributaries of the Ohio River were inadequate to make them viable for steamboat navigation, thus a slackwater system of locks and dams arose in the 1830s. Dams were placed across a stream at intervals insuring a minimum depth of water year-round. Each dam had a lock through which vessels passed. By the mid-1840s, such systems were in operation on the lower Kentucky River, and also on the Green and Barren, the Licking, the Muskingum, and the Monongahela Rivers.

Once in place, these early slackwater systems faced financial and technical problems. Inadequate capital for repairs and maintenance delayed completion and limited their operation. Poor engineering and construction, as well as flooding and icy conditions, limited service. Revenues did not meet expectations, such as the Kentucky River project, suspended in 1842 after the building of five dams. A Muskingum project, completed in 1842, showed profit for a decade, but then faced financial problems. The most successful project was one on the lower Monongahela River, which benefitted from coal shipments to Pittsburgh. Even state-supported open-channel projects had financial problems. On the Kanawha River, smaller boats took advantage of the improvements, but large coal barge tows could not.[83]

Until 1852, relatively little river work was carried out by the U.S. Army Corps of Engineers or the Corps of Topographical Engineers.[84] Even with the 1852 Rivers and Harbors Act, which provided funds for dike repair and construction on the Ohio and for building new snagboats, navigation improvement was sporadic at best for several years.[85] The Democrats won the 1852 election, kept power until 1860, and consistently opposed internal improvements, so that Congress did not pass another general rivers and harbors bill until after the Civil War.[86]

Post War Navigation and the Ohio River: 1866–1885

The years after the war witnessed a shift from specific open-channel improvements—especially the elimination of obstructions or bypassing them with canals—to elaborate plans for slackwater systems and storing flood waters in large reservoirs on the headwaters.[87] Pressure for federal involvement intensified especially because states' rights interests had been quelled. And the Republican Party, strongly committed to federal public works, was in control. Despite the rising competition from the railroads, the government focused on the Mississippi River because of its commercial importance.

In June, 1866, Congress appropriated approximately $3.7 million for about 50 projects and almost 40 examinations and surveys across the country. In the 1870s alone, total appropriations reached almost $54 million. Between the end of the Civil War and 1882, U.S. Presidents signed 16 river and harbor bills and federal appropriations for river and harbor projects totaled over $91 million. In 1882, Congress provided $18.7 million for 371 projects and 135 surveys.[88]

As a result of the 1866 Rivers and Harbors Act, William Milnor Roberts was appointed to oversee improvements and to conduct surveys of the Ohio. In his 1869 report, he provided a "radical" plan for a slackwater, lock, and dam canalization. Colonel William E. Merrill, who replaced Roberts in 1870, supported the proposal because of its relatively successful use on a number of tributaries.[89]

Ironically, those coal shippers who were dominating Ohio river commerce and who stood to benefit from a deepened channel and year-round navigation, argued that the dams would obstruct the channels, require breaking tows to pass through the locks, and require heavy tolls. In addition, the flatboat and rafting trade objected because the proposal would sacrifice the natural navigation of the river for ten months of the year to gain two additional months of navigation for larger vessels. Others warned of possible increased flood heights, stagnant slackwater pools, and silting of river channels.

In the wake of such criticisms, Merrill began exploring alternatives. He sent his deputy to examine movable dams in Europe where 124 movable dams had been completed. Such dams could be raised to increase depths during shallow periods and then lowered when the water was high.[90] As a result, Merrill recommended, in 1874, that a series of movable dams be utilized in the canalization of the Ohio River with the first experimental movable dam and lock to be built at Davis Island near Pittsburgh. While critics complained loudly, he received support from the Ohio River Commission, a variety of shipping interests, and the Grange—a farmer group that hoped to undercut railroad costs. With the additional support of the Senate Committee on Transportation Routes, Congress appropriated funds for the project in 1875. Work began in 1878, and eventually was completed in 1885.[91]

For his movable dam, Merrill chose the design of Frenchman Jacques Chanoine. Invented in 1852, the Chanoine wicket consisted of a line of timbers bolted together into a rectangular panel hinged to a concrete foundation placed on the river bottom. Upon completion, the movable dam was 1,223 feet in length and contained 305 wickets. The wickets were raised by a grapple on a maneuver boat and supported by an iron pole sloping downstream. When the river was high, the pole was removed and the wickets returned to the river bottom. The lock itself was 110 feet wide by 600 feet long. Both dam and lock were the largest of their kind in the world. The lock also was one of the first in the United States to use concrete instead of masonry. Although the new system, as a prototype, faced some problems, the critics were silenced since the Pittsburgh harbor increased in depth and large tows could be assembled there.[92]

Completion of the Davis Island Dam opened a new era in the improvement and navigation of the western rivers. It also marked the modern era of lock design in the United States. While only a 174-mile section between Pittsburgh and Marietta was completed by 1896, a series of about 50 dams extended slackwater navigation along the Ohio by 1929.[93]

The Upper Mississippi and the Headwater Dams: 1866–1899

In 1866, after many attempts to channelize the upper Mississippi, Congress appropriated $400,000 for a 4-foot-deep channel between Minneapolis and St. Louis. In 1878, before the channel project was completed, Congress authorized the Corps to seek a 4 ½-foot depth to the channel through the use of wing and closing dams.[94] As a result of the deepening project, the banks gradually moved inward thus constricting the river and also changing the landscape.[95] The Mississippi River Commission, established in 1879, had set a goal of a minimum year-round channel depth of 6 to 4 ½ feet from St. Louis to St. Paul, the results of which would fundamentally change the physical character of the river. Methods proposed included low-water dams to concentrate the flow of water in the main channel, spur dikes or wing dams to narrow the channel in places where the river was too wide, protection of the river banks from erosion, and occasional dredging. In addition, the Congress authorized reservoirs on the headwaters of the upper river to store surplus water during the wet season.[96]

In fact, as early as 1870, Brevet Major General G. K. Warren recommended construction of 41 reservoirs on the St. Croix, Chippewa, Wisconsin, and Mississippi Rivers. In 1878, Representative William D. Washburn of Minnesota raised the issue of the reservoirs again, in part to benefit his own flour mills at St. Anthony and also to counter the growing railroad challenge. In 1880, in spite of opposition from St. Paul, Congress made its first of several appropriations for these headwaters dams, thus beginning a project that would be one of the earliest large-scale systems of reservoirs constructed in the United States.[97]

Congress initially authorized five headwater dams. The first and largest project resulted in an experimental dam at Lake Winnibigoshish (completed 1883; reconstructed 1899), followed by dams at Pokegama Falls (completed 1885; reconstructed 1904), Leech Lake (completed 1884; reconstructed 1903), Pine River (completed 1886; reconstructed 1907), and Sandy Lake (completed 1895; reconstructed 1911). A sixth dam was completed in 1912 at Gull Lake. They all were located upriver from St. Paul on the main stem and tributaries of the upper Mississippi near the river's source at Lake Itasca. The Corps built all of them at lake outlets in remote areas with no existing roads and few settlements. The isolation of the sites led to initial construction with timber. At the turn of the century, the dams were reconstructed with concrete.

Although historically important as a reservoir system—and for developing an efficient method of constructing a series of standardized dams—the headwater dams project did not utilize unique technology. Each site had an earthen embankment and a timber outlet structure footed on timber piles. The cores of the embankments were filled with puddled clay and contained a timber diaphragm. The length of the dam determined the number of discharge sluices, but each was controlled by a timber gate. All the dams had log sluices. At Sandy Lake a navigation lock was added to serve steamboats between Aitken and Grand Rapids.[98]

Initially, civic leaders and businessmen in St. Paul had opposed the reservoirs for fear that they would give Minneapolis an unfair economic advantage, and lumbermen in northern Minnesota worried that the dams would constrain their logging activities. Predictably, the railroads also had opposed the project. While it was anticipated that improved steamboat navigation would be the primary beneficiary of the project, commercial interests in Minneapolis-St. Paul benefitted the most, particularly lumber, flour milling, and waterpower.[99]

In addition to political and economic rivalries, the construction of the headwater dams also highlighted social and environmental problems that would plague dam projects in the twentieth century. For example, the land to be inundated by the construction of the Lake Winnibigoshish and Leech Lake Dams belonged to approximately 1,300 Chippewa Indians. Constructing the dams required taking a substantial amount of timber from the area. Also, opening the dams damaged or destroyed the Chippewa's wild rice fields, some of their fisheries, and tamarack and cedar tree stands.

A commission in 1884 authorized to determine damages, recommended $10,000 in property damage and an annual additional payment of $26,800, but by 1886 even that paltry award was not paid. In 1890, the commission authorized a meager appropriation of $150,000 as full payment for damages. Some of the overflowed acres were ceded to the United States government, and all of the lands likely to be damaged were subject to construction and building of new dams and reservoirs.[100]

The improved ability to transport lumber by water, aided by the construction of the headwater reservoirs and dams, increased water pollution along the upper Mississippi. Sawmill refuse, already a serious problem in the Minneapolis area by the late 1870s, obstructed river navigation. The Corps and many river interest groups favored a refuse act to prohibit such dumping. Lumber interests, however, fought such action, in part at least because they were not the only culprits. Minneapolis dumped approximately 500 tons of refuse into the Mississippi each day.[101]

Changes in federal law were meant to address in some fashion pollution problems like those faced on the upper Mississippi. With the 1899 Rivers and Harbors Act, especially section 13 (the Refuse Act), loopholes were closed and the law made illegal the casting of "any refuse matter of any kind or description" into navigable waters without permission of the Secretary of War. In time, the 1899 act would be regarded as a seminal piece of legislation in the recognition of water pollution as a major problem. It did not, however, seriously reduce pollution along the upper Mississippi or other rivers.[102]

As settlement increased along the nation's great rivers and their tributaries, a wide array of environmental issues complicated the use of the waters, including sewage and industrial pollutants and urban and agricultural runoff. Also, the impact of river improvements themselves in the form of dredging, canalizing, and dam and reservoir construction would raise serious concerns about silting, land inundation, flooding, and threats to fisheries.

The U.S. Army Corps of Engineers would find themselves in the midst of these controversies. The Corps' status as the lead federal bureau in water resources development was challenged at the turn of the century by the creation of a new federal bureau focused on the arid West. The story of large federal dams thus will involve at its center the political and technical lives of the Corps and the Reclamation Service of 1902 (renamed the Bureau of Reclamation in 1923).[103]

WATER IN THE WEST: ORIGINS OF THE RECLAMATION SERVICE

The West Before the Nineteenth Century

Cultivation of irrigated crops in the West predates the arrival of both Spanish and Anglo-American settlement. For instance, the Hohokam and their predecessors used canal irrigation on the Santa Cruz River in the Tucson area as early as 1,200 b.c., and canal irrigation was well established in the Phoenix area by about 500 a.d. Some of these canals were quite large even by modern standards. In addition, the Pueblo Indians of the Rio Grande Valley were using canal irrigation at the time of Spanish contact in 1540. By the time Europeans first explored the Southwest in the sixteenth century, Hohokam culture had vanished, a victim of unknown environmental or cultural forces. But their canals survived largely intact; in the 1860s they were cleaned out and re-excavated by Anglo-American settlers who transformed them into irrigation canals that still lie at the heart of Phoenix's hydraulic infrastructure. But prehistoric Indian irrigation did not sustain the bulk of native food production in the West; many tribes made no attempt to use riverflow for agriculture, and they had little or no impact on the riparian environment.[104]

In the seventeenth century, the Spanish took control over what later became the southwestern United States, bringing with them an understanding of agricultural techniques suitable for an arid environment. That knowledge supplemented Pueblo irrigation skills. Their settlements in the Southwest involved some development of irrigation, most notably at San Antonio, in the pueblos settlements of the upper Rio Grande, and at the Franciscan missions of California. On the whole, however, Spanish irrigation initiatives were limited in scope and did not involve the construction of large storage dams.[105] Not until large numbers of pioneers from the eastern United States began moving westward in the mid-nineteenth century did interest in large-scale development of western water resources become manifest.

Water and Mormon Migration

The first Anglo-Americans to embrace the possibilities of irrigation technology were Mormon refugees who emigrated to Utah's Salt Lake Valley in the late 1840s. To survive in the wilderness, they quickly began diverting creeks that flowed from the Wasatch Mountains, using the water for crops. This was first accomplished at City Creek in Salt Lake City and quickly spread along the mountains of the "Wasatch Front" that form the eastern edge of the Great Basin.

Mormon settlements centered around the small streams were able to erect numerous irrigation systems that did not depend on large dams or lengthy canals. Extending less than five miles on average, Mormon canals typically supported small communities comprised of farms less than 30 acres in size.[106] Early irrigation systems in Utah comprised a relatively rudimentary technological achievement, but they proved successful in supplying food, and the communal settlements helped to inspire western agrarian development. The Mormon's success in building irrigation-based communities set a precedent for later pioneers seeking to colonize the West.[107]

Most early non-Mormon irrigation development did not depend upon a strong social mission tying together settlers. For example, Anglo-American agricultural settlement in Arizona's Salt River Valley dates to the late 1860s and represents a much more prosaic endeavor. In 1867 "Jack" Swilling, a former Confederate Army officer formed the Swilling Irrigation Canal Company and quickly cleared out an ancient Hohokam canal. Swilling's canal extended a mile-and-a-half across the desert and then curved back toward the Salt River; farm land "under the ditch" could now be cultivated using water from the river and resulting crops could be sold to the army outpost at Fort McDowell. By 1870, the townsite that became Phoenix had been laid out and the Anglo-American settlement of Central Arizona began to grow slowly as it met the needs of the local Army encampment.[108]

California Water Development

In central California agriculture became a major economic activity as early as the 1850s when crops were cultivated for sale in the gold mining camps of the Sierra Nevada. In addition, the fertile lands of the Sacramento River Valley were developed as large farms to export wheat through the busy port of San Francisco. These wheat fields depended upon nutrients and moisture that had accumulated in the soils over hundreds of years and, initially, they did not require irrigation.

As the soils were depleted, the attractiveness of non-irrigated agriculture began to fade while commercial interest in irrigation development increased, especially in the drier lands of the San Joaquin River Valley that lay to the southeast of San Francisco.

By 1886, there were 21 irrigation colonies in the Fresno region, covering 45,000 acres and supporting 7,500 residents. Real estate speculators and large landholding syndicates promoted these colonies, which drew water from the Kings River or (less frequently) from the San Joaquin River.[109]

Irrigation development in the San Joaquin Valley also centered around large tracts of land in the low-lying areas adjoining the river in the region north of Fresno and south of Stockton. By the late 1860s, much of this rich riparian land was under the control of a consortium of influential San Francisco businessmen headed by William Ralston of the Bank of California. Ralston and his partners soon formed the San Joaquin & Kings River Canal and Irrigation Company. In the early 1870s, the company planned a valley-wide irrigation system capable of watering hundreds of thousands of acres of land.[110] Although the company was willing to invest its own money in those parts of the system that would divert water on to the low-lying lands that it directly controlled, they were hesitant to underwrite any broader scheme without government assistance.

Rebuffed by the state legislature, Ralston approached the federal government with hopes of obtaining a large land grant and associated canal rights-of-way that would make the project economical. This plan also failed. However, in early 1873, Congress authorized $6,000 for a Board of Irrigation Commissioners to study the water resources of Central California. Their report did nothing to further Ralston's efforts to obtain federal help, but it was a precursor of large, federally sponsored projects that were implemented in California in the 1930s. Furthermore, the report enhanced the nation's political consciousness that western agriculture represented a potentially major segment of the burgeoning national economy.

The Exploits of John Wesley Powell

By the mid-1880s, most of the small streams in the West had been diverted for irrigation and other uses. It was becoming clear that larger dams on the major rivers would be needed to expand water supplies. In 1888, a Senate resolution called for the Department of the Interior to identify possible reservoir sites and to protect them for future development. Later that year, Congress passed another resolution designating the U.S. Geological Survey as the body to examine the arid region, determine the capacity of streams and where irrigation could be practiced, and arrive at the cost of construction and the capacity of

23

possible reservoirs.[111] It was the 1879 publication of John Wesley Powell's *Report on the Lands of the Arid Region of the United States* that had opened the eyes of Americans to the significance of irrigation development.[112] With Powell as a primary proselytizer, the notion that "reclaimed" land in the West might serve a larger national purpose began to assume momentum.

Born in New York in 1834, Powell moved to Illinois as a young man and informally explored the Mississippi River and its tributaries in the 1850s. He had attended Illinois College, Oberlin, and Wheaton, which helped him develop a sense of self-sufficiency and a facility for self-education. Powell served with the Union Army during the Civil War, losing his lower right arm in the Battle of Shiloh. After mustering out as a Major in 1865, his attention focused on the far West; in 1868 he voyaged down the Green and Colorado Rivers.[113] In 1869 he and a crew of nine men loaded up four small wooden boats and set out down the Colorado River proper. Upon its completion, the expedition marked the first recorded Anglo-American trip through the Grand Canyon.

The success of Powell's journey prompted Congress to support a second expedition through the Grand Canyon two years later. Powell subsequently published and lectured on a variety of western scientific topics centered on the geology of the Colorado River watershed, substantially boosting his stature both within the scientific community and among the public at large as a folk hero. After his successful navigation of the Colorado River, Powell set out to become as knowledgeable as possible about the topography and geology of western America. His *Report on the Lands of the Arid Region* called for creation of a government bureau to explore and classify western lands; this soon led to formation of the Geological Survey within the Interior Department.

Appointed Director of the Geological Survey in 1881, Powell quickly became a major spokesman for development of the West's water resources. Because of his stature as a "scientist," Powell found an audience in Congress and with the American people for his claim that areas of desert land could be "reclaimed" for agriculture by impounding flood waters for use throughout the year. Powell's charismatic lecturing and persuasive writing began a crusade for opening the West to agricultural development through irrigation.[114]

Socially concerned citizens intent on countering the seemingly baneful effects of industrial development in eastern cities saw irrigation as a way for population growth in rural settlements. Recalling Madison's idea of a continental nation and Jefferson's agrarianism, advocates of this irrigation crusade considered western reclamation an ideal means for small family farms to foster the American ideal that life rooted to the soil was better than life despoiled by the "evil" city. Powell became prominently associated with this idealistic crusade.

In 1888, Congress authorized Powell to head the Irrigation Survey to explore the potential for developing western water resources. Among the many young engineers who participated in this work were Frederick Haynes Newell and Arthur Powell Davis, both of whom would later serve as Chief Engineer and Director of the U.S. Reclamation Service.[115] The legislation creating the Irrigation Survey offered no indication that the bureau was conceived as a direct antecedent for federal sponsorship or financing of irrigation projects. However, the legislation did give Powell and his staff the power to "withdraw" public land from entry to prevent private ownership claims from impeding economical construction of a storage reservoir.[116] In essence, the withdrawal authority was to preclude speculators from using information gathered by survey personnel to purchase, at cheap prices, choice public lands that might later be sold for large profits.

Speculative exploitation by large landowners ran counter to the ideal of western irrigation for small-scale, independent farmers. But with no stated mechanism for using the Irrigation Survey's data, the idea that reservoir sites could be closed indefinitely to private development aroused alarm among many westerners who already owned land in the arid region. This consternation found political expression in the person of U.S. Senator William Stewart of Nevada who, in his capacity as chairman of the Senate Select Committee on Irrigation and the Reclamation of Arid Lands, prompted Congress to cut off funding for the Irrigation Survey in the summer of 1890.[117] Furthermore, Congress enacted an 1891 law allowing individuals and private irrigation companies to file claims to reservoir sites on public lands, providing that construction work begin within five years.[118]

The Sentimental and Practical during the 1890s

Powell had come to realize that most of the best irrigable land had already fallen into private hands by the 1890s. He created an uproar by announcing this fact at the 1893 National Irrigation Congress in Los Angeles.[119] The following year, Powell resigned from the Interior Department, becoming a victim of the "triumph of sentimentalism" among those who championed the social primacy of the small family farm.[120]

Opening the West to a new generation of yeoman farmers was popular even among American citizens at large, many of whom had only the vaguest concept of western reclamation. In the early twentieth century, sentiment for promoting family farms in the arid West found political expression in a national reclamation program; but while this federal program would rhetorically espouse the ideals of the irrigation crusade, it would also provide benefits to large tracts of land that had long since been removed from the public domain.

Practical political efforts to promote reclamation also continued. Francis G. Newlands, Nevada's member of the U.S. House of Representatives, Senator Francis E. Warren of Wyoming, and Senator Joseph M. Carey of Wyoming, persisted in seeking support for western agriculture during the 1890s.[121] While it remained politically infeasible to advocate a federally controlled reclamation program, in 1894 Congress did pass the Carey Act, which authorized the federal government to cede up to a million acres of public land to states on their assurance that the acreage would be developed through viable irrigation projects. Eventually, these projects proved important in some northern states such as Idaho, where the act helped fund the Milner Dam and the Twin Falls Canal that irrigated more then 300,000 acres in the Snake River Valley. However, the Carey Act proved too cumbersome, failing to open the public domain to widespread irrigation.[122]

Western reclamation in the 1890s has no simple narrative trajectory leading inevitably to the U.S. Reclamation Service in 1902. Rather, a variety of private interests as well as state and local politicians promoted initiatives to increase agriculture and land values. Among the most prominent of these advocates of western irrigation was George Maxwell, a California lawyer who, in 1897, organized the National Irrigation Association to call for federal legislation benefitting western agriculture. But also important in the adoption of the Reclamation Act of 1902 were several broader factors including: the depression of the 1890s that crippled construction of private irrigation projects in the West and drove down the value of irrigable land; the rapid disposal of public land to grazing interests and land speculators after 1889; and the desire of leading railroads to boost their traffic in the West and to sell farmland.[123]

The Chittenden Survey of 1897

In another strategy to attract federal support for western irrigation, Senator Warren took the model of "river and harbor improvement" so successfully used by eastern states in garnering government assistance. In 1896, he called for a survey of reservoir sites in Wyoming and Colorado to help reduce floods in the Missouri/Mississippi River basin. Warren justified federal support because the Missouri was an interstate river, and because Wyoming, Colorado, New Mexico, Arizona, Utah, and Nevada previously had received no benefits from "river and harbor" improvements authorized by Congress.[124]

Warren won approval in May 1896 for the survey. At his request, the U.S. Army Corps of Engineers appointed Captain Hiram Chittenden to direct it. He traveled throughout the West inspecting irrigation systems, and he examined reservoir sites at the headwaters of the Platte and Laramie Rivers in Wyoming and Colorado. While he held little enthusiasm for flood control from reservoirs placed in the lower Missouri/Mississippi river basin, Chittenden nonetheless recommended the construction of dams at five reservoir sites (two in Colorado, three in Wyoming) that were well-suited to support irrigation development. Avowing that such reservoir construction could "properly be carried out only through public agencies," he also advised that all water stored behind government-built reservoirs be "absolutely free to the people forever, just as the canals, harbors, and other public works are free for general use without toll or levy of any kind."[125]

Chittenden's December 1897 report attracted the interest of western irrigation advocates, but the attention of the nation as a whole was rapidly becoming absorbed by the impending war with Spain. Senator Warren could not entreat Congress to consider an expansion of Chittenden's work until early 1899. In addition, his attempts to utilize a "rivers and harbors" appropriation for dam and reservoir construction were thwarted by eastern congressmen who argued that the Constitution gave a clear mandate for government control over interstate rivers, but the water stored behind upstream reservoirs was not to be owned or controlled by the federal government. Despite intense lobbying and political maneuvering on Warren's part, the 1899 federal "rivers and harbors" appropriation authorized no expenditure for western storage dams, and the issue of federal support for reclamation remained unresolved.[126]

While Congress vacillated in the 1890s, privately financed projects continued to open land for settlement. Some proved viable, others failed.[127] The most prominent state initiative for locally controlled irrigation districts experienced only limited success. That initiative began in 1887, when C. C. Wright, a state legislator in California, won approval for legislation allowing formation of irrigation districts authorized to issue bonds and tax all landowners within a district in order to pay for water supply improvements. Scores of "Wright Act Districts" were established in California during the late 1880s and early 1890s, but in the face of drought, the nationwide financial Panic of 1893, and the difficulties in preventing financial fraud, the Wright Act came to symbolize the generic problems of locally controlled irrigation development.[128]

Newell, Roosevelt and the Move to Reclamation

Although nineteenth-century irrigation settlements in the West proved the feasibility of building diversion dams and distribution canals, the financial risks associated with building large remote storage reservoirs and lengthy feeder canals through rough terrain discouraged private financiers.[129] By the turn of the century, the possibility of "capturing the floods" for widespread irrigation remained much more a vision than a reality. Because huge quantities of flood water were lost to the ocean or dissipated in desert lakes, political support for federal intervention began to grow stronger and found its leading advocate in Frederick H. Newell.[130]

Born in Pennsylvania, Newell graduated as a mining engineer from the Massachusetts Institute of Technology in 1885. He helped Powell administer the Irrigation Survey and, after its demise, became the U.S. Geological Survey's Chief Hydrographer in 1894. Newell proudly associated himself with America's scientific elite.[131] As Chief Hydrographer, his mandate was to measure flows in America's rivers; initially he could not advocate a large-scale federal reclamation program. However, in the late 1890s, this changed with the foundering of the Carey and Wright acts. By 1900, Newell was interacting with key western businessmen (such as George Maxwell) and politicians (such as Francis Newlands) to promote a federal role in irrigation among the Capital's political leadership. For example, in January 1901 Newell gave an evening lecture at Newlands' home where he "showed lantern slides and talked irrigation" to 16 guests, including the Secretary of the Interior and several members of the U.S. Congress.[132]

Newell had attended the contentious 1893 Irrigation Congress which had reacted vehemently to Powell's report on the paucity of prime irrigable land in the public domain. In fact, Newell even gave a speech to the Irrigation Congress echoing Powell's remarks.[133] Despite an awareness of land ownership patterns in the West, Newell projected a romantic image of yeoman farmers populating arid public lands in his 1902 book *Irrigation in the United States*: "Home-making is the aim of this book. . ... The dead and profitless deserts need only the magic touch of water to make arable lands that will afford farms and homes for the surplus people of our overcrowded Eastern cities. . ... The national government, the owner of these arid lands, is the only power competent to carry this mighty enterprise to a successful conclusion."[134]

In 1901, Congressman Newlands—working closely with Newell—submitted a proposal jointly with Senator Henry Hansbrough of North Dakota to fund federal irrigation projects from proceeds derived from public land sales.[135] In addition, Newlands received assistance from George Maxwell who, through his leadership of the National Irrigation Association, continued to lobby on behalf of a national reclamation program. While political support for federal irrigation appeared to be rising within Congress, President McKinley showed little interest.

Everything changed when Theodore Roosevelt became president following McKinley's assassination in September 1901. Roosevelt, the popular "rough rider" of the Spanish-American War, was an irrepressible outdoorsman and an ardent conservationist who would not be afraid to wield the power of the federal government in directly influencing and promoting the nation's economic life. Roosevelt believed that government should play a major role in conserving, and efficiently using, the nation's natural resources. Conservation of water in the West soon became a high priority for Roosevelt. In his first formal message to Congress in December 1901, Roosevelt explicitly endorsed federal support for irrigation by stressing that the construction of "great storage works . . . [had] been conclusively shown to be an undertaking too vast for private effort." Roosevelt proclaimed that "it is as right for the national government to make the streams and rivers of the arid region useful by engineering works for water storage as to make useful the rivers and harbors of the humid region by engineering works of a different kind."[136]

With the new president's vigorous support, Newlands' bill sailed through Congress and Roosevelt signed it into law on June 17, 1902. Advocating a national perspective and a scientific approach to natural resource management, the National Reclamation Act (which is sometimes referred to as the Newlands Act) provided that:

> All moneys received from the sale of public lands in Arizona, California, Colorado, Idaho, Kansas, Montana, Nebraska, Nevada, New Mexico, North Dakota, Oklahoma, Oregon, South Dakota, Utah, Washington and Wyoming . . . are hereby reserved, set aside, and appropriated as a special fund in the Treasury to be known as the "reclamation fund" to be used in the examination and survey for and the construction and maintenance of irrigation works for the storage, diversion and development of waters for the reclamation of arid and semiarid lands in the said states and territories.[137]

As originally enacted, the Act directed the Secretary of the Interior to select irrigation projects without any further review or authorization by Congress. Construction would be undertaken directly by the Department of the Interior, acting through the newly formed U.S. Reclamation Service.[138] Upon completion of each project, the farmers benefitting from increased water supply were to repay all construction costs to the federal government. This was to be accomplished in annual payments made during the first ten years after construction was completed. Theoretically, the repaid money (not to include any interest) would then be available to fund other federal reclamation projects.

Unlike the Carey Act, state governments would play no role in the program's implementation. The National Reclamation Act contained provisions for "reserving" public lands served by irrigation projects to insure that speculators would not take advantage of planned improvements, and it stipulated that farms benefitting from the irrigation be 160 acres or smaller. Tacitly recognizing

John Wesley Powell's observation at the International Irrigation Congress in Los Angeles nine years earlier, the Act also allowed land already in private ownership to receive water from the federal projects without prejudice. But it also reflected the enduring legacy of the "irrigation crusade" by specifying that no farmer operating an irrigated tract larger than 160 acres could benefit from water supplied by a Reclamation Service project. The details of how individual projects were to be developed (and how issues of large-scale landownership within authorized projects would be resolved) were left in the hands of the Secretary of the Interior.

In the summer of 1902, responsibility for administrating the fledgling Reclamation Service fell to Newell who, not surprisingly, became the Reclamation's first Chief Engineer. After years of lobbying and proselytizing, Newell now faced the challenge of actually implementing large-scale dam and water supply projects. Although he possessed no real experience in the construction of major engineering works, Newell could not afford the luxury of slowly learning the skills required to plan, design, construct, and operate reclamation projects. The political circumstances that fostered the establishment of the Reclamation Service also encouraged–and in fact almost demanded–that the new federal bureau prove its service to the nation by building large storage dams in the West as quickly as possible. Only a dozen years had passed since Powell's Irrigation Survey had been curtailed in response to pressure from western interests; but by 1902 the federal government was widely perceived as a savior possessing the technical skills and financial resources necessary to "make the desert bloom" and to open a new era of regional growth.[139]

The stage was set for the great confrontation between the professional and military Corps and the amateur and civilian Reclamation Service. But the amateurs, with strong presidential backing, would quickly become professional just as the Corps, with continuing congressional support, became increasingly civilian.

In the nineteenth century, fresh water was a commodity and rivers needed to be channeled or otherwise modified to 'improve' them. These notions were not abandoned in the early twentieth century, but they were certainly modified by the efficiency concepts imbedded in progressivism, by early stirrings about the multiple uses of rivers, and by growing interest in hydropower. By the late 1920s, the stage was set for the arrival of the Big Dam Era commencing in the 1930s.

PROGRESSIVISM AND WATER RESOURCES

Progressivism, Conservation, and Efficiency

Conservation became an important national issue in the Progressive Era. Proponents were increasingly dismayed by the wanton waste and destruction of natural resources in the name of economic progress. Some, like John Muir, viewed preservation of public lands and pristine waterways as the only way to stave off the worst impulses of the industrial age. But many other conservationists favored the management of natural resources and their efficient use. In his 1907 conservation message to Congress, President Theodore Roosevelt stated:

> As a nation we not only enjoy a wonderful measure of present prosperity but if this prosperity is used aright it is an earnest of future success such as no other nation will have. The reward of foresight for this nation is great and easily foretold. But there must be the look ahead, there must be a realization of the fact that to waste, to destroy, our natural resources, to skin and exhaust the land instead of using it so as to increase its usefulness, will result in undermining in the days of our children the very prosperity which we ought by right to hand down to them amplified and developed.[140]

Proponents did not want to undermine development per se, but questioned short-term private gain at the expense of long-term public benefit. Progressive Era government regulation challenged the notion of unfettered private exploitation of resources by asserting a utilitarian ethic based on "the greatest good for the greatest number." But more than some generalized communal ideal was a commitment to efficient use of those natural resources. Problems could be solved, they believed, if well-trained experts armed with the techniques of applied science and located within the government were the spearheads of change. These experts came from a variety of fields, including hydrology, forestry, agrostology, geology, anthropology, and civil engineering. In the Progressive Era, governmental technical expertise addressed forest depletion through selective harvesting and planting techniques; ranching problems through new forage mixtures, fencing, and the introduction of pure-bred stock; and water use through dam building and new irrigation systems.[141]

Multipurpose Stirrings

During the early years of the Theodore Roosevelt presidency, problems associated with forestry received central attention. But the evolution of American conservation policy depended upon more than the application of scientific forestry practices.¹⁴² To conservationists, issues concerning timber and grass were directly linked to water. Roosevelt often stated that water conservation had to be associated with forest reserves, which preserved watersheds in timbered regions. For his part, Chief Forester Gifford Pinchot supported a concept of management of forest reserves that integrated the protection of watersheds and grazing rights with timber management.¹⁴³ It was western water development in particular that shaped the burgeoning conservation movement in the early twentieth century. The promotion of a federal irrigation program, debate over water rights, the problem of speculation, and concern over siltation "gave rise to extensive ideas about water conservation." Historian Samuel Hays also argued that these issues "became crystallized into an overall approach and by 1908 emerged as a concept of multiple-purpose river development," although that conclusion may exaggerate the actual commitment to multiple-purpose development by more than a decade.¹⁴⁴ The promotion of hydroelectric power—both in the East and the West—also was crucial to the rise of the multiple-purpose movement, but not until the end of World War I.¹⁴⁵

Beginning in the late nineteenth century, hydraulics data and new theories of natural resource development and control helped bring into question water as a single-purpose resource. Interest in irrigation, flood waters, new sources of urban water supply, hydroelectric power, and navigation stimulated promotion of broader economic development plans for whole river basins. Such plans included the protection of watersheds, headwater reservoirs, and coordination of the various water uses.¹⁴⁶ The U.S. Geological Survey is credited with advocating the idea of water as a resource with many uses. The Reclamation Service, which was constructing reservoirs for irrigation purposes, saw the possibility of combining irrigation storage with hydroelectric power production. However, the Reclamation Act (1902) made no provisions for hydropower, and Congress did not authorize the bureau to take up its general development and sale until 1906.¹⁴⁷

Conservation leaders within the Roosevelt administration faced an array of problems raised by various water uses and proposed water uses, but also began to envision the possibilities of basin-wide river development. An emerging viewpoint was to avoid opportunities lost. Pinchot echoed these sentiments: "To develop a river for navigation alone, or power alone, or irrigation alone, is often like using a sheep for mutton, or a steer for beef, and throwing away the leather and the wool."¹⁴⁸

The multiple-purpose approach ultimately reinforced the notion that the federal government needed to take the lead on river development because of the complexity of the issues and because of the many jurisdictions involved. (This was not always popular with state governments, however.) From a practical perspective, the multiple-purpose approach was not only meant to deal with whole river basins, but with such matters as the size, type, design, and purpose of dams.[149]

Attention to inland waterways navigation proved an opportunity for federal officials to implement the multiple-purpose viewpoint. Waterways associations and related groups, particularly in the Mississippi basin, called for federal aid to increase navigable depths along the rivers, but appeared to have little concern for a broader approach. However, a common interest in a deep channel navigable by ocean-going vessels—from the Gulf of Mexico to Chicago—seemed to offer a chance to promote such a plan.[150] Combining the development of hydroelectric power with the navigation goals, the argument went, could provide revenue to pay for the desired river improvements.[151]

Standing in the path of the deep channel was the U.S. Army Corps of Engineers, which blocked efforts at acquiring construction funds for the basin-wide plan. Some believed that the Corps was stubbornly clinging to the single-use philosophy of the past, but the Corps had good reason to regard the multiple use idea as impractical at the time. Hydroelectric power had yet to compete on the open market with other forms of energy. River transportation was facing stiff competition from railroads. And the idea of building dams large and inexpensive enough to be practical had not been tested.[152]

Alternately, Representative Newlands concluded that congressional statutes imposed clear limits on Corps functions and that the Corps itself narrowly interpreted the functions assigned to it by the Congress. Of course, the Corps may have simply been protecting its long-standing leadership role in determining waterways policy, fending off all other contenders. However, several members of Congress also were impediments to multipurpose development. They opposed efforts by the administration to coordinate the activities of agencies concerned with water resource policy because they did not want to have their influence eroded.[153]

W. J. McGee, a geologist and anthropologist, an associate of John Wesley Powell, and a former member of the Geological Survey, was the primary architect and promoter of the new waterways movement connected to the Roosevelt administration. To circumvent the traditionalists in the Corps and the Rivers and Harbors Committee, McGee urged the president to appoint a commission to examine possibilities for integrated river basin development.[154] In 1907, Roosevelt appointed the Inland Waterways Commission (IWC), stating that the time had come to merge "local projects and uses of inland waters in a comprehensive plan designed for the benefit of the entire country."[155] This clearly placed Roosevelt behind multiple-purpose river development. Beginning in

April, the commission devoted much of its time to problems of navigation, but it also appointed a subcommittee to examine the water power issue.[156]

In February, 1908, the commission issued its report recommending that future plans "shall take account of the purification of the waters, the development of power, the control of floods, the reclamation of lands by irrigation and drainage, and all other uses of the waters or benefits to be derived from their control."[157]

Resistance from the Corps to the multiple-purpose approach arose at several junctures. The Corps opposed the recommendation of the Geological Survey's Chief Hydrographer, Marshall O. Leighton, to regulate streamflow with reservoirs. Brigadier General William H. Bixby believed that the hydrographer's data was too limited to make such claims and that the economic feasibility of the idea was questionable. While Bixby's position did not demonstrate overt hostility to the multiple-purpose approach, it did reflect extreme caution in abandoning basic Corps principles and historic practices.[158]

The commission also recommended that a "National Waterways Commission" be established to coordinate the work of the Corps, the Reclamation Service, the Department of Agriculture's Forest Service and Bureau of Soils, and other federal agencies.[159]

But the Corps objected to bureaucratic changes proposed in the report that would undermine its authority and stressed the primacy of navigation in federal river development. This viewpoint carried significant weight in Congress.[160] When Newlands presented a bill to carry out the recommendations of the Inland Waterways Commission—particularly to centralize all water-resource issues under a single agency—it received a frosty reception in the Senate and the bill eventually died.[161]

Ultimately, a joint congressional commission was created by the 1909 Rivers and Harbors Act. While not the vehicle for multiple-purpose river development that advocates hoped, it called for several navigation improvements, regulation of wharves and terminals, prevention of deforestation near mountain streams, and legislation promoting water power development. It also recommended a federal reservoir system for flood control based on multiple-purpose benefits. At this stage, the Corps remained unconvinced that the multiple-purpose approach had broad applicability, although by World War I Congress expanded its program to include flood control along with navigation.[162]

A spirited controversy over the damming of Hetch Hetchy Valley in Yosemite National Park further intensified the debate over water use, and in so doing, also drove a wedge among conservationist groups. In San Francisco, disagreements surrounding the franchise held by Spring Valley Water Company led to provision in 1900 for a municipal water system. Reform mayor James D. Phelan applied to the Secretary of Interior for dam construction permits along the Tuolumne River running through the Hetch Hetchy Valley in the northern part of Yosemite National Park. The secretary denied Phelan's request, but the new Secretary of the Interior, James R. Garfield, accepted the application in 1907 because he was not very interested in guarding resources for aesthetic purposes and because he felt that the 1906 earthquake and fire in San Francisco demonstrated a real need.[163]

San Francisco officials were ecstatic, but opposition mounted. Spring Valley Water Company voiced its objections, as did farmers in Modesto and Turlock, who claimed the water of the Tuolumne. However, the opposition of preservationist John Muir and a throng of wilderness advocates turned the dispute into a national debate[164] The effort to invade the Hetch Hetchy Valley infuriated Muir. "Dam Hetch Hetchy!" he declared, "As well dam for water-tanks the people's cathedrals and churches, for no holier temple has ever been consecrated by the heart of man."[165]

Muir failed to attract support from President Roosevelt, who was torn between his relationship with Muir, concern about the reaction of Californians, and his sympathy for resource conservationism. Muir then began a public campaign to win support for protecting the Hetch Hetchy. Approval of the Hetch Hetchy project was successfully blocked in Congress in 1909, but a bill was passed in 1913 to transfer the proposed site to San Francisco. In 1923, the O'Shaughnessy Dam on the Tuolumne was completed.[166]

The Hetch Hetchy controversy not only shattered Muir's vision of the protection of the Sierra Nevada, but also divided the conservation movement. In the hearing before Congress over the Raker Bill to approve the project, Muir and his allies squared off against Pinchot and supporters of resource conservation. It was a bitter squabble. Pinchot and Muir had been friends and allies in several conservation battles. With Hetch Hetchy, Muir clearly divorced himself from the utilitarian approach that Pinchot had come to represent.[167]

Hetch Hetchy also was about hydroelectric power and to some degree multiple-use. San Francisco had turned to the valley for water, but also identified three hydroelectric sites for future development. To defenders of the Hetch Hetchy, hydroelectric power was "the Trojan Horse of the whole fight" since dam advocates had been cool to seriously consider alternative sites that could provide water but little prospect of hydropower. An amendment to the 1913 act required the city to distribute hydropower from the valley directly to consumers.

This action put the private utility, Pacific Gas and Electric (PG&E), on the same side with Muir, but for different reasons. The city's efforts, however, to contract with PG&E as an agent for Hetch Hetchy power met with resistance from supporters of private power. The debate was settled in 1945 when San Francisco leased a transmission line from PG&E to deliver its power to the city. Public power was defeated at this site, but not the desire for multiple-use.[168]

The Hydroelectric Challenge

From a national perspective, hydropower was a key component in the evolution of multiple-purpose river development and hence in the construction of large federal dams. With respect to the latter, the recurrent use of storage reservoirs to increase capacity is linked to the use of hydropower.[169] The generation of power is one of the prime benefits of running or falling water, and thus an essential resource to be conserved through wise use. It also was considered by proponents of multiple-use as a means to underwrite the cost of dam building and river development in general.

Prior to the advancement of hydroelectric power in the late nineteenth century, almost 66 percent of the waterpower in use in the United States was concentrated in the North Atlantic States (primarily New York and New England). The amount of water horsepower in use by eastern manufacturers far outstripped similar use in the rest of the country. By 1920, however, demand for the distribution of waterpower potential shifted to the Rocky Mountain and Pacific Coast States thanks largely to electricity. By 1890, hydroelectric power had been successfully applied in Europe and was making inroads in the United States.[170]

While a plant in Appleton, Wisconsin, was the first to utilize falling water to generate electricity in the early 1880s, the harnessing of Niagara Falls in the mid-1890s brought major national attention to hydroelectric power.[171] Since the mid-nineteenth century, there had been strong interest in utilizing the water of Niagara Falls for power production. The falls were an excellent choice because of their steady flow and their proximity to large populations. Until the advent of alternating current (AC) and efficient dynamos, the project was impractical. However, as the technology changed and the market for electricity increased, the development of the falls became more practical. In 1895, the first of three 5,000-horsepower AC generators was installed. The completion of the plant marked the beginning of large-scale hydroelectric generation in the United States. With less fanfare, hydroelectric power generation began in the West–as early as 1889 in Oregon, followed by similar ventures in California, Washington, and Montana.[172]

Governmental regulation related to hydroelectricity evolved with the technology. Private hydroelectric dams on waterways in the East and Midwest increasingly interfered with navigation. Urged on by the U.S. Army Corps of Engineers, Congress attempted to regulate dam construction through the Rivers and Harbors Acts of 1890 and 1899, requiring that dam sites and plans for dams on navigable rivers be approved by the Corps and the Secretary of War before construction. Regardless, between 1894 and 1906 Congress issued 30 permits for private dams, mostly along the Mississippi River.[173]

Prior to the twentieth century, waterpower sites in the public domain were claimed by private companies without any effort by the federal government to reserve those sites or regulate their use. Part of the reason was ambiguity over federal jurisdiction, and part was the lag in identifying waterpower as a central feature of water conservation and wise use. The first step toward waterpower conservation occurred in 1901, with passage of the Right-of-Way Act. Although primarily intended as a way of facilitating reclamation and irrigation programs adjacent to public lands, it was broadened to cover many utility functions. The Secretary of the Interior could grant rights-of-way over public lands for dams, reservoirs, waterpower plants, and transmission lines.

Private companies continued to fight for more favorable legislation, but they accepted the permit system. By 1916, power facilities in the national forests represented 42 percent of the total developed power in the western states. President William Howard Taft's appointment of Richard Ballinger as Secretary of the Interior in 1909 weakened the new regulatory scheme. Ballinger refused to apply the Forest Service permit system to waterpower sites on public domain. In 1911, Walter L. Fisher succeeded Ballinger and decided to follow Garfield's policy. He had to contend with the General Land Office, which regarded the permit system as illegal and thus he gave the Geological Survey the responsibility for administering it. The revised permits included a fifty-year limited grant and imposed a waterpower fee ("conservation charge").[174]

These efforts did not resolve the problem of waterpower development on navigable streams. An important issue was the relationship between the multiple-purpose development of waterways and the question of financing such development. In the 1903 veto of private construction of a dam and power stations on the Tennessee River at Muscle Shoals, Alabama, Roosevelt protected the site for later government development, but he also helped to establish the principle of national ownership of resources previously considered only of local value. In this particular case, Roosevelt recommended using revenue from power production to finance navigation improvements.

The General Dam Act of 1906 standardized regulations concerning private power development, requiring dam owners to maintain and operate navigation facilities—without compensation—when necessary at hydroelectric power sites. The act helped to clarify the role of the federal government in safeguarding river navigability, and, in a general way, also strengthened federal regulatory authority in the area of water that already had been established with respect to forest land. A 1910 amendment to the 1906 act more closely linked hydropower to plans for waterway improvements by requiring the Corps to take hydropower development into account when evaluating dam construction permits. The emphasis of the amendment was on hydropower as a financing mechanism for navigation and flood control projects.[175]

The 1906 act and the 1910 amendment, however, engendered strong disagreements of interpretation of water development. Traditionally, the Corps viewed power dams as obstructions to expanding navigation and only slowly was moving toward a broader viewpoint. The Taft administration, like the Corps, looked at dams essentially as obstructions to navigation and was no more supportive of a multiple-purpose approach, approving hydropower franchises that required neither a limited permit nor compensation. Prior to World War I, hydroelectric power development continued to remain a private venture.[176]

In general, the Woodrow Wilson administration showed little interest in conservation issues. For example, Democrats usually had not favored expanding government power to withdraw public lands from use, which limited equal access to resources. However, the issue of private development of hydroelectric power on navigable rivers in the public domain remained a lively issue in the Wilson years.[177]

The desire to improve rivers through human technology had not disappeared in the United States by the end of World War I. Indeed, the demand on water resources had become greater. Large dams, rather than artificial canals and steamboats, would become the primary tool to harness rivers. And during the course of the next several decades, dam building greatly accelerated.

This early twentieth century increase in dam building brought forth a new interest in the structural analysis and design of these huge works. Interest focused especially on the new material of that century, structural concrete, which was beginning to assume a major role in buildings and bridges. Thus, as the politics of dams became more complex, so did the engineering, and it is, therefore, crucial to understand these new technical ideas and to recognize that they can be as controversial as political ideas.

Endnotes

1. Joseph M. Petulla, *American Environmental History* (San Francisco: Boyd and Fraser, 1977), 24.
2. William L. Graf, "Landscapes, Commodities, and Ecosystems: The Relationship Between Policy and Science for American Rivers," in Water Science and Technology Board, National Research Council, National Academy of Sciences, *Sustaining Our Water Resources*,(Washington, D.C.: National Academy Press, 1993). 1-2. See also E. J. Nygren, *Views and Visions: American Landscapes Before 1830* (Washington, D.C.: Corcoran Gallery of Art, 1986).
3. John Seelye, *Beautiful Machine: Rivers and the Republican Plan, 1175-1825* (New York: Oxford University Press, 1991), 8-9. See also Graf, "Landscapes, Commodities, and Ecosystems . . .," 12.
4. Theodore Steinberg, *Nature Incorporated: Industrialization and the Waters of New England* (Amherst: University of Massachusetts Press, 1991), 16.
5. Martin V. Melosi, *Coping with Abundance: Energy and Environment in Industrial America* (New York: Alfred A. Knopf, 1985), 17.
6. Ernest S. Griffith and Charles R. Adrian, *A History of American City Government, 1775-1870* (Washington, D.C.: University Press of America, 1983), 218.
7. Zane L. Miller and Patricia M. Melvin, *The Urbanization of Modern America: A Brief History*, 2nd edition (San Diego: Harcourt Brace Jovanovich, 1987), 31-2, 48-9, 57.
8. Charles E. Brooks, *Frontier Settlement and Market Revolution: The Holland Land Purchase* (Ithaca, New York: Cornell University Press, 1996), 7-9.
9. Ben Moreell, *Our Nation's Water Resources–Policies and Politics* (Chicago: Law, School, University of Chicago, 1956), 15-6.
10. Todd Shallat, "Water and Bureaucracy: Origins of the Federal Responsibility for Water Resources, 1787-1838," *Natural Resources Journal* 32 (Winter, 1992), 5-6.
11. Richard N. L. Andrews, *Managing the Environment, Managing Ourselves: A History of American Environmental Policy* (New Haven, Connecticut: Yale University Press, 1999), 140.
12. Petulla, *American Environmental History*, 28-31. Ellis L. Armstrong, Michael C. Robinson, and Suellen M. Hoy, editors, *History of Public Works in the United States, 1776-1976* (Chicago: American Public Works Association, 1976), 7-8. Andrews, *Managing the Environment, Managing Ourselves*, 71. See also John Opie, *The Law of the Land: Two Hundred Years of American Farmland Policy* (Lincoln: University of Nebraska Press, 1987). Frank Gregg, "Public Land Policy: Controversial Beginnings for the Third Century," in Michael J. Lacey, editor, *Government and Environmental Politics* (Washington, D.C.: Woodrow Wilson Center Press, 1989), 143.
13. Andrews, *Managing the Environment, Managing Ourselves*, 79-88. Peter S. Onuf, *Statehood and Union: A History of the Northwest Ordinance* (Bloomington: Indiana University Press, 1987), xiii-xvii, 40-2.
14. Armstrong, Robinson, and Hoy, editors, *History of Public Works in the United States, 1776-1976*, 7. Andrews, *Managing the Environment, Managing Ourselves*, 88-9.
15. John Lauritz Larson, "A Bridge, A Dam, A River: Liberty and Innovation in the Early Republic," *Journal of the Early Republic* 7 (Winter 1987), 354.
16. Martin Reuss and Paul K. Walker, *Financing Water Resources Development: A Brief History* (Washington, D.C.: Historical Division, Office of Administrative Services, Office of the Chief of Engineers, July 1983), 4-5. See also, Petulla, *American Environmental History*, 111-3.
17. Seelye, *Beautiful Machine*, 8-10.
18. Armstrong, Robinson, and Hoy, editors, *History of Public Works in the United States, 1776-1976*, 23-30. See also Petulla, *American Environmental History*, 114-8. Carol Sheriff, *The Artificial River: The Erie Canal and the Paradox of Progress, 1817-1862* (New York: Hill and Wang, 1996).
19. See Andrews, *Managing the Environment, Managing Ourselves*, 41-2.
20. The Industrial Revolution brought steampower to supplant humans and animals. However, the origins of large-scale manufacturing, factory production, and resource extraction at first depended on fuelwood for iron production, and waterpower for integrated New England textile mills. Not until the 1880s did steam exceed water as an energy source for industrial expansion in the United States. Louis C. Hunter, *A History of Industrial Power in the United States, 1780-1930 Volume One: Water in the Century of the Steam Engine* (Charlottesville: University Press of Virginia, 1979), 139. Martin V. Melosi, *Coping with Abundance*, 22-3.
21. Morton J. Horwitz, *The Transformation of American Law, 1780-1860* (Cambridge, Massachusetts: Harvard University Press, 1977), 34.
22. Hunter, *Water in the Century of the Steam Engine*, 54-5, 140.
23. Hunter, *Water in the Century of the Steam Engine*, 115, 141, 151.
24. Steinberg, *Nature Incorporated*, 141.

25. However, since surface water was abundant, few conflicts over water rights needed to be resolved in the eastern state courts in the nineteenth century. See William F. Steirer, Jr., "Riparian Doctrine: A Short Case History for the Eastern United States," in John W. Johnson, *Historic U.S. Court Cases, 1690-1990: An Encyclopedia* (New York: Garland Publishing, 1992), 312.
26. Hunter, *Water in the Century of the Steam Engine*, 33. See also 140-2, 145.
27. Horwitz, *The Transformation of American Law*, 1780-1860, 35-6.
28. The greatest departure from riparian principles occurred in Massachusetts, where "mill acts" had granted privileges to mill owners since the colonial period.
29. Horwitz, *The Transformation of American Law*, 37, 40-2. See also Steinberg,, *Nature Incorporated*, 146-7.
30. Horwitz, *The Transformation of American Law*, 42.
31. Horwitz, *The Transformation of American Law*, 43. See also Hunter, *Water in the Century of the Steam Engine*, 141. Donald J. Pisani, *To Reclaim a Divided West: Water, Law, and Public Policy 1848-1902* (Albuquerque: University of New Mexico Press, 1992), 11-2.
32. See Opie, *The Law of the Land*, 108-9.
33. Samuel P. Hays, *Conservation and the Gospel of Efficiency: The Progressive Conservation Movement, 1890-1920* (New York: Atheneum, 1972; originally published 1959), 16-7. See also Robert G. Dunbar, *Forging New Rights in Western Waters* (Lincoln: University of Nebraska Press, 1983), 59-60.
34. Richard White has argued that appropriative rights in the West were "a local implementation of a larger national modification of the law, not a western innovation." This is a useful refinement, but does not contradict the point that the application of appropriative rights in the West had local roots. See his *"It's Your Misfortune and None of My Own": A History of the American West* (Norman: University of Oklahoma Press, 1991), 401. See also Donald J. Pisani, "State vs. Nation: Federal Reclamation and Water Rights in the Progressive Era," *Pacific Historical Review* 51 (August 1982), 266-7. Patricia Nelson Limerick, *The Legacy of Conquest: The Unbroken Past of the American West* (New York: W. W. Norton, 1987), 72.
35. Donald J. Pisani, *From the Family Farm to Agribusiness: The Irrigation Crusade in California and the West, 1850-1931* (Berkeley: University of California Press, 1984), 30-1.
36. Walter Prescott Webb, *The Great Plains* (Boston: Ginn and Company, 1931), 3-44.
37. See Lawrence B. Lee, *Reclaiming the American West: An Historiography and Guide* (Santa Barbara, California: ABC-Clio Press, 1980). Donald J. Pisani, "Deep and Troubled Waters: A New Field of Western History," *New Mexico Historical Review* 63 (October 1988), 311-31. See also Lawrence B. Lee, "Water Resource History: A New Field of Historiography?" *Pacific Historical Review* 57 (November 1988), 457-67. Pisani, *From the Family Farm to Agribusiness*. Pisani, *To Reclaim a Divided West*. Donald Worster, *Rivers of Empire: Water, Aridity and the Growth of the American West* (New York: Pantheon, 1985). Norris Hundley, *Water and the West: The Colorado River Compact and the Politics of Water in the American West* (Berkeley: University of California Press, 1975). Norris Hundley, *The Great Thirst: Californians and Water, 1770-1990* (Berkeley: University of California Press, 1992). Stanley Davison, *The Leadership of the Reclamation Movement: 1875-1902* (New York: Arno Press, 1979). and James Earl Sherow, *Watering the Valley: Development Along the High Plains of the Arkansas River* (Topeka: University Press of Kansas, 1990).
38. Useful histories include Worster, *Rivers of Empire*. Marc Reisner, *Cadillac Desert: The American West and Its Disappearing Water* (New York: Viking Press, 1986). William A. Warne, *The Bureau of Reclamation* (New York: Praeger Press, 1973). Michael Robinson, *Water for the West* (Chicago: Public Works Historical Society, 1979). Karen Smith, *The Magnificent Experiment: Building the Salt River Project, 1870-1917* (Tucson: University of Arizona Press, 1986). Joseph Stevens, *Hoover Dam: An American Adventure* (Norman: University of Oklahoma Press, 1988). In addition, the role of the federal government in supporting the construction of the Los Angeles project in the Owens Valley has been described in William L. Kahrl, *Water and Power: The Conflict over Los Angeles' Water Supply in the Owens Valley* (Berkeley, University of California Press, 1982). Abraham Hoffman, *Vision or Villainy: Origins of the Owens Valley-Los Angeles Water Controversy* (College Station, Texas: Texas A&M Press, 1981).
39. The need for capital in hydraulic mining helped to promote the success of prior appropriation by the mid-1850s.
40. Those states are Alaska, Arizona, Colorado, Idaho, Montana, New Mexico, Nevada, Utah, and Wyoming. See Hundley, *The Great Thirst*, 69-73. Dunbar, *Forging New Rights in Western Waters*, 61-3.
41. Hays, *Conservation and the Gospel of Efficiency*, 16-7.
42. Hundley, *The Great Thirst*, 84-5.
43. Hundley, *The Great Thirst*, 88.
44. Pisani, *From the Family Farm to Agribusiness*, 246-7. For a thorough discussion of the legal context for western water law from 1848 to 1902, see Pisani, *To Reclaim a Divided West*, 11-68.

45. According to Donald J. Pisani, "After the Desert Land Act was passed in 1877, it was impossible to assert riparian rights–at least not if someone had claimed water under prior appropriation before a final patent was issued to that land." Donald J. Pisani to David P. Billington, February 15, 1998.

46. Hundley, *The Great Thirst*, 95-7. Dunbar, *Forging New Rights in Western Waters*, 67-72. Pisani, *From the Family Farm to Agribusiness*, 246-7. Pisani, *To Reclaim a Divided West*, 35-6.

47. Pisani, *From the Family Farm to Agribusiness*, 248. White, *"It's Your Misfortune and None of My Own,"* 402. Dunbar, *Forging New Rights in Western Waters*, 78-98. In Utah, however, appropriative rights developed by the Mormons were unique. Based on the Mormon's "ecclesiastical brotherhood," they divided rights into primary and secondary categories. Within each category water was divided proportionately among the users. See Dunbar, *Forging New Rights in Western Waters*, 82.

48. Dunbar, *Forging New Rights in Western Waters*, 99-112. White, *"It's Your Misfortune and None of My Own,"* 402.

49. Andrews, *Managing the Environment, Managing Ourselves*, 60-1.

50. Pisani, *To Reclaim a Divided West*, 31-2, 37-8.

51. Arthur A. Ekirch, Jr., *Man and Nature in America* (New York: Columbia University Press, 1963), 81, 83, 88. G. Michael McCarthy, *Hour of Trial: The Conservation Conflict in Colorado and the West, 1891-1907* (Norman: University of Oklahoma Press, 1977),15. Petulla, *American Environmental History*, 217-8. Roderick Frazier Nash, editor, *American Environmentalism: Readings in Conservation History* (New York: McGraw-Hill Publishing Company, 1990), 36, 45. Craig W. Allin, *The Politics of Wilderness Preservation* (Westport, Connecticut: Greenwood Press, 1982, 24-36. See also Roderick Nash, *Wilderness and the American Mind* (New Haven: Yale University Press, 1967).

52. Petulla, *American Environmental History*, 221-5. Ekirch, *Man and Nature in America*, 87. Nash, editor, *American Environmentalism*, 59. Allin, *The Politics of Wilderness Preservation*, 19-23. McCarthy, *Hour of Trial*, 15.

53. See Carolyn Merchant, editor, *Major Problems in American Environmental History* (Lexington, Massachusetts: D.C. Heath and Company, 1993), 338. Robert Gottlieb, *Forcing the Spring: The Transformation of the American Environmental Movement* (Washington, D.C.: Island Press, 1993), 19-21. Petulla, *American Environmental History*, 217. Nash, editor, *American Environmentalism*, 10-11. McCarthy, *Hour of Trial*, 3-4, 12.

54. See Ekirch, *Man and Nature in America*, 82. See also Elmo R. Richardson, *The Politics of Conservation: Crusades and Controversies, 1897-1913* (Berkeley: University of California Press, 1962), 2-3, 6.

55. W. Stull Holt, *The Office of the Chief of Engineers of the Army: Its Non-Military History, Activities, and Organization* (Baltimore: Johns Hopkins Press, 1923), 2.

56. Todd Shallat, *Structures in the Stream: Water, Science, and the Rise of the U.S. Army Corps of Engineers* (Austin: University of Texas Press, 1994), 2. Terry S. Reynolds, "The Engineer in 19th-Century America," in Terry S. Reynolds, editor, *The Engineer in America* (Chicago: University of Chicago Press, 1991), 7.

57. Prior to 1750, the state was the primary employer of engineers throughout Europe. The growth of commercial operations and the acceleration in industrial development, however, enabled self-trained civilian engineers—especially in Great Britain—to practice their occupation. See Reynolds, "The Engineer in 19th-Century America," 7-9. Shallat, *Structures in the Stream*, 14-5, 18.

58. Reynolds, "The Engineer in 19th-Century America," 8-9.

59. Holt, *The Office of the Chief of Engineers of the Army*, 2. See also Forest G. Hill, *Roads, Rails & Waterways: The Army Engineers and Early Transportation* (Norman: University of Oklahoma Press, 1957), 5-7.

60. Shallat, *Structures in the Stream*, 3, 92-3.

61. Reynolds, "The Engineer in 19th-Century America," 16. Todd Shallat, "Building Waterways, 1802-1861: Science and the United States Army in Early Public Works," *Technology and Culture* 31 (January 1990), 25-6. Edwin Layton, "Mirror-Image Twins: The Communities of Science and Technology in 19th-Century America," *Technology and Culture* 12 (October 1971), 570. Hill, *Roads, Rails & Waterways*, 12-9. See also Stephen E. Ambrose, *Duty, Honor, Country: A History of West Point* (Baltimore: Johns Hopkins Press, 1966), 22-3, 62-8, 90-2, 97-100, 100-3.

62. Shallat, "Building Waterways, 1802-1861," 22.

63. Martin Reuss and Charles Hendricks, "The U.S. Army Corps of Engineers: A Brief History," 1-2. The history is available under "history" on the U.S. Army Corps of Engineers web site, www.usace.army.mil/.

64. Louis C. Hunter, *Steamboats on the Western Rivers: An Economic and Technological History* (Cambridge: Harvard University Press, 1949). See also Melosi, *Coping with Abundance*, 20-2.

65. Shallat, "Water and Bureaucracy," 5.

66. Reuss and Walker, *Financing Water Resources Development*, 6. Hill, *Roads, Rails & Waterways*, 25-26, 154-61. Isaac Lippincott, "A History of River Improvement," *Journal of Political Economy* (1914), 634.

67. Reuss and Walker, *Financing Water Resources Development*, 3. Allan H. Cullen, *Rivers in Harness: The Story of Dams* (Philadelphia: Chilton Company, 1962), 35.

68. Livingston and Fulton had died by that time.

69. Francis N. Stites, "A More Perfect Union: The Steamboat Case," in Johnson, *Historic U.S. Court Cases*, 187-93.

70. Stites, "A More Perfect Union," 190.

71. See Beatrice Hort Holmes,, *A History of Federal Water Resources Programs, 1800-1960* (Washington, D.C.: Economic Research Service, U.S. Department of Agriculture, Miscellaneous Publication No. 1233, June 1972), 3. Reuss and Walker, *Financing Water Resources Development*, 8. Shallat, "Water and Bureaucracy," 14-5. Shallat, *Structures in the Stream*, 125-7.

72. Michael Robinson, "The United States Army Corps of Engineers and the Conservation Community: A History to 1969" (draft manuscript, November 1982), (Files, Office of History, Headquarters, U.S. Army Corps of Engineers, Alexandria, Virginia), 3. See also Frank E. Smith, *The Politics of Conservation* (New York: Pantheon Books, 1966), 71.

73. Holmes, *A History of Federal Water Resources Programs*, 3.

74. Robert W. Harrison, *History of the Commercial Waterways & Ports of the United States,* Volume I (Fort Belvoir, Virginia: U.S. Army Engineer Water Resources Support Center, 1979), 5. Reuss and Walker, *Financing Water Resources Development*, 6. Michael C. Robinson, *History of Navigation in the Ohio River Basin* (Alexandria, Virginia: U.S. Army Corps of Engineers, Water Resources Support Center, 1983), 12. Holmes, *A History of Federal Water Resources Programs*, 3. Charles F. O'Connell, Jr., "The Corps of Engineers and the Rise of Modern Management, 1827-1856," in Merritt Roe Smith, editor, *Military Enterprise and Technological Change: Perspectives on the American Experience* (Cambridge, Massachusetts: MIT Press, 1985), 97.

75. Holmes, *A History of Federal Water Resources Programs*, 4. Aubrey Parkman, *History of the Waterways of the Atlantic Coast of the United States* (Washington, D.C.: Institute for Water Resources, January 1983), 43. Reuss and Walker, *Financing Water Resources Development*, 6. Reuss and Hendricks, "The U.S. Army Corps of Engineers," 3.

76. Shallat, *Structures in the Stream*, 98. See also Reuss and Walker, *Financing Water Resources Development*, 6-7.

77. The board was in existence for eight years until the Corps of Topographical Civil Engineers were placed in a separate bureau. Its responsibilities included determining general possibilities for improvement of roads, canals, and railways on a national scale. See Hill, *Roads, Rails & Waterways*, 63-6.

78. Holt, *The Office of the Chief of Engineers of the Army*, 6-7. Reuss and Hendricks, "The U.S. Army Corps of Engineers," 3. Hill, *Roads, Rails & Waterways*, 56-57.

79. Robinson, *History of Navigation in the Ohio River Basin*, 13-4. Hunter, *Steamboats on the Western Rivers*, 193-8, 202. See also Hill, *Roads, Rails & Waterways*, 164. Lippincott, "A History of River Improvement," 639.

80. According to Louis Hunter, "Bars acted as dams to hold back and conserve the water supply during the dry seasons. Their elimination would simply result in the stabilization of the depth of water at a proportionately lower level, to the prejudice of navigation. To deepen the channel over a bar without increasing the 'expense' of water and draining the pool above, the [Board of Engineers] proposed the construction of dikes of timber and stone to concentrate the flow of water at a lower stage within a limited space. The flow of water thus concentrated would cut a deeper channel for the benefit of navigation. This was the hope, at any rate, and the act of 1824 provided for experiments on certain bars to determine the practicality of the scheme." See Hunter, *Steamboats on the Western Rivers*, 201.

81. The Italians and the French had experimented many years earlier with structures that would enlarge a channel's carrying capacity by increasing a stream's velocity. See Shallat, "Building Waterways, 1802-1861," 41-2. Shallat, "Water and Bureaucracy," 19-20.

82. Robinson, *History of Navigation in the Ohio River Basin*, 12-6. Hunter, *Steamboats on the Western Rivers*, 200-3.

83. Hunter, *Steamboats on the Western Rivers*, 205-7. See also Robinson, *History of Navigation in the Ohio River Basin*, 17-8.

84. In 1838, the Topographical Bureau was given the assignment of directing all river and harbor work. All new work was assigned to it and some works in progress also were transferred. However, the Corps of Topographical Engineers remained an independent unit only until 1863. See Hill, *Roads, Rails & Waterways*, 177. See also Frank N. Schubert, editor, *The Nation Builders: A Sesquicentennial History of the Corps of Topographical Engineers, 1838-1863* (Fort Belvoir, Virginia: Office of History, U.S. Army Corps of Engineers, 1988).

85. Robinson, *History of Navigation in the Ohio River Basin*, 16-7. Hill, *Roads, Rails & Waterways*, 173. Holt, *The Office of the Chief of Engineers of the Army*, 8-10. Hunter, *Steamboats on the Western Rivers*, 203.

86. Parkman, *History of the Waterways of the Atlantic Coast of the United States*, 47. See also Holt, *The Office of the Chief of Engineers of the Army*, 10-1. Shallat, *Structures in the Stream*, 106. Hill, *Roads, Rails & Waterways*, 181-98.

87. Hill, *Roads, Rails & Waterways*, 205. Robinson, *History of Navigation in the Ohio River Basin*, 22.

88. Parkman, *History of the Waterways of the Atlantic Coast of the United States*, 49. Reuss and Walker, *Financing Water Resources Development*, 12.

89. Robinson, *History of Navigation in the Ohio River Basin*, 23-25. Hunter, *Steamboats on the Western Rivers*, 210.

90. Robinson, *History of Navigation in the Ohio River Basin*, 25. Hunter, *Steamboats on the Western Rivers*, 209. Leland R. Johnson, *The Davis Island Lock and Dam, 1870-1922* (Pittsburgh: U.S. Army Engineer District, 1985), 34-5.

91. Robinson, *History of Navigation in the Ohio River Basin*, 26. Hunter, *Steamboats on the Western Rivers*, 212.

92. Robinson, *History of Navigation in the Ohio River Basin*, 26-27. Hunter, *Steamboats on the Western Rivers*, 211-2. Reuss and Hendricks, "U.S. Army Corps of Engineers," 5. Johnson, *The Davis Island Lock and Dam, 1870-1922*, 37-8. See also Francis H. Oxx, "The Ohio River Movable Dams," *Military Engineer* 27 (January-February 1935), 49-58.

93. Hunter, *Steamboats on the Western Rivers*, 212. See also John P. Davis, "Locks and Mechanical Lifts: State of the Art" (Paper prepared for the National Waterways Roundtable, Norfolk, Virginia, April 22-24, 1980), 6.

94. Roald D. Tweet, *History of Transportation on the Upper Mississippi & Illinois Rivers* (Washington, D.C.: Government Printing Office, 1983), 51-52. William Patrick O'Brien, Mary Yeath Rathbun, Patrick O'Bannon, and W Christine Whitacre, *Gateways to Commerce - The U.S. Army Corps of Engineers' 9-foot Channel Project on the Upper Mississippi River* (Denver: National Park Service, 1992), 19. For more information on the debate over channelization, see Hunter, *Steamboats on the Western Rivers*, 213-5. John R. Ferrell, "From Single- to Multi-Purpose Planning: The Role of the Army Engineers in River Development, 1824-1930" (Draft manuscript, Historical Division, Office of the Chief of Engineers, February 1976), 11-25. Martin Reuss, "Andrew A. Humphreys and the Development of Hydraulic Engineering: Politics and Technology in the Army Corps of Engineers, 1850-1950," *Technology and Culture* 26, No. 1 (January 1985), 1-2. Luna B. Leopold and Thomas Maddock, Jr., *The Flood Control Controversy: Big Dams, Little Dams, and Land Management* (New York: Ronald Press Company, 1954), 98-9.

95. John O. Anfinson, "Commerce and Conservation on the Upper Mississippi River," *The Annals of Iowa* 52 (Fall 1993), 387-8.

96. Hunter, *Steamboats on the Western Rivers*, 214.

97. Tweet, *History of Transportation on the Upper Mississippi & Illinois Rivers*, 54.

98. Jane Lamm Carroll, "Mississippi River Headwaters Reservoirs" (*Historic American Engineering Record*, Report No. MN-6), Prints and Photographs Division, Library of Congress, 2-8. See also Jane Lamm Carroll, "Lake Winnibigoshish Reservoir Dam" (*Historic American Engineering Record*, Report No. MN-65). Jane Lamm Carroll, "Leech Lake Reservoir Dam" (*Historic American Engineering Record*, Report No. MN-67). Jane Lamm Carroll, "Lake Pokegama Reservoir Dam" (*Historic American Engineering Record*, Report No. MN-66). Jane Lamm Carroll, "Pine River Reservoir Dam" (*Historic American Engineering Record*, Report No. 68). Jane Lamm Carroll, "Sandy Lake Reservoir Dam and Lock" (*Historic American Engineering Record*, Report No. MN-69). Jane Lamm Carroll, "Gull Lake Reservoir Dam" (*Historic American Engineering Record*, Report No. MN-70). See also Raymond H. Merritt, *The Corps, the Environment, and the Upper Mississippi River Basin* (Washington D.C.: Historical Division, Office of the Chief of Engineers, 1981), 1.

99. Carroll, "Mississippi River Headwaters Reservoirs," 11-5. See also Anfinson, "Commerce and Conservation on the Upper Mississippi River," 389-90.

100. Merritt, *The Corps, the Environment, and the Upper Mississippi River Basin*, 3-10. See also Carroll, "Mississippi River Headwaters Reservoirs," 23-30.

101. Merritt, *The Corps, the Environment, and the Upper Mississippi River Basin*, 13-9.

102. *Ibid.*, 19-20. See also Philip V. Scarpino, *Great River: An Environmental History of the Upper Mississippi, 1890-1950* (Columbia, Missouri: University of Missouri Press, 1985), 153.

103. See Ferrell, "From Single- To Multi-Purpose Planning," 307. For figures on the amounts of congressional appropriations for river and harbor improvements between 1820 and 1894, see I. Y. Schermerhorn, "The Rise and Progress of River and Harbor Improvement in the United States," *Journal of the Franklin Institute* 139 (January-June 1895), 261.

104. Early irrigation by Native Americans is described in R. Douglas Hurt, *Indian Agriculture in America: Prehistory to the Present* (Lawrence: University Press of Kansas, 1987), 19-26. See also Hundley, *The Great Thirst*, 16-24.

105. For more on Spanish irrigation in the Southwest, see Michael C. Meyer, *Water in the Hispanic Southwest: A Social and Legal History, 1550-1850* (Tucson: University of Arizona, 1984). For information on Old World irrigation, see J. A. Garcia-Diego, "The Chapter on Weirs in the Codex of Juanelo Turriano: A Question of Authorship," *Technology and Culture* 17 (April 1976), 217-34. and J. A. Garcia-Diego, "Old Dams in Extramadura" *History of Technology* 2 (1977), 95-124.

106. Mormon irrigation is discussed in detail in Charles H. Brough, *Irrigation in Utah* (Baltimore: The Johns Hopkins University Press, 1898). See also George Thomas, *The Development of Institutions under Irrigation with Special Reference to Early Utah Conditions* (New York: MacMillan Company, 1920). See Leonard J. Arrington, *Great Basin Kingdom: Economics History of the Latter-Day Saints* (Lincoln: University of Nebraska Press, 1958) for a complete overview of Mormon economic development in the nineteenth century.

107. Leonard J. Arrington and Dean May, "A Different Mode of Life: Irrigation and Society in Nineteenth Century Utah," *Agricultural History* 49 (January 1975), 3-20. See also Brough, *Irrigation in Utah*, 75. Additional discussion of small-scale Mormon agricultural colonies can be found in Charles Penrose, "Utah Colonization Methods," *Proceedings of the Twelfth National Irrigation Congress* (Galveston, Texas, 1905), 361-9.

108. For information on nineteenth-century irrigation colonies (and irrigation settlements in general) that were based upon private financing, see Pisani, *To Reclaim a Divided West*, 69-104.

109. See Karen Smith, *The Magnificent Experiment*, 4-5.

110. See Pisani, *From the Family Farm to Agribusiness*, 121-8.

111. W. Turrentine Jackson, Rand F. Herbert, and Stephen R. Wee, introduction, reprint of *Engineers and Irrigation: Report of the Board of Commissioners on the Irrigation of the San Joaquin, Tulare and Sacramento Valleys of the State of California, 1873* (Engineer Historical Studies Number 5) (Fort Belvoir, Virginia: Office of History of the U.S. Army Corps of Engineers, 1990), 10-2.

112. Thomas G. Alexander, "The Powell Irrigation Survey and the People of the Mountain West," *Journal of the West* 7 (January 1968), 48-9.

113. Historically the Colorado River started at the confluence of the Green and Grand Rivers. In 1921 the name of the Grand was officially changed to the Colorado River above that confluence so that the Colorado extended to its current headwaters in Rocky Mountain National Park.

114. John Wesley Powell, *Report on the Lands of the Arid Region in the United States* (Washington, D.C.: Government Printing Office, 1879).

115. Historiographical analysis of the irrigation movement in this era can be found in Lee, *Reclaiming the American West*, 6-15. Important early studies include John T. Ganoe, "The Beginnings of Irrigation in the United States," *Mississippi Valley Historical Review* 25 (1938), 59-78. See also John T. Ganoe, "The Origins of a National Reclamation Policy," *Mississippi Valley Historical Review* 18 (June 1931), 34-52.

116. The work of the Irrigation Survey is documented in the *Annual Reports of the U.S. Geological Survey from 1888-89 through 1892-93*. For example, see John Wesley Powell, *Tenth Annual Report of the United States Geological Survey, Part II - Irrigation* (Washington, D.C.: Government Printing Office, 1890). See also Pisani, *To Reclaim a Divided West*, 143-68. Davison, *Leadership of the Reclamation Movement*. See also and Everett Sterling, "The Powell Irrigation Survey, 1888-1893," *Mississippi Valley Historical Review* 27 (1940), 421-34.

117. *Tenth Annual Report of the United States Geological Survey, Part II - Irrigation*, 22-9, describes the process of withdrawing lands from "sale, entry, settlement or occupation."

118. Pisani, *To Reclaim a Divided West*, 150-65.

119. S. T. Harding, *Water Rights for Irrigation: Principles and Procedures for Engineers* (Palo Alto, California: Stanford University Press, 1936), 130-1, describes and analyzes the requirements and legal procedures stipulated by the Act of March 3, 1891.

120. Davison, *Leadership of the Reclamation Movement*, 127-63.

121. For the views of a major proponent of sentimentalism, see William E. Smythe, *The Conquest of Arid America* (New York, privately printed: 1900). The 1905 edition of this book was reprinted by the University of Washington Press in 1969.

122. Pisani, *To Reclaim a Divided West*, 169-272.

123. Discussion of the Carey Act can be found in Lee, *Reclaiming the American West*, 12-3. See also Pisani, *To Reclaim a Divided West*.

124. Davison, *Leadership of the Reclamation Movement*, 245. Pisani, *To Reclaim A Divided West*, 285-94.

125. Pisani, *To Reclaim a Divided West*, 275.

126. For more on Chittenden's work in reclamation planning see Gordon Dodds, *Hiram Martin Chittenden: His Public Career* (Lexington: University Press of Kentucky: 1973), 29-41. His report to Congress was formally distributed as: U.S. Congress, House, "Preliminary Examination of Reservoir Sites in Wyoming and Colorado," prepared by Hiram Martin Chittenden, 55th Congress, 2nd session, 1897, House Document 141, series 3666. Government Printing Office. Pisani, *To Reclaim a Divide West*, 276-8, provides good discussion of Chittenden's project.

127. See Pisani, *To Reclaim a Divided West*, 279-85, for extended discussion of the politics that underlay the rejection of Warren's efforts to utilize "rivers and harbors" expenditures for irrigation work.

128. The collapse of the Bear Valley Irrigation Company in Southern California is noted in Davison, *Leadership of the Reclamation Movement*, 188-9.

129. For extended discussion of the Wright Act and the difficulties that accompanied its implementation, see Pisani, *From the Family Farm to Agribusiness*, 252-80.

130. Smith, *The Magnificent Experiment*, 6. Undeniably, there had been noteworthy successes. For example, by the beginning of the twentieth century, over 110,000 acres were under irrigation in the greater region of Phoenix-Tempe-Mesa in Arizona Territory; in slightly more than thirty years the population of greater Phoenix had grown from almost nothing to more than 20,000 settlers. Nonetheless, no major storage dam had been built to capture the flood flow of the Salt River, and the limits of irrigation based upon "natural flow" long since had been reached.

131. For biographical information on Newell see his professional obituary: "Frederick Haynes Newell," *Transactions of the American Society of Civil Engineers* 98 (1933), 1597-1600.

132. Newell's relationship with Gifford Pinchot, the first Director of the U.S. Forest Service and a leader in the conservation movement, is documented in M. Nelson McGeary, *Gifford Pinchot: Forester - Politician* (Princeton: Princeton University Press, 1960).

133. Frederick Haynes Newell Papers, diary entry for January 8, 1901. Manuscript Division, Library of Congress, Washington, D.C.

134. Davison, *Leadership of the Reclamation Movement*, 150-1.

135. Frederick Haynes Newell, *Irrigation in the United States* (New York, 1902), 1, 406. The original edition of this book appeared before passage of the 1902 Reclamation Act. A second edition was published in 1906 that included a small amount of new material related to activities of the Reclamation Service.

136. Newlands has become prominently associated with federal reclamation thanks to his co-sponsorship of the National Reclamation Act. However, his contributions to the irrigation movement as a whole appear limited. For more on his career as a lawyer and politician, see William D. Rowley, *Reclaiming the Arid West: The Career of Francis G. Newlands* (Bloomington: University Press of Indiana, 1996).

137. The complete text of Roosevelt's first annual message to Congress sent on December 3, 1901, can be found in Theodore Roosevelt, *State Papers as Governor and President, 1899*, in *The Works of Theodore Roosevelt, National Edition 15* (New York: C. Scribner and Sons, 1926), 81-135.

138. The full text and all amendments to the National Reclamation Act prior to 1919 appear in: Institute for Government Research, *The U.S. Reclamation Service: Its History, Activities and Organization* (New York: D. Appleton and Company, 1919), 104-124. The placement of the federal reclamation program in the Department of the Interior represented a victory for Newell and the Interior Department over the bureaucratic forces of Elwood Mead and others involved in irrigation investigations for the Department of Agriculture. See Lee, *Reclaiming the American West*, 22-3. Pisani, *To Reclaim a Divided West*, 306-10

139. Between 1902 and 1907 the Reclamation Service operated as a part of the U.S. Geological Survey. In 1907 it became an independent bureau within the Interior Department.

140. Cited in Leon Fink, editor, *Major Problems in the Gilded Age and the Progressive Era* (Lexington, Massachusetts: D.C. Heath and Company, 1993), 412.

141. See Hays, *Conservation and the Gospel of Efficiency*, 1-2. In the cities, government officials passed anti-pollution laws, scientists established laboratories for testing water purity, engineers built elaborated sanitation systems, and reformers promoted programs of "civic responsibility." See Martin V. Melosi, editor, *Pollution and Reform in American Cities, 1870-1930* (Austin: University of Texas Press, 1980).

142. Ekirch, *Man and Nature in America*, 94. Guy-Harold Smith, editor, *Conservation of Natural Resources* (New York: John Wiley & Sons, Inc., 1971; 4th edition), 7.
143. Petulla, *American Environmental History*, 267, 271-2.
144. Hays, *Conservation and the Gospel of Efficiency*, 5.
145. Hays, *Conservation and the Gospel of Efficiency*, 5.
146. Reuss and Hendricks, "U.S. Army Corps of Engineers," 11-2. Ekirch, , *Man and Nature in America*, 94.
147. John R. Ferrell, "From Single- to Multi-Purpose Planning," 91.
148. Gifford Pinchot, *The Fight for Conservation* (Seattle: University of Washington Press, 1967; originally, 1910), 54.
149. Hays, *Conservation and the Gospel of Efficiency*, 100-2.
150. The Lakes-to-the-Gulf-Waterway.
151. Ferrell, "From Single- to Multi-Purpose Planning," 96-7.
152. For an interesting parallel argument about the Corps and its views on stream flow, see Gordon B. Dodds, "The Stream-Flow Controversy: A Conservation Turning Point," *Journal of American History* 56 (June 1969), 59-69.
153. Ferrel, "From Single- to Multi-Purpose Planning," 103, 125-6.
154. Smith, *The Politics of Conservation*, 89, 101, 106. Hays, *Conservation and the Gospel of Efficiency*, 102-5, 108-9. Ferrell, "From Single- to Multi-Purpose Planning," 105-6. Donald C. Swain, *Federal Conservation Policy, 1921-1933* (Berkeley: University of California Press, 1963), 96-8.
155. Quoted in Ekirch, , *Man and Nature in America*, 94.
156. Hays, *Conservation and the Gospel of Efficiency*, 105-8. Ekirch, , *Man and Nature in America*, 94-5. Ferrell, "From Single- to Multi-Purpose Planning," 115-7. Moreell, *Our Nation's Water Resources—Policies and Politics*, 37.
157. "Report from the Inland Waterways Commission, February 26, 1908," in Frank E. Smith, editor, *Conservation in the United States: A Documentary History* [Five volumes] (New York: Chelsea House, 1971), 91.
158. See Ferrell, "From Single- to Multi-Purpose Planning," 125-30. Paul K. Walker, "Developing Hydroelectric Power: The Role of the U.S. Army Corps of Engineers, 1900-1978," (Washington, D.C.: Department of the Army, Office of the Chief of Engineers, draft, February 1979), 29.
159. Holmes, *History of Federal Water Resources Programs*, 6.
160. "Supplementary Report of Commissioner General Alexander Mackenzie," in Smith, editor, *Conservation in the United States*, 93-94. See also Reuss and Walker, *Financing Water Resources Development*, 15. Ferrell, "From Single- to Multi-Purpose Planning," 118-9.
161. Hays, *Conservation and the Gospel of Efficiency*, 109-14. Newlands was able to get his legislation for a new Inland Waterways Commission passed in 1917. However, President Wilson was too deeply involved in the war effort to appoint commissioners, and conservationists were too severely split to force the appointments. See Swain, *Federal Conservation Policy, 1921-1933*, 97-8.
162. Holmes, *History of Federal Water Resources Programs*, 6. Ferrell, "From Single- to Multi-Purpose Planning," 123.
163. Hundley, *The Great Thirst*, 170-4. James C. Williams, *Energy and the Making of Modern California* (Akron: University of Akron Press, 1997), 250-1. Stephen Fox, *The American Conservation Movement: John Muir and His Legacy* (Madison: University of Wisconsin Press, 1981), 139-40. Petulla, *American Environmental History*, 282.
164. Michael P. Cohen, *The Pathless Way: John Muir and American Wilderness* (Madison: University of Wisconsin Press, 1984), 280-1. Douglas H. Strong, *The Conservationists* (Menlo Park, California: Addison-Wesley Publishing Company, 1971), 91-104. Hundley, *The Great Thirst*, 178-80. Peter Wild, *Pioneer Conservationists of Western America* (Missoula, Montana: Mountain Press Publishing Company, 1979), 1-17.
165. Quoted in Cohen, *The Pathless Way,* 330.
166. Strong, *The Conservationists*, 110, 114-5. Hundley, *The Great Thirst*, 176-84. Williams, *Energy and the Making of Modern California,* 251-2. Richardson, *The Politics of Conservation*, 44-5. See also Ben Robert Martin, "The Hetch Hetchy Controversy: The Value of Nature in a technological Society," (PhD dissertation, Brandeis University, 1982), 232-305. Smith, *The Politics of Conservation,* 134-7. Eric B. Kollgaard and Wallace L. Chadwick, editors, *Development of Dam Engineering in the United States* (New York: Pergamon Press, 1988), 49, 84, 86.
167. Richard White, "It's Your Misfortune and None of My Own," 413. See also Williams, *Energy and the Making of Modern California*, 251-2. Strong, *The Conservationists*, 104, 111. Fox, *The American Conservation Movement: John Muir and His Legacy*, 111, 115, 130, 140-6.

168. Fox, *The American Conservation Movement*, 141. Williams, *Energy and the Making of Modern California*, 252-3. Hundley, *The Great Thirst*, 187-90. Petulla, *American Environmental History*, 282-3. Strong, *The Conservationists*, 114-5..

169. Hunter, *Waterpower in the Century of the Steam Engine*, 513.

170. Hunter, *Waterpower in the Century of the Steam Engine*, 137-9, 388, 512.

171. Previous advances in equipment and materials for dam building made possible the construction of the high dams necessary for power production. See Reuss and Walker, *Financing Water Resources Development*, 36.

172. Reuss and Walker, *Financing Water Resources Development*, 138. See also Martin V. Melosi, *Coping with Abundance*, 64, 81.

173. Reuss and Hendricks, "U.S. Army Corps of Engineers," 11-2. Reuss and Walker, *Financing Water Resources Development*, 36. Walker, "Developing Hydroelectric Power: The Role of the U.S. Army Corps of Engineers, 1900-1978", 8-9.

174. Harold T. Pinkett, *Gifford Pinchot: Private and Public Forester* (Urbana: University of Illinois Press, 1970), 61-62. Andrews, *Managing the Environment, Managing Ourselves*, 142-4. Melosi, *Coping with Abundance*, 81-2. McGeary, *Gifford Pinchot: Forester-Politician*, 73, 75, 77-8.

175. Reuss and Hendricks, "U.S. Army Corps of Engineers," 12. Reuss and Walker, *Financing Water Resources Development*, 36-37. Hays, *Conservation and the Gospel of Efficiency*, 114-9. Holmes, *A History of Federal Water Resources Programs, 1800-1960*, 7. Pinkett, *Gifford Pinchot: Private and Public Forester*, 63. Melosi, *Coping with Abundance*, 81-2. Walker, "Developing Hydroelectric Power: The Role of the U.S. Army Corps of Engineers, 1900-1978," 15, 21-2.

176. The Corps did install a power station substructure at Lock and Dam #1 on the upper Mississippi, however. See Reuss and Hendricks, "U.S. Army Corps of Engineers," 12. Reuss and Walker, *Financing Water Resources Development*, 36-7.

177. Smith, *The Politics of Conservation*, 126, 141. Petulla, *American Environmental History*, 281. Arthur S. Link, *Wilson: The New Freedom* (Princeton: Princeton University Press, 1956), 128.

CHAPTER 2

THEORIES AND COMPETING VISIONS FOR CONCRETE DAMS

In this chapter, we describe the history of ideas about design and analysis of concrete dams. These include stone masonry dams as well. We reserve for a later chapter (chapter 6) the presentation of embankment dams where they can be put in the context of the largest federal embankment dams ever built. The design of embankment dams is part of the introduction of the twentieth century field of soil mechanics (Geotechnical Engineering).

This chapter, therefore, presents engineering principles for concrete (and stone masonry) dam design and describes two contrasting traditions of dam design. Dams in the massive tradition rely on their weight for safety while those in the structural tradition rely more on their form. From these two there evolved analytic techniques that attempted to integrate the two traditions, and that the Bureau of Reclamation defined as the trial-load method. This chapter provides the engineering context for considering economic implications of these designs and gives insight into how federal agencies chose the forms for these dams. The U.S. Army Corps of Engineers, a leader in embankment dams, did not enter the early debate about these two traditions because it began to design high dams in concrete in the 1930s. Its contributions in concrete dam design will be considered in later chapters (chapters 5 and 8) in the context of major dams in the Columbia and Ohio basins.

PRE-TWENTIETH CENTURY THEORIES: THE MASSIVE TRADITION

Massive or Structural

For thousands of years, societies have stored water, altered river flow, and transformed environments to increase food production or achieve other social and economic goals. The oldest known dams, small earthen structures built 6,000 years ago in present-day Jordan, were designed to capture rainfall for agricultural and domestic use before the water evaporated or sank into the desert sands.[1]

Compared with intricate technologies such as an automobile engine or a "spinning jenny" in a textile mill, dams represent relatively simple structures designed to control water, which is relatively the same around the world. Thus, the art of dam-building exhibits patterns that often transcend particular features of individual cultures or environments. This is evident in two distinct dam design traditions–the massive and the structural–that extend back to ancient times and found expression in the United States during the nineteenth century. They represent two relatively distinct approaches to the problem of storing water.

In its most elementary form, a dam in the massive tradition consists of a mass of material that, by its weight alone, holds back a volume of water. Such structures are known as gravity dams, an appropriate name because it is the force of gravity pulling vertically down on the dam that provides resistance against pressure exerted horizontally by water in the reservoir.[2] Designs adhering to the massive tradition can be based upon sophisticated engineering analysis, but the basic principle underlying the tradition is simple: accumulate as much material as economically or physically possible, thus ensuring that the dam will not tip over, slide, or rupture; in turn, the massive dimensions will increase the likelihood that the dam can achieve long-term stability in holding back a reservoir.

In many locales, earth and rock are available in large quantities, and often they do not require any complicated machinery for excavation or transportation. Although massive embankments comprised of earth and loose rock (often called rockfill) are susceptible to erosion or washouts, dams made of these materials can function quite successfully in impounding water; in fact, many dams built in the ancient world were embankment dams of earth and/or rock.[3] However, such dams require some kind of relatively impervious barrier—for example, a layer of dense earthen clay, a surface of timber planks, or a concrete slab—either on top of the upstream face or within the dam's interior. Otherwise, seepage and percolation through the structure can undermine the dam and cause collapse. To combat seepage, early dam builders filled the space between loose rock or masonry blocks with mortar to create a waterproof barrier well suited for a dam's upstream face. Massive designs of this latter type are usually termed masonry gravity dams, a name that refers to solid structures comprised of stone blocks, concrete, or some mixture of these two.

Structural dams have existed for centuries with Roman engineers being credited for the first arch dam and several buttress dams. After the collapse of the Roman Empire, the structural tradition (along with large-scale dam building in general) waned in western civilization; but during the thirteenth and fourteenth centuries, Ilkhanid Mongols in Persia constructed arch dams including the 190-foot-high Kurit Dam that—although unrecognized until recent times—stood as the world's tallest dam for 500 years.[4]

Gravity Dam Design Theory

Dam building flourished in late sixteenth century Spain, which was then the richest country in Europe. These early dams included the Almans, a curved

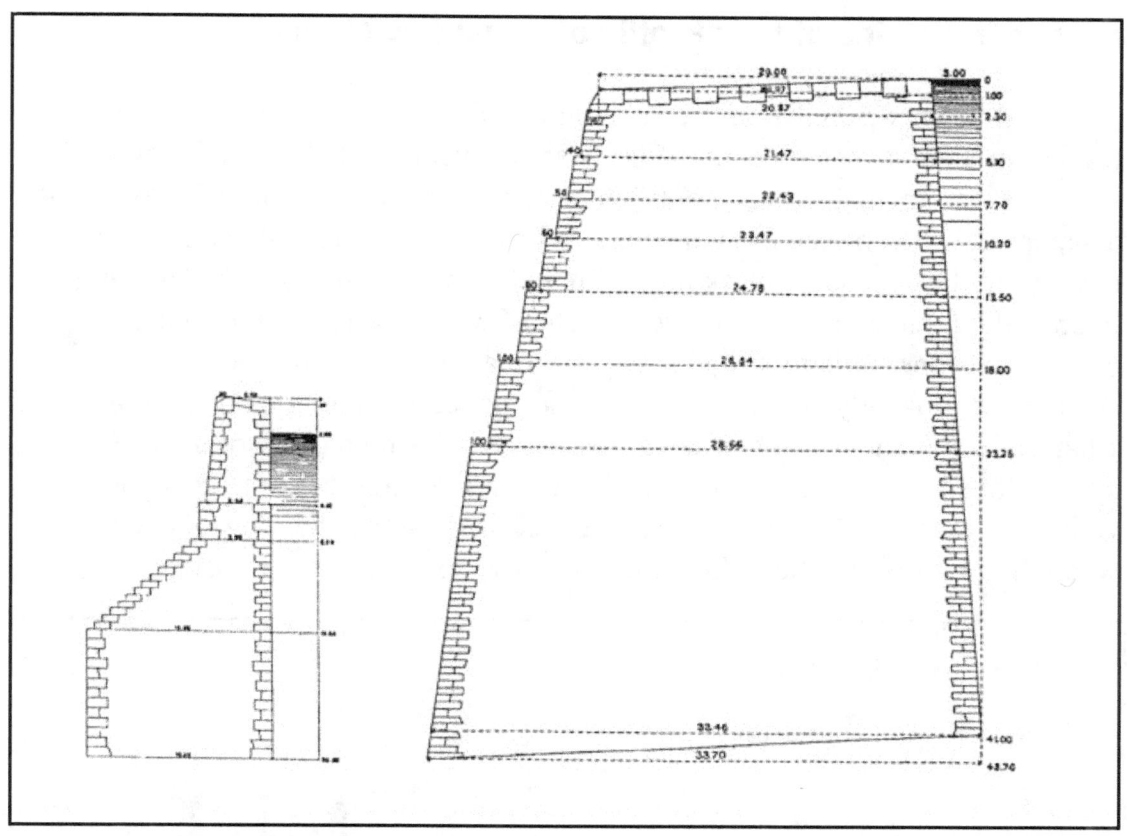

Figure 2-1: Spanish dams of the sixteenth century began the modern era. The most important are shown in these cross sections of the fifty-foot high Almans curved gravity dam of 1586 (left) and the one hundred forty-foot high Alicante Dam of 1594 (right). Source: Edward Wegmann, Jr., *The Design and Construction of Masonry Dams: Giving the Method Employed in Determining the Profile of the Quaker Bridge Dam,* (New York: John Wiley and Sons, 1889), 2nd edition revised, plate 21 (left) and plate 22 (right).

gravity dam (50 feet high, in operation by 1586), the Alicante gravity dam (140 feet high, 1594), the Elche arch dam (76 feet high, ca. 1650), and the Rellue arch dam (105 feet high, ca. 1650).[5] During this period, Spanish engineers codified the construction principles, and by 1736, Don Pedro Bernardo Villa de Berry (a Basque nobleman) had outlined geometrical rules that pointed toward a less intuitive approach to proportioning dams.[6]

Prior to the late eighteenth century, dam builders had not utilized mathematics to help calculate dimensions. Gradually, this began to change, as techniques of physical logic, promulgated by Isaac Newton and Robert Hooke, found their way into engineering practice.[7] By the early nineteenth century, several engineers in France and England had published treatises on gravity dam theory.[8] Although these works did not have any immediate or dramatic effect, they established useful precedents for adapting mathematical theory to the practice of dam design.

European Origins of the "Profile of Equal Resistance"

Masonry gravity dams can be built without any reliance upon mathematics, but in the nineteenth century European engineers realized that this type of structure was amenable to a quantifiable approach to design. In the early 1850s, a paper published by the French engineer J. Augustin Tortene de Sazilly set the course for all subsequent work in this area of gravity dam design.[9] Knowing the hydrostatic force exerted by a given height of water (which weighs about 62.5 pounds per cubic foot) and the approximate weight of masonry used in dam construction (usually about 140-150 pounds per cubic foot), de Sazilly conceived what he termed the "profile of equal resistance." Using basic formulas of statics, he developed a cross-section in which compressive stresses at the upstream face when the reservoir is empty equal compressive stresses at the downstream face when the reservoir is filled. In taking these two extreme conditions, he

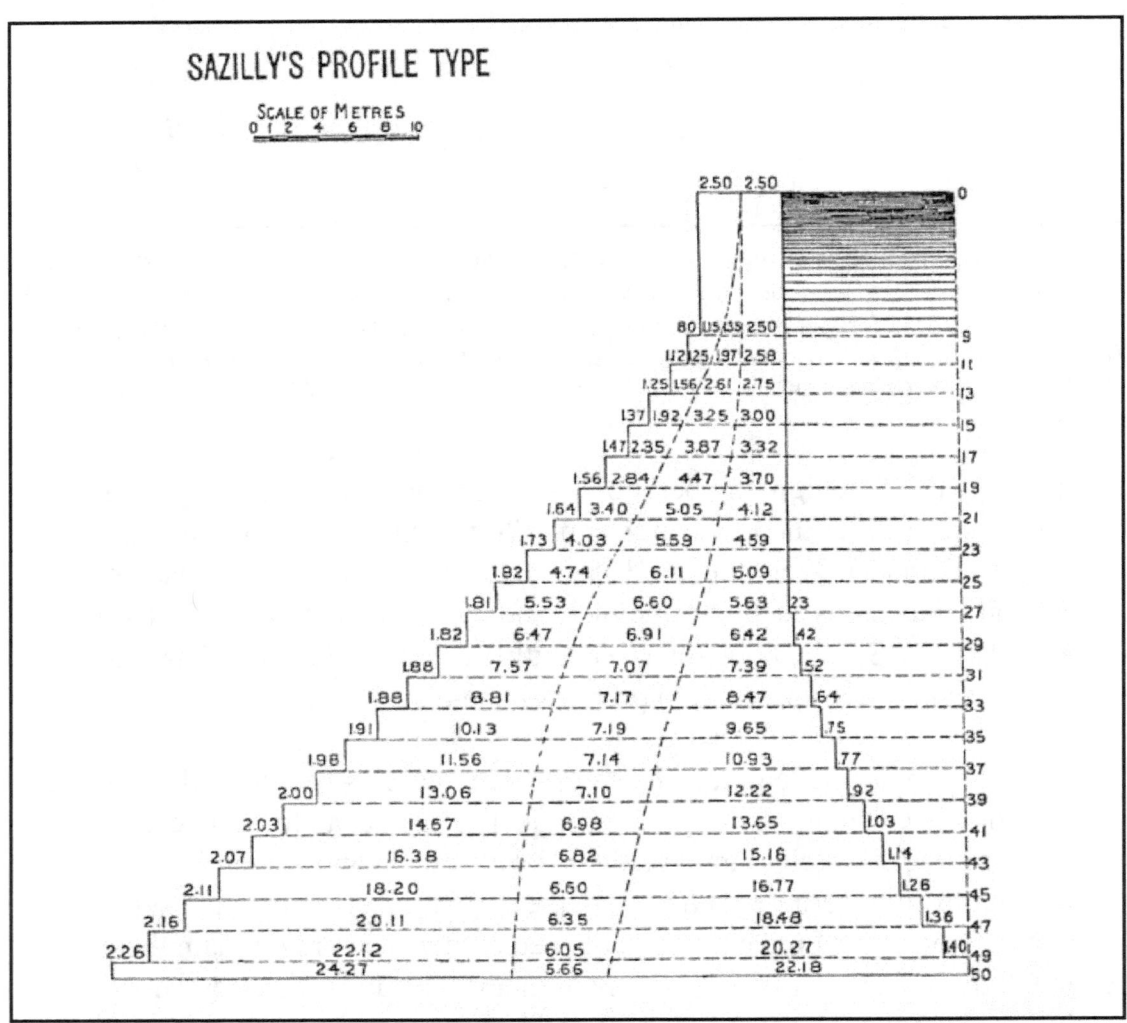

Figure 2-2: The course for designing masonry gravity dams was set in 1850 by French engineer J. Augustine DeSazilly who conceived the "profile of equal resistance". In 1858 F. Emile DeLocre applied this theory to the 183-foot high Furens Dam across the Loire River. Source: Wegmann, *Design and Construction of Masonry Dams,* 2nd edition, Plate 1.

hypothesized a design that, at least in cross-section, would minimize the material necessary to erect a stable masonry gravity dam.

In 1858 the French engineer F. Emile Delocre utilized de Sazilly's theory to develop a "profile of equal resistance" for the 183-foot-high Furens Dam across the Loire River. In formulating his design, Delocre empirically analyzed

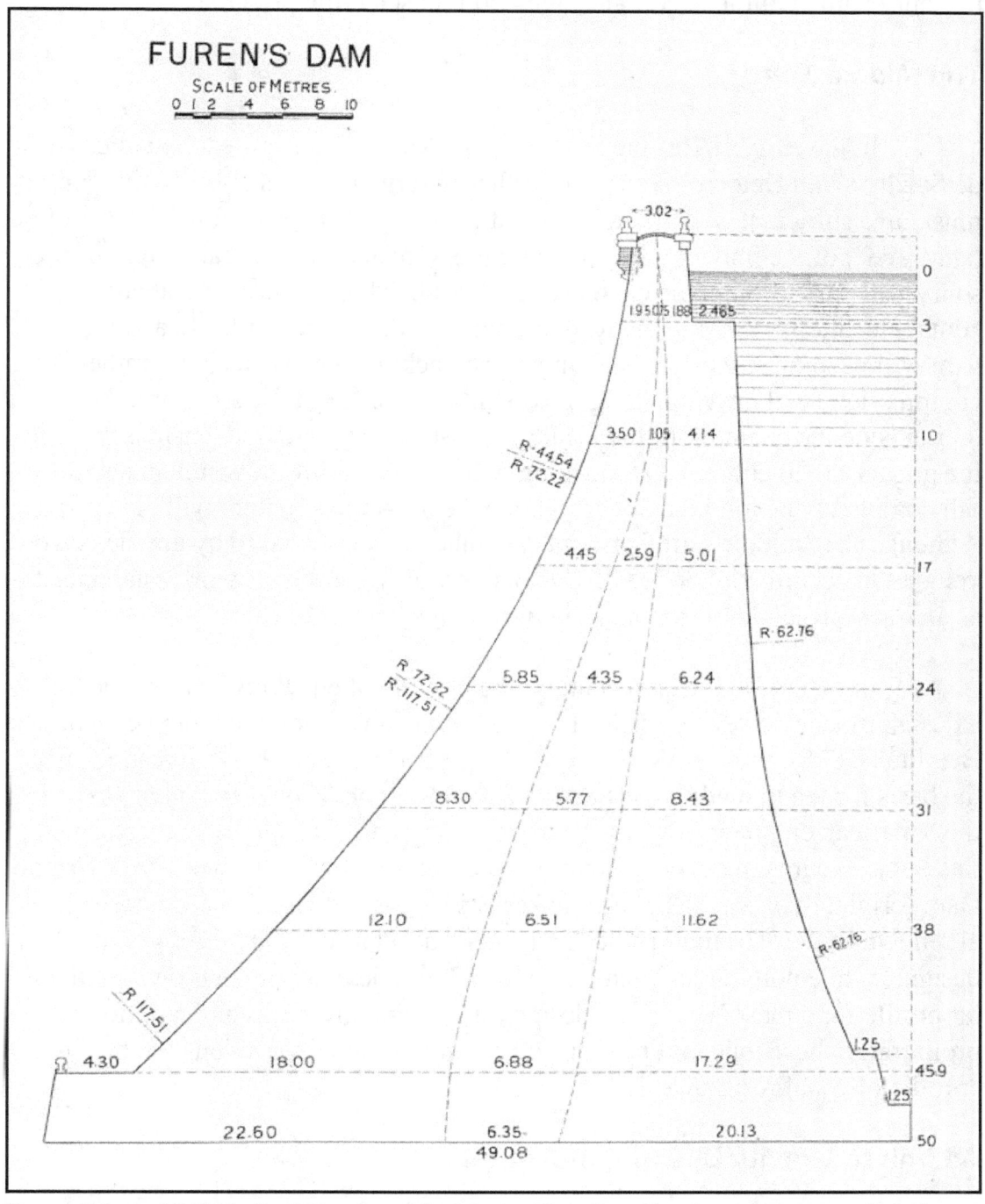

Figure 2-3: In 1858 Emile DeLocre applied de Sazilly's "profile of equal resistance" design method to the Furen's Dam across the Loire River. Source: Wegmann, *Design and Construction of Masonry Dams,* 2nd edition, plate 36.

several long-standing dams (including Almansa and Alicante in Spain) and calculated that a compressive stress of 86 pounds per square inch (psi) could be safely withstood by a masonry dam; in turn, this provided him with a compressive stress that he believed he could safely use for his own design.[10] Completed in 1866, the curved gravity Furens Dam contained over 52,000 cubic yards of masonry. Within a few years, textbooks on gravity design heralded it as the first masonry dam "built in accordance with correct scientific principles."[11]

The Middle Third

In the early 1870s, the Scot W. J. M. Rankine confirmed the validity of de Sazilly's and Delocre's work; he further observed that a stable gravity dam must have sufficient cross-section so that the combined vector force (or "resultant force") of the horizontal hydrostatic pressure and the vertical weight of masonry will pass through the center (or middle) third of the structure at any horizontal elevation.[12] Should the resultant fall outside the center third, a gravity dam will become susceptible to dangerous cracking because tension (rather than compression) will develop along the downstream edge of the structure; the farther outside the center third the resultant passes, the greater the tensile stress and the greater the likelihood that cracking will occur. And if the resultant should fall completely beyond the downstream edge, then the structure will "overturn." Although the "middle third" precept was inherently adhered to by any design developed in accord with de Sazilly/Delocre profiles, Rankine's work established it as an overt principle of masonry gravity design.[13]

In the late nineteenth century, the "profile of equal resistance" served as a design model for several major European dams including the Gileppe Dam in Belgium (1875), and the Vyrnwy Dam in Great Britain (1890).[14] It also formed the basis for the first edition (1888) of *The Design and Construction of Dams* by the American engineer Edward Wegmann. Using the design method presented in this book, Wegmann developed a cross-section for New York City's New Croton Dam (originally it was to be the Quaker Bridge Dam) that achieved international renown as the "Croton Profile" and served as a basic standard for gravity dam design. While nothing prevented engineers from developing their own particular profile for a masonry gravity design (and seemingly profound variation did proliferate), these innovations represented only minimal variations on the basic "profile of equal resistance."

A Limit to Gravity Design Innovation

In 1897, E. Sherman Gould authored a monograph entitled *High Masonry Dams* in which he praised: "the masterly treatise of Mr. Edward Wegmann," and observed that "the mathematical researches have established a vertical section, the base basis of which is a right-angled triangle of base equal to two-thirds or three-quarters of its height. . . . The most refined calculations will inevitably bring us back to the neighborhood of this form." He proposed that we should "start our designs by first laying down such a triangle, surmounting it by a practical top width instead of its own sharp apex, and, if its height exceeds 80 to 100 feet, giving a flare to the lower part of the part of its inside face to expand the footing on that side."[15]

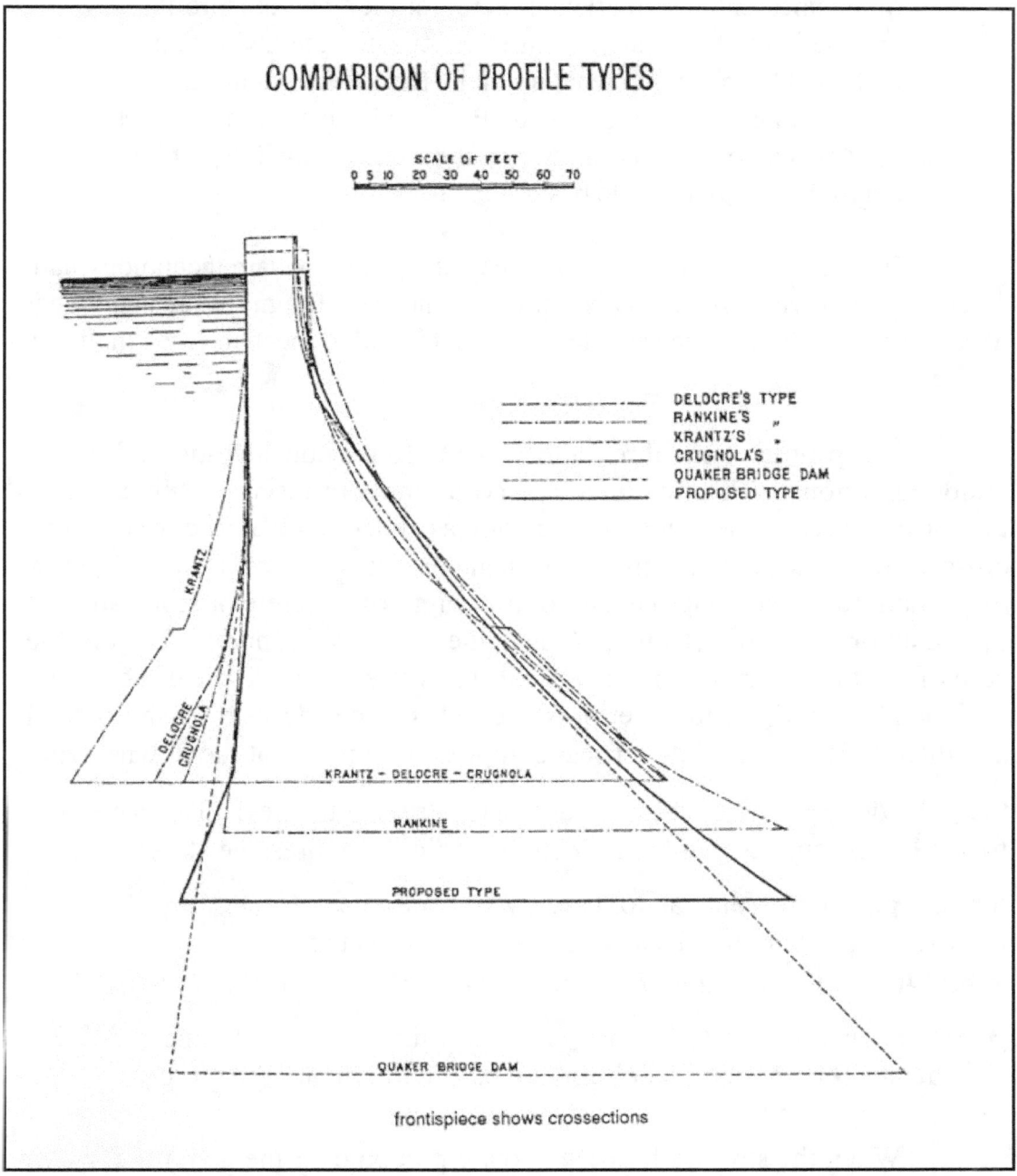

Figure 2-4: Edward Wegmann's first edition of Design and Construction of Masonry Dams appeared in 1888 with 109 pages, 59 plates, 18 figures, and 14 pages of advertising for cement, drills, hoses, steam pumps, etc. Wegmann's expanded eighth edition in 1927 contained 740 pages, 11 drawing plates, 291 figures, and no advertising. Source: Wegmann, *Design and Construction of Masonry Dams,* 2nd edition, frontispiece.

More than 15 years after Gould's observations, the character of gravity dam design was reiterated by George Holmes Moore in a 1913 *Engineering News* article that describes practically all masonry gravity dam profiles as featuring "the two-three triangle" in which the ratio of height-to-thickness is at least 3:2. As Moore further observed:

> In possibly no other branch of dam design is the amplification of the unessential so marked as here, for the 'theoretic profile' and the 'hyperbolic-curve' nonsense [of some gravity design methods] heaped upon what might be termed the standard gravity section is astounding indeed. Pages, chapters, even volumes are devoted to a discussion of gravity profiles which depart but negligibly from a simple basic section.[16]

By the early part of the twentieth century, gravity dam technology had reached a point where relatively empirical methods of design (based upon what Moore termed a "two-three triangle") were sufficient to meet any real engineering needs.

The profile of equal resistance came from a consideration of two major conditions of dam loading: reservoir empty or reservoir filled. For the former case, the dead load of the dam, assumed to be a pure triangle in cross section, caused a maximum vertical compressive stress f_1^h at the heel of the dam (upstream edge) equal to the weight of concrete or stone above that point or $f_1^h = Hwc$ (Height H times the density of concrete ^wc). For the case of the full reservoir, to the vertical stress one must add the effect of the horizontal force F due to water pressure. This force causes the dam to bend and thus creates maximum vertical compressive stresses at the downstream toe $f_2^t = Hw_w \frac{H^2}{B^2}$ with equal vertical tensile stress at the heel. The criterion for least resistance is that the maximum vertical compressive stress for case one be the same as for case two, hence $Hw_c = Hw_w \frac{H^2}{B^2}$ or $\frac{H^2}{B^2} = \frac{w_c}{w_w}$. For example, where the density of concrete w_c is taken to be 140 pounds per cubic foot and the density of water w_w to be 62.5 pounds per cubic foot, then $\frac{H^2}{B^2} = \frac{140}{625} = 2.25$ so that $\frac{H}{B} = \sqrt{2.25} = 1.5$ or about 3/2. This means for a dam 60 feet high, the base width would be 40 feet.

When the stresses for case two are plotted over the dam base, we find that they form a triangle with the maximum value at the downstream toe and the minimum (equals zero) at the heel. The centroid of that pressure lies at B/3 from the downstream toe. Likewise, for the reservoir empty in case one, the centroid lies at B/3 from the heel. Thus, the centroids of all loading cases between one and two lie between those two positions or within the middle third of the dam width B. When the stresses for case two are plotted over the dam base, we find that they form a triangle with the maximum value at the toe and the minimum (equals zero) at the heel. The centroid of that pressure lies at B/3 from the toe.

Of course, interest in other issues relating to gravity design did not remain stagnant and this is best reflected in concern over the influence of "uplift" on the safety of gravity structures. Uplift is a phenomena resulting from water seeping under/through the foundation (or into the interior of the dam proper) that—because of pressure exerted by water in the reservoir—pushes upward and increases the likelihood that the structure will slide horizontally downstream. Uplift attracted the attention of engineers in the early twentieth century and encouraged both the use of thicker profiles as well as the development of grouting and drainage techniques that would mitigate its occurrence and possible effect.

The 1911 failure of a gravity dam in Austin, Pennsylvania, led the American engineering profession to look more closely at the influence of uplift on dam safety, especially as it related to sliding. In addition to the force of the water and weight of the dam, the water pressure underneath the dam produces uplift while the cohesion between dam and rock resists sliding. In addition, the friction between dam and foundation (usually rock) will resist sliding in proportion to the vertical force W less the uplift. Neglecting cohesion and assuming full uplift on a dam where $B/H = 2/3$, we find the safety factor against sliding to be less than one. This result helps explain the Austin Dam failure, where $B/H = 0.6$, and investigations after failure led to the conclusion of substantial uplift. Part of the solution was to increase B/H and also to drain the base to relieve the pressure and hence reduce the uplift force to 0.5 or less.[17]

The most significant drawback to gravity designs involved their high cost. While the "profile of equal resistance" offered a mathematically rational basis of design, this did not mean that gravity dams would necessarily be cheap to build. For major municipalities, the economic benefits that accompanied an increased water supply might easily justify the huge expenditures required to build large masonry gravity designs. But once cities such as Boston (with the Wachusetts Dam completed in 1904) and New York (with the New Croton Dam completed in 1907) erected masonry gravity structures as part of major civic improvement projects, the technology came to represent—at least in many people's eyes—the most conservative, the most appropriate, and, if at all economically feasible, the most desirable type of dam.

PRE-TWENTIETH CENTURY THEORIES - THE STRUCTURAL TRADITION

Gravity Dams versus Structural Dams

A dam in the structural tradition in contrast to gravity designs does not rely exclusively upon bulk; rather, it depends upon its shape—and not simply its mass—to resist hydrostatic pressure. For example, an arch dam in a narrow canyon with hard rock foundations allows a significant amount of the hydrostatic pressure to be carried by arch action horizontally into the canyon walls. Because of this arch action, the thickness (and hence bulk) of the dam's profile can be much less than a gravity dam of the same height. In essence, the amount of material in (or the mass of) a structural dam is a less important attribute than it is for a massive dam; in a dam adhering to the structural tradition, it is more important to develop a design that takes advantage of shape and not just weight.

In addition to thin arch dams, the structural tradition includes designs that feature buttresses built perpendicularly to the downstream side of a relatively thin masonry or concrete wall.[18] Buttress structures which utilize a flat surface for the upstream face are called flat-slab dams. Those featuring a series of arches are known as multiple-arch dams. In contrast to massive gravity designs, flat-slab and multiple-arch buttress dams are not solid monoliths that present a continuous, solid cross-section that extends the length of the structure. Often called 'hollow dams' because of the empty space that lies between adjacent buttresses (which can stretch out to distances of more than 60 feet), buttress dams require much less material than gravity dams of comparable height. Moreover, unlike thin arch designs, buttress dams do not require relatively narrow canyons capable of absorbing the horizontal thrust of a single, large arch.

In the abstract, the key distinction between massive and structural dams is easy to formulate. However, the line separating the two traditions can become blurred, especially when a structure contains enough material to function as a gravity dam, but is built along a curved axis like an arch dam. Known as curved gravity dams, the cross-section of such dams is sufficient for them to function as gravity structures; it is appropriate that a curved gravity dam be considered a part of the massive tradition because—if it were somehow straightened out and the curve eliminated—the cross-section would still be ample to impound water without tipping over.

In contrast, the profile of a thin arch dam is insufficient to resist a hydrostatic load simply by the force of gravity; in other words, if an arch dam were somehow straightened, then the cross-section would prove insufficient to resist the pressure exerted by a full reservoir. In acknowledging that thin arch dams must be curved in order to be safe, we also need to recognize that all arch dams also act to some degree as gravity designs. Phrased another way, those parts of arch dams closest to the foundations actually resist the water pressure by gravity action. How to analyze the relationship between "gravity" and "arch" action in curved dams became an important component of early twentieth-century concrete dam design history.

In their basic form, massive gravity dams are relatively easy to conceptualize, but they require large amounts of construction material and are often expensive to build. In contrast, thin arch and buttress dams require relatively small amounts of material but they can also entail more complicated design and construction techniques. Both traditions can foster safe designs and it is not a question of the massive being correct and the structural incorrect. For centuries they have coexisted within the evolving art of dam-building as different engineers within different cultures (or within different parts of a larger culture) championed particular types of designs.

Arch Dam Theory: European Origins and the Cylinder Formula

During the late nineteenth century, gravity dam technology attracted the most professional and public attention, but thin arch dams also began to be built utilizing new, mathematically-based methods of design. The seventeenth century Spanish dam at Elche and the Ponte Alto Dam in Italy featured profiles too thin to stand as gravity structures, but it took several score more years before any other prominent arch dams were built.[19] In the early 1830s, Lieutenant Colonel John By of the British Army supervised construction of an arch dam as part of the Kingston Canal connecting Lake Ontario with the Ottawa River. The slender profile of the Jones Falls Dam (58 feet high, maximum thickness 19 feet) reduced the amount of masonry necessary to build the arch, and apparently this prompted By to adopt the design in order to speed up construction.[20]

In the case of Francois Zola and the Zola Dam built in France starting in the late 1840s, the role of mathematical theory in developing the design is well known.[21] In the 1830s, Zola began developing a plan to increase the water supply of Aix-en-Provence. As part of this, he devised an arch dam design with basic dimensions that were calculated using a simple equation known as the "cylinder formula."

For absolute accuracy, the cylinder formula must assume that the arch is infinitely thin and supported on completely rigid foundations. Despite being based upon such idealized assumptions, the formula can still provide a useful means of estimating arch stresses that, while never absolutely precise, can represent reasonable approximations. In designing his dam, Zola conceived the structure as a stack of horizontal arches. At the dam's crest—where water pressure is low—the stack is relatively thin. As the water pressure increases, the thickness of Zola's design increases proportionally; significantly, the rate at which the arch becomes thicker is based upon the cylinder formula. The Zola Dam was completed in 1854 (after its designer's death) and attracted attention among engineers through the next decade. However, de Sazilly's and Delocre's gravity designs apparently eclipsed any strong interest in arch dams among European engineers, and, aside from one small structure in Australia, no other major arch dams are known to have been built until the 1880s.[22]

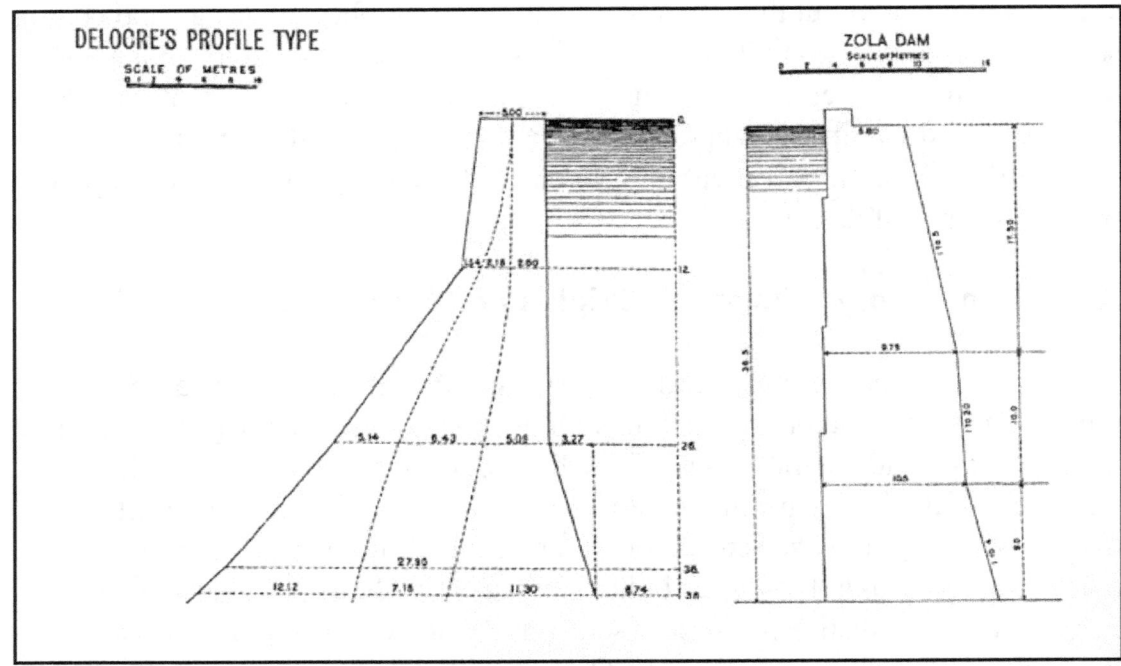

Figure 2-5: The DeLocre profile (left) is shown cut off at 38 meters to approximate the Zola Dam (right). The cylinder formula was the basis of design for the 36.5 meter high arch dam. It was completed in 1854 after the death of its designer, engineer Francois Zola, and was named the Zola Dam. Source: Wegmann, *Design and Construction of Masonry Dams*, 2nd edition, plates 2 (left) and 35 (right).

American Arch Dams

In 1884, the American engineer Frank E. Brown completed a masonry arch dam in Southern California with a profile so slender that the resultant force did not simply pass outside the "middle third," but fell well beyond the downstream edge of the foundation.[23] Designed using the cylinder formula, Brown's 64-foot-high Bear Valley Dam featured a maximum thickness of only 20 feet and dramatically demonstrated that this mathematical theory could help create designs that—in comparison with gravity designs—dramatically reduced the material necessary to build a solid masonry dam. In 1886, Brown began building another arch structure (50 feet tall, base width 10 feet) on the Sweetwater River near San Diego. He erected only a portion of the dam before being replaced by James D. Schuyler as the engineer in charge of construction. Schuyler revised Brown's design so that it featured a thicker profile, but the 90-foot high, 46-foot thick Sweetwater Dam, completed in 1888, still comprised a notable arch dam design that further demonstrated the value of utilizing the cylinder formula in dam design.[24]

Constant Angle Designs

Brown's and Schuyler's accomplishments reflected a growing interest in the cylinder formula among dam engineers who were attracted to the structural tradition. Within a few years, this led to a significant design variation that represented a final important innovation derived from the theory. In the late 1870s, the French mathematician Albert Pelletreau published a theoretical study demonstrating his understanding that it was not necessary to employ the same horizontal radius for all vertical elevations of the arch.[25] Because the cylinder formula postulates that the arch thickness is directly proportional to the radius, the thickness of any particular arch slice can be reduced simply by making the radius smaller. Because most canyons are narrower at the bottom than at the top, it is easy to conceptualize the construction of arch dams consisting of a stack of arches with progressively smaller radii. Referred to as either constant-angle arch dams (i.e., the angle generally remains constant while the radius gets smaller) or variable-radius arch dams (i.e., the radius is variable in length rather than constant) structures of this type can visually resemble downward-pointed cones. Following Pelletreau's conception of this type of dam, the idea was discussed by the American engineers Gardiner Williams in 1904 and John S. Eastwood in 1910.[26] Although actual construction of a constant angle arch dam did not occur until 1913, when the Danish-born engineer Lars Jorgensen (who subsequently moved to California) designed the Salmon Creek Dam for a hydroelectric plant near Juneau, Alaska, the possibilities of utilizing the cylinder formula for a constant angle design extend back into the late nineteenth century.[27]

American and European Dam Design Practice

In assessing the state of dam design theory at the beginning of the twentieth century, it is apparent that European engineers and mathematicians led the way in hypothesizing the key design methodologies for both gravity dams and arch dams. American engineers picked up on the basic character of these design innovations within a short period of time and quickly developed designs in both the massive and structural traditions comparable to European practice. Just as European engineers innovated in both the massive and structural tradition, so too did American engineers. Of course, particular engineers often—if not usually—focused their design energies on specific types of structures, and they were not necessarily inclined to view design types outside their particular area of expertise with great favor. But, taken as a whole, at the beginning of the twentieth century, American engineers were interested in—and capable of developing—the full range of design possibilities inherent in both the massive and structural traditions. Decisions about particular designs were based on a variety of factors, including the topography and geology of a site, the availability of construction materials, the availability of labor, financial constraints imposed by the patron or client (whether corporate, governmental, or individual), the professional experiences of the design engineer, and the social importance or prominence of the project. But there was no single or distinctive American style of dam building that defined what would come to be built after the turn of the century.

THE STRUCTURAL TRADITION AND THE RATIONAL DESIGN OF CONCRETE STRUCTURES

The Beginning of Rational Design

Structural engineering, as a modern profession, begins with the building of iron bridges in the late eighteenth century in Great Britain. It began because of the desire for lighter bridges that could nevertheless be as strong as or even much stronger than those built of stone or wood. Starting with the French schools, the *Ponts et Chaussées* established in 1748 and the *Ecole Polytechnique* established in 1794, structural engineering by the early nineteenth century began to be put onto a scientific basis where mathematical theory could help predict performance and be, therefore, a guide to designing new forms.

Bridges were the primary focus of early structural theory because they were pure structure, they had the longest spans, and they also had the most dramatic failures. During the last half of the nineteenth century, structural theory became formalized, began to be used extensively for buildings, and was taught systemically in the Polytechnic Institutes of Western Europe. By contrast with bridges and buildings, dams did not receive the same intensive attention in schools or in the technical literature. This was so because most dams were low and were built of earth or rock and, thus, remained part of a preindustrial technological culture. Throughout the nineteenth century, dams received little attention either in the technical literature or in schools of engineering. But at the end of the century, four major events in the United States brought dams into the forefront of engineering: first, cities were expanding at an unprecedented rate and they could not grow without new sources of water; second, the new electric power industry moved rapidly into hydroelectric stations; third, the closing of the frontier raised strong social pressures to develop the West in large part through irrigation; and fourth, the 1889 Johnstown dam disaster dramatically increased public sensitivity to dam safety issues.

Those social pressures combined with the advanced state of structural theory to produce the desire for a more scientific treatment of dams. Engineers believed that dams could be more rationally, hence more economically and more safely, designed. Just at this time the use of the new and prototypical twentieth century material, structural concrete, came into general practice and encouraged designers to abandon stone masonry, and sometimes embankment dams, for dams built using the new material. But even where earth or rock dams seemed still preferable, concrete became widely used in spillways, powerhouses, and diversion works.

As the twentieth century unfolded, major dam building in the United States and elsewhere began to take a new direction, a direction characterized by high multipurpose dams, huge reservoirs, and the search for rational methods of analysis as a basis for design. Almost all the large dams that are the focus of this book reflect these trends, and, in addition, they tend to involve such a strong restructuring of the environment that their planning required adjudicating among competing objectives. Of these objectives, the one that concerns us in this chapter is the competing vision of structural form, characterized by the structural tradition versus the massive tradition, which we can rephrase as the battle between form and mass.

Form and Mass in Structure

In the preindustrial European world, with the notable exception of the high gothic cathedrals, there was an implicit belief that great works went together with massive structures which were primarily of stone. This aesthetic of mass connoted permanence, opulence, and power; it stood in opposition to the ephemeral wooden structures of peasants and the urban poor. To be monumental was to be safe and handsome. Skeletal metal bridges of the nineteenth century often were banned from urban settings, and when concrete entered practice in the 1890s, it had to be covered in, or formed to look like, stone to be accepted. Building in the capital city of Washington reflects well this attitude. The light iron dome of the U.S. Capitol, built by the U.S. Army Corps of Engineers, is covered to look like masonry, and the Washington Monument, also largely built by the Corps, is the largest stone obelisk ever erected.[28] The Arlington Memorial Bridge represents a heroic attempt to make concrete—and even steel in one span—look like cut stone. The heaviness of modern buildings in the city reflect this belief in mass over form.

It is, therefore, not surprising that when large dams entered modern America in the twentieth century, they would reflect that context, especially those dams designed by large municipalities and agencies of the federal government. And yet, right from the start of federal dam building in concrete with the founding of the Reclamation Service in 1902, the conflict between form and mass was immediately present, and it would remain as a continuing issue, never fully resolved, throughout the century. This story begins with the twentieth century, when for its first series of major dams, this new Reclamation Service sponsored a detailed analytic study of dams from the structural perspective.

Arch and Cantilever Behavior in Dams

A dam is really a wall or barrier that resists the pressure of water stored in the reservoir. Consider a wall that runs straight across a valley. The wall, when rigidly fixed into the valley floor, will hold back water by acting as a cantilever. In engineering terms, a cantilever is a structure rigidly attached at one end and free of any restraint at the other. Therefore, the free end (top) of a straight gravity dam will move horizontally as the cantilever bends downstream under water pressure. In this way, the water load is carried down to the foundation (on the valley floor) by bending.

Now the dam, if curved into an arch form between the sides of the valley, will also carry water load to the vertical canyon walls, by compression forces calculated from the cylinder formula. As these horizontal arches carry compression, they will become shorter and hence move in the horizontal direction downstream. Thus, a curved arch dam can carry loads both vertically as a cantilever and horizontally as an arch. The challenge to the engineer is to determine how much of the load goes to the canyon floor and how much to the canyon walls.

This issue is crucial to design because much more material is required for safe cantilever behavior than for safe arch action. For example, designers proportioned gravity dams (those assumed to act as cantilevers alone) with a base thickness equal to about 2/3 of the dam height. Where the height is 60 feet and the width 40 feet, the amount of concrete required per foot of dam length would be $V = 60 \times 40 \times \frac{1}{2} = 1200$ cubic feet. By contrast, an arch dam with a height of 60 feet would require a base thickness of about 7.5 feet from the cylinder formula (for $f = 350$ pounds per square inch or psi) and, hence, a total volume of $60 \times 7.5/2 = 225$ cubic feet or less than 20 percent of the material required for the gravity or massive dam.

As a result, some engineers, seeing this great advantage of arch dams, had a strong incentive to find some rational way to determine analytically how much load was carried by the arching action and thereby to justify designing a safe dam with far less material than a gravity dam carrying load by cantilever action. Engineers consulting with the newly established Reclamation Service began this process of analysis as early as 1903.

In September 1903, the Reclamation Service held a conference of engineers at Ogden, Utah, where their newly appointed (March 1903) consulting engineer, George Y. Wisner (1841-1906), presented a paper which called for a thorough study of stresses in high masonry (stone or concrete) dams to ensure safety and achieve minimum construction cost.[29] F. H. Newell, the Chief Engineer of the Reclamation Service, asked a select committee of four, including Arthur Powell Davis (1861–1933), later to become Director of the Reclamation Service, to make him a recommendation, which it did formally on October 5, 1904. Its letter spoke of the two high dams proposed for Wyoming (Pathfinder and Buffalo Bill) and of the fact that "no thorough analysis has ever been made of the relative economy and stability of reinforced concrete dams as compared with similar dams of gravity sections. . . ." They suggested that such an analysis be commissioned by the Reclamation Service and they recommended Mr. E. T. Wheeler, of Los Angeles, for the job.[30] Under the supervision of Wisner, Wheeler began work in January of 1905. In March, the Board of Consulting Engineers met to review the work and they decided to revise the dam dimensions used by Wheeler up to then. Wheeler submitted his final report on May 5, 1905, and Wisner sent that report, preceded by a lengthy discussion of his own, to Newell on May 16. Its importance was considered to be so great that the Wisner-Wheeler paper was published in the August 10, 1905, issue of *Engineering News*. Since this report inaugurated the structural tradition of large-scale dam design within the federal government of the United States, it is essential to explain its substance and its impact.[31]

Although Wisner proposed the study in the light of the Reclamation Service's new big dams—Roosevelt, Buffalo Bill, and Pathfinder, he and Wheeler actually focused only on Pathfinder, it being the first one to be completed (1909). Wisner described how an arch dam in a narrow valley (he called it "of short span") carried water loads and also how it behaved under wide swings of temperature both with the reservoir full and with it drawn down. He then gave Wheeler's report, which consisted of the sets of formulas for water loads: one of which assumed that the dam carried the water pressure as a series of horizontal arches supported by the side walls of the canyon. He then computed the horizontal deflection of these arches at their crowns—essentially, only the vertical centerline of the dam.

Wheeler next took a vertical slice of the dam at this centerline, and assuming it carried all the water pressure as a cantilever, supported only on the floor of the canyon, he computed its horizontal deflection at various points from

Figure 2-6: Upstream face of Pathfinder Dam on the North Platte River near Casper, Wyoming. This is a masonry arch with a gravity section built 1905-1909 by the Bureau of Reclamation. The 214 foot height was unprecedented at the time of construction in 1909. Source: Bureau of Reclamation

base to top of the dam. Clearly, the arch deflections and the cantilever deflections must be the same at the same points on the dam, but this two-part calculation will not give such results. Thus, Wheeler had to make a second calculation by adjusting the amount of load taken by the arches and the amount taken by the cantilevers. The first calculation shows that the free cantilever deflects far more than the arches do in the top portion of the dam, while the reverse is true at the bottom. Thus, the arches should carry more load at the top and the cantilevers more at the base. This redistribution of load would eventually be called the "trial-load method of analysis." Moreover, Wheeler found that the Pathfinder Dam could carry all the water load as a series of arches with compressive stresses under 200 psi for a material (stone masonry) whose compressive strength is well over 2,000 psi.

Next, Wheeler studied temperature stresses in Pathfinder Dam. Here, he assumed that the temperature dropped 15°F at the top with the reservoir filled only to 100 feet from the top and that the temperature drop decreased linearly to zero at 120 feet below the top. This drop causes the arches to bend and deflect in the downstream direction, causing vertical cracks in the upper arches; and the deflection of the arches above, relative to the undeflected cantilevers below, will cause vertical bending in the lower parts of the dam and, hence, horizontal cracks there. This qualitative description helps explain where reinforcing steel needs to be placed (if it were a concrete dam), but it does not give a good quantitative measure. However, by iteration, he was able to make a more reasonable estimate of the temperature stresses, which he then combined with the water load to give one design condition.

The significance of this admittedly quite approximate approach is that, for the first time, the engineers were seeking to take advantage of the true behavior of a curved dam in a narrow canyon to the end of making it more

Figure 2-7: Pathfinder Dam was constructed with huge blocks of stone. Source: Bureau of Reclamation

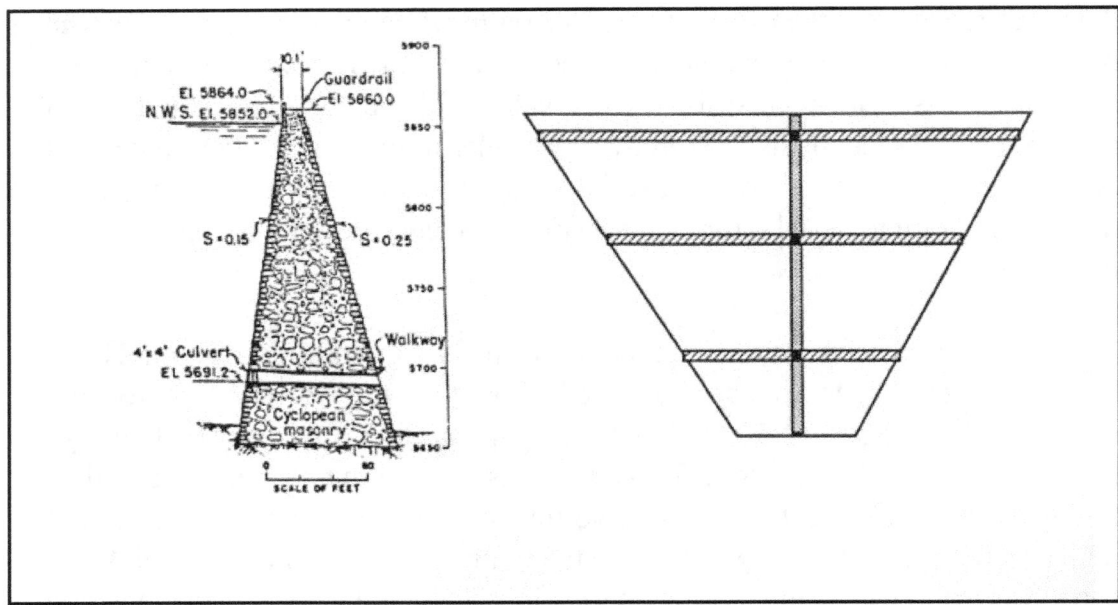

Figure 2-8: In 1905 E. T. Wheeler analyzed Pathfinder Dam for the Bureau of Reclamation under the supervision of George Y. Wisner. He took a vertical slice of Pathfinder as a cantilever and analyzed the deflections. Then he analyzed horizontal arch section to determine deflections. The calculated deflections did not match, but this analysis provided the basis for the trial-load method of analysis. Source: Bureau of Reclamation.

economical through saving material. But many engineers did not trust this approach. Indeed, when Wisner suggested the year before that the Lake Cheesman Dam, near Denver, could have been designed with substantially less material had arch action been considered, the designer replied that "the suggestions for a different design are for a lighter section than the one used . . . [but] when the consequences of failure are very great, the engineer should build abundantly strong."[32] Wisner was seeking to show strength does not necessarily mean more mass.

The Constant-Angle Arch Dam

In spite of the belief that neglecting arch actions would make dams "abundantly strong," up to 1914 there had been no public record of any arch dam failure,[33] whereas numerous gravity dams had failed. In that same year, the Danish immigrant engineer Lars R. Jorgensen (1872-c.1937) presented a paper on the constant-angle arch dam in which he argued that counting on arch actions realistically for high, narrow canyons would "show a savings of material of 33 percent or more over an ordinary gravity dam, and at the same time it will possess a factor of safety more than twice as great as that of the gravity dam."[34]

Jorgensen showed how to achieve an arch dam with a minimum thickness at each level by using the cylinder formula and by reducing the radius of curvature in the lower regions where the water pressure is higher. He found that

Figure 2-9: In 1910 the Bureau of Reclamation completed Buffalo Bill Dam on the Shoshone River near Cody, Wyoming. This constant radius concrete arch dam is 325 feet high. The dam delivers irrigation water and has a powerplant. Source: Bureau of Reclamation.

as the radius decreased the volume of material needed would be a minimum if the arc angle remained the same and had a value of about 133.6°—hence his dam designs are called constant-angle dams.

Here we have the classic design problem well known in elevated water tanks made in the shape of cones to keep the forces in the wall roughly constant.

These forces come from the cylinder formula and are the product of the water pressure and the wall radius. Thus, the design goal here is to decrease the radius as the water pressure increases.

Jorgensen proposed the same idea for concrete dams in narrow V-shaped canyons. He showed that the thickness for his Salmon Creek Dam would have to have been over twice as thick as he made it if the arc radius had been kept constant rather than the arc angle. Clearly, the thicker dam becomes a gravity dam in sections with a width to height of nearly 2/3.[35]

Jorgensen was well aware of the arch-cantilever analysis made by Wisner and Wheeler, and he presented a similar analysis in his 1915 paper, but his goal was to show that by having a changing radius, i.e., one which decreased toward the canyon floor, the arches lower down would, thereby, become stiffer and carry more of the water load than in the case of the designs where the radius did not so diminish. Furthermore, the arch behavior allows the structure to be thinner because there is less bending than with the cantilever behavior and thus the thinner sections near the base reduce further the cantilever action.

Jorgensen did more than present a clever design idea, he actually was partly responsible for constructing the first such dams: the Salmon Creek Dam in Alaska (1914) and the Lake Spaulding Dam in California (1919). In 1920, Jorgensen reported that 25 gravity dams had failed, of which 19 were built during the preceding 30 years, while no arch dams had ever failed. By 1931, he would state that over 40 constant-angle arch dams had been built in the last 16 years—beginning with Salmon Creek.[36]

Noetzli and the Curved Dams

Strictly speaking, the analysis of Wisner and Wheeler was a trial-load method because it assumed a distribution of loads between arches and cantilevers and then, after various other trials, it based design on a final iteration. Fred Noetzli (1887-1933), a Swiss trained engineer, summarized the situation in a landmark 1921 paper in which he reviewed the practice of arched dams, gave relatively simple formulas for calculating the cantilever and the arch actions in horizontally curved dams, and then applied his formulations in detail to the Pathfinder Dam. This last part is the heart of his paper in which he compares his semigraphical approach to the purely analytical calculations presented by Wisner and Wheeler in 1905. He concludes that his "distribution of load between cantilever and arches compares very favorably with that obtained analytically by Mr. Wheeler."[37]

Noetzli then proceeded to discuss the central issues in concrete dam design that went beyond the statics of water-pressure loading: stresses due to temperature change, to shortening of the arches under water pressure, and to shrinkage of the concrete as well as the influence of cracks in the concrete. He showed by simple calculations that these effects were at least as important as those due to the statics of water pressure loading.

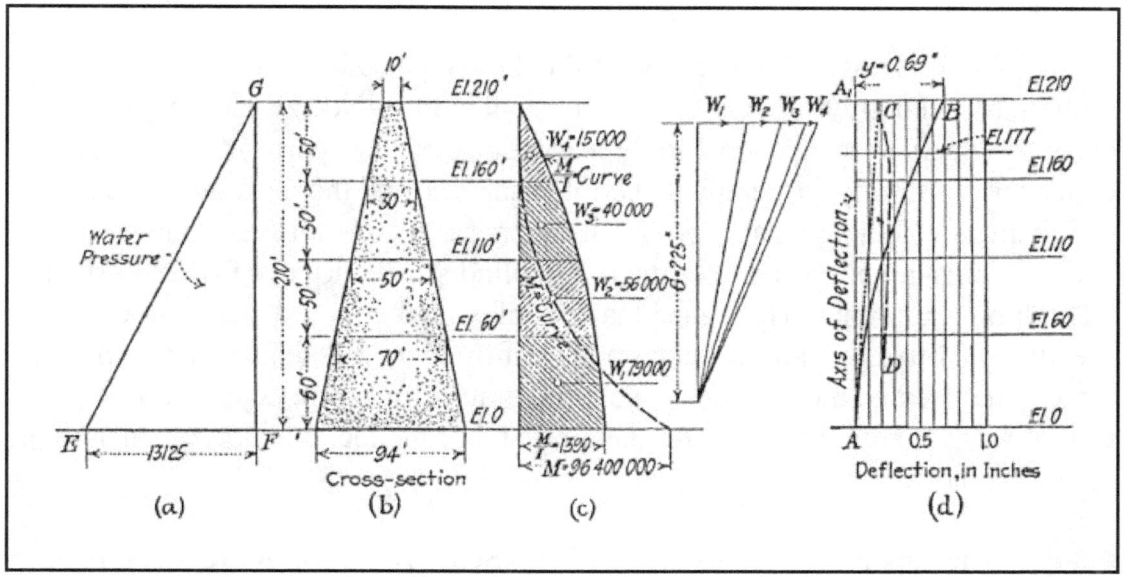

Figure 2-10: In 1921 Fred A. Noetzli calculated the loading on Pathfinder Dam and showed these graphic results. Source: Noetzli, "Gravity and Arch Action in Curved Dams," 18.

He then discussed two ideas crucial to dam design: vertical arching and the rigid-body overturning of curved dams. In the former case, a gravity dam in a V-shaped canyon will surely not carry the vertical loads directly to the valley floor but rather the lower section will act like an arch, transferring the loads to the side walls of the valley. Noetzli cautioned that this behavior could lead to very high compression stresses in the side wall near the floor where vertical arching (some vertical load for the dam weight is carried to the canyon sides by arch action in the vertical plane) adds greatly to the compression calculated solely by gravity and the cantilever action due to water pressure. Furthermore, he went on to point out that pure gravity dams rarely have a safety factor against overturning of over 2.0 and usually (where full uplift is assumed) it is close to 1.0. This surprising claim allowed him to make a strong criticism of such dams, i.e., "no other engineering structure of acknowledged good design has such a small factor of safety as a pure gravity dam."[38]

Noetzli then described the rigid-body overturning of curved dams using his own term of "Curved Dams as Cylinder Hoofs." He considered here the curved ground plan of the dam and noted that, as the structure tends to overturn, the moment of inertia of that plan is considerably greater than that of a straight

71

dam ground plan of the same base width. Therefore, he argued, even if there were no horizontal arching behavior (as there might not be in a wide valley) the curved dam will have a far greater resistance to overturning than the straight dam, especially as the curvature increases. Noetzli is here rendering a severe criticism of the procedure which simply takes a vertical slice from a curved gravity dam and studies it alone for overturning. Such a procedure fails to show any difference between a curved and a straight gravity dam.

Finally, after a brief review of the three types of curved dams—gravity section, reduced gravity sections, and constant angle—Noetzli gave eight conclusions which we can group into three categories: the superiority of curved dams, the possibility for relatively simple analyses, and the need to consider special features. He noted that curved dams were safer and more economical than straight dams largely because of the hoof-cylinder behavior even when horizontal arching is minimal. He argued that complex mathematical analysis could be greatly simplified by graphic statics and confirmed by full-scale measurements. Finally, he stressed the significance of temperature and shrinkage, of crack control, of vertical arching in narrow canyons, and of the fact that horizontal arching may often not act.

The paper of 60 pages in the 1921 ASCE *Transactions* drew vigorous discussions stretching out to 75 pages with 14 discussers, including such major figures of the period as Lars Jorgensen, B. F. Jakobsen, A. J. Wiley, and Edward Wegmann (all of whom, save Wiley, were like Noetzli, from Europe).[39] The paper established the Swiss engineer as a leading theoretician for dams, and the discussion largely confirmed Noetzli's reputation. Running through Noetzli's writing was the two-part theme, typically Swiss, that good design implies form over mass and that analysis—often graphically done—can be greatly simplified to improve understanding as well as to encourage designers to think in terms of form over mass. He was at great pains to stress the historical fact that mass did not mean safety but that form, properly conceived, did so and with greater economy as well.

Much of the discussion revolved about the relative simplicity of the graphical approach as compared to the complexity of the mathematical one.[40] One factor in the form versus mass debate was the perception that lighter forms needed more rigor in solution.

The writings on curved dams continued throughout the 1920s as the nation was beginning to move into the largest program of dam building ever attempted. In the 1922 ASCE *Transactions*, for example, there were eight major articles (out of 20) related to dams; Noetzli wrote one and contributed discussions to five others. In his 1922 article, Noetzli developed formulas for use with field measurements of displacements to calculate stresses that may arise from "water pressure, temperature, shrinkage, swelling, lateral deformation, etc. or a combination of any or all of them."[41] He was still focused on simplified mathematics and the centrality of measurements, which also elicited much discussion. Curiously, the three dams for which measurements were then available were all outside the continental United States (two in Australia and one in Alaska).

The Trial Load Method

The articles and discussions up to 1929 discussed both arch and cantilever behavior and, hence, qualify as trial-load methods. However, not until the 1929 article by C. H. Howell and A. C. Jaquith does the method acquire publicly the name of "trial load."[42] Both authors had worked for the Bureau of Reclamation in Denver where they had begun to study the method in 1923; and when the full mathematical framework appeared in 1938, Reclamation identified seven concrete dams designed on the basis of trial load analysis: Boulder (now, and hereafter in this study known as Hoover),[43] Owyhee, Parker, Seminoe, Gibson, Deadwood, and Cat Creek (actually designed by the Navy Department).[44]

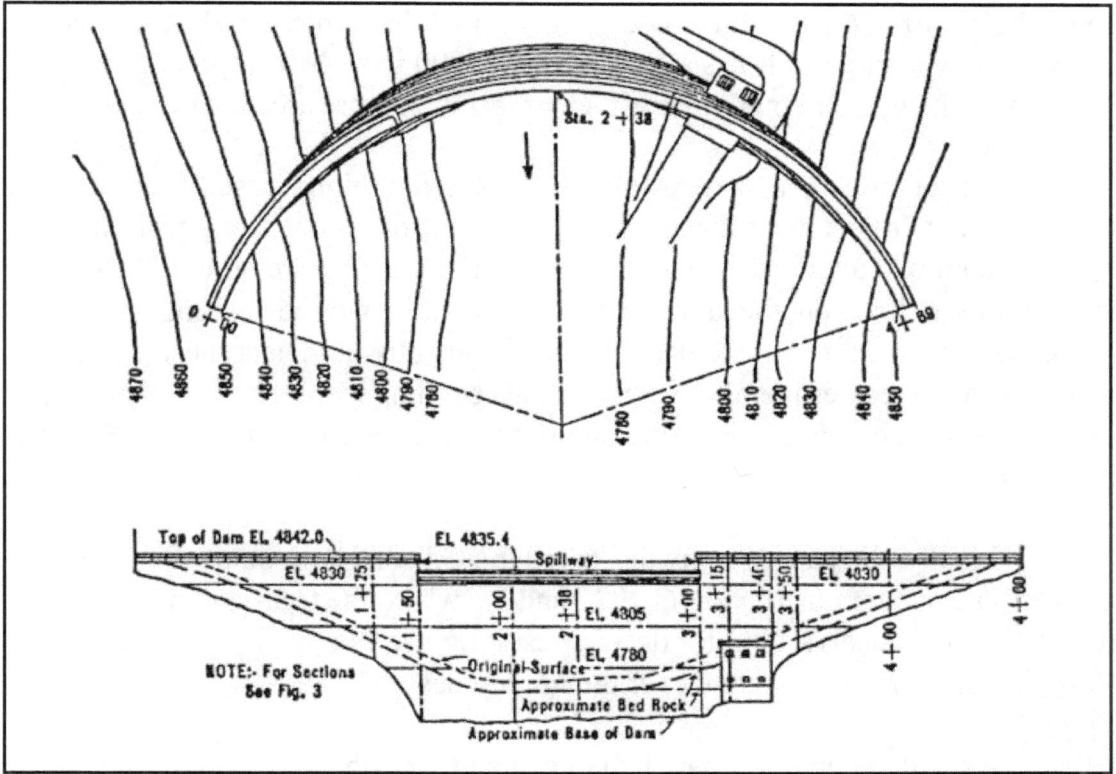

Figure 2-11: C. H. Howell and A. C. Jaquith studied both cantilever and arch action in developing what is called the "trial load method" of dam analysis. Variations in the shape of the canyons were such that two shapes were considered necessary for realistic analysis. This is Dam No. I. Source: C. H. Howell, and A. C. Jaquith. Paper No. 1712. "Analysis of Arch Dams by the Trial Load Method." *Transactions of the American Society of Civil Engineers* 93 (1929), 1194.

Howell and Jaquith had completed a first draft in January 8, 1925, and it was approved for publication on January 16.[45] But it did not appear until four years later. In the final version, they defined the method as one which considers the dam to be made up of a series of horizontal arches and a series of vertical cantilevers, with part of the water load carried by the dam considered as arches and part by the dam considered as cantilevers. The arch loads and the cantilever loads are adjusted so that the deflection of the arches are nearly the same as the deflections of the cantilevers at the same points. They distinguish the trial load method from previous similar methods by the fact that they were considering more than the one single cantilever, which is what Wheeler, Noetzli, and others had done. By considering a series of cantilevers, rather than one cantilever only at the centerline of the dam, Reclamation engineers created an immensely complex procedure that took the entire 266 page Bulletin (ref. 45) to explain, without even going into any numerical calculations.

In their 1929 paper, the authors began by noting the variations in the shapes of canyons in which dams appear and thus they established the need to use more than one cantilever for more realistic analyses. They then proceeded to define two designs identified only as Dam No. I and Dam No. II. The first has a center height of 75 feet and a base thickness of 18.75 feet, and the second has a center height of 256 feet with a base width of 48.4 feet. In No. I, the crest length is 469 feet, whereas No. II has a crest length of 900 feet. Not surprisingly, No. II, with its greater length of crest, exhibits much more arch action than No. I.

The authors then gave results for the 271 foot high Horse Mesa Dam, on the Salt River Project in Arizona, where the base width is only 40.2 feet; therefore, the dam will act predominantly as an arch in the upper regions. When discussing this arch action, the authors showed that there will be bending in the arches, especially lower in the dam and predominantly both near the valley side walls and at the centerline of the arches. This bending results in some tension, causing the concrete to crack, reducing the arch section and hence increasing the compression stresses.

Finally, the paper briefly discussed the Gibson Dam, on the Sun River in Montana, under construction in 1927 and for which the trial-load method was used not just for analysis but for design, resulting in a "savings of more than 41,000 cubic yards of concrete over the gravity design."

As with Noetzli's paper, the Howell and Jaquith paper brought forward much substantial discussion: a 34-page paper resulted in 90 pages of discussion. Noetzli and Jakobsen both observed that Alfred Stucky had used the trial load method for a Swiss dam in 1922, although the method was not so named. In fact, Robert Maillart had used the same idea in 1902 for a water tank also in

Switzerland.[46] Probably the most significant discussion from the point of view of federal dams came from John Savage and Ivan Houk, both of Reclamation. They gave a more detailed discussion of Gibson Dam and gave also results from their analysis of the 405 foot high Owyhee Dam in eastern Oregon. Savage had assumed a dominant role in Reclamation dam design and was already in 1929 deeply involved with the Boulder Canyon Project. But as the dams got higher and higher, Reclamation recognized the need to develop not just mathematical analyses but also physical model testing and the instrumentation of full scale dams.

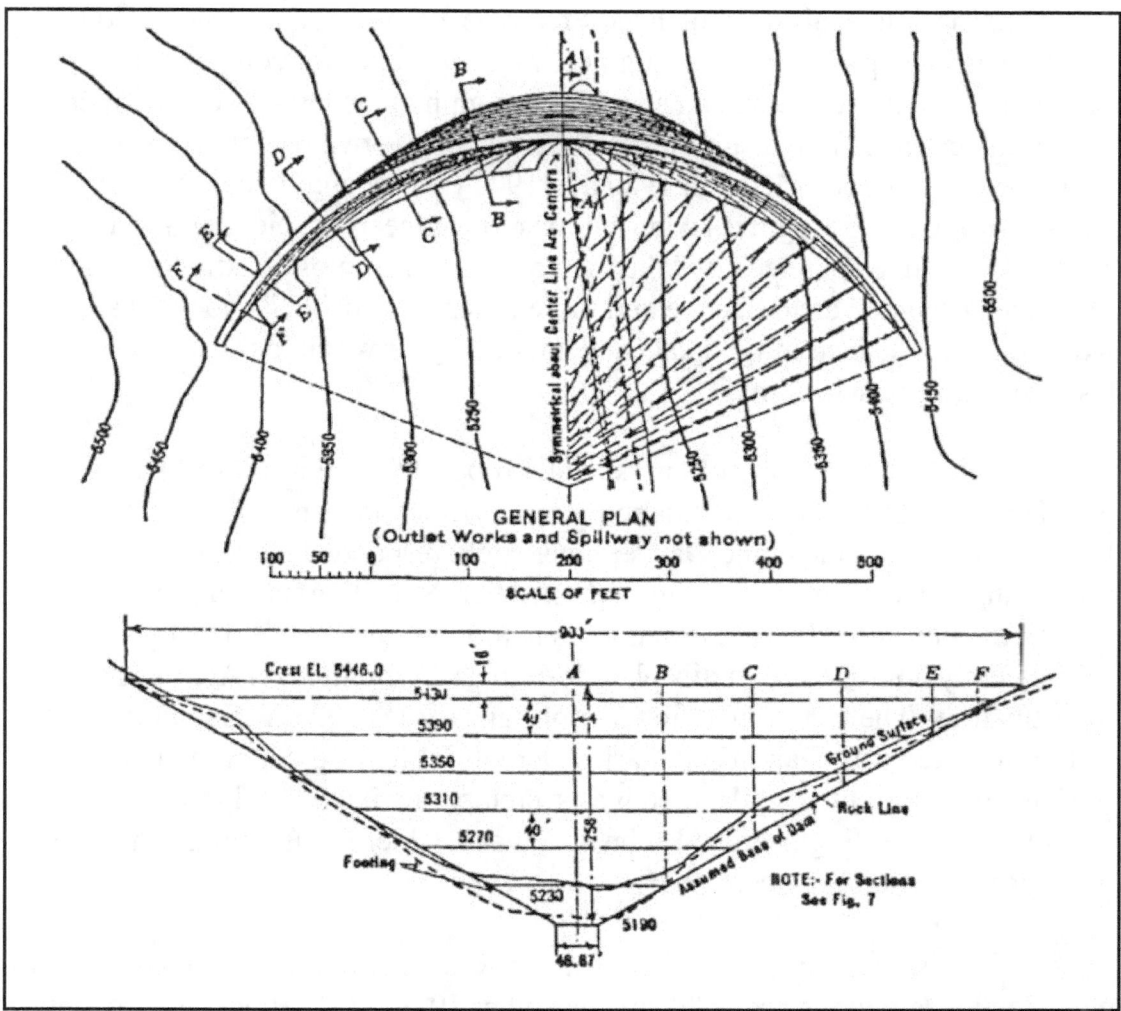

Figure 2-12: This is Dam No. II used by Howell and Jaquith in development of the trial-load method of analysis. Source: C. H. Howell, and A. C. Jaquith. Paper No. 1712. "Analysis of Arch Dams by the Trial Load Method." *Transactions of the American Society of Civil Engineers* 93 (1929), 1199.

The Stevenson Creek Test Dam

During the first three decades of the twentieth century, engineers focused intently on concrete arches, creating numerous designs for bridges as well as dams, and stimulating more mathematically complex analytic schemes. In 1924, four of the twenty ACSE *Transactions* papers dealt with concrete arches, and many of these pages were filled with formulas and tables. The 1925 ASCE *Transactions* contained two extensive articles on arch analysis: one 98 pages long with 73 pages of discussion; the other 40 pages with 61 pages of discussion. The two articles total about 20 percent of the entire volume.

But already by 1922, some engineers became uneasy with so much abstraction and began to worry about field performance as opposed to office abstractions. Particularly, engineers in the western states saw the need for a different approach to analysis, which led Fred Noetzli, one of their leaders, to request financial support from the Engineering Foundation for collecting performance data on existing arch dams and for designing new tests and experiments.[47] Noetzli noted the national significance of arch dams—two recent papers on the subject won the Croes Medal in 1920 and 1921 (the second highest award given by the ASCE, it recognizes contributions to engineering science)—and urged aid for physical testing because "the methods by which most existing arch dams have been designed are defective and more or less unreliable." Noetzli had been worried about the lack of field data and later that year would publish a paper on test results in full size dams.[48]

Noetzli sent a copy of his skeletal proposal to Arthur Powell Davis, then Director of the U.S. Reclamation Service, to request his help with the Engineering Foundation, and Davis immediately responded by sending it a supporting letter. The Foundation requested that Noetzli prepare a program statement which could be presented to its board in May 1922.[49] The board subsequently approved Noetzli's plan, forming a Committee on Arch Dam Investigation. The Committee first met on January 18, 1923, in San Francisco. This committee was chaired by Charles Derleth Jr., a professor at the University of California–Berkeley, with Noetzli as secretary and including Frank E. Weymouth, Chief Engineer, and John L. Savage, Chief Design Engineer, both of the Reclamation Service.

It was becoming clear, then, that the Reclamation Service would have to play a major role in the project.[50] In December 1923, W. A. Brackenridge, Senior Vice President of the Southern California Edison Company, proposed the building of a large scale concrete arch dam designed expressly for research, and he further offered to provide a large amount of the money for it as well as the use of his company's facilities. Located on Stevenson Creek, a tributary of the San Joaquin River about 60 miles east of Fresno, California, this test dam was approved by the committee and construction began in August of 1925.[51]

The test arch design was startlingly thin. The physical structure was built in a V-shaped canyon, to be 60 feet high with a thickness throughout the top half of only 2 feet, tapering from mid-height to the base from 2 feet to 7.5 feet, respectively. The arch is of a constant 100 foot radius throughout.[52] The tests used mechanical strain gages, and, from these measurements, stresses were calculated. Deflection and temperatures were also measured.

Meanwhile, engineers had been collecting measurements from existing dams as part of the overall program, and they had found discouraging results because of the difficulties in relating strains and displacements to loading and temperature changes. They debated the data from the test dam construction and instrumentation at a meeting in Fresno in early December, 1925.[53] Reclamation was becoming convinced that the test dam alone would no be sufficient and that a series of small scale models ought also to be included in the program.[54] In early 1926, the Commissioner of Reclamation, Elwood Mead, approved funding for part of the work with small scale models.[55] A report on all this work appeared in November 1927 and on December 8, 1928, a concrete model of the Stevenson Test Dam was loaded to destruction.[56]

Figure 2-13 John (Jack) L. Savage, U. S. Reclamation Service/Bureau of Reclamation, Chief Design Engineer (1916-1945). Source: Bureau of Reclamation.

We can summarize the conclusions reached by the committee in late 1927 under three categories: first, the great strength of the arch dams; second, physical experiments have given data useful to engineers developing mathematical analyses; and third, arch dams may be designed more economically (by being thinner) in the future.[57]

The full report included an analysis by Noetzli following his 1921 paper and a summary of data on all major arch dams built in the world up to 1927 (57 in the USA, 39 abroad) and all major multiple arch dams (38 in the USA, 16 abroad). After a brief discussion on the construction, there followed an

77

extensive detailing of the test results by Willis A. Slater, followed by brief reports on the properties of concrete (by Raymond Davis of University of California-Berkeley), on models (by Savage and Houk), on a celluloid model of Stevenson Creek Dam (by Professor George Beggs of Princeton University), a theoretical report (by H. M. Wester-gaard of Illinois University), and a brief discussion of loads and concrete deterioration by Alfred Flinn of the Engineering Foundation. An extensive bibliography ends the report. Thus, by 1927, substantial data from both physical testing and mathematical calculations existed, but there were conflicting views about the results.

Fred Noetzli, whose primary aim had been to use tests and calculations, predicted in 1924 that gravity dams would be replaced by thin arch structures; he quoted several engineers saying that "the gravity dam is a thing of the past" and "the gravity dam is an economic crime." He held the belief common to many in the 1920s that "engineering science is advancing" and that a more rational analytic base would lead to thinner and less costly structures.[58] But Noetzli did not imply that more rational would necessarily mean more complex. He worked with graphical methods typified by his education at the Federal Technical Institute in Zurich. He did not publish the detailed mathematical formulations that had begun to appear in the 1920s and would culminate in Reclamation's 1938 report (ref. 45).

This type of mathematical complexity was criticized sharply by one of the most famous structural engineering teachers, Hardy Cross of Illinois University. In discussing a highly mathematical 1925 paper on concrete arches, Cross noted the uncertainties of loadings, of actual stress, and of foundations, none of which were dealt with in the paper which "having swallowed these 'camels' only the 'gnat' of mathematical analysis remains. The 'gnat' should be an *hors d'oeuvre* and engineers are giving abnormal gustatory attention to it." He goes on to proclaim that "the theories of arch analysis which are now being elaborated in engineering literature are distinctly 'high brow' in that their elaborateness camouflages with erudition uncertainties and inaccuracies which are inevitable."[59]

In spite of Noetzli's hope and Cross's warning, the profession charged ahead with complexity of analysis and the result was that in the Bureau of Reclamation the dams did not get thinner but thicker. A 1988 summary of this period stated that:

> Many arch dams built at the time showed a tendency for increasing thickness. On the one hand the failure of St. Francis Dam in California in 1928 had raised questions regarding the safety of any proposed dam of large size. On the other hand, it

seemed that the excellent results obtained at Stevenson Creek, including a verification of the trial load method, were not carried forward with these arch gravity-type dams.[60]

While many of these ideas and methods focused on high, single-arch dams, another form of structural dam had appeared in the early twentieth century, and received considerable attention in the technical literature. This form, the buttress dam, consists of either slabs or arches supported by buttresses within the valley. It, too, held out possibilities for lighter, more economical dams, and we next turn to these to have a fuller picture of the choices available to the dam designer in the 1920s.

Buttress Dams

Early in the twentieth century, two types of buttress dams appeared in the United States: the flat slab and the multiple arch. Still in service today are 44 flat slab and 24 multiple arch dams over 30 feet in height, built up to 1927. None of these are federal dams, although the Bureau of Reclamation did complete one flat slab in 1928 (Stony Gorge) and one multiple arch in 1939 (Bartlett).[61] What kept Reclamation from designing such dams?

Of the two types of "hollow" dams, as they were called, the flat slab dam, while more often used, received far less attention in the civil engineering literature than did the multiple arch dam, and for good reasons. The slab designs came largely from one man, Nils F. Ambursen and his company (Ambursen Hydraulic Construction Company), which built the great majority of the more than 350 such dams built in the United States.[62] By choosing such a design, a federal bureau was almost precluding competition for construction. In addition, the intellectual rage of the 1920s was for concrete arches both for bridges and dams. Between 1918 and 1929, the highest ASCE awards for papers went to writings on concrete arches four times. In the next two highest awards, arch studies also won four times.[63] No such recognition went for slab design or analysis. These slab dams were usually relatively small scale—over 80 percent were 30 feet or less in height.

On the other hand, the multiple arch dams represent a quite different approach and, with few exceptions, have been well built and exist today in good condition. Of the 30 dams over 30 feet high listed by Noetzli and completed before 1928, at least 22 are still in service.[64] The leading practitioner of these dams was John S. Eastwood (1857-1924), who, after studying engineering at the University of Minnesota (he never graduated), went west in 1880 to work on railroads. By 1883, he had settled in Fresno, California, and after several civil works jobs he became an early pioneer in the development of hydroelectric power for California.

In 1906, while working on a plan at Big Creek to develop hydroelectric power from the mountain valleys of central California for Los Angeles and San Francisco, Eastwood struck upon the idea of abandoning the normal earthen dam with a concrete corewall for a purely concrete multiple arch dam. Apparently, he came to the idea on his own, even though Henry Goldmark in a paper of 1897, which won the ASCE Thomas Fitch Rowland Prize, had proposed a 105 foot high multiple arch dam for an electric company in Utah. It was never built and its design was relatively bulky compared to the designs later developed by Eastwood.[65] Eastwood lost the Big Creek project, but two years later he succeeded in building the first multiple arch dam; it is located at Hume Lake in the Sierra Nevada Mountains and was built for the Hume-Bennett Lumber Company.

Before his death in 1924, Eastwood designed a total of seventeen such dams that were put into construction.[66] These dams, and some designed by others, consisted of relatively short span arches carrying water loads horizontally to buttresses which take those forces vertically to the foundations. Eastwood used simple mathematics on which to base his design, but later engineers began to introduce more complex procedures in an effort to account for more detail. In a summary treatise of 1927, Fred Noetzli presented a more rigorous approach and used Eastwood's Lake Hodges Dam of 1917 as a numerical example. Noetzli had shown that the arch compression stresses would be 300 psi at 100 feet below water level (the total height is 136 feet) with the simple cylinder formula used by Eastwood. When including effects of rib shortening and temperature change, he found the maximum arch compression stress to be 640 psi and a possible tension stress of 69 psi.[67]

The concrete compression strength in the 1:2:4 mix concrete of that period was about 3,000 psi, which means that the stresses rose from 10 percent to about 21 percent of the strength by more rigorous calculations. The small tension stress is well below that which would cause cracking (about 10 percent of the strength). Thus, the additional rigor, while of considerable engineering interest, had nothing to do with the safety of the dam. No such structure, designed in that simple way and built by an experienced engineer like Eastwood, ever failed because of lack of rigor in the stress calculations.

In discussing the history of dams or any other large-scale structures, it is essential to be clear on the uses for analysis. It is a tool of design, not a means to design, and the numerical values for stresses are always suspect in concrete structures because they cannot be measured directly. In the Stevenson Creek test dam studies, engineers measured only deflection and strains (which are also physical movements). From the strains they could compute stresses only by using an assumed relationship between them, defined as the modulus of elasticity, which, however, is not a constant and must be estimated.

In the Stevenson Creek report, Willis Slater described results as follows:

> At the 30 feet elevation the stress is 650 pounds per square inch and is seen to be larger than the observed total stress for all points except for a few feet close to the abutment. This stress is about 180 percent in excess of the observed stress at the center line, while at the abutment it is about 35 percent less than the observed stress. At the 50-feet elevation the stress by the cylinder formula is 217 pounds per square inch compression, whereas the observed stress was a tension of about 125 pounds per square inch at the crown, and a compression of about 290 pounds per square inch at the abutment. About 20 feet from the abutment the observed stress was about 575 pounds per square inch compression, or about 165 percent greater than that given by the cylinder formula. It is quite obvious that the cylinder formula is entirely inadequate to represent conditions found in a elastic arch of this kind.[68]

It is a correct conclusion that the cylinder formula gives inadequate stress results, but, as a guide to arch dam design, it does not seem to have been misleading. No concrete arch dam designed on the basis of that formula has been known to fail. But in the 1920s, the ideal was for more rigor and simplified methods were considered inadequate. For the multiple arch dams, the main problem with the cylinder formula was the neglect of bending in the fixed arch. Eastwood recognized this, and in several of his dams he made the arches three hinged; that is, he built the arches in two halves connected to each other and to the abutments by rounded edges. These hinge lines, running along the sloping upstream sides of the dam, permit the arches to rotate and thereby eliminate the bending due to rib shortening and temperature. When Fred Noetzli studied one of these dams (Mountain Dell from 1917) in 1925, he found no evidence of rotation in the hinges and hence one could conclude that there was no noticeable evidence of bending. In other words, the cylinder formula appeared to have been validated.[69]

Costs in Form and Mass

Eastwood was able to build multiple arch dams primarily because they were the least expensive designs, an important factor for private companies and local public authorities. Expense was not necessarily a factor for the

Bureau of Reclamation. In 1928, Reclamation reported on cost estimates for two major arched gravity dams: Gibson and Owyhee.[70] For Gibson, the cost was $2,116,745. For Owyhee, Reclamation chose a heavy arch even though it was 19.3 percent more expensive than a light arch and in spite of the good results from the thin Stevenson Creek test dam. The Gibson Dam valley has the same general shape as the valley in which Eastwood built the 1923 Palmdale Dam, but the latter is about two-thirds the size of Gibson. One would expect the cost of Palmdale to be about two-thirds the cost of Gibson i.e. $1,400,000. Palmdale actually cost less than $550,000.[71] Thus, in both cases, Reclamation chose more expensive solutions than it apparently needed if it had thoroughly embraced the structural tradition

For the Mountain Dell Dam, the owner solicited bids for three different types of concrete dams. The results are shown in Table 2.1.

Type of dam	Bid	Concrete
Gravity	$230,000	45,000 cu. yds.
Ambursen (flat slab)	$217,000	21,300 cu. yds.
Multiple Arch (Eastwood)	$139,000	14,700 cu. yds

The owner had no difficulty in choosing the Eastwood multiple arch design, which was completed in 1917.[72] The cost was 60 percent of that bid for the gravity dam. Reclamation chose massive designs because it thought them safer, it believed them to appear safer, and it was not under the same constraints as private industry to keep costs low.[73]

Concrete Form and Masonry Mass

By 1927, there had emerged well documented traditions of massive and of structural dams. The structural tradition brought forth new methods of analysis, both by physical test and by mathematical calculation. The goal had been to build lighter, less expensive, and safer dams. But as the methods of analysis became more complex, there seemed to grow an anxiety about uncertainties in the analysis itself, and the federal agencies addressed these worries by designing heavier structures, which they believed to be safer even though the lighter ones were performing at least as well.

It seems to be a natural result of centralized agencies that they seek to avoid risks, to question innovations, and to justify heavy expenditures by invoking the specter of failure. But behind this apparent criticism, there lay a deep cultural ideology that was characterized by the new and prototypical building material of the twentieth century. American society, and indeed western society as a whole, reacted to reinforced concrete in a profoundly ambiguous way.

Modern concrete clearly stimulated the search for new forms that would carry loads with less material and at least as much safety as heavier designs. But many engineers, not seeing these possibilities or not valuing them, sought to discredit this search for innovation. They saw concrete rather as a mere substitute for stone masonry—indeed, the term masonry today still connotes both stone and concrete. Webster's 2005 Collegiate Dictionary defines masonry as "something built, as by a mason, of stone, brick, concrete etc."

Modern concrete, cast monolithically, made the building of integrated structures possible, leading to great savings of materials and weight. But the idea of monolithic structures set the academic mind off on a search for new mathematics that left most of the practitioners bewildered. The counter-intuitive result was an increased anxiety within the profession over these new forms that seemed to deliver primarily new formulas. The performance of equations seemed to replace the performance of structure. It was much easier simply to put the new cast material into old familiar forms. Thus, it could be given a modern look by some surface treatment or added filigree.

American society had, by 1927, passionately embraced completely new forms of twentieth century engineering: the automobile, the airplane, and the radio; but it was much more difficult to accept the new forms of pre-twentieth-century engineering works for bridges, vaulted roofs, and dams. These structures were already part of the culture long before reinforced concrete came to encourage their transmutation: stone arches, stone vaults, and stone dams. Had there been no such models built into the cultural retina, the concrete arches of Robert Maillart, the concrete thin shells of Anton Tedesko, and the concrete dams of John Eastwood would no doubt have found widespread acceptance during the first third of the twentieth century.

In spite of this reluctance to abandon the old forms of stone design, society did embrace the great transforming event of electricity and its late nineteenth-century innovation of power at a distance. Dams had, for a long time, been used for mill power, but the power was local; it had to be used by mechanical transmission. In the East, new industrial towns, such as Lowell, Massachusetts, had sprung up in the early nineteenth century to produce textiles by water power, but the dams were low and the power dependent upon the vagaries of river flow. With the introduction of electric power, new possibilities arose and mainstream dams could now become multipurpose structures. Furthermore, with the development of the West, irrigation dams could also become multipurpose through power production. Most significantly, the sale of power could now justify the cost of high dams on non-navigable rivers. Two dams illustrate this change from single to multipurpose dams—Roosevelt Dam in Arizona built by the Reclamation Service and Wilson Dam in Alabama built by the U.S. Army Corps of Engineers. These two serve also as the prelude to the big dam era that would follow soon after.

Endnotes

1. See S. W. Helm, "Jawa, A Fortified Town of the Fourth Millennium B.C." *Archaeology* 27 (1974), 136-7.
2. The concept of traditions" as applied to dam design is discussed in Donald C. Jackson, *Building the Ultimate Dam: John S. Eastwood and the Control of Water in the West* (Lawrence: University Press of Kansas, 1995), 18-21.
3. Nicholas J. Schnitter, *A History of Dams: The Useful Pyramids* (Rotterdam, Netherlands: A. A. Balkema, 1994), includes numerous descriptions of early embankment dams.
4. Schnitter, *A History of Dams,* 55-80.
5. Schnitter, *A History of Dams,* 123-7.
6. See J. A. Diego-Garcia, "The Chapter on Weirs in the Codex of Juanelo Turriano: A Question of Authorship," 217-34. Norman Smith, *A History of Dams* (London Peter Davies,1971), 117-20. Also see Smith, 192-3, for a discussion of Simon Stevin's 1586 treatise "De Beghinson des Waternichts."
7. J. E. Gordon, *Structures: Or Why Things Don't Fall Down* (New York: Penguin Books, 1978), 33-44, provides good discussion on how the late seventeenth century work of Newton and Hooke influenced the theoretical development of structural mechanics.
8. See Peter Molloy, "19th Century Hydropower: Design and Construction of Lawrence Dam, 1845-1848," *Winterthur Portfolio* 15 (Winter 1980), 315-43, for references to dam design publications by Belidor, Bossut, Smeaton, Coulomb, and Moseley.
9. J. A. T. de Sazilly, "Sur un type de profil d'egale resistance propose pour les murs des reservoirs d'eau," *Annales des Ponts et Chaussees* (1853), 191-222. M. Delocre, "Memoire sur la forme du profil a adopter pour les grande barrs ages en maconnerie des reservoirs," *Annales des Ponts et Chaussee*, Memoires et Documents (1866), 212-72. Both of these references are taken from Smith, *History of Dams*.
10. Smith, *History of Dams*, 197-200. For later structures, such as the Ternay Dam (1865-68) and the Bon Dam (1867-70), Delocre selected maximum compressive stresses of 100 psi and 114 psi, but by modern standards, even these were extremely conservative.
11. Edward Wegmann, *Design and Construction of Dams*, 3rd edition (New York: John Wiley & Sons, 1900), 69.
12. W. J. M. Rankine, "Report on the Design and Construction of Masonry Dams," *The Engineer* 33 (January 5, 1872), 1-2.
13. In his article on the Lawrence Dam, Molloy reports that Moseley recognized the importance of the "middle third" in the early nineteenth century. However, when Rankine discussed the significance of the middle third in the 1870s, there is no evidence that anyone considered it a reiteration of previously discussed ideas. Smith makes no mention of Moseley in his *History of Dams*.
14. See Smith, *History of Dams*, 205-6, for a discussion of how the Gileppe Dam was designed with extremely conservative proportions. See Wegmann, *Design and Construction*, 3rd edition, 81-90, for descriptions of the Gileppe and Vyrnwy.
15. E. Sherman Gould, *High Masonry Dams* (New York: D. Van Nostrand, 1897) 3-4.
16. George Holmes Moore, "Neglected First Principles of Masonry Dam Design," *Engineering News* 70 (September 4, 1913), 442-5.
17. C. L. Harrison, "Provision for Uplift and Ice Pressure in Designing Masonry Dams," *Transactions of the American Society of Civil Engineers* 75 (1912), 142-5, with discussion, 146-225. For a modern treatment see Max A. M. Herzog *Practical Dam Analysis* (Thomas Telford, Ltd.: London, 1999), 80.
18. The upstream face of buttress dams can also be built using wood planks or steel sheets; however, the most common types of large-scale buttress dams feature upstream faces built of reinforced concrete.
19. Schnitter, *A History of Dams*, 124-7 and 144-5.

20. Robert F. Leggett, "The Jones Falls Dam on the Rideau Canal," *Transactions of the Newcomen Society* 31 (1957-59), 205-218. Unfortunately, any theoretical calculations that may have been used to proportion the design remain unknown, and, as such, it cannot be considered a structure that used mathematical theory as a basis.

21. Smith, *History of Dams*, 181-3.

22. In examining the profile of the Zola Dam, it is evident that the cylinder formula was not used as a rigid means of proportioning the structure. Rather, Zola used it as a guide to assist him in increasing the thickness of the arch as it descended into the canyon. For the eclipse of arch dams, see Schnitter *A History of Dams*, 195.

23. James Dix Schuyler, *Reservoirs for Irrigation, Water-Power, and Domestic Water-Supply* (New York: John Wiley and Sons, 1902), 246-256 provides a good description and photograph of the original Bear Valley Dam. Also see F. E. Brown, "The Bear Valley Dam," *Engineering News* 19 (June 23, 1888), 513-4.

24. Schuyler, *Reservoirs*, 213-37. As designer of this important dam, Schuyler provided an extensive description of it in his book. Also see James D. Schuyler, "The Construction of the Sweetwater Dam," *Transactions of the American Society of Civil Engineers* 19 (1888), 202-3.

25. Albert Pelletreau, "Barrages cintres en forme de voute," *Annales des Pont et Chaussees* (1879). This reference is included in Schnitter, "Arch Dams."

26. Gardiner Williams, discussion on "Lake Cheesman Dam and Reservoir," *Transactions of the American Society of Civil Engineers* 53 (1904), 183. John S. Eastwood, "An Arch Dam Design for the Site of the Shoshone Dam," *Engineering News* 63 (June 9, 1910), 678-80.

27. Lars Jorgensen, "The Constant-Angle Arch Dam," *Transactions of the American Society of Civil Engineers* 78 (1915), 685-733.

28. Louis Torres, *"To the Immortal Name and Memory of George Washington:" The United States Army Corps of Engineers and the Construction of the Washington Monument* (Washington, D.C., 1985). Albert E. Cowdry, *A City for a Nation: The Army engineers and the Building of Washington, D.C. 1790-1967* (Washington, D.C., 1978).

29. Wisner reviews this history in his report, George Y. Wisner, "Investigation of Stresses Developed in High Masonry Dams of Short Span." Report submitted to F. H. Newell, Chief Engineer, U.S. Geological Survey, Washington, D.C., May 16, 1905, 22 pages.

30. Letter from Wisner et al., in Montrose, California, to Newell in Washington, D.C., October 5, 1904. Response from Newell to Wisner, Detroit, October 12, 1904. The recommendation of Wheeler comes in a letter from Wisner and J. H. Quinton to Newell on November 2, 1904. Presumably the final approval came quickly thereafter. For details on A. P. Davis, see Charles H. Bissell and F. E. Weymouth, "Memoir for Arthur Powell Davis," *Transactions of the American Society of Civil Engineers* 100 (1935), 1582-91.

31. Letter report from Wheeler in Los Angeles to Wisner in Los Angeles, May 5, 1905, 9 pages. Covering letter from Wisner in Los Angeles to Newell in Washington, D.C. May 16, 1905 with the complete report (ref. 31). The published paper appeared as George Y. Wisner and Edgar T. Wheeler, "Investigation of Stresses in High Masonry Dams of Short Spans," *Engineering News* 54, No. 60, (August 10, 1905), 141-4. Earlier publications did deal with the arch designs, but they appeared in obscure publications. For example, L. Wagoner and H. Vischer, "On the Strains in Curved Masonry Dams." *Proceedings of the Technical Society of the Pacific Coast* 6 (December 1889). Such articles did not have the influence of Wisner and Wheeler.

32. Discussions by Wisner of "Lake Cheesman Dam and Reservoir" *Transactions of the American Society of Civil Engineers* 53 (1904), 170-172 and reply by C. L. Harrison, 203-4.

33. H. Hawgood, "Huacal Dam, Sonora, Mexico." *Transactions of the American Society of Engineers* 78 (1915), 565.

34. Lars. R. Jorgensen, "The Constant-Angle Arch Dam," 686.

35. If radius ®) is to decrease as pressure (P) increases, then the horizontal angle (2q) that defines the arc of each arch slice should remain constant to follow the V-shape of the canyon walls whose width (W) is decreasing linearly. Since W= 2Rsinq, the radius will decrease linearly with W so long as q remains a constant, hence the constant angle arcH. Since arch stiffness is a function of $1/R^2$, the arch will lose stiffness near the base if R is not reduced. Since cantilever stiffness depends upon T^3, it will greatly diminish for thinner dams, especially near the base.

36. Lars Jorgensen, "The Record of 100 Dam Failures," *Journal of Electricity*, San Francisco, 44, No. 6 (March 15, 1920). Lars Jorgensen, "Memorandum on Arch Dam Developments," *Proceedings American Concrete Institute* 27 1931.

37. Fred A. Noetzli, "Gravity and Arch Action in Curved Dams," *Transactions of the American Society of Civil Engineers*, 84 (1921), 23.

38. Noetzli, "Gravity and Arch Action in Curved Dams," 41. The low factor of safety can be calculated for overturning as follows: For H/B = 1.5 and for no uplift S.F. = 2.00 (assuming a triangular dam section) while for full uplift S.F. = 1.1.

39. Lars Jorgensen, "Improving Arch Action in Arch Dams," *Transactions of the American Society of Civil Engineers* 83 (1919-1920), (New York: 1921), 316-31. His discussion appears in 84, 68-71. Jorgensen had already reported on the measurements in the Salmon Creek Dam especially related to temperature, and he had referred to observations on his Lake Spaulding Dam in discussing Noetzli's paper.

40. William Cain, *Discussion*, 84, 71-91. His analysis was expanded and published as "The Circular Arch Under Normal Loads," *Transaction of the American Society of Civil Engineers* 85 (1922), 233-48. Noetzli wrote a discussion to the paper in 85, 261-4. Also in the discussion, William Cain, a professor of mathematics at North Carolina University, compared Noetzli's graphical method to a purely analytic approach given the previous year by B. A. Smith and noted their close correspondence.

41. Fred Noetzli, "The Relation Between Deflections and Stresses in Arch Dams," *Transactions of the American Society of Civil Engineers* 85 (1922), 306. Noetzli also spoke of time effects and tended to discount them in dams; today we recognize that as *creep* and would agree that in most cases it is not crucial for dams unless the compression stresses are high.

42. C. H. Howell and A. C. Jaquith, "Analysis of Arch Dams by the Trial Load Method," *Transactions of the American Society of Civil Engineers* 93 (1929) 1191-225.

43. According to Ray Lyman Wilbur and Northcutt Ely in the book Hoover Dam Documents (Washington, D.C., Government Printing Office, 1948) discuss the name of Hoover Dam on pages 80-2. "On September 17, 1930, on initiating construction, Secretary Wilbur issued the following order [to Commissioner Mead]: . . ."This is to notify you that the dam which is to be built in the Colorado River at Black Canyon is to be called the Hoover Dam . . ." in December 1930, Congressman Taylor . . . called the attention of the House (of Representatives) to the fact that the committee (Interior Department Subcommittee on Appropriations) had designated the dam as Hoover Dam in the appropriation bill, saying" . . . we unanimously and very gladly wrote into this action those words making the naming of that great dam the Hoover Dam by the action of Congress . . . so that the dam is now officially named by both the Secretary of the Interior and by Congress." The bill became law on February 14, 1931. In the next four succeeding appropriations acts, in 1932 and early 1933, the dam was designated as "Hoover Dam." "After Mr. Hoover left office, the Interior Department, although failing to take any formal action, avoided the use of the name 'Hoover Dam,' and publicized the names "Boulder Canyon Dam' or 'Boulder Dam. . . ." Early in the Eightieth Congress, a number of bills were introduced to restore the name of Hoover Dam. . . ." On April 30, 1947, President Harry S. Truman signed a resolution restoring the name Hoover Dam. Though the name "Boulder Dam" was used by Reclamation and others during the planning and authorization phases, and later (1933 until 1947) the dam was popularly and unofficially known as Boulder Dam or Boulder Canyon Dam, Hoover Dam was ever officially named Boulder Dam.

44. U.S. Department of the Interior, Bureau of Reclamation, *Boulder Canyon Project Final Reports, Part V - Technical Investigations, Bulletin 1, Bureau of Reclamation, Trial Load Method of Analyzing Arch Dams* (Denver Colorado: 1938), 13, 23. This 266-page report filled with mathematical formulations does not give any numerical results and hence no interpretations of the method as to is relationship to dimensioning. The 1929 paper (ref. 44) does give some data on Horse Mesa and the Gibson dams.

45. Letters from R. F. Walter (Acting Chief Engineer) in Denver to Commissioner Mead in Washington, January 16, 1925 and from Commissioner Elwood Mead in Washington to the secretary of the American Society of Civil Engineers in New York, January 16, 1925. NARA, Denver, Record Group 115 [hereinafter Record Group will be designated with RG].

46. R. Wuczkowski, "Flussigkeitsbehälter," in Friedrich Ignaz Edler von Emperger, editor, *Handbuch für Eisenbetonbau* (Berlin: Wilhelm Ernst & Sohn, 1911edition), 3, 348-51, 407-13.

47. Letter from Fred Noetzli in San Francisco to the Board of Trustees of the Engineering Foundation in New York City, March 3, 1922, 2 pages.

48. Fred Noetzli, "Arch Dam Temperature Changes and Deflection Measurements," *Engineering News-Record* 89, No. 22 (November 30, 1922), 930-2.

49. Letters from Noetzli in San Francisco to Davis in Washington, D.C., March 4, 1922, and from Davis in El Paso to the Engineering Foundation in New York, March 13, 1922. See also a letter from Alfred Flinn (secretary of the Engineering Foundation) in New York to Davis in Washington, March 17, 1922.

50. Letters from F. E. Weymouth in Denver to Davis in Washington, November 10, 1922, and from Morris Bien (Acting Director) in Washington, D.C., to Weymouth in Denver, November 20, 1922. In the fall of 1922, Weymouth asked Davis if Reclamation could not give financial support to the project; the Acting Director agreed to some support. NARA, Denver RG115.

51. "Report on Arch Dam Investigations," 1, November 1927, *Proceedings of the American Society of Civil Engineers* (May 1928), 6, 43. In Bulletin No. 2 (June 1, 1924), the Engineering Foundation noted that the Edison Company had put up $25,000 and another $75,000 was needed.

52. "A Progress Report on the Stevenson Creek Test Dam," *Bulletin No. 2,* Engineering Foundation, December 1, 1925, 8 pages. On December 1, 1925, the Engineering Foundation issued a progress report giving the dam design and the list of contributors, which was dominated by private industry. The Bureau of Reclamation was not listed although the Bureau of Standards was, largely through the assignment of W. A. Slater, its engineer-physicist.
53. Letter from Julian Hinds, in Denver, to R. F. Walter, in Denver, December 17, 1925. NARA, Denver, RG115.
54. Letter from R. F. Walter, in Washington, D.C., to Elwood Mead in Washington, D.C., December 16, 1925. NARA, Denver, RG115.
55. Letters from R. F. Walter, in Denver, to Elwood Mead, in Washington, D.C., February 11, 1926, and from Mead, in Washington, D.C., to Walter in Denver, February 19, 1926.
56. John L. Savage, "Arch Dam Model Tests - Progress Report," December 17, 1928.
57. "Report," November 1927, *Proceedings of the American Society of Civil Engineers* (May 1928), 8-9.
58. Fred Noetzli, "Improved Type of Multiple-Arch Dam," *Transactions of the American Society of Civil Engineers* 87 (1924), 410, Closure.
59. Hardy Cross, "Discussion of Design of Symmetrical Concrete Arches by C. S. Whitney," *Transactions of the American Society of Civil Engineers* 88 (1925), 1075-7. In 1932, Cross would introduce the most significant simplified analysis procedure ever presented for reinforced concrete frame structures. Noetzli worked from a solid math basis but simplified it by graphic statistics (a Swiss tradition). Cain (see ref. 42) was a mathematician, not an engineer, and his work is very complex, but Noetzli could understand it easily. Reclamation carried on from Noetzli's simplified approach to develop an elaborate and highly complex mathematical method. Cross criticized all such complexity and sought to develop methods that the average practicing structural engineer could easily use.
60. Kollgaard and Chadwick, editors, *Development of Dam Engineering in the United States*, 269.
61. Kollgaard and Chadwick, editors, *Development of Dam Engineering in the United States,* 537, 539.
62. Kollgaard and Chadwick, editors, *Development of Dam Engineering in the United States,* 535-8.
63. Jorgensen won the 1918 Norman Medal for his paper on multiple arch dams, Noetzli won the 1922 Croes Medal for his paper on arch cantilever dams (no Norman Metal awarded that year), and Jacobsen won the 1924 and 1927 Norman Medals for his papers on multiple arch dams and thick arch dams respectively. B. A. Smith won the 1920 Croes Medal, William Cain won the 1926 Croes Medal, and Charles Whitney won the 1925 Croes Medal, all three awards were for arch analyses. H. de B. Parsons won the 1925 Rowland Prize for a paper on a multiple arch dam. See *American Society of Civil Engineers Official Register* (1982), 182-3, 186.
64. Report, 1927, *Proceedings of the American Society of Civil Engineers* (May 1928), 38-41 and Kollgaard and Chadwick, editors, *Development of Dam Engineering in the United States*, 539.
65. Jackson, *Building the Ultimate Dam*, Chapters 3 and 4. For Goldmark's design, see Henry Goldmark, "The Power Plant, Pipe Line and Dam of the Pioneer Electric Power Company at Ogden, Utah," with Discussion and Correspondence. *Transactions of the American Society of Civil Engineers* 38 (December 1897), 246-314.
66. Jackson, *Building the Ultimate Dam*, 4.
67. Fred Noetzli, "Multiple-Arch Dams;" Edward Wegmann, *The Designs and Construction of Dams*, 8th edition (New York: John Wiley & Sons Inc., 1927), 446-50.
68. Report, 1927, *Proceedings of the American Society of Civil Engineers* (May 1928) 189.
69. Noetzli, "Multiple-Arch Dams," 1927, 482.
70. "Summary of Cost, Gibson Dam," *Final Report*, April 1931, 1, 11-3. Letter from R. F. Walter, in Denver, to Elwood Mead, in Washington, March 13, 1928, which transmitted a report of the Board of Engineers for the Owyhee Dam, February 25, 1928. NARA, Denver, RG115.
71. Noetzli, "Multiple-Arch Dams," 1927, 486-90. Jackson, *Building the Ultimate Dam*, 207 gives a cost of $435,000.
72. Noetzli, "Multiple-Arch Dams," 478-82. Jackson gives a value of 17,750 cubic yards of concrete for the multiple arch design but the same cost. See Jackson, *Building the Ultimate Dam*, 147-8.
73. Noetzli, "Multiple-Arch Dams," see pages 187-92 for a more detailed discussion of the attitude of the Reclamation Service (changed to the Bureau of Reclamation in 1923).

CHAPTER 3

EARLY MULTIPURPOSE DAMS: ROOSEVELT AND THE RECLAMATION SERVICE, WILSON AND THE U.S. ARMY CORPS OF ENGINEERS

FROM SINGLE TO MULTIPURPOSE DAMS

Roosevelt and Wilson Dams

Throughout the nineteenth century, people built dams either for municipal storage, for water power, for flood control, or for irrigation. Many were privately built, some by local governments, but none by the federal government except for river navigation. With only a few late nineteenth-century exceptions, all dams were low and brought forth little modern engineering. By the end of that century, the problems associated with erecting masonry dams gave rise to their study, which Edward Wegmann summarized in the first edition of his classic text on *The Design and Construction of Dams* (1888). He considered only masonry gravity dams, and it was the Quaker Bridge Dam design, over 100 feet higher than any previous masonry dam, that stimulated Wegmann to write his first edition. He noted only 14 American dams, none by the federal government, and most were for municipal water supply. He did not write about earth embankments, although many had been built and the U.S. Army Corps of Engineers had long been engaged in such work for levees.[1]

It was not until early in the twentieth century that dams designed to serve more than one purpose began to be built, and, then, largely because of the growing demand for electrical power and its natural source in the flow and fall of water. That demand, in itself, did not bring the federal government into the building of multipurpose dams; rather, it arose from the mission that the two agencies had: navigation for the U.S. Army Corps of Engineers and irrigation for the Reclamation Service. No structures better show the origins of dams built for more than one purpose than Roosevelt Dam on the Salt River and Wilson Dam on the Tennessee River. These dams will form the central part of this chapter.

In both cases the governmental agencies adopted forms that had been worked out by nonfederal designers. For the Reclamation Service, the structures were for storage, but for the Corps, they were run-of-the-river dams. Municipalities had been building large storage dams of masonry in the late nineteenth century, and mainstream dams began to be commonly used early in the twentieth century. The storage dams are usually marked for their height and often appear in narrow valleys, while the mainstream dams are low but often built across wide stretches of river.

We have already seen how the Reclamation Service dealt with storage dams by designing curved masonry dams, which led engineers to develop the trial load method of analysis. Roosevelt Dam was such a design, but its initial design neglected completely any arch action. However, its 400 foot radius and relatively thin upper regions certainly resulted in substantial load being carried by arching. After completion of the design, it was analyzed both as an arch (with no cantilever action) and as a cantilever (with no arch action) to estimate conservative values for its margin of safety.[2]

By contrast, Wilson Dam is a pure gravity structure. Its crucial problems were with the foundations and were centered on uplift, sliding, and overturning. In addition, much of the dam is a spillway passing that part of the river flow not used for power. A critical issue studied since the late nineteenth century was the downstream shape of the spillway cross section, designed to avoid separation between the water flow and the concrete surface. This separation can lead to a partial vacuum on the concrete, pulling it loose, and thereby forming holes by cavitation.

The U.S. Army Corps of Engineers had studied this question of shape in connection with dams on the Cumberland River, and in 1908 Major William Harts directed a survey of overflow masonry weirs; i.e., spillways and other dam cross sections for relatively low overflow dams. Carried out in detail by the experienced John S. Walker, who had been an engineer for the Corps since 1872, the study appeared in summary form in a later edition of Wegmann's famous book.[3] Wegmann had described the general problem with water pressure, uplift pressure, and the S-shaped curve on the downstream face. It was this type of design that would engage the engineers of the Corps as they moved toward multi-purpose mainstream dams.[4]

But the movement toward such dams was controversial within the Corps. One prime illustration came from the same William Harts in 1909. President Roosevelt, in 1908, had articulated clearly to the Congress the multi-purpose nature of river development by linking navigation to power. But the following year, as Nashville's district engineer, Colonel Harts published eight criticisms of the high dams and storage reservoirs required by multipurpose river development. He even predicted that "it seems improbable that it [such development] will ever be extensively used." He was not alone in the Corps, even West Point taught (until 1938) that such developments would be prohibitively expensive.[5]

Because of the naturally competing uses for dams, these objections were not outlandish. The most serious one, referred to in the West Point text, was the conflict between flood control and power generation. The reservoir behind the dam needs to be kept low to be able to catch flood waters; whereas the reservoir needs to remain full to generate maximum power. Moreover, navigation requires the reservoir to be drawn down during periods of low water runoff so that the downstream water level can be kept high enough for shipping and so that irrigation can proceed. On the other hand, that loss of water makes recreation less attractive and can hurt the value of lake side property. It is far easier to manage a single purpose dam where the reservoir is dedicated to one or the other use. For the Corps, this meant navigation and for Reclamation, irrigation.

The multipurpose debate is most strikingly characterized in the 1920s by a strong opponent of Colonel Harts. Harts had become division engineer in 1920 and his opponent was Major Harold C. Fiske, then the district engineer for both Nashville and Chattanooga Districts. Fiske, rather than Senator George Norris, has even been called the father of the Tennessee Valley Authority. More accurately, Fiske was a primary stimulus for the well-known 308 reports and "perhaps the greatest proponent of multipurpose and comprehensive water resource development in the nation during the 1920s." But he would achieve that distinction at a price. In the Corps, one did not publicly dispute a superior like Harts without it damaging one's career.[6]

Fiske was carrying on that major function of the Corps to survey rivers, but the Corps' river surveys had, by the early twentieth century, expanded to river basin surveys rather than just the navigable streams. So, both the Reclamation and the U.S. Army Corps of Engineers were encouraged by the Progressive Era President Theodore Roosevelt to view their historic missions more broadly and to think specifically of dams as part water resources. No part of the system would characterize that multipurpose concept better than large-scale multipurpose dams, such as the two named for Presidents, Roosevelt and Wilson, whose administrations helped to define the era in which such major restructuring of the nation began.

Roosevelt Dam, along with Pathfinder and Buffalo Bill Dams, was a precursor to Hoover Dam, the first major multipurpose structure from Reclamation; similarly, Wilson Dam was the precursor to Bonneville Dam, the first major multipurpose structure designed and built entirely by the U.S. Army Corps of Engineers. Roosevelt Dam was followed by Arrowrock, Elephant Butte, and Owyhee Dams, all impressively large for their time, but none was conceived as multipurpose to any significant degree. By contrast, the McCall Ferry, Keokuk, and Wilson Dams were clearly designed for the generation of power with the latter two being truly multipurpose, having both navigation and power as primary objectives.

The two agencies moved into a broader concept for their dams during the period leading up to Hoover and Bonneville. Irrigation required study of arid regions around and sometimes at great distance from the reservoirs. The experience of the Corps with Wilson led them to think about electrical power distribution at a distance. Even before those dams, the Corps had, for a long time, been concerned about flooding and levees, which required consideration of more than just navigation.

In this chapter, we shall explore the origins and developments of the Roosevelt and Wilson Dams and end with the manner in which the Tennessee Valley studies led Major Fiske to propose river basin studies that would lead directly to the 308 reports. These reports laid the basis for the great federal dams that began in the 1930s amid the turmoil of the depression. But there were other factors, both political and technological, that would influence dam building after 1932. American democracy pits the executive branch against the legislative branch, which meant that the 308 concept, however similar it made the centralized organization of river basins seem, still had to pass the more localized scrutiny of Congress. Fiske called for a Tennessee Valley plan, to be directed by the U.S. Army Corps of Engineers. President Franklin Roosevelt took the idea and decided that it should be a model for other major river basins, but the Congress, after approving the Tennessee Valley Authority, would go no further.

Thus, the story of major dams in the other basins exhibits continued conflict between the President and Congress and between these two federal agencies themselves. The resolution of the conflicts, or more accurately the physical results of these conflicts, illustrate how the American system of politics operated in the 1930s and still operates even at the beginning of the twenty-first-century. Because of the uniqueness of each major basin--the Colorado, the Columbia, the Missouri, the Central Valley of California, and the Ohio-upper Mississippi--the resolutions would turn out to be unique to each basin. The major dams we have chosen to study tend to be significantly different from each other, both in their forms and in their associated politics.

YEARS OF TURMOIL: FROM RECLAMATION SERVICE TO BUREAU

Newell, Davis and Wisner

When the Reclamation Service was authorized in the summer of 1902, Frederick H. Newell had worked for the U.S. Geological Survey for almost 15 years. He began with Powell's Irrigation Survey in the late 1880s, and, after Congress withdrew support for this initiative, he repositioned himself as the Geological Survey's Chief Hydrographer in charge of measuring the flow of rivers throughout the United States. Based in Washington, D.C., Newell cultivated relationships with political and cultural leaders through participation in non-government organizations (such as the National Geographic Society); by the start of the new century, he was acknowledged nationally as a prominent irrigation advocate. Nonetheless, at the time of his appointment as Chief Engineer for the Reclamation Service, his professional experience had been almost entirely that of an administrator. He had not been involved in the design, construction or operation of any functioning irrigation projects, nor had he supervised the design or construction of any large dams or water storage structures. In this light, it is not

Figure 3-1: Frederick H. Newell, U.S. Reclamation Service: Chief Engineer (1902-1907); second Director (1907-1914). Source: Bureau of Reclamation.

surprising that Reclamation Service records contain little evidence of Newell's involvement in the bureau's technical work.[7]

In Newell's place, Arthur Powell Davis functioned as the Reclamation Service's senior official responsible for matters of engineering; in 1902, he was appointed Assistant Chief Engineer, and, after Newell's formal designation as Director in 1907, he assumed the title of Chief Engineer. A nephew of John Wesley Powell, Davis received an engineering degree from Columbian University (later renamed George Washington University) in Washington, D.C. After graduation in 1888, he joined Powell's Irrigation Survey and remained

Figure 3-2: Arthur Powell Davis, U. S. Reclamation Service: Assistant Chief Engineer (1902–1907), Chief Engineer (1907-1920), Director (1914-1923). Source: Bureau of Reclamation.

through the 1890s as an employee of the U.S. Geological Survey.[8] Davis was active in undertaking field work—for example in 1897 he authored a detailed Geological Survey publication outlining possible dam and irrigation projects on the Gila and Salt Rivers in Central Arizona[9]—but, like Newell, he had not been substantively involved in the actual operation of irrigation systems or the construction of dams prior to formation of the Reclamation Service.

As a means of facilitating Reclamation's engineering work, Newell instituted a system of consulting "engineering boards." These boards usually consisted of three men who were either high-level, full-time Reclamation Service employees (known as supervisory engineers) or part-time consultants (designated consulting engineers). These boards—which sat atop a hierarchy of district engineers (who were originally assigned responsibility for a state or territory and later given jurisdiction over districts encompassing more than one state), planning and construction engineers for specific projects, engineering assistants, and engineering aides—were authorized to prepare, review or approve plans related to various projects.[10] The need for a fairly elaborate system of project administration was largely dictated by the geographically-diffuse character of Reclamation's work; for example, a 1904 letter from Charles Walcott (Director of the U.S. Geological Survey and Newell's immediate superior) to the Secretary of the Interior explaining Reclamation's organization reported that:

Figure 3-3: (Theodore) Roosevelt Dam on the Salt River Project, May 15, 1941. Source: Bureau of Reclamation.

> The plan of organization is necessarily different from that which would be adopted if all of the work were concentrated in one state or locality. It would be a comparatively small matter to supervise the work under these circumstances. The conditions are such that plans must be prepared and executed almost simultaneously in 13 states and 3 territories and in localities hundreds of miles apart. Hence it becomes necessary to have at each point men located to carry on certain work and other men so situated that they can travel from point to point and give expert information and advice.[11]

As a result, each Reclamation project was overseen by a separate and distinct "engineering board," but many supervising and consulting engineers served on more than one board.

When selecting an engineer to provide advice on large-scale dam design, Newell turned to George Y. Wisner, a 63-year-old hydraulic engineer from Detroit, Michigan. Prior to 1903, Wisner's professional interests focused on issues such as sanitary engineering, harbor development, and water transportation

Figure 3-4: Engine room and cable towers on the east side of Roosevelt Dam during construction (1903-1911). Source: Bureau of Reclamation.

along the Great Lakes and the Mississippi River.[12] Wisner may have lacked experience in irrigation, storage dam construction, and western water projects in general, but this did not deter Newell from appointing him as a senior member of engineering boards for several projects. In addition, Wisner was also called upon to take responsibility for mathematically analyzing arch dam designs for the Shoshone and Pathfinder reservoir sites.[13] Wisner's participation in the early work of the Reclamation Service was unquestionably important, but he remains a problematic figure because, after his unexpected death in July 1906, his name practically never appears in regard to any ongoing bureau projects; in addition, neither Newell, Davis, nor anyone else associated with the Reclamation Service prepared a "memoir" (or obituary) recounting his career for publication in the *Transactions of the American Society of Civil Engineers,* something that was a common practice for senior members of the profession.

The Salt River Project

Under Newell's leadership, the early Reclamation Service planned and built several irrigation systems. These included the Salt River Project in central Arizona, the Truckee-Carson Project in Nevada (later named the

Figure 3-5: This November 1, 1909, photograph shows the large blocks of stone used in construction of Roosevelt Dam. Note the powerhouse in the lower right corner of the picture.
Source: Bureau of Reclamation.

Newlands Project in honor of Francis Newlands), the Milk River Project along the Canadian border in northern Montana, the Shoshone Project near Cody in Northern Wyoming, and the North Platte Project in southeastern Wyoming and western Nebraska. These early projects were spread out to encompass as many different states as possible. This action reflected both the legal requirement that monies accrued from public land sales be expended on projects in those states or territories where the land was sold and Newell's desire to obtain widespread political support for the newly established Reclamation Service. But, in attempting to plan and implement several major reclamation projects within a time span of only a few years, great strain was placed on Newell, Davis, and the entire bureau staff. The pressure on the nascent organization was further exacerbated by Newell's arguments, expressed prior to the bureau's formation, that extolled the scientific expertise and skill of the federal government in constructing projects beyond the capacity of privately-financed enterprise.

Newell's bureau never suffered from any cataclysmic dam failure and—within this context—avoided any disasters related to technical competence. But, in an economic context, Reclamation experienced shortcomings related to cost overruns as well as lengthy delays in project completion. One of the most prominent of Reclamation's early efforts was Roosevelt Dam, built as part of the Salt River Project in central Arizona. A brief review of this project offers insight into the problematic success of the Reclamation Service.

Roosevelt Dam

At the time of its authorization in 1902, the Reclamation Service was intended purely as a means of providing federal support for irrigation in 16 western states. Electric power production was not perceived as a primary or essential purpose of the Reclamation Service, and projects initially were not intended to serve "multiple purposes," but rather were to focus on supplying water for irrigation. However, once construction commenced on a dam and irrigation system, purposes other than irrigation could become integrated into it.

Anticipation of the benefits to be accrued by Reclamation Service endeavors did not derive from any emphasis that they might serve "multiple purposes." Rather, they were promoted by Frederick Newell as providing an efficient means of financing large-scale public works that would avoid wasteful spending and offer value to American society beyond the scope of what privately financed irrigation projects were capable of. Efficiency (by which they meant primarily minimizing cost)—and not "multiple-purpose"—would become the hallmark by which early Reclamation Service projects were judged by the water users who would assume responsibility for repaying the costs of these projects.

To better understand how the early Reclamation Service grappled with the problems and opportunities that confronted it in the decade or so after 1902, the following discussion focuses on Roosevelt Dam in central Arizona. Selected by the Secretary of the Interior as one of the Reclamation Service's first five authorized projects, Roosevelt Dam was erected in concert with land owners in the greater Phoenix region (known as the Salt River Valley Water Users' Association) in what became known as the Salt River Project. The Salt River Project eventually came to include a significant hydroelectric power component. And it came to embroil Newell and the Reclamation Service in a struggle with the Water Users' Association over repayment of—and control over—a project that dramatically exceeded initial cost estimates and engendered skepticism over the actual "efficiency" of the Reclamation Service. This struggle was not unique to Roosevelt Dam and the Salt River Project, but, in fact, it reflected difficulties that encumbered Newell and the Reclamation Service on a region-wide basis.

Modern irrigation development in the Phoenix area commenced in the 1860s, when Jack Swilling (an ex-Confederate soldier) and other early Anglo settlers cleaned out an ancient canal originally built by the Hohokam Indians.[14] In the 1870s and 1880s, several privately owned canals began to draw water from the Salt River, fostering concern that the river's natural flow could not serve all the competing ditches.[15] In reaction to this concern, an 1889 survey expedition, sponsored by the City of Phoenix and Maricopa County, located a large reservoir site about 60 miles east of Phoenix, just downstream from where Tonto Creek enters the Salt River.[16] Known initially as the Tonto Dam site, it became the site of Roosevelt Dam (named after the sitting President) when the Reclamation Service initiated plans to build one of its first large dams at the location.

In the early 1890s, the privately owned Hudson Reservoir and Canal Company filed an application to build a large dam at the Tonto site.[17] Exactly how close the Hudson Company ever came to actually building a dam at the site remains obscure, and certainly no construction work of any scale or significance ever occurred. But plans for the Hudson Company's Tonto Dam were widely disseminated and they received prominent notice in the first edition of James D. Schuyler's book, *Reservoirs for Irrigation, Water-Power and Domestic Water Supply*, published in 1901. Among the more striking aspects of the Hudson proposal was its inclusion of a 6,768 horsepower hydroelectric power component in the project, or, as Schuyler described the plan, "a combined irrigation and electric-power project, the same water being used for both purposes."[18] The revenue derived from the sale of electric power to businesses and consumers throughout the region was envisaged as playing a key role in paying the bonds necessary to finance construction.

At the start of the twentieth century, Phoenix's civic leaders began lobbying for federal assistance to build a storage dam at Tonto. At first, this lobbying focused on gaining approval for the issuance of publicly-backed bonds to help support dam construction (federal approval was necessary because Arizona was a territory and did not become a state until 1912); but once Theodore Roosevelt became president in September 1901, Phoenix leaders changed strategies and started advocating direct federal involvement in sponsoring the work.[19] In 1903, their lobbying bore fruit when the Secretary of the Interior selected the Salt River Valley for one of the Reclamation Service's first major projects.[20]

Although financing for early Reclamation Service projects was intended to come from the proceeds of public land sales, almost all of the almost 200,000 acres of land to be served by the Salt River Project was in private hands.[21] Nonetheless, the Salt River Project offered the Reclamation Service an opportunity to develop one of the West's premier reservoir sites. And this opportunity appeared to Newell as too attractive to pass by. For Phoenix boosters, the federal government was a welcomed supporter for their dreams of storing the flood waters of the Salt River behind a large dam, especially because it appeared that the private investment market had no interest in assuming the risk of financing such an endeavor. Reclamation Service Chief Engineer Newell's promise of an efficiently and economically prudent plan to build Roosevelt Dam under federal auspices generated excitement and anticipation on the part of Phoenix area farmers and residents. But, as will be seen, this same excitement and anticipation eventually spawned resentment when Newell could not deliver completion of the Roosevelt Dam project in a timely and economical manner.

In 1903, Newell directed his Assistant Chief Engineer, Arthur Powell Davis, to take the lead in organizing work on Roosevelt Dam and urged him to:

> Concentrate your energies as far as practicable on the pushing forward of the Salt River Project . . . and laying out a scheme of further work, involving the purchase of lands, rights of way, etc. . . .[22]

In selecting a design for Roosevelt Dam, Davis adopted a masonry curved gravity design that depended upon its huge mass for stability. In the late nineteenth century, masonry gravity dams had become the standard for metropolitan water systems (i.e., New York's Croton Dam and Boston's Wachusett Dam), and Newell and Davis apparently considered it appropriate that their bureau's most visible dam project an aura of permanence and stability similar to dams built by major municipalities.[23] Although the Reclamation Service never explicitly stated that Roosevelt Dam was to mimic major structures of the Eastern

United States, there is no evidence that alternatives to a masonry curved gravity design were ever contemplated.[24]

In the 1890s, the Hudson Reservoir and Canal Company publicized a masonry curved gravity dam design for the Tonto site.[25] Later, in 1902, Davis proposed a structure that closely resembled the Hudson Company's plan, and this was subsequently adopted for construction.[26] As Newell later wrote:

> We [the Reclamation Service] have been inclined to adhere to the older, more conservative type of solid dam, largely perhaps because of the desire not only to have the works substantial but to have them appear so and recognized by the public as in accordance with established practice.[27]

With storage dams, Newell's concern for appearance and its effect on public perception was particularly acute:

> Plans for the construction of storage works, while they must be prepared with regard to reasonable economy, must be [undertaken] with a view to being not merely safe but looking safe. People must not merely be told that they are substantial, but when the plain citizen visits the works he must see for himself that there is every indication of the permanency and stability of a great storage dam . . . he must feel, to the very innermost recesses of his consciousness, that the structure is beyond all question.[28]

Given such sentiments as expressed by the top official in the early Reclamation Service, the choice of a masonry curved gravity design for Roosevelt Dam–without serious consideration of possible alternatives–becomes more understandable.

The selection of a basic design represented one of the least problematic aspects of early planning for the Roosevelt Dam project. Because of the Tonto site's remote location (more than 60 miles from a railhead in the Phoenix/Tempe area and 40 miles from the railroad that reached the mining town of Globe), a special effort was needed to build the arduous "Apache Trail" supply road connecting Phoenix to the site. Essentially, all supplies necessary to build the dam–other than those procured locally–would have to be hauled in by mule teams over the 60-mile-long Apache Trail. This would prove logistically cumbersome from the time that early construction work commenced in 1904 until formal completion of the masonry dam in 1911. The other key component of the project necessary to complete before work on the dam proper could start was the construction on the upper Salt River of a diversion dam connected to the damsite by a 19-mile long canal. This canal was designed to provide water to a hydroelectric powerplant that would generate the electricity necessary to operate equipment (including the aerial cableway) at the dam site.

The masonry blocks comprising the bulk of the dam were quarried from the site's canyon walls and placed via the aerial cableway strung across the site. The concrete used to bond the masonry blocks into a monolithic mass was formed using cement produced at a cement mill built by the Reclamation Service a few hundred feet upstream from the site. The cement mill represented an attempt to limit the amount of construction material that would have to be hauled in over the Apache Trail. But while the cement mill certainly eliminated the need to haul in cement, it required large quantities of fuel oil to be imported to operate the high-heat cement kilns, and securing a reliable fuel oil supply proved to be a troublesome task.[29]

To allow excavation of the site down to bedrock, an initial diversion tunnel was driven through the southern canyon wall, starting in 1905. Later supplemented with a second diversion tunnel drilled through the north abutment, this initial diversion tunnel proved too small to handle the heavy floods that descended down the Salt River in 1905. Eventually the dam site was successfully excavated down to bedrock, and, commencing in 1906, masonry blocks began to be placed as part of the dam proper. But construction constituted a long slow process that taxed the patience of the Reclamation Service staff, the citizens of greater Phoenix and U.S. citizens as a whole, who wondered how large a federal expenditure the project would entail before reaching completion.

Rather than take direct responsibility for hiring and supervising the labor necessary to build Roosevelt Dam, Newell and Davis sought to contract out as much work as possible. As Davis explained:

> The Department and the Director, as well as most of the consulting engineers, are strongly in favor of doing work by contract wherever this is practicable, even if in advance it may not seem the most economical method.[30]

At Roosevelt Dam, the contract method proved disastrous in building the 19-mile-long "power canal" system designed to supply hydroelectric power during the construction of the dam proper. Originally estimated to cost $215,260, this power canal (along with its diversion dam and associated hydroelectric powerplant) eventually cost over $1.4 million to complete.[31] Reliance upon outside contracts also led to the selection of John M. O'Rourke & Company of Galveston, Texas, as the main contractor for the dam although the firm was new to Arizona and had no experience in dam construction.[32] In 1905, O'Rourke came in with a low bid of $1,147,000 to complete the main structure of the dam up to the 150-foot-level in less than two years.[33]

Because of heavy flooding that wreaked havoc with the construction site, it may have been impossible for any company or organization to have built Roosevelt Dam in a timely, cost effective manner. However, O'Rourke never came close to meeting the terms of the original contract. By the spring of 1907, when the vast majority of the structure was to have been complete, O'Rourke had placed only five percent of the total masonry required.[34] In fact, the dam did not reach the 150-foot level until November 1909, more than two and one-half years past the deadline stipulated in the original contract. Final completion of the structure occurred in February 1911.

As exemplified by the experience with O'Rourke, the use of a contract system did not prove effective in insuring that expenditures by the Reclamation Service could be easily controlled. At the same time, a prominent attempt by Newell and Davis to limit costs by avoiding the use of the contract system also proved problematic. Roosevelt Dam consists primarily of large sandstone blocks bonded together by concrete, requiring large quantities of cement. In his original 1902 design proposal, Davis estimated that purchasing cement and delivering it to the Roosevelt site would cost $9.00 per barrel; at the same time, Davis reported that cement using local limestone and clay deposits could be produced for about $2.00 per barrel.[35] As a result, the Reclamation Service opted to build and operate its own cement mill at Roosevelt despite subsequent offers by private companies to deliver cement on-site for as low as $4.51 per barrel.[36]

After construction of the cement mill commenced in March 1904, it took the Reclamation Service a year to get the cement mill operating because "the inaccessibility of the dam site caused long delay in securing the necessary equipment for economical and rapid work."[37] And after operations began, the need to fire the cement kilns with imported fuel oil largely offset any advantages that accrued by using locally available limestone at the plant. Fuel oil came to the Salt River Valley from California by railroad and was then hauled overland sixty miles to the cement plant. Because of a railroad tank car shortage and problems with the haulage contract, the cement plant suffered from periodic shutdowns throughout the construction process.[38] The cement plant ultimately worked, but the original cost estimates proved to be too optimistic. Instead of $2.00 per barrel, the per unit cost came to $3.14 per barrel; and this figure was reduced because an additional 60,000 barrels (out of a total of 338,000) were produced for the Granite Reef Diversion Dam.[39] The Reclamation Service's effort to manufacture cement at Roosevelt Dam was hardly a disaster, but it suffered from significant cost overruns that exceeded original estimates and demonstrated that avoidance of the contract system also posed pitfalls for the Reclamation Service.

Costs for the Roosevelt Dam exceeded original estimates and the overall project proved much more expensive than anticipated in 1903. Whereas the dam project was originally estimated to cost about $1.9 million (including the power canal, powerplant, cement mill, hydraulic gates, tunnels, roads, and "damage to private lands"), it ultimately cost more than $3 million dollars–and this figure excluded more than $1 million for land purchases, placement and repair of hydraulic gates, tunnel excavation and road construction, as well as more than $2.3 million for the entire electric power system.[40] Hopes for financing the Salt River Project with monies derived from public land sales in Arizona Territory quickly faded, and initial project costs eventually exceeded $10 million.[41] Some of this increase resulted from economic inflation that afflicted all aspects of the U.S. economy, and some of it resulted both from raising the height of Roosevelt Dam and from the Reclamation Service taking responsibility for building a new diversion dam at Granite Reef. But, much of it also resulted from the Reclamation Service's failure to efficiently build facilities such as the power canal and to develop plans related to the dam proper that were feasible for the contractor to implement in a timely manner.

In the case of the contract that O'Rourke signed to complete the main bulk of the dam in 24 months, the unrealistic expectations–and the inexperience–of the Reclamation Service were particularly in evidence. Even at the time, there should have been little reason for Newell or Davis to have believed that–even under the best of conditions at the remote Tonto site–O'Rourke could successfully meet the original terms of his contract. Consider the following contemporaneous situation. It took Boston's Metropolitan Water District 49 months (June 1901–July 1905) to construct the Wachusett Dam and place 273,000 cubic yards of masonry; and this was at a site directly accessible to a railroad and in the midst of a major industrial region.[42] In contrast, O'Rourke had offered to erect approximately the same quantity of masonry in only 24 months in a remote, harsh wilderness with a minimal supply of local skilled labor.

The Reclamation Service's problems did not arise unexpectedly near the project's end; even during the earliest phases of construction, Newell knew that expenditures were escalating out of control. By the end of 1905, almost $3.5 million had been contractually obligated for the Salt River Project, prompting Newell to express concern over "the enormous expenditures which have been made in Arizona," while bemoaning that "we have already allotted to this project a very large sum, more than can easily be defended, and the allotment has been so liberal that I did not suppose that we should come to an end so quickly. The end, however, has been reached. . . ."[43]

This financial predicament did not abate after 1905, and, as O'Rourke fell further behind schedule, the notion of the Reclamation Service as an efficient engineering organization became difficult to sustain. But Newell could not easily confront Reclamation's failures because this brought the skills and progressive attributes of the Reclamation Service into question. Meanwhile, residents of the Salt River Valley participated in the effort to reduce the repayment requirements of the 1902 Reclamation Act, and, especially after the Roosevelt Dam began storing water, they did not hesitate to express dissatisfaction with the Reclamation Service.[44]

Based on the legislation authorizing the program, Newell had insisted that beneficiaries of Reclamation Service projects pay their full costs (excluding interest charges). But in the wake of financial overruns exemplified by what transpired in the case of the Salt River Project, Reclamation's ability to pursue such a policy faltered. In his 1901 message to Congress, President Roosevelt had set a high standard for federal reclamation in assuring the American people that:

> No reservoir or canal should ever be built to satisfy selfish personal or local interests, but only in accordance with the advice of trained experts. . . . There should be no extravagance, and the believers in the need of irrigation will most benefit their cause by seeing to it that it is free from the least taint of excessive or reckless expenditure of the public moneys.[45]

Unfortunately, cost overrun problems were endemic with Reclamation Service projects, and Newell, one of Roosevelt's "trained experts," could not easily justify what opponents might term "excessive or reckless expenditure of the public moneys."[46] While Newell held to the view that the original repayment terms should be enforced, the Westerners who were to bear the brunt of increased construction costs expressed outrage and sought politically expedient ways to reduce their financial liability.

A few months after the completion of Roosevelt Dam, Newell expressed dismay at how members of the Salt River Valley Water Users' Association were attempting to evade payment as stipulated under the original terms of the National Reclamation Act:

> It is not wise to let it be understood that the project is completed and that the Valley has nothing more to expect. This has been emphasized by a number of the citizens who have implied that now that there is nothing more to be had out of Uncle Sam they

> can concentrate their energies on securing deferred payments,
> or even the repudiation of a whole or a part of the debt.
>
> I have been astonished at the way this feeling has apparently spread. . . .[47]

As reflected in his concern for monumentality in dam design, Newell was drawn to the construction of huge structures as symbols both of safety and of his bureau's ability to accomplish great things. And–despite the fact that it did nothing to open up the public domain to new farms–Roosevelt Dam represented an opportunity to develop one of the West's best reservoir sites. But when the Reclamation Service floundered in economic difficulties, his enthusiasm for this and other bureau endeavors waned. He expressed his disappointment in 1910 by acknowledging that:

> The outlook is very dubious, and we [the Reclamation Service]
> do not know from day to day what will occur. I am keeping
> the work going as well as I can under the circumstances, but, of
> course, there is not the feeling of satisfaction or of enthusiasm
> which formerly existed.[48]

Newell possessed no personal financial cushion to fall back on (in contrast to his colleague and contemporary Gifford Pinchot who left his post as Chief Forester of the U.S. Forest Service in 1910 after clashing with the Taft Administration), and he lingered on as nominal leader of the Reclamation Service until 1914, when he was finally dismissed by Franklin Lane, Woodrow Wilson's Secretary of the Interior. But from 1910 onwards, the financial problems of the Reclamation Service were so manifest that Newell's original hopes and ambitions for the bureau were forever lost.

Of course, this does not mean that structures such as Roosevelt Dam disappeared from the landscape or that the condition of western water development reverted back to a pre-1902 world. Taking Roosevelt Dam as an example, we can see what transpired in a more general context for Reclamation Service projects as a whole. Although the cost for the original Salt River Project that the Salt River Valley Water Users' Association was responsible for repaying remained at the substantial figure of $10.2 million, the federal government adopted a forgiving posture as to the terms for this repayment; it was not completely paid off until 1956—thirty-five years later than it should have been paid under the original terms of the Reclamation Act, but within the terms of that Act as amended. Perhaps even more importantly, control over the operation of the entire Salt River Project—including Roosevelt Dam and all associated electric power facilities built as part of the project—was transferred from the federal government to the Salt River Water Users' Association in 1917.[49] At the time that control of the Roosevelt facilities was transferred to the Water Users' Association, the powerplant at Roosevelt had a rated generating capacity of 9,500 kW.[50]

From 1917 on, the Reclamation Service (and later the Bureau of Reclamation) certainly remained involved in activities associated with the development of the Salt River watershed. But this development did not depend on the work or planning of Reclamation staff. In fact, the construction of the three large hydroelectric power dams on the Salt River below Roosevelt Dam (Mormon Flat, Horse Mesa, and Stewart Mountain Dams) that were constructed in the 1920s was handled by the Water Users' Association with funding provided by bonds sold on the private investment market.[51] In essence, the history of Roosevelt Dam and the Salt River Project highlights how tenuous the Reclamation Service's role in western water development became before the authorization of the Boulder Canyon Project in the late 1920s. The Reclamation Service/Bureau never lost all relevancy, but compared to the heady days of 1904--05, the Reclamation Service/Bureau of the early 1920s had lost much of its luster and influence. It would take the construction of Hoover Dam to bring it back; and it would take Hoover—combined with the economic effects of the Great Depression—to establish that multiple purpose dams dependent upon hydroelectric power revenues could become a central component of federally financed water projects in the West.

Crises and Rebirth

After Newell's formal resignation in late 1914, Arthur Powell Davis was appointed Director of the Reclamation Service. Under Davis's leadership the bureau completed some prominent large-scale concrete dams, including the 348-foot-high Arrowrock Dam in southern Idaho and the 306-foot-high Elephant Butte Dam in New Mexico, but its financial problems did not abate. In the early 1920s, severely depressed agricultural prices prompted Congress to enact repayment "moratoriums" that relieved project beneficiaries from meeting their annual obligations until prices rebounded. This was welcome news for farming interests, but it also highlighted Reclamation's financial difficulties. By 1923, Reclamation had expended over $135 million dollars while repayments totaled less than $10 million. In reaction to these problems, Davis was fired as Director in 1923 and the name of the bureau changed to the Bureau of Reclamation.[52]

Elwood Mead and Reclamation

After Davis's dismissal, Secretary of the Interior Hubert Work established a special Fact Finding Commission to examine the financial and operational problems that plagued the federal reclamation program.[53] Formed in

September 1923, this commission included several prominent men involved in western development and agriculture, including Thomas Campbell (a former governor of Arizona), James Garfield (the Secretary of the Interior under Theodore Roosevelt), John Widstoe (a highly regarded agricultural scientist and former president of Utah State University), and Elwood Mead. Mead had been a longtime champion of western irrigation, dating back to the 1890s when he served as Wyoming's first state engineer. He later became head of irrigation investigations for the Office of Experiment Stations in the U.S. Department of Agriculture and–prior to the formation of the Reclamation Service in 1902– clashed with Frederick Newell over issues involving the proper role of the federal government in promoting irrigation.[54]

Mead advocated a smaller-scale, more community-oriented approach to federal reclamation than Newell, and–rather than promote the idea of increasing water supplies by building large-scale storage dams–Mead considered it more important to teach farmers better techniques of irrigation that would help eliminate problems such as overwatering and salt accumulation. After Newell's ascension to the leadership of the Reclamation Service, Mead remained with the Agriculture Department for a few years while also teaching part-time at the University of California as a Professor of Irrigation; in 1903 he published a well-regarded book on *Irrigation Institutions* that described in detail the practical and legal character of irrigation in various western states.[55] In

Figure 3-6: Elwood Mead, Commissioner of the Bureau of Reclamation from 1924 to 1936. Source: Bureau of Reclamation.

1907, Australian officials invited him "Down Under" to take charge of Victoria's State Rivers and Water Supply Commission. Originally planning to stay only a year, he remained in Australia until 1915 and oversaw the settlement of several irrigation communities sponsored by the state government. After Australia's entry into World War I, he returned to the United States, becoming Professor of Rural Institutions at the University of California, Berkeley. Drawing upon his experiences in Australia, he wrote *Helping Men Own Farms*, a book that championed the ideal of the small-scale farming community.[56] He also assumed leadership of the California Land Settlement Board and supervised the establishment of two state-sponsored irrigation colonies in northern California, beginning in 1918. The Durham and Delhi colonies proved unsuccessful (in fact, by the early 1930s, both were abandoned), but Mead's reputation survived intact.[57]

At the time Mead joined the Department of the Interior's Fact Finding Commission in 1923, he was considered a leading authority on irrigation. In addition, he had been unsullied by any association with the Reclamation Service during the previous two decades. As a result, when the Commission wrapped up its investigation, Mead was considered a logical person to take charge of the Bureau of Reclamation and revitalize the federal government's reclamation program. Appointed Commissioner of the Bureau of Reclamation in April 1924, Mead assumed responsibility for putting the bureau on a solid financial foundation in which repayment obligations would be met by project beneficiaries. Congress also addressed this issue by establishing 40 years as the standard time period for reimbursing the government for project costs (although in some cases payments could stretch out for more than 100 years).[58] As a key part of his plan for stabilizing Reclamation, Mead established a policy of completing and developing existing reclamation projects instead of initiating new projects; the effect of this policy can be seen in the construction of new dams such as Stoney Gorge (part of the Orland Project) and Gibson (part of the Sun River Project) that contributed to existing federal projects. At times, Congress authorized new initiatives (such as the Owyhee Project on the Oregon-Idaho border), but in the early years of Mead's leadership most Reclamation work focused on already established projects.

In his speeches and writings, Mead continued to champion the ideal of small-scale, community-based irrigation systems.[59] But during his time in office, a completely different justification for federally sponsored water projects emerged. The catalyst for this new type of project came from Southern California's desire to tap into the water resources of the Colorado River. The idea of building a large storage dam across the Colorado River at Boulder Canyon (located on the Arizona-Nevada border) derived from a desire both to protect California's Imperial Valley (located just north of the Mexican border) from floods and to provide the valley with additional water supply; by the early

1920s, this idea had been picked up by civic boosters in greater Los Angeles who perceived such a dam as a key element in their plans to increase municipal water supply by building a long aqueduct across the Mohave Desert.[60] Promoted as a multipurpose structure that would provide water for irrigation and municipal development, hydroelectricity, and flood control, Hoover Dam represented a far different project from what Mead was promoting in terms of small-scale rural settlement. Justified largely in terms of hydroelectric power generation (the sale of power–and not revenue derived from agricultural production–would comprise the financial foundation of the Boulder Canyon Project), Hoover Dam came to represent a whole new type of project. Authorized in December 1928, the Boulder Canyon Project came into existence well before the onset of the Great Depression. But the model it established—especially in regards to the generation and sale of hydroelectric power—became a powerful and prominent part of large-scale dam building during the Great Depression of the 1930s.

YEARS OF INDECISION: FROM NAVIGATION TO POWER

Floods and Politics

Congress, having given the Corps responsibility for river navigation in the nineteenth century, found early in the twentieth century that river floods were commanding increasing political attention. If irrigation brought the Reclamation Service to life, floods gave new direction to the Corps as the century began, and both bureaus would quickly face the third great river issue, hydroelectric power. Early in the century, there seemed to be little overlap because the Reclamation Service had to stay in the West and the Corps' big problems lay to the east—especially in the gigantic watershed of the Mississippi, Ohio, and Tennessee Rivers.

The devastating Pittsburgh flood of 1907 badly damaged the steel city, and the Ohio floods of 1913 (especially in the Miami Valley region) killed 467 people and cost nearly $150 million. Finally, with more floods raising political pressure, the representatives agreed to form a House Committee on Flood Control in 1916 and then passed the Flood Control Act the next year. Although limited to the lower Mississippi and the Sacramento Rivers, this Act was a major landmark in government because, for the first time, Congress openly allocated funds for flood control. It began long-range planning, it included the requirement for local cost sharing, and it directed the Corps that whenever it undertook flood control studies it had to include a comprehensive assessment of the watershed or watersheds.[61]

After World War I, Congress turned to river issues in 1920 with passage of the Water Power Act, but it failed to address the issue of combined usage until 1925. Then, on February 25, 1925, the House Rivers and Harbors Committee asked the federal government to estimate costs for a comprehensive survey of navigable rivers, to which the Corps replied in 1926 with House Document 308. This major report identified the 180 rivers and numerous tributaries to be studied with navigation and water power in mind. Congress authorized the studies in 1927, and then disaster struck. It was the Mississippi flood which Herbert Hoover, then Secretary of Commerce called the "greatest disaster of peace times in our history."[62] Between 250 and 500 people were killed, more than sixteen million acres flooded (an area larger than Rhode Island, Delaware, Connecticut, New Jersey and Massachusetts combined), 41,000 buildings destroyed, 162,000 homes flooded, and 325,000 people tended by the Red Cross in temporary camps. This mammoth flood proved that the "levees only" policy of previous years was an enormous error.[63]

That error had been introduced by Andrew A. Humphreys, Chief of Engineers of the Corps, in his 1861 book written with Henry L. Abbot, *Report on the Physics and Hydraulics of the Mississippi River*, and the error had been the policy of the Corps until the 1927 flood forced it to face the river's reality and to think about such measures as storage dams on tributaries.[64] Thus, the rise of hydroelectric power and ravaging floods brought the Corps into the big dam business in a major way.

Still, the Corps had already been drawn in by war, and that experience helped prepare it for the huge surge in dam building that would characterize the next half century. For the origin of the Corps' work on multipurpose dams, we turn to the work of one primary figure, Hugh Cooper, and to the origins and completion in 1925 of one huge structure, Wilson Dam, on the Tennessee River at Muscle Shoals, Alabama.

Muscle Shoals: The Battle for Control of Hydropower

Hydroelectric power development in the early twentieth century American West touched only lightly on the issue of public power. However, federal ownership and operation of hydroelectric facilities in the rural South, especially at Muscle Shoals on the Tennessee River, led to more vigorous interest in the issue. World War I was primarily responsible for turning Muscle Shoals into a national controversy over the issue of public power and economic development.

With the prospect of American participation in the war, Congress appropriated $20 million in 1916 for the production of nitrate (a necessary ingredient in explosives). President Wilson chose Muscle Shoals as the site for a nitrate plant because of the area's potential for generating abundant, inexpensive electric power—an essential factor in extracting nitrogen from the atmosphere. In 1918, two nitrate plants were built and work began on what later was named Wilson Dam (completed in 1925). The total government investment for the project eventually came to approximately $145 million.

Figure 3-7: Wooden formwork was used to cast the arches over the spillways at Wilson Dam. Source: U.S. Army Corps of Engineers.

Figure: 3-8: Wilson Dam at Muscle Shoals on the Tennessee River was constructed under contract by the U. S. Army Corps of Engineers. Source: U.S. Army Corps of Engineers.

Figure 3-9: On July 13, 1923, open gates on the north end of Wilson Dam produced turbulent flow. Source: U.S. Army Corps of Engineers.

The development of Muscle Shoals raised many questions about the government's role in projects affecting navigation, flood control, economic rehabilitation, conservation of agricultural lands, regional planning, development of natural resources, and the generation of power. The last issue took priority in the postwar years after passage of the Federal Water Power Act of 1920 and the agitation of midwestern progressives who sought stricter regulation of water power.[65] For many years, the Federal Water Power Act of 1920 proved to be relatively weak in practice. While the Act permitted federal supervision of hydroelectric facilities on both public lands and navigable streams and established the Federal Power Commission (FPC), the Commission was generally limited to licensing and site location. Flood control and irrigation were not included, power revenues were not linked with multiple-purpose dam construction, and federal revenues derived from hydroelectric facilities proved to be small. And, both the Tennessee River and the Boulder Canyon Project were excluded from the purview of the FPC.[66]

Immediately after World War I, Secretary of War Newton D. Baker attempted to turn the nitrate plants over to private companies for production of fertilizers. Attracting no takers, a bill was introduced in Congress to create a government corporation for that purpose, but it failed. In March 1921 the new Republican Secretary of War, John W. Weeks, announced that the government

113

would accept bids for the facility, which it would be willing to sell for a reasonable price.

A bid from Henry Ford to buy the Wilson Dam and generating plant set off a major dispute over control of Muscle Shoals. The bid attracted support from Secretary of Commerce Herbert Hoover, Thomas Edison, the Farm Bureau, several key southern politicians, and local developers in the Tennessee Valley. Power progressives fought the bid vigorously, as did southern power companies who feared the competition, and southern manufacturers who were skeptical of Ford's motives. Locally, the distaste for land speculators reinforced the notion that the Ford offer was exploitative. With the groundswell of opposition, especially championed by Senator George Norris of Nebraska, Ford withdrew his offer. The Muscle Shoals Inquiry Commission, appointed by President Calvin Coolidge, recommended in 1925 that the properties be leased to a private operator for fertilizer production and only incidentally for power production. Lukewarm interest in the recommendation resulted in no lessee being secured. And while Norris continued to push for public operation of the site, the time was not right for public power.[67]

Hugh Cooper, McCall Ferry, and Keokuk

Electric power characterizes as well as does any technology the tension so typical in the United States between individual freedom and government regulations, between private industry and public works. This tension, intensified by the debate over power from Niagara Falls, led to the Burton Bill of 1906 that brought government into Niagara River and Great Lakes water regulation while angering private power companies.[68] This issue was to become far more important with a dam planned by the Corps a decade later at Muscle Shoals.[69] This far larger project would demand far better engineering, and for that the Corps reached out to the leading designer of such works, Hugh Cooper.

Hugh Lincoln Cooper (1865–1937) left home after graduating from high school in 1883 determined to become an engineer. In 1885, he began to work on bridges, especially on the construction of steel bridges. By 1894, he had decided to leave bridge engineering and to focus on hydroelectric power projects, something quite new. He soon became an expert in the design of such plants and worked on design and construction in the United States, Canada, Brazil, and Mexico. He surveyed, designed, and built a 100,000 hp powerplant for the Electrical Development Company of Ontario above Horseshoe Falls in Niagara, Canada. His reputation grew, and in 1905 he opened his own office in New York City. He was shortly to take up his best known work to date, the McCall Ferry hydroelectric powerplant on the Susquehanna River.[70] The McCall Ferry Power Company had positioned the dam at the center of a circle whose 70 mile radius included Philadelphia, Baltimore, Wilmington, and Harrisburg so that the market for power existed and the new technology of high voltage transmission could easily send it to profitable centers. The company also explicitly insisted that the work be "absolutely first class" in the light of the many poorly built facilities like Hales Bar that had sprung up.[71] Cooper designed the dam and power house and also oversaw the construction from 1906 to 1908.

With the McCall Ferry Dam, Cooper established a structural type that became characteristic of main stem dams over the next 75 years. It is a low gravity dam made of unreinforced concrete, roughly triangular in section. Internal concrete stresses were not critical. Rather, the studies focused on foundation pressures and overturning safety. This latter required that the dam's vertical weight be large enough to prevent the horizontal water pressure load from tipping the dam over by rotating it about the toe. (See Wilson Dam discussion below.) Far more technically challenging than stresses and overturning was the construction process in the 2,700-foot-wide fast-flowing river. Cooper devised a system whereby half the river was blocked by a cofferdam to allow construction there while the river ran through the open half. The river was then directed through the partly completed dam while the second half was cast within a new cofferdam. This was a major undertaking and it prepared Cooper well for his next step, one which brought him into close contact with the Corps and the Mississippi River.

When Cooper moved from the Susquehanna to the Mississippi, he left the provincial and entered onto the world stage: the Susquehanna with a maximum flow of 50,000 cubic feet per second compared to the Mississippi with over 370,000 cubic feet per second. The mightiest American river had never been dammed below Minneapolis, even though the Corps had been working on the river for almost a century.[72] The big change, of course, was hydroelectric power, and the early 1890s saw formation of the Keokuk and Hamilton Water Power Company named for the two towns facing each other in Iowa and Illinois. In 1901, Congress approved a power project for the company that involved a wing dam (really a slanted jetty) and a power canal but no river dam.[73] The company could not finance it, but, the following year, the Rivers and Harbors Act called for a survey at Keokuk to study a possible dam.

The survey report by Montgomery Meigs (son of the Union Quartermaster-General during the Civil War) was favorable, and, in 1905, the Congress authorized the power company to proceed. Even the railroads called for better river navigation, and, in the spring of 1908, President Roosevelt, with typical Teddy flourish, sailed from Keokuk to Memphis in what was acclaimed as the "largest steamboat parade in history. . . ."[74]

Meanwhile, in 1907, Cooper began to study the Keokuk Dam project and to raise funds for its construction. He went to the nation's leading consulting electrical power engineers, Stone and Webster of Boston, who that year had formed Stone and Webster Management Association to handle powerplant construction.[75] Together with Cooper, they formed the Mississippi River Power Company of Boston with Edwin Webster as president and Hugh Cooper as vice president and chief engineer. Cooper designed the project, hired his own workers, and supervised all the construction. It was the largest hydroelectric plant in the world.[76]

What began to develop early in the century were huge dam projects, like McCall Ferry and Keokuk across wide rivers and Pathfinder and Buffalo Bill within high narrow canyons. The wide river dams were only justified because of hydroelectric power, whereas justification for the high narrow dams lay in water storage and irrigation. In the former type, design interest focused more on the power house and penstocks than on the dam structure; whereas in the latter, the structure caught the imagination of engineers. This contrast paralleled that contemporaneous structural development in bridge design where some engineers built long concrete viaducts with little intrinsic structural interest while others imagined wide spanning suspension forms of breathtaking daring.

But if Cooper's low wide dams were not structurally innovative, his means of construction were pioneering. Indeed, at Keokuk the brilliance of his plan would catch the attention not only of the international profession but also of the local citizenry. As the *Engineering News* reported in 1911, two years before completion,

> Engineering works rarely receive much attention on the part of the public, but the work at Keokuk forms a local attraction, the people of the surrounding country having a popular interest in "the big dam across the Mississippi," as the project has been in the public eye for so many years. In fact, excursions are run from nearby points to Keokuk, with the dam as the main attraction. To meet this condition and still provide for excluding the public from the works, the power company has erected on each side of the river a covered pavilion or observation platform, which is provided with seats and from which there is a very good view of the work as well as of the scenery along this part of the river.[77]

Essential features of the performance were the steel travelers, huge truss bridges that moved out over the river from both shores to place the total of 540,000 cubic yards of concrete, the most ever used in a dam up to then.[78] That immense mass would, however, be more than doubled in Cooper's next major work in which the Corps would play a much larger role than it had at Keokuk.

From Muscle Shoals to Wilson Dam

Nothing illustrates better the Corps' ambiguity toward river development than their activities on the Tennessee River during the first two decades of the twentieth century. This uncertainty and at times hostility toward multipurpose dams mirrored a great national debate during the progressive era about the role of the federal government in the development of natural resources, a debate that extended from river basins to oil trusts.

The story of Muscle Shoals characterizes this debate while also centering on the world's largest dam built up to that time. If the Keokuk project "marked a serious shift of direction for navigation improvement," the Tennessee River dam brought the Corps into multipurpose dam building in a major way.[79] The 58 steel gates above the concrete dam permitted a maximum flow at Muscle Shoals of 950,000 cubic feet per second, or over 2.5 times the flow at Keokuk.[80]

Before the turn of the century, the Corps had proposed development at Muscle Shoals, but not until 1909 did any serious action begin when a special board of engineers concluded a study with the cautious observation that " . . . any partnership relation between the United States and a private corporation is necessarily to be closely scrutinized as the results in the past have been that the government, as a party to such agreements, has usually suffered thereby."

Nevertheless, the Board admitted that times were changing and that water power may "require a new departure in governmental policy."[81] After an abortive 1914 attempt to get Congressional approval, Major Harry Burgess of the Nashville District took charge and produced a monumental 1916 report that lay the technical basis for the development at Muscle Shoals. Before any project could begin, however, the war intervened, and, in the fall of 1917, President Wilson chose Muscle Shoals as one site for a large nitrate plant to make munitions and ordered the Corps to begin work on a hydroelectric facility to power the plant. This was a completely new venture for the Corps, and they called on the acknowledged leader in such works, Hugh Cooper.

Meanwhile, two weeks after Woodrow Wilson's second inauguration in early March, German U-boats sank three American ships, and, after much agonizing, the President told a special session of Congress on April 2 of the necessity for war. Congress confirmed the state of war by April 6. Immediately, Hugh Lincoln Cooper, aged 52, volunteered to serve, and in May he received a commission as a major of engineers. By July, he was in France planning base-port facilities; in October he was promoted to Lt. Colonel.[82]

The Corps, back home, was struggling with its huge new project in Alabama, and in March of 1918, at the urging of the Chief of Engineers of the Army, Cooper was transferred to Muscle Shoals, where he made a careful study, many recommendations, and redesigned the entire project. He was quickly promoted to colonel, but in May, he was transferred back to France. He never knew why he had been transferred.[83] Cooper himself, in a 1922 testimony to the Military Affairs Committee of the House, confessed that he "never knew exactly how that happened." He also displayed his consternation that, although he was the designer of the entire project and responsible for its construction, the government refused to pay him anything.[84] Cooper apparently had failed to appreciate that his time as a U.S. Army Corps of Engineers officer could not be compensated beyond the normal remuneration given to an officer of his rank.

Just before Armistice Day, on November 9, 1918, construction resumed on the dam while the Corps made more subsurface tests and, following Cooper's recommendations, began to make design sketches. Cooper, meanwhile, had gone back into private practice and then, on May 21, 1920, the Corps signed a contract with Hugh L. Cooper & Co. that put the company in charge of design, construction, and inspection of the entire project. Cooper began work, and by election day 1920 he had completed numerous drawings laying out the dam and powerhouse.[85] From then until 1924, Cooper produced drawings and supervised construction.

The politics of how the dam came to be used raged through Congress during the 1920s while the Corps moved toward large scale dams of which Wilson was its first major effort. The dam itself represented the largest of its type; the overflow masonry weir dam. There are 58 openings formed by a gravity dam 95 feet high and 101 feet wide at its base surmounted by 18 foot high steel control gates and all flanked by buttresses that support a continuous arch bridge. The shape of the gravity section followed from those at McCall Ferry and Keokuk; this shape had evolved from late nineteenth-century dams, such as that at Holyoke across the Connecticut River which was completed in 1899.[86] The principal problems in design for these mainstem, or run-of-the-river, dams is sliding and uplift instability as well as foundation scour at the downstream toe. For these reasons, such dams are wider than they are high, have a counter curvature on the downstream face, and are supplied with drains to control water pressure under the dam base.

Cooper's design solved these problems, and the dam has not had any significant structural difficulties since its completion in 1925 after all

1,350,000 cubic yards of concrete were in place. Cooper was particularly meticulous in his concrete control, a major factor in the dam's satisfactory performance.[87] Power generation began on September 12, 1925, and by June of 1926 six generators were operating to produce an average of about 112,000 hp, or 83,500 kW of power.[88] The full power planned was about 600,000 hp from four 30,000 hp units and fourteen 35,000 hp units.[89] This immense project moved newly elected Franklin Roosevelt to state in a 1933 address at Montgomery, Alabama, "My friends, I determined on two things as a result of what I have seen today. The first is to put Muscle Shoals to work. The second is to make of Muscle Shoals a part of an even greater development that will take in all of that magnificent Tennessee River. . . ."[90]

Before the president created TVA and took Wilson Dam away from the Corps, another major political event would firmly fix the Army engineers in the multipurpose mode.

From the Tennessee River to House Document No. 308

Long before Franklin Roosevelt expressed his goal of developing the entire Tennessee basin, the Chief of Engineers of the Corps, General Lansing Beach, had ordered a study of that potential.

The Rivers and Harbors Act of June 5, 1920, authorized the U.S. Army Corps of Engineers to make preliminary examinations and surveys of the Tennessee River and tributaries, and General Beach ordered the survey by letter on June 30, 1920. Beach had verbally explained:

> the intention of Congress to include studies of present or potential hydroelectric developments, the mineral and industrial resources of this region, drainage, flood protection, and such other allied subjects as may reasonably appear to have an appreciable influence on the project that may be finally recommended for adoption for the improvement of navigation.[91]

In short, Beach asked for a comprehensive report with navigation included but not the central issue. He assigned that task to a young officer, Major Harold C. Fiske, commander of the Nashville District.

Because of limited funds, Fiske decided to use the new technique of aerial photo topographical mapping. Developed only during the World War, the method served Fiske well as he began in 1921 to take photos from a flimsy De Havilland airplane at a 12,500-foot altitude.[92] Gerard Matthes concluded his 1923 paper on the survey by stating that "the general plan . . . was first conceived by Major Fiske. It is only fair to state that it was due to his resourcefulness and keen personal interest in the survey and the development of the details, that it was possible to accomplish so much with the small funds available."[93]

Fiske submitted his preliminary report on January 15, 1921, with the recommendation that a full survey be carried out and that it include all the aspects mentioned by General Beach in June of 1920.[94] On January 29, 1921, Colonel Harts, Fiske's superior, wrote to the Chief of Engineers that, "For the foregoing reasons I do not feel that I can consistently recommend the survey that is proposed by the district engineer." His reasons came down to the image of the Corps as a bureau for navigation only. He claimed that a complete survey had been made in 1909, when he was district engineer, solely for navigation; so no new one was needed. Harts criticized Fiske's preliminary report for being "clearly an investigation into the water-power possibilities, mostly on the tributaries, with no explanation as to how it is expected that navigation will be benefitted thereby." Moreover, he continued, "the cost of the proposed survey is so far beyond what seems reasonable that it should, in my opinion, not be commenced. . . ."[95]

General Beach supported Fiske, but instead of the over $500,000 requested, Fiske got a mere $20,000. Undaunted, he kept up his preliminary work over the next year and submitted another, briefer report on March 15, 1922, recommending a series of specific studies for a reduced cost of $250,000.[96] A new division engineer, Colonel C. W. Kutz, again objected to the report on the grounds of slighting navigation in favor of power and of being far too costly.[97] On April 4, the Board of Engineers for Rivers and Harbors reported that "This purpose is commendable but Congress has never sanctioned an inquiry of this kind and therefore no authority appears to exist for making it [the Fiske survey]."[98]

In spite of division disapproval again, General Beach still supported Fiske and recommended "that an appropriation of $250,000 be made for continuing the work and that the full amount for completing the survey, viz., $515,800, be authorized." On September 22, 1922, the Congress agreed that Fiske's detailed survey could continue.[99] By early 1924, it had proceeded far enough that the House Committee on Rivers and Harbors on March 31, and April 1, could subject the survey to a thorough critique, resulting in the observation that, "Major Fiske's plan [is] 'astounding and amazing' and [it was] at this session [that] the suggestion was first made that similar surveys should be initiated on other rivers of the United States." The result of this review and recommendation appeared in the Rivers and Harbors Act of March 3, 1925 (section 3), which authorized and directed the U.S. Army Corps of Engineers and the Federal Power Commission jointly to prepare cost estimates for comprehensive surveys of all navigable streams and their tributaries, except the Colorado River, where hydroelectric power appears to be practical.[100]

Fiske's report in 1926 recommended public and private cooperation, and he set the precedent for the national survey that was soon to follow. However, his enthusiasm got him into deep trouble with his superiors, and, because he directly proselytized members of Congress, the Chief of Engineers reprimanded him. Nevertheless, his irregular behavior was crowned with success when Congress, "astounded" by the excellence of his surveys, authorized similar surveys throughout the nation.[101]

The 308 Reports

In a letter report of April 7, 1926, the Chief of Engineers, responding to the 1925 Act, gave detailed costs for studying "navigable streams upon which power developments appear to be feasible." This is House Document No. 308 which, in just over four pages, laid out a national program of immense scope for which the surveys would cost $7,322,400.[102] Clearly, this implied that the Corps would enter the multipurpose dam business, but it did not settle the issue of public versus private development. The document referred to private activity. This issue would not be fully resolved until after 1932, but the direction for power and navigation had been set and now both federal agencies; Reclamation and the Corps, were to begin a new adventure that would lead to clash and compromise.

The 308 document set the stage for multipurpose dams by its focus on river basins with a combined use for navigation and power. The document itself was signed by the Chief of Engineers and the executive secretary of the Federal Power Commission.

For the major rivers, the surveys were to determine: discharge, locations and capacities of reservoir sites, location and practicability of dam sites, capacities of power sites, present and prospective power markets available, best plan of improvement for all purposes, preliminary cost estimates, and feasibility of the best plan. Relationship to navigation was to be identified, and, where the benefits were sufficient, the federal government could share in the cost. But power costs were assumed to be the responsibility of private companies.

There were twenty-four separate surveys ranging from the Raritan River ($19,400) to streams (except the Mississippi) that drained into the Gulf of Mexico ($909,000). For the Tennessee, the cost estimate was $300,000; for the Columbia, $734,100; for the Missouri, $425,000; for the Ohio, $393,100; and for California, $420,000. The surveys, which were also to consider flood control, were written into law by acts of Congress on January 21, 1927, and May 15, 1928.

With the 308 surveys before Congress, Wilson Dam complete, and Reclamation planning its most ambitious project, at Hoover Dam below Las Vegas, the landscape was set for major restructuring. The state-of-the art in dam design and construction appeared in 1927 in the eighth and last edition of Wegmann's treatise on dams. Coupled with the great flood on the lower Mississippi, this treatise and the 308 document marked the end of an era and, soon thereafter, the beginning of a time of social trauma and technological achievement.

In his introduction to the eighth edition, Edward Wegmann noted two major changes from his 1888 first edition: one was the inclusion of "a mathematical discussion of multiple-arch dams" and the other was the immense growth of the field so that his first treatise of 109 pages, 59 plates, and 18 figures had grown in 39 years to "740 pages of text, 191 plates, and 291 figures in the text."[103] Things had gotten more complex and dams more numerous. These changes were characteristic of all engineering and, indeed, of all society.

On June 1, 1927, the locks opened at Wilson Dam, and commercial transportation began. Ten days earlier, Charles Lindbergh touched down at Le Bourget Aerodrome, Paris, to symbolize dramatically a new pathway for transportation. In June, France returned to America the draft Kellogg-Briand Treaty in which the two governments renounced war with each other, while earlier, in January, the Allies abolished their control commission for supervising German disarmament.[104] This was a year largely of optimism as grand plans were laid for peace and prosperity. Among them were the plans for river basins in the United States.

We move now to the individual river basins and their most significant multipurpose dams, beginning with the Boulder Canyon Project planned, with no reference to any 308 report, during the prosperous 1920s. The project's greatest symbol is Hoover Dam, the third[105] of these presidentially-named structures and, by far, the best known. When completed, it was the highest dam ever built, used the most concrete, generated the most power, and impounded the largest reservoir. It stands as a great monument to American engineering, but it has also a great story to tell about its political, economic, and urban history. To that story and that monument we now turn.

Endnotes

1. Edward Wegmann, *The Design and Construction of Dams* 1st edition, (New York: John Wiley & Sons, 1888).

2. "Theodore Roosevelt Dam," *Development of Dam Engineering in the United States*, Kollgaard and Chadwick, editors, 368.

3. Edward Wegmann, *The Design and Construction of Dams* 6th edition, (New York: John Wiley and Sons, 1911), plates CII and CIII.

4. Wegmann, *Design and Construction of Dams*, 6th edition, 403.

5. Leland R. Johnson, *Engineers on the Twin Rivers* (Nashville: U.S. Army Corps of Engineers, 1978), 175-8. See Gerard H. Matthes, "Discussion," *Transactions of the American Society of Civil Engineers* 100 (1935), 919-23.

6. Leland R. Johnson, *Engineers on the Twin Rivers*, 178.

7. For example, in 1903, Newell directed his Assistant Chief Engineer, Arthur P. Davis, to "concentrate your energies as far as practicable on the pushing forward of the Salt River Project. . . . and laying out a scheme of further work, involving the purchase of lands, rights of way, etc., also [on] action which may be taken regarding the Arizona Canal and the construction of the new dam in the river. I should like to have, in brief, a consistent plan for future operations covering these important matters." This letter can be interpreted as a sign that Newell, the administrator, felt comfortable leaving a plethora of technical problems to a trusted subordinate. But it also reflects how Newell separated himself from Reclamation's engineering work. See F. H. Newell to A. P. Davis, June 23, 1903, National Archives and Records Administration, Denver, Colorado, Records of the Bureau of Reclamation, RG115, Entry 3, Administrative and Project Records, 1902-1919. Salt River 24D-26. Box 845. Folder 26: Salt River Project. Consulting Engineer Reports thru 1906. This source refers to the General Administrative and Project Records (1902-1919) for the Reclamation Service's Salt River Project retained in the National Archives.

8. Davis's career is described in his professional memoir: "Arthur Powell Davis" *Transactions of the American Society of Civil Engineers* 100 (1935), 1582-91.

9. For example, see Arthur P. Davis, *Irrigation Near Phoenix, Arizona* [U.S. Geological Survey Water Supply Paper No. 2] (Washington D.C., 1897).

10. The "engineering board" system of project review is described in the *Third Annual Report of the Reclamation Service, 1903-4* (Washington, D.C., 1905) 41.

11. See Charles D. Walcott to Secretary of the Interior [E. A. Hitchcock], March 15, 1904, National Archives, College Park, Maryland, Records of the Secretary of the Interior, RG48, Entry 632. Newell initialed this letter and he was almost certainly responsible for drafting its contents. Miscellaneous Projects, Records 1901-07, Box 3, File 197-1904.

12. Wisner's career is synopsized in Frederick Haynes Newell, compiler, *Proceedings of the First Conference of Engineers of the Reclamation Service: With Accompanying Papers* (Washington, D.C.: Government Printing Office, 1904), 350. He graduated as a civil engineer from the University of Michigan in 1865 and spent most of his professional life working on harbor and transportation projects involving the Great Lakes, the Illinois River, the Mississippi River and ports on the Gulf of Mexico. He also served with the Lighthouse Service and the International Waterways Commission.

13. Wisner's role in analyzing designs for Shoshone Dam (renamed Buffalo Bill Dam) and Pathfinder Dam is documented in the file labeled: "Discussion Related to Dams," National Archives, Denver, RG115, Entry 3, Box 289. Among others, he served on engineering boards for the Salt River Project (Arizona), the Carlsbad Project (New Mexico), the Klamath Project (Oregon/California), the Milk River Project (Montana), and the Minidoka Project (Idaho). His role in reviewing the Reclamation Service's project plans for the lower Colorado River in California is noted in Pisani, *From the Family Farm to Agribusiness*, 311-3.

14. The early history of the city is documented in Geoffrey P. Mawn, "Phoenix, Arizona: Central City of the Southwest, 1870-1920" (Ph.D. dissertation, Arizona State University, 1979).

15. The valley's early irrigation canals are described in Arthur P. Davis, *Irrigation Near Phoenix, Arizona*, U.S. Geological Survey Water Supply and Irrigation Paper No. 2 (Washington, D.C., 1897). See Earl A. Zarbin, *Roosevelt Dam: A History to 1911* (Phoenix: Salt River Project, 1984), 19-28, for contemporary newspaper references documenting legal disputes over water supply in the 1870s and 1880s.

16. The 1889 reservoir survey led by Maricopa County Surveyor William Breakenridge is described in Mawn, "Phoenix, Arizona," 220-1. Also see James H. McClintock, *Arizona*, 3 volumes (Chicago: S. J. Clarke Publishing Company, 1916), 431-2. This locally-sponsored expedition occurred contemporaneously with John Wesley Powell's Irrigation Survey, but did not draw upon federal sponsorship.

17. The Hudson Company's filings on the Tonto Dam site are dated April 15, 1893 (and posted April 20, 1893), as recorded in "Secretary of State, Appropriation of Water, Dam and Reservoir Sites, 1893-1910," Arizona Department of Library, Archives and Public Records, Phoenix, Arizona.

18. Schuyler, Reservoirs, 340-8, quote on 343. The Tonto Dam was also described in James D. Schuyler, "Water Storage and Construction Of Dams," *USGS Eighteenth Annual Report* (Washington, D.C.: Government Printing Office, 1897), 715-7.

19. Phoenix's efforts to obtain federal support for a large-scale Salt River storage reservoir is detailed in Smith, *The Magnificent Experiment*, 20-42.

20. Karen Smith, "The Campaign for Water in Central Arizona, 1890-1903," *Arizona and the West* 23 (Summer 1981), 127-48.

21. This statistic was calculated by Karen Smith from information relating to the original subscribers to the Salt River Valley Water Users' Association in the spring of 1903. it is somewhat misleading because it excludes consideration of thousands of acres of land owned by Dwight Heard and others. The final legal determination of land to be served by the project also stipulated that acreage which had been irrigated before 1903 (which was all privately owned) would have priority rights to water stored by Roosevelt Dam. See Smith, *The Magnificent Experiment*, 43-8, 125-135. Earl A. Zarbin, "The Committee of Sixteen," *The Journal of Arizona History* 25 (Summer 1984), 129-54.

22. F. H. Newell to A. P. Davis, June 23, 1903, National Archives, Denver, RG115, Entry 3, General Administrative and Project Records, 1902-1919. Salt River 24D-26. Box 845. Folder 26: Salt River Project. Consulting Engineer Reports thru 1906.

23. These structures are described in Wegmann, *The Design and Construction of Dams*, 8th edition.

24. In the letter officially recommending selection of the Salt River Project for construction by the Reclamation Service, the Director of the U.S. Geological Survey made explicit reference to recent work on the New York and Boston water supply systems. The Salt River Project was the only initial Reclamation Service project intended to serve the region surrounding a burgeoning city. As such, it is not so surprising that the Roosevelt Dam utilized a technology very similar to that employed for prominent eastern dams. See Charles Walcott to E. A. Hitchcock [Secretary of the Interior], March 7, 1903, National Archives, Denver, RG115, Entry 3, Salt River Project. Specific comparison of the proposed Salt River dam with the Croton and Wachusett Dams is made in Arthur P. Davis, "Investigations in Arizona," in Newell, compiler, *Proceedings of the First Conference of Engineers of the Reclamation Service*, 129.

25. The Hudson Reservoir and Canal Company's proposed curved gravity dam design is illustrated in Schuyler, "Water Storage and Construction Of Dams," 715-7.

26. The first design description of Roosevelt Dam appeared in Arthur Powell Davis, *Water Storage on Salt River, Arizona* (Washington D.C.: Government Printing Office, 1903). This publication disseminated a report submitted by Davis in mid-1902 that was partially financed by Phoenix irrigation interests. *The First Annual Report of the Reclamation Service* (Washington D.C.: Government Printing Office, 1903) essentially repeats the description presented in Davis's 1902 report. As one of his first contributions to the Reclamation Service, consulting engineer George Wisner visited the Tonto site in April 1903 and shortly afterwards inspected the San Mateo Dam (a concrete curved gravity dam) in the company of Davis. In reporting upon his trip, Wisner stated that "a curved masonry dam, such as designed by Mr. Davis, is particularly well adapted for the [Tonto] site." George Y. Wisner to F. H. Newell, May 4, 1903, National Archives, Denver, RG115, Entry 3, Box 845, General Administrative and Project Records, 1902-1919, Salt River 24D-26. Folder 26. Salt River Project. Consulting Engineer Reports. Thru 1906. The engineering board was later consulted on plans that raised the Roosevelt design from a height of 240 feet to 280 feet. However, the enlarged gravity design was not modified except to proportionally increase the thickness of the base. For verification that Davis was primary designer of the Roosevelt Dam, see C. R. Olberg to A. P. Davis, March 6, 1911, National Archives, Denver, RG115, Entry 3, Salt River Project.

27. F. H. Newell to A. H. Dimock, April 16, 1912, National Archives, Denver, RG115, Entry 3, General Administrative and project Records, 1902-1919. 910.6A-910.7. Box 289. Folder 910.7: Straights. Technical Discussions. Discussion Relative to Dams. 1911-1914.

28. Quote taken from F. H. Newell, "Irrigation: An Informal Discussion," *Transactions of the American Society of Civil Engineers* 62 (1909), 13. Newell's preoccupation with visual appearance is also reflected in his complaint that "the general appearance of the grounds at the Roosevelt headquarters could be improved . . . [because] our work is judged by the general public more by its outward appearance than by its real merit . . ." See F. H. Newell to L. C. Hill, May 8, 1909, National Archives, Denver, RG115, Entry 3, General Administrative and Project Records, 1902-1919. Salt River 24D-26. Box 845. Folder 26: Salt River Project. Consulting Engineers Reports. January 1, 1907-December 31, 1912.

29. Davis, *Water Storage on the Salt River*. These cost estimates for purchasing and making cement are repeated in the *First Annual Report of the Reclamation Service*, 99-102.
30. A. P. Davis to Louis C. Hill, February 27, 1906, National Archives, Denver, RG115, Entry 3, General Administrative and project Records, 1902-1919. Salt River 24D-26. Box 845. Folder 26: Salt River Project. Consulting Engineers Reports. Thru 1906. Hill was supervising engineer for the Salt River Project.
31. Initial estimates for the construction of the power canal and hydroelectric plant are given in the *First Annual Report of the Reclamation Service*, 97. Final costs are provided in the *Ninth Annual Report of the Reclamation Service* (Washington, D.C., 1911), 69.
32. O'Rourke's firm had recently completed the concrete seawall protecting Galveston's gulf coast shoreline and this apparently qualified them for the Roosevelt contract.
33. Construction bid data is published in the *Fourth Annual Report of the Reclamation Service* (Washington, D.C.), 69-71.
34. *Sixth Annual Report of the Reclamation Service* (Washington, D.C., 1908), 64.
35. Davis, *Water Storage on the Salt River*. These cost estimates for purchasing and making cement are repeated in the *First Annual Report of the Reclamation Service*, 99-102.
36. Report on "Expense of Operation and Maintenance, Roosevelt Cement Plant, May 1905 to July 1910, Inclusive," National Archives, Denver, RG115, Entry 3, Salt River Project. Also see F. H. Newell to A. P. Davis, March 4, 1904; and E. A. Hitchcock to Charles Walcott, March 4, 1904.
37. The quotation is taken from G. Y. Wisner and W. H. Sanders to F. H. Newell, March 3, 1905, National Archives, Denver, RG115, Entry 3, General Administrative and Project Records, 1902-1919. Box 874. Salt River 639-741. Problems in erecting the cement plant are outlined in "Folder 733: Salt River. Operation of Cement Mill," (no author) March 3, 1905.
38. See L. C. Hill to F. H. Newell, July 5, 1905; and L. C. Hill to F. H. Newell, August 8, 1905; and A. P. Davis to James Kruttsschnitt, July 17, 1907, National Archives, Denver, RG115, Entry 3, Salt River Project.
39. Final costs for the cement production are given in "Expense of Operation and Maintenance, Roosevelt Cement Plant, May 1905 to July 1910, Inclusive." Granite Reef Dam is located more than 40 miles downstream from Roosevelt Dam. It is a concrete overflow structure that diverts water out of the Salt River and into distribution canals serving the valley. The dam's construction was not planned as part of the original Salt River Project but became necessary after the privately owned Arizona Dam failed in the heavy flooding of 1905. The 60,000 barrels of cement used in its construction helped reduce the unit costs of production at the Roosevelt cement mill by allowing fixed costs to be distributed over a greater number of barrels. Because of its close proximity to Phoenix, the costs of supplying cement to the Granite Reef Dam from private contractors would have been much less than offers made to supply cement to Roosevelt.
40. Cost of the Salt River Project, as of the end of 1910, is given in the *Ninth Annual Report of the Reclamation Service*, 69-70. Work on Granite Reef Dam and distribution canals in the valley came to a little more than $2 million.
41. Determination of the Salt River Project's official initial cost of approximately $10.2 million is documented in Smith, *The Magnificent Experiment*, 140.
42. Construction of Wachusett Dam is described in Wegmann, *Design and Construction*, 6th edition, 185-94.
43. F. H. Newell to L. C. Hill, December 26, 1905, National Archives, Denver, RG115, Entry 3, Salt River Project.
44. Detailed discussion of how the Salt River Valley Water Users' Association sought to extend the project repayment period and reduce its financial responsibility is presented in Smith, *The Magnificent Experiment*, 92-124.
45. Quoted in Newell, *Irrigation*, 396.
46. Another dramatic cost overrun afflicted the Strawberry Valley Project in Utah. In 1905, cost estimates were $1.25 million, but by the time of project completion almost a decade later costs had reached almost $3.5 million. See "Strawberry Valley Project," HAER UT-26, Prints and Photographs Division, Library of Congress.
47. F. H. Newell to L. C. Hill, November 25, 1911, National Archives, Denver, RG115, Entry 3, General Administrative and Project Records, 1902-1919. Salt River 24D-26. Box 845. Folder 26: Salt River Project. Consulting Engineers Reports. January 1, 1907-December 31, 1912. In 1917 the federal government gave the locally-controlled Salt River Valley Water Users' Association complete operational control over the Salt River Project system. However, the federal government retained legal ownership of the Roosevelt Dam and other major structures. Final payment for the Association's initial $10.2 million project debt to the government was tendered in the mid-1950s.
48. F. H. Newell to J. B. Lippincott, February 11, 1910, Newell Papers, Library of Congress Manuscripts Collections. Red thesis binder devoted to correspondence with Lippincott.

49. Smith, *The Magnificent Experiment*, 184-95.
50. The powerplant at Roosevelt is described in Arthur P. Davis, *Irrigation Works Constructed by the U.S. Government* (New York: John Wiley and Sons, 1917), 18-9.
51. The history of hydroelectric power development along the Salt River by the Water Users' Association is described in Donald C. Jackson, "History of Stewart Mountain Dam," Historic American Engineering Record Collection, Prints and Photographs Division, Library of Congress, Washington, D.C. Also see T. A. Hayden, "Salt River Project, Arizona, Irrigation and Hydroelectric Power Development by the Salt River Water Users' Association-Six Major Dams," *Western Construction News* 5 (June 25, 1930), 298-300.
52. Robinson, *Water for the West*, 37-43, provides a sympathetic, yet generally realistic, treatment of the financial problems that beset the early Reclamation Service. Newell remained associated with the Reclamation Service as "Chief of Construction" until May 1915, when he formally left the federal government. See also Jackson, *Building the Ultimate Dam*, 187-190.
53. For an excellent description and analysis of this commission's work see Brian Q. Cannon, "'We Are Now Entering a New Era': Federal Reclamation and the Fact Finding Commission of 1923-24," *Pacific Historical Review* 66 (May 1997), 185-211. After A. P. Davis left the Reclamation Service, David W. Davis (a former governor of Idaho who possessed no background in engineering) was appointed the first Commissioner of the Bureau of Reclamation.
54. Mead's early career is well described in James R. Kluger, *Turning on Water with a Shovel: The Career of Elwood Mead* (Albuquerque: University of New Mexico Press, 1992). See 27-35 for discussion of his rivalry with Newell and his opposition to the National Reclamation Act that established the Reclamation Service.
55. Elwood Mead, *Irrigation Institutions: A Discussion of the Economic and Legal Questions Created by the Growth of Irrigated Agriculture in the West* (New York: The Macmillan Company, 1903).
56. Elwood Mead, *Helping Men Own Farms: A Practical Discussion of Government Aid in Land Settlement* (New York: The Macmillan Company, 1920).
57. Kluger, *Turning on Water with a Shovel*, 57-101, describe Mead's activities from 1907 through the early 1920s.
58. Robinson, *Water for the West*, 45-6.
59. Kluger, *Turning on Water with a Shovel*, 115-26.
60. An alternate spelling is Mojave Desert.
61. Joseph L. Arnold, *The Evolution of the 1936 Flood Control Act*, Office of History, U.S. Army Corps of Engineers (Fort Belvoir, Virginia: 1988), 11-5.
62. Arnold, *Evolution of the 1936 Flood Control Act*, 17-9.
63. Martin Reuss, "The Army Corps of Engineers and Flood-Control Politics on the Lower Mississippi," *Louisiana History*, 23, No. 2 (Spring 1982), 131-2.
64. Martin Reuss, "Andrew A. Humphreys and the Development of Hydraulic Engineering: Politics and Technology in the Army Corps of Engineers, 1850-1950," 1-33.
65. Melosi, *Coping with Abundance*, 122-3. Walker, "Developing Hydroelectric Power: The Role of the U.S. Army Corps of Engineers, 1900-1978," 34-5 Preston J., Hubbard, *Origins of the TVA: The Muscle Shoals Controversy, 1920-1932* (New York: Norton, 1968 [originally published in Nashville: Vanderbilt University Press, 1961]).
66. Hays, *Conservation and the Gospel of Efficiency*, 118-21. Swain, *Federal Conservation Policy, 1921-1933*, 98, 113-4. Melosi, *Coping with Abundance*, 81-2. Walker, "Developing Hydroelectric Power: The Role of the U.S. Army Corps of Engineers, 1900-1978," 47. See also Judson King, *The Conservation Fight: From Theodore Roosevelt to the Tennessee Valley Authority* (Washington, D.C.: Public Affairs Press, 1959), 45-58.
67. Melosi, *Coping with Abundance*, 123-4. Swain, *Federal Conservation Policy, 1921-1933*, 115-20. Hubbard, *Origins of the TVA*, 138. Reuss and Walker, *Financing Water Resources Development*, 37; Walker, "Developing Hydroelectric Power: The Role of the U.S. Army Corps of Engineers, 1900-1978," 38-9. See also Hays, *Conservation and the Gospel of Efficiency*, 192-5.
68. Gail E. Evans, "Storm Over Niagara: A Catalyst in Co-shaping Government in the United States and Canada During the Progressive Era," *Natural Resources Journal* 32, No. 1 (Winter 1992), 45-6.
69. Leland R. Johnson, *Engineers on the Twin Rivers*, U.S. Army Engineer District - Nashville, 1978, 163-9.
70. Valentine Ketcham and M. C. Tyler, "Hugh Lincoln Cooper Memoir," *Transactions, American Society of Civil Engineers* 103 (1938), 1772-7.
71. "The McCall Ferry Hydro-Electric Powerplant on the Susquehanna River," *Engineering News* 58, No. 11 (September 12, 1907), 267.
72. Roald D. Tweet, *A History of the Rock Island District U.S. Army Corps of Engineers: 1866-1983* (Rock Island, Illinois: U.S. Army Engineer District, Rock Island, 1984), 245.

73. "Water Power Development on the Mississippi River at Keokuk, Iowa." *Engineering News* 66, No. 13 (September 28, 1911), 355.
74. Tweet, *History of the Rock Island District*, 244.
75. Thomas P. Hughes, *Networks of Power: Electrification in Western Society: 1880-1930* (Baltimore: The Johns Hopkins University Press, 1983), 390.
76. Hugh L. Cooper, "The Water Power Development of the Mississippi River Power Company at Keokuk Iowa," *Journal of the Western Society of Engineers* 17 (January to December 1912), 213. This article came from the paper presented on October 18, 1911.
77. Cooper, "Water Power Development on the Mississippi River at Keokuk, Iowa," 356.
78. Cooper, "Water Power Development on the Mississippi River at Keokuk, Iowa," 364.
79. Tweet, *History of the Rock Island District*, 245.
80. Hugh L. Cooper and Co., "Electric Power for the Tennessee River, Muscle Shoals, Alabama," *Interim Construction Progress Bulletin* (November 1, 1923), 7.
81. Johnson, *Engineers on the Twin Rivers*, 169.
82. Ketcham and Tyler, "Cooper Memoir," 1774.
83. Harold Dorn, "Hugh Lincoln Cooper and the First Détente," *Technology and Culture* 20, No. 2 (April 1979), 327-8. The quotation came from "Conflict in House Reports on Col. Cooper's Muscle Shoals Work," *Engineering News-Record* 84, No. 22 (1920), 1082. In 1920 two views surfaced in a report from the House of Representatives. The Republican majority, critical of Wilson's plans for Muscle Shoals, held that "anyone who opposed or sought to modify the plans . . . was either disregarded or disciplined . . ." and that Cooper was one who was so disciplined. The minority, Democrat report, denied that charge and explained that "this eminent officer was needed in the building of an immense dock project in France. . . ."
84. Dorn, "Cooper and the First Détente," 328. Most likely, this meant that his personal time was uncompensated. His company produced substantial design work which must surely have been paid for.
85. Fred A. Noetzli, , "With Mathematical Discussion and Description of Multiple Arch Dams," Edward Wegmann, *The Design and Construction of Dams*, 8th edition. For Wilson Dam see 630-5. For Cooper's work, see for example, drawings Numbers 3450-3452 dated November 1, 1920, and Number 3453 dated October 23, 1920. The Corps continued to make drawings into October 1920, but Cooper's work had by then become the official record.
86. Wegmann, *Design and Construction*, 8thedition, 292. For a summary of such overflow masonry weirs see also plates CII and CIII.
87. John W. Hall, "The Control of Mixtures and Testing of Wilson Dam Concrete," Paper delivered at a meeting of the American Concrete Institute, Chicago, February 24, 1926. Hall was resident engineer for Hugh Cooper.
88. Johnson, *Engineers on the Twin Rivers,* 173. He reports that in the month of June 1926, the dam produced 60,226,100 kilowatt hours of work, which implies an average production of 60,226,100 /(30 x 24) = 83,500 kW. of power.
89. Cooper, "Water Power Development on the Mississippi River at Keokuk, Iowa," 1923, 5.
90. Quoted in Johnson, *Engineers on the Twin Rivers,* 173. Roosevelt delivered the speech on January 21, 1933, after a visit to Wilson Dam.
91. Harold C. Fiske, "Preliminary Examination of Tennessee River and Tributaries," January 15, 1921, 9, in *Tennessee River and Tributaries North Carolina, Tennessee, Alabama, and Kentucky,* House of Representatives, 6 Congress, 2nd Session, Document No. 319, May 19, 1922.
92. Johnson, *Engineers on the Twin Rivers,* 181.
93. For a full description of the Tennessee River survey see Gerard H. Matthes, "Aerial Photography as an Aid in Map Making, with Special Reference to Water Power Surveys," *Transactions of the American Society of Civil Engineers* 86 (1923), 779-802.
94. Johnson, *Engineers on the Twin Rivers,* 9-82.
95. William W. Harts, "To the Chief of Engineers, United States Army," in Johnson, *Engineers on the Twin Rivers,* 182-4.
96. Harold C. Fiske, "Partial Survey of Tennessee River and Tributaries," March 15, 1922, in Johnson, *Engineers on the Twin Rivers*, 148, 165.
97. C. W. Kutz, "To the Chief of Engineers, United States Army," March 21, 1922, in Johnson, *Engineers on the Twin Rivers*, 165-7.
98. H. Taylor, for the Board of Engineers for Rivers and Harbors, "To the Chief of Engineers, United States Army," April 4, 1922, in Johnson, *Engineers on the Twin Rivers,* 8.

99. Lansing H. Beach, "To the Secretary of War," May 16, 1922, in Johnson, *Engineers on the Twin Rivers*, 1-3. The Congress authorized an expenditure of $200,000 on September 22, 1922. See *Tennessee River and Tributaries, North Carolina, Tennessee, Alabama, and Kentucky,* House of Representatives, 71st Congress, 2nd Session, Document 328, March 24, 1930, 29. This is the 308 Report. The total allotted to the general survey by 1927 was $910,075.51. *Tennessee River and Tributaries, North Carolina, Tennessee, Alabama, and Kentucky,* House of Representatives, 67th Congress, 2nd Session, Document 319, 3.

100. *House Committee on Rivers and Harbors,* 68th Congress, 1st Session, March 31 and April 1, 1924, and the Rivers and Harbors Act of March 3, 1925, 43 Stat. 1186, Section 3.

101. Johnson, *Engineers on the Twin Rivers*, 181-4.

102. H. Taylor and O. C. Merrill, "Estimate of Cost of Examinations, etc. of Streams where Power Development Appears Feasible," House of Representatives, 69th Congress, 1st Session (December 7, 1925, to November 10, 1926), *Document No. 308*, Washington D.C., 6 pages.

103. Wegmann, *Design and Construction*, 8th edition, vii-viii.

104. Raymond J. Sontag, *A Broken World: 1919-1939* (New York: Harper & Row, 1971), 133-4.

105. (Editor's note) This is the third dam named for a president discussed in this book. By this time, however, a fourth dam was already in-place – Coolidge designed and built by the Indian Irrigation Service on the Gila River to the southeast of Phoenix.

CHAPTER 4:

THE BOULDER CANYON PROJECT: WATER DEVELOPMENT IN THE COLORADO RIVER BASIN, AND HOOVER DAM

THE COLORADO RIVER: IRRIGATION AND FLOOD

The River

With an average flow of about 14 million acre-feet per year, the Colorado River does not stand as an American giant in terms of the water volume it carries. The mighty Columbia River, in the Pacific Northwest, carries almost ten times as much water, and many rivers in the humid East (such as the Susquehanna, the Delaware, the Hudson, and the Connecticut) are comparable to the Colorado in terms of annual flow. But the Colorado drains one of the driest regions in North America, and the water that passes through its channel is a rare and precious resource; the muddy, turbulent stream stands in stark, dramatic contrast to the arid terrain of the southwestern landscape. In absolute terms, the Colorado River may not be a large river, but within the context of its surrounding environment it offers possibilities of social and economic development that imbue it with enormous significance. And this significance is reflected in the political battles that accompanied efforts by government officials, businessmen, boosters, engineers, and the citizenry as a whole to take control of the Colorado and utilize its water for the purposes they thought most advantageous.

The construction of dams, powerplants, canals, and aqueducts to effect control over the Colorado unquestionably represents a story of technological development and—over time—advancement. But the story of where, when, and how these technological artifacts came to be built is, of necessity, a political story—a story of how American political institutions were utilized by the nation's citizenry (acting through a dynamic matrix of individuals, organizations, companies, and interest groups) to foster the implementation of specific engineering systems serving the desires of particular groups of people. The Colorado River flows through seven western states, and each of these states is populated by people who see (and who have long seen) the river as an appropriate source of economic benefit for their state and their projects.

In terms of how Hoover Dam came to be built and how the dam relates to the overall development of the Colorado River, the driving force behind the project is easily traced to political and business interests tied to Southern California. The technological and legal initiatives central to the construction of the river's first major storage dam all emanated from California and were designed to facilitate: (1) the agricultural development in the Imperial Valley, and (2) the municipal growth of greater Los Angeles. The irony of this circumstance is that California provides very little water to the flow of the Colorado; nonetheless, California occupies a geographical and topographical relationship to the river that facilitated its ability to utilize the flow before any other states in the Southwest could develop projects of comparable scale or economic importance.

Because the total flow is so limited (at least in terms of the amount of land that can conceivably benefit from it) and the possible uses so vast, it did not take long for citizens of the various states to begin to perceive other states as

Figure 4-1: The Colorado River Basin showing the Upper Basin and Lower Basin states.
Source: Bureau of Reclamation.

potential competitors. For many years, this competition remained hypothetical and abstract, but things began to change in the early 1920s when serious and eminently plausible initiatives were advanced to erect what became the first major storage dam across the stream. With the completion of Hoover Dam, utilization of the river's resources would pass from the realm of the possible to the realm of reality. And because of how American society (acting through state laws, federal statutes, and constitutional rulings by the judiciary) had formulated the legal structure of water rights on a national level, the placement of this huge technological construct into the riparian landscape of the lower Colorado River represented (and would effect) a huge political ordering of how the river's water resources would be allocated and used for generations to come.

Thus, while the story of dam building along the Colorado River might appear, at first glance, to constitute a tale of developing engineering expertise brought to bear on problems involving such things as diversion tunnel blasting or concrete placement, in aggregate it represents something much more complicated (although the process of dam-building was nothing if not complicated in its own terms). The political character of dam-building is certainly not a unique characteristic of the Colorado River basin, but the history of the river's development offers a particularly engaging example of how dam-building is inextricably intertwined into the political fabric of American life.

The Bureau of Reclamation and the West

By the beginning of the 1920s, the feasibility of building large-scale water projects in the West could no longer be dismissed as fanciful dreaming. Several projects, some sponsored by the federal government and some underwritten by private capital, had demonstrated the possibility of transforming the arid western environment through the control and diversion of regional water supplies. Most of these involved agricultural development, while some focused on hydroelectric power production or municipal water supply. The organization, funding, and implementation of these various endeavors may have differed, but they all provided evidence that increased utilization of water resources could foster increased economic growth.

While not the only player in the game of big dam construction, during the first two decades of the twentieth century, the federal government made major contributions to the art of hydraulic engineering. The Reclamation Service may have failed in its efforts to achieve the high (and perhaps unrealistic) standards of managerial efficiency and financial success that Frederick Newell heralded at the time of its original authorization; nonetheless, it had proved beyond cavil its ability to plan and complete big construction jobs in remote and difficult locations. While the ideal of promoting the family farm never disappeared from official bureau rhetoric, during the 1920s and into the 1930s, Reclamation gradually reshifted its energies toward projects with economic goals extending far beyond simple agricultural production.

In its first two decades, the Reclamation Service included hydroelectric power production as part of some projects (most notably the Roosevelt Dam in Arizona). However, in the first two decades of the Reclamation Service's existence, electric power production always remained ancillary to irrigation and was never pursued as a central goal, tenet, or objective unto itself. Eventually identifying (and promoting) its work in a more broad-based manner that extended beyond an agricultural focus, by the mid-1920s the renamed bureau began to more directly focus on fostering a range of benefits that could include flood control, hydroelectric power, municipal water supply, and—last but not least—irrigation. Because of the scope and scale that multipurpose projects both encompassed and required, Reclamation's interests gradually expanded beyond efforts to build individual dams and water supply systems. Instead, it began to conceive its mission as one dedicated to planning—and implementing—the hydraulic development of entire river basins.

The ascension of federally sponsored multipurpose dams did not derive from some master plan conceived in the Reclamation Service conference rooms in Washington, D.C., or Denver, Colorado. Rather, it slowly evolved out of a longstanding tradition (dating as far back as the U.S. Supreme Court's 1824 ruling in *Gibbon v. Ogden*) that the federal government maintained a special constitutionally-derived authority over navigable waterways. During the late nineteenth and early twentieth centuries, the legal orientation of the U.S. judicial system was not one that encouraged an active (some might say intrusive) role for the federal government in the economic affairs of the United States. From a vantage point at the start of the twentyfirst century, it might seem perfectly natural—if not inevitable—that the federal government would take on responsibility for erecting huge water supply and electric power systems. But prior to the 1930s, during which the lingering devastation of the Great Depression fostered a complete rethinking of how government should interact with the national economy, it was not at all obvious that such projects represented initiatives properly undertaken by a federal bureau.

Certainly, the passage of the National Reclamation Act in 1902 and subsequent actions such as Theodore Roosevelt's convening of a National Conservation Convention at the White House in 1908 offer evidence that–at least in the minds of many progressive conservationists–there existed a role for the federal government in economic and natural resource development. Furthermore, the national debate over control and regulation of water power that extended from the Roosevelt Administration through the establishment of the

Federal Power Commission in 1920 reflected a willingness on the part of the American electorate to accord a federal role in overseeing hydroelectric power production by privately-financed electric power companies. Nonetheless, the notion that the federal government would assume direct responsibility for financing and building dams and water control systems dedicated to generating electric power for public consumption and to supporting non-agricultural (i.e., municipal/urban) water supply systems was not at all obvious prior to the 1930s.

Any telling of the story of how the federal government came to embrace multipurpose projects focused around river basin development cannot be completely comprehensive if focused on only a single river or a single dam project. For example, the controversy over how (or whether) the federal government should become involved in the business of electric power production came to the forefront of the national political arena in the 1920s during debate over how Muscle Shoals (Wilson) Dam should be integrated into the economy of the Southeastern United States.[1] A narration of the federal/river basin/hydroelectric power story might simply take Wilson Dam and analyze its history as representative of how a federal presence in such affairs—while originating prior to the Franklin Roosevelt Administration—did not become fully manifest until the coming of the New Deal.[2] Certainly, there would be some truth to such a portrayal and the importance of the Wilson Dam/TVA connection in regard to the ascendance of federally owned hydroelectric generating plants cannot be discounted. But such a story would leave a mistaken impression that the Roosevelt Administration's New Deal activism—or special defense-related motivations—comprised a necessary factor in the implementation of federal multipurpose projects.

A key initiative that complicates any historical interpretation positing the New Deal as an essential factor in the rise of federally sponsored river basin development concerns the authorization and construction of Hoover Dam. Located across the Colorado River about 150 miles downstream from the Grand Canyon (and only about 25 miles east of Las Vegas, Nevada), this massive curved gravity concrete dam was formally authorized by the Boulder Canyon Project Act, which was signed into law by President Calvin Coolidge in December 1928. Clearly, the approval of the project by a Republican president famed for his view that "after all, the chief business of the American people is business" (and approved long before the stock market collapse of October 1929) reveals that Hoover Dam must derive from something much more than a simple "reaction" to the economic downturn of the Great Depression.

In addition, the fact that it was named Hoover Dam by President Hoover's Secretary of the Interior, Ray Lyman Wilbur in 1930, also speaks to its origins as something quite different from a New Deal project (a note on

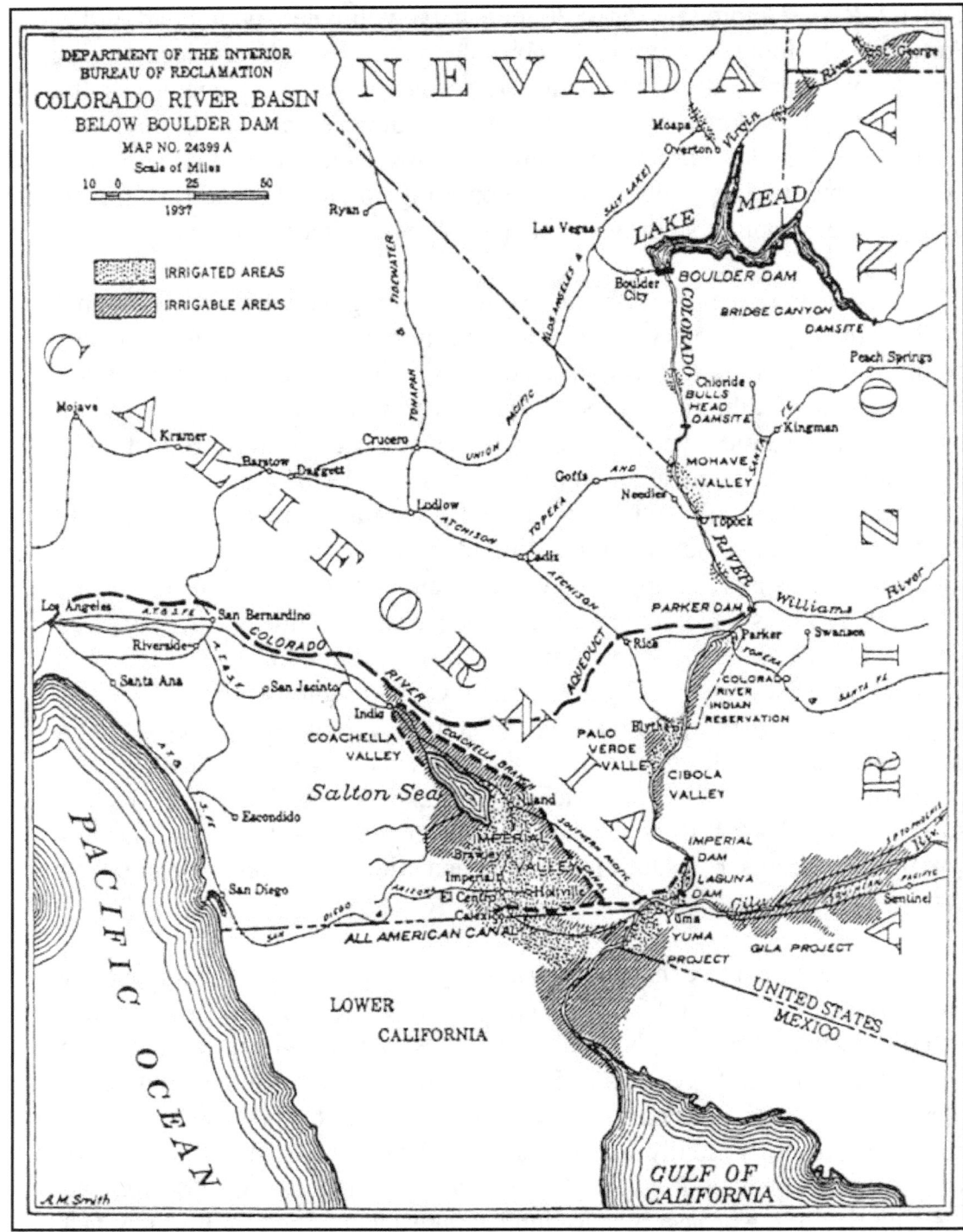

Figure 4-2: This early 1930s map shows the relationship of the Imperial Valley to the Colorado River, including the site of Hoover (Boulder) Dam and the delta lands of northern Mexico. Source: Bureau of Reclamation.

nomenclature: early planning perceived Boulder Canyon as a logical site for a dam across the lower Colorado River and for this reason proposed legislation authorizing the dam was titled "Boulder Canyon Project Act"). However, as early as 1924, it was recognized that nearby Black Canyon offered a better site for a large dam, and—despite being characterized as Boulder Dam—the structure, as it stands today, is located in Black Canyon).

The first of what came to be characterized as multipurpose dams–and associated river basin initiatives–involved plans to create a huge reservoir on the lower Colorado River to serve agricultural and municipal interests in southern California. The huge concrete gravity dam responsible for creating this reservoir revolutionized the way the federal government participated in water control projects.

Early Developments and the Imperial Valley

The origins of Hoover Dam lay in an ambitious, privately financed project to irrigate southern California's Imperial Valley with water from the Colorado River. As conceived by the Colorado Development Company in the late 1890s, this scheme diverted water from the river to a huge tract of desert land just north of the California/Mexico border. Significantly, much of this land lies below sea level, which makes it relatively easy for water to flow to the valley; at the same time, this distinctive topographical condition also makes the valley susceptible to flooding. In the absence of human interference, the flooding of the valley occured in cycles of hundreds or thousands of years depending upon silt accumulation in the river's delta flood plain. With human manipulation of the river in order to foster irrigated agriculture, the possibility of flooding assumed a new dimension, not because settlers wanted to flood the Imperial Valley but because of unintended consequences that could result from the construction and clearing of canals connected to the mainstem of the Colorado. To understand better how floods affected development of the Imperial Valley and the eventual construction of storage dams on the Colorado River, it is necessary to consider the character of the river's entire watershed.

The tributaries of the Colorado River drain over 200,000 square miles of land varying in elevation from over 14,000 feet to sea level (and even lower).[3] The main river is primarily fed by a few large tributaries flowing west and south out of the Rocky Mountains. The most important of these include the Green River flowing out of Wyoming and through eastern Utah; the Yampa, White and Gunnison Rivers flowing out of western Colorado; and the San Juan River flowing out of northwestern New Mexico. In aggregate, these streams contribute almost 90 percent of the river's annual flow, and they constitute what is commonly called the Upper Basin. The above-named tributaries feed into the main stem of the Colorado River before it reaches the forbidding canyon lands of southwestern Utah where the streambed flows through a series of gorges and canyons lying hundreds to thousands of feet below the surrounding mesas and plateaus. After crossing the Utah/Arizona border and passing the famous river crossing point

135

known as Lee's Ferry, the Colorado flows westward through the Grand Canyon. Soon afterwards, it reaches the Boulder and Black Canyons that straddle the Arizona/Nevada border. Turning southward once it reaches the vicinity of Las Vegas, Nevada, the Colorado River soon forms the 250-mile-long border between the states of California and Arizona. Finally, at a distance of 1,450 miles from its headwaters in Colorado, the Colorado River enters–at least it did prior to the erection of large storage dams–the state of Sonora, Mexico, and disperses across an expansive delta. Only then does it drain into the shallow arm of the Pacific Ocean known as the Gulf of California.

Figure 4-3: A newly planted grapefruit orchard in the Imperial Valley, circa 1920. Source: Bureau of Reclamation.

With a flow that can range in intensity from 2,500 to over 300,000 cubic feet per second, the annual capacity of the Colorado River is not particularly remarkable if compared with rivers in the humid region of the eastern United States. But the river drops thousands of feet in its journey seaward and offers the possibility of developing significant amounts of hydroelectric power, especially if seasonal floods can be captured behind high storage dams and gradually released through turbine/generators. The power potential of the stream attracted little attention prior to the twentieth century because the possibility of harnessing its energy seemed remote and impractical. Instead, during the nineteenth century, the lower reaches of the stream from Mexico up along the California/Arizona border–which constituted a relatively flat stretch of river unencumbered by rapids or rocky shoals–was initially utilized to support steamboat traffic serving mining districts in the region.[4] Within the larger scope of the national economy, this steamboat traffic was of minor importance. It quickly dissipated by the early 1880s, after the Southern Pacific Railroad completed its southern transcontinental line through Arizona and New Mexico. In a strictly economic context, the Colorado's short-lived steamboat trade would barely rate a footnote in the history of the American Southwest. But within a legal context, the existence of this steamboat traffic demonstrated that the stream was unquestionably "navigable" and, thus, subject to federal jurisdiction based upon long-standing legal precedents involving interstate commerce.

Figure 4-4: An Imperial Valley strawberry field, c. 1920. Source: Bureau of Reclamation.

In contrast to the lower Colorado River (which had been known to Spanish explorers as far back as the sixteenth century), the Upper Basin of the river's watershed (i.e., the territory upstream from the Grand Canyon stretching into Utah, Colorado, and Wyoming) constituted one of the last great "unknown" regions of the North American Continent. Characterized by a rugged terrain in which the river proper often lay far below the level of the surrounding countryside, exploration of this region was the focus of John Wesley Powell's famous Colorado River expeditions of 1869 and 1871.[5] By the end of the nineteenth century, the upper Colorado River watershed was long past being "unknown" by Anglo-American society at large. But, aside from a relatively small number of irrigation diversion ditches serving communities such as Grand Junction,

Colorado, the water resources of the Upper Basin remained largely unexploited in terms of economic development. Similarly, by 1900, only a relatively small amount of irrigation development had taken place in the Lower Colorado Basin near Blythe, California, where a small portion of the river flow was diverted onto bottom lands paralleling the river in what came to be called the Palo Verde Valley (although in truth it is more properly characterized as encompassing part of the lower Colorado River Valley). Aside from this (and a few other minor irrigation diversions along the length of the stream), the Colorado still flowed free and unfettered in its journey through the delta to the Gulf of California. With the coming of the twentieth century, this would quickly change.

As far back as the 1850s, Oliver Wozencroft, a pioneering Anglo-American who traveled through Southern California in the wake of the Gold Rush, had come to appreciate the agricultural possibilities that were afforded by the distinctive topography of the lower Colorado River Delta. Specifically, Wozencroft and his engineer colleague, Ebenezer Hadley, realized that an ancient channel of the river (long-since filled in with sediments) had once carried water directly into the Imperial Valley.[6] The reason for subsequent change in the location of the main course of the river away from this channel related to the huge amount of sediment carried by the river. Prior to the construction of major storage dams, the river deposited, on average, approximately 130,000 acre-feet of sediment atop its delta every year. As the river neared the Gulf of California, the streambed flattened out, the rate of flow decreased, and the sediment "load" gradually settled out–thus, slowly but surely raising the level of the streambed. As with all river deltas (including the mouths of the Nile River in Egypt or the Mississippi River in Louisiana), this buildup of silt eventually proves so great that the river will overflow its banks and naturally "discover" a new, less silt-clogged, steeper, and hence more physically advantageous route to the sea. In the case of the lower Colorado River, the Pacific Ocean did not represent the only possible outlet; in fact, because much of the Imperial Valley lies below sea level, it offers an even more "logical" destination for the river than the Pacific Ocean because it can allow for a steeper, faster flow. In fact, the region now known as the Imperial Valley had once been a part of the Gulf of California; it was only due to the accumulation of sediment in the Colorado Delta that it became separated from the Gulf and was allowed to become dry land below sea level. As Wozencroft discovered, in ancient times the river had carried fresh water into the Imperial Valley as a result of shifting distributor channels. And there was no reason why—with a little human assistance—it could not do so again.

Wozencroft died before any serious effort was made to develop the Imperial Valley as an irrigation settlement, but not before he had attracted congressional interest in surveying and assessing the proposed scheme. By the 1890s, his basic idea was picked up by the engineer, Charles Rockwood, the irrigation promoter and developer, George Chaffey (who had previously been involved successfully in establishment of the irrigation colony at Ontario, west of San Bernardino); and other investors in the California Development Company. Rockwood, Chaffey, and their company are generally credited for popularizing the name Imperial Valley in place of the much less evocative term "Colorado Desert" used previously to denote the region.[7] But beyond helping coin a more attractive name for the valley, the California Development Company also undertook practical and vital engineering work beginning in 1896.[8] Most importantly, this entailed cleaning out silt from the ancient channel (usually called the Alamo River) that had once flowed into the valley. Rockwood and Chaffey intended to rehabilitate the channel as the right-of-way for a major irrigation canal.

Just north of the Mexican border on the California side of the river, the company "cut" a short canal connecting the existing riverbed to the ancient channel. Fitted up with wooden headgates intended to control the amount of flow allowed into their canal, this deceptively simple system provided a successful and relatively inexpensive means for diverting water into the Imperial Valley. In the short term, the company's plans to make the desert bloom proved surprisingly easy to implement—nature had accomplished most of the "excavation" work hundreds of years earlier—and, by 1902, thousands of acres of prime agricultural land was in process of being irrigated and made economically productive. However, two potential problems, one political and one environmental-technological, threatened the endeavor's long term success.

The first of these problems derived from the political reality that the ancient river channel used by the California Development Company crossed over the international boundary and ran for about fifty miles through Mexican territory before reentering the United States at the southern edge of the Imperial Valley. Although the Mexican Government allowed the canal to traverse its territory, under the jurisdiction of a separate Mexican company associated with the California Development Company, there remained concern that international political action—not to mention the possible action of *bandidos*—might cut off the valley's water supply. In addition, the company's agreement with Mexican authorities stipulated that Mexico reserved the right to draw half of the water flowing through the canal for use on land lying outside the United States.

As it turned out, most of the Mexican acreage eventually watered by the Alamo Canal was controlled by Los Angeles businessman and *Los Angeles Times* publisher Harry Chandler. Clearly, the politics of operating an international canal to serve the Imperial Valley were complicated by the fact that the financial interests of some highly influential American businessmen did not stop at the border. Nonetheless, there existed strong feelings among many westerners that a canal lying entirely within U.S. territory would be in the national interest. By the early 1920s, this sentiment had coalesced into a movement to win federal support for what came to be known as an "All-American Canal," entirely within the bounds of U.S. territory.

The second potential problem facing the California Development Company system was inherently environmental (and hence technological) in character. Specifically, it related to the short "cut" excavated between the ancient Alamo River channel and the main channel of the Colorado as it existed at the start of the twentieth century. After this cut was made into the river bank, water quickly began flowing into the valley and agricultural and economic growth rapidly followed. By October of 1903, 100,000 acres were under cultivation and the valley supported a population of 4,000.[9] As a result of this remarkable growth, demand for water also grew, but the company had difficulty keeping the Alamo Canal free of silt and thus capable of sustaining its maximum potential carrying capacity. As it turned out, silt accumulation proved particularly troublesome in the excavated section of the canal closest to the river, thus prompting the company to excavate two new (and larger) openings into the river a short distance downstream from the original "cut." As before, water flow was to be controlled by wooden "headgates" designed to ensure that uncontrolled heavy floods could not pour into the Alamo Canal and down toward the Imperial Valley.[10]

In the late spring of 1905, the possibility of diverting too much water through this new diversion system became reality. In June of that year, heavy floods washed out the new headgates and huge—essentially uncontrolled—quantities of water began surging into the Imperial Valley. As more water flowed out of the "old" channel of the Colorado River, the "new" Alamo River canal deepened and widened. In turn, this process of erosion allowed more water to be diverted and the "new" channel continued to increase in size. The Southern Pacific Railroad (whose trackage passed through the valley and which—as a freight carrier—maintained a strong economic interest in the valley's agricultural production) worked valiantly, in concert with the California Development Company, to dump trainload-after-trainload of rock to close off the canal entrance with a rock embankment.[11]

For months this effort had little effect as the Colorado reached flood stages unprecedented in the short time period that Anglo-Americans had come to know and study the stream. Appeals were made to the federal government to aid in staunching the flow through the Alamo Canal, but, while sympathetic to farmers in the valley, President Theodore Roosevelt refused to directly aid or interfere in what he and his Administration perceived as the affairs of a private corporation.[12] The California Development Company undertook its work as a private, non-governmental initiative and—after the project had encountered severe difficulties—Roosevelt saw no reason that U.S. taxpayers need be drawn into a costly effort to rescue the company. In a January 1907 message addressed to the Senate and House of Representatives, Roosevelt specifically responded to pleas for direct government intervention:

> The California Development Company began its work by making representations to possible settlers of the great benefits to be derived by them by taking up this land. A large amount of money which might have been used in needed works was expended in advertising and in propounding the enterprise. The claims were not only extravagant, but in many cases it appears that willful misrepresentations was made . . . the money thus obtained from settlers was not used in permanent development, but apparently disappeared either in profits to the principal promoters or in the numerous subsidiary companies. . . . At the present moment there appears to be only one agency equal to the task of controlling the river, namely the Southern Pacific Company, with its transportation facilities, its equipment, and control of the California Development Company and subsidiary companies. The need of railroad facilities and equipment and the international complications are such that the officers of the United States, even with unlimited funds, could not carry on the work with the celerity required. . . .[13]

Whether the direct involvement of the federal government would have actually accelerated the process of staunching the flow into the Imperial Valley remains uncertain. Given the enormous effort made by the Southern Pacific on its own account, it is difficult to imagine that federal action would have made any dramatic difference in the practical work to close the breach. As it turned out, above average flooding in the Colorado River watershed during this period clearly exacerbated the problem of protecting the valley; eventually, the flooding through the Alamo Canal was brought under control, but it took almost two years–and an expenditure of two million dollars–before the "cut" was closed. In terms of the physical environment, the flood permanently inundated thousands of acres of land under a newly created lake that came to be known as the Salton Sea. Even today, the Salton Sea (which is primarily fed by subsurface irrigation drainage rather than surface flow) remains an enduring, essentially permanent, part of the Southern California landscape that bears prominent witness to the effect of the 1905-07 Colorado River floods and the ability of humans to dramatically (if unwittingly) transform the hydraulic landscape.

As a result of the tremendous disaster attending these floods, the California Development Company entered bankruptcy in 1909, but not before most of its assets had been absorbed by the Southern Pacific Railroad in the wake of efforts to finance the flood control campaign.[14] In 1911, landowners in the valley who had previously relied upon the California Development Company for their water supplies formed the Imperial Irrigation District. This locally administered governmental authority was designed to operate the regional water supply system to promote the political and economic interests of those people who had invested in the valley. In 1916, the Imperial Irrigation District formally purchased the water supply system from the railroad (for about $3 million dollars in irrigation district bonds) and assumed all responsibility for delivering water to the valley's farmers.[15]

Once the flooding had been checked in 1907, the valley resumed agricultural production (minus thousands of acres of low-lying land now permanently inundated by the Salton Sea). But, understandably, fear that another uncontrolled "break" might occur remained very real and acted to reduce land prices in the valley. Although the Imperial Irrigation District did not place any limitations on the size of individual land holdings within the district, the formation of this public entity (directors of the district are chosen by public election) helped downplay the notion that irrigation in the valley was simply a private endeavor that should depend upon private resources for its growth and development. And soon the district and its boosters began clamoring for federal support to aid them in their desire for flood protection and for protection from Mexican interference.

THE FEDERAL INITIATIVES

The Fall/Davis Report

Even before the conclusion of World War I, the Imperial Irrigation District sought assistance from the federal government to help plan an engineering project that would excavate a completely new canal entirely within U.S. territory. After the end of the war came in late 1918, the district also began to seek federal support for construction of a flood control and storage dam somewhere in the lower Colorado River Valley—most likely in the Boulder Canyon region—that would capture annual floods and thus protect the valley from any possible reoccurrence of the disastrous inundation of 1905-07. In holding back flood water, such a dam could also serve to increase the amount of water available for irrigation in the watershed below the dam, including the vast expanse of the greater Imperial Valley.

As early as 1902, Arthur Powell Davis, who was then the assistant chief engineer of the Reclamation Service, initiated preliminary studies of how to develop the resources of the Colorado River. Although serious efforts to construct a major storage dam across the stream were not pursued at the time (the scale and potential cost of the endeavor paled in comparison to the perceived short-term benefits to irrigation development), one of the Reclamation Service's early projects involved building the Laguna Diversion Dam to irrigate riparian lands surrounding the settlement of Yuma, Arizona, a short distance north of the Mexican border. Between 1902 and 1919, the issue of lower Colorado River development never dropped from consideration by the Reclamation Service, but it was overshadowed by the myriad other projects that the bureau was constructing throughout the West. However, at the end of World War I the Reclamation Service had completed many of the large projects that had occupied its attention during the previous decade (including, most recently, Arrowrock Dam in southern Idaho and Elephant Butte Dam in New Mexico), and the ability of Reclamation to devote major time, energy, and resources to a lower Colorado River storage dam became more feasible.

In 1915, Davis had taken over command as director of the Reclamation Service. Based upon his earlier studies, he was familiar with engineering issues related to building a big dam on the lower Colorado. He also understood that such a project would potentially involve construction of one of the largest, most prominent, and most visually dramatic dams in the world. Thus, when it became clear that boosters in the Imperial Valley were willing to push the U.S. Congress to support federal financing for development of the lower Colorado River Basin, Davis made clear that the Reclamation Service would be willing to assist in developing plans for such work. In the words of California water historian Norris Hundley:

> The proposed legislation [for an All-American Canal] immediately caught the eye of Arthur Powell Davis...who saw it as a perfect opportunity to raise anew his dream of harnessing the Colorado River.... The canal made sense, concluded Davis, but only if it were part of a larger design. To build such an aqueduct without also constructing dams to control "the flood menace" would doom the canal to a short life.... Davis told all who would listen [that the Imperial Valley problem] "is inseparably linked with the problem of water storage in the Colorado Basin as a whole."[16]

As a result of political activism on the part of the Imperial Irrigation District and encouragement from Davis, in May 1920 Congress approved a study that authorized the Reclamation Service to develop preliminary plans for an All-American Canal and a Colorado River storage dam.[17] Known as the Kincaid Act (in recognition of its sponsorship by the chairman of the House Committee on Irrigation, Moses Kincaid of Nebraska), this action represented the beginning of practical planning for what came to be the Hoover Dam.[18]

When formally issued in 1922, the study authorized by the Kincaid Act was known as the Fall/Davis Report because it was officially prepared under the auspices of Secretary of the Interior Albert Fall and Reclamation Service Director Davis. In the report, Davis strongly advocated constructing a large storage dam that would do much more than simply store floods and protect the Imperial Valley; he proposed that hydroelectric power production be considered a key part of the project and that construction costs be underwritten by the sale of hydroelectric power made possible by the dam.

This proposal made sense from a strictly practical point of view because there was no question that huge amounts of power could be generated by a dam extending to a height of over 500 feet and holding back a reservoir of more than 20 million acre-feet of water. From a political perspective, the notion of using electric power revenues as the primary means of financing the dam was much more problematic because it raised questions as to the proper place of the federal government in the generation and sale of electricity. The privately-financed electric power industry controlled most of America's electric power grid in the 1920s, and it lobbied on the local, state, and federal level for favorable legislation that would reinforce its ability to retain that control. A huge federally financed dam on the Colorado River that was to be paid for by hydroelectric power revenues represented a threat to private control. As such, Davis's plan spurred opposition—or at least serious concern—among business interests that wished to limit the role of government in America's economic life.

In the political environment of the pro-business 1920s, when the Republican Party controlled both the White House and Congress, the "public power" issue was always a source of contention regarding the proposed Boulder Canyon Project. In terms of constructing a "high storage" dam, it proved impossible for anyone to devise a practical alternative scheme that could pay for the dam in the assured and reliable manner that proponents of hydroelectricity could claim. Hydroelectricity offered the only economically feasible means of building a "high storage" dam. As a result, during subsequent debates about the project, the possibility of building a smaller-scale (and hence less expensive) "flood control" dam on the lower Colorado was often projected by "private power" advocates as a more reasonable alternative. Such a dam could be erected without reliance upon hydroelectric power sales. In essence, it represented less of a threat to the private electric power industry. But a smaller dam devoted simply to "flood control" and irrigation would not allow full storage and use of the Colorado River's flow. Davis and the engineers of the Reclamation Service saw this as a needless and undesirable inefficiency. In this context, Davis, local Congressman Phil Swing, and others interested in the economic development of Southern California, proved unwilling to accept a smaller "flood control" dam as a true alternative to a "high dam."

As part of the Fall/Davis Report, Davis took on the task of developing a basic plan for the design and construction of a high dam and hydroelectric powerplant in the vicinity of Boulder Canyon. Although the Reclamation Service recognized that there were other possible storage dam sites along the length of the Colorado River (in fact, such sites as Glen Canyon would eventually be developed after World War II), Reclamation quickly focused on Boulder Canyon because of its large storage capacity and its proximity to prospective water users and electric power consumers in Southern California.

Both Boulder Canyon and the nearby Black Canyon (which lies about 20 miles further downstream) offered dramatic, narrow gorges with steep walls extending upwards from the riverbed for hundreds of feet. Investigations initially focused on Boulder Canyon (hence the name chosen for the project) but, early in the process, studies were also carried out at Black Canyon in order to find the best possible site.

The Design of Hoover Dam

During the early planning stages for what became the Boulder Canyon Project, Arthur Powell Davis and his staff made an effort to consider a range of possibilities for the design of the big storage dam on the lower Colorado. Based

Figure 4-5: The Black Canyon damsite, circa 1921. Source: Bureau of Reclamation.

upon the Reclamation Service's experiences with the Roosevelt, Elephant Butte, and Arrowrock Dams, it is not surprising that a massive masonry gravity design attracted the interest of Davis, his Chief Engineer Frank Weymouth, Chief Designing Engineer John L. Savage, and Project Engineer Walker Young. The Reclamation Service had experience building massive embankment dams (such as Belle Fourche in South Dakota and Strawberry Valley in Utah) as well as thin arch concrete masonry dams (Pathfinder and Shoshone (Buffalo Bill), both in Wyoming). In this context, the decision to utilize a curved gravity concrete design did not come without some consideration of alternative designs. However, the selection did come quickly and without public review of alternative designs.[19]

In late 1920, Davis initiated correspondence with Lars Jorgensen, a European-trained engineer who had become a prominent advocate of thin arch dam design (especially constant angle arch dams), for the purpose of discerning whether a storage dam of this type might be feasible to build across the lower Colorado.[20] While previously Davis had been prominently associated with massive gravity dams such as Roosevelt and Arrowrock, he retained an interest in thin arch designs, and his interaction with Jorgensen testifies to this point.[21]

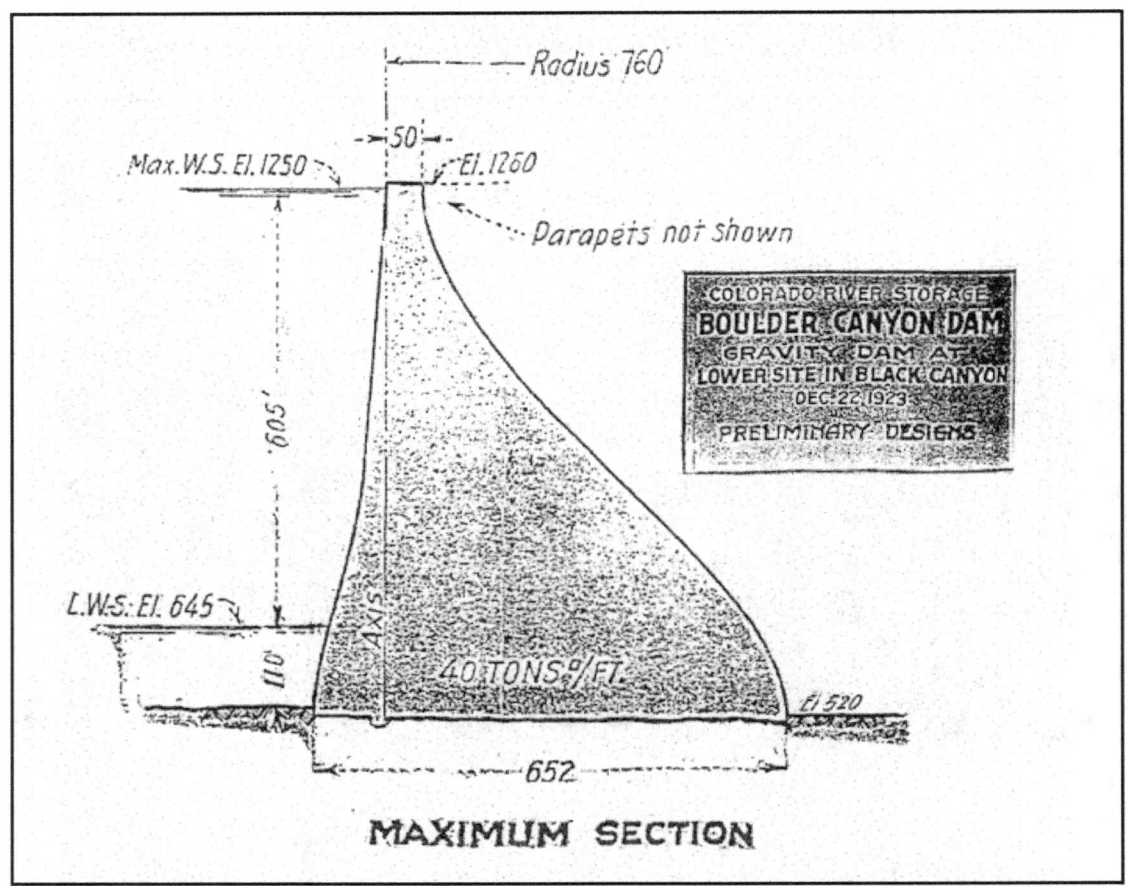

Figure 4-6: A preliminary design (from a blueprint) for a gravity dam at the lower site in Black Canyon, December 22, 1923. Source: Bureau of Reclamation.

During the next year, the use of a thin arch design (either constant radius or constant angle) officially remained a possibility but little action to forcefully promote such a design is evident in available records.[22]

In contrast, the notion that the Reclamation Service would rely upon a massive design was publicly expressed by Davis as early as October 1920, even before he corresponded with Jorgensen, when he wrote the Chief Engineer of the Los Angeles County Flood Control District in response to a "request for some information concerning tentative plans made for a dam in Boulder Canyon." At that time, Davis indicated that "studies have been made for a section of masonry or concrete of the gravity type, and a rockfill and earth section, the latter, however, not being regarded as certainly feasible."[23] In early 1924 (after Davis had been displaced as Director of the Reclamation Service and immediately prior to the appointment of Elwood Mead as Commissioner of the Bureau of Reclamation), Weymouth submitted a "Report on the Problems of the Colorado Basin"—oftentimes simply referred to as the "Weymouth Report"—in which Volume Five focused on "Boulder Canyon: Investigations, Plans and Estimates." In this report, no mention is made of any thin arch designs that may have been

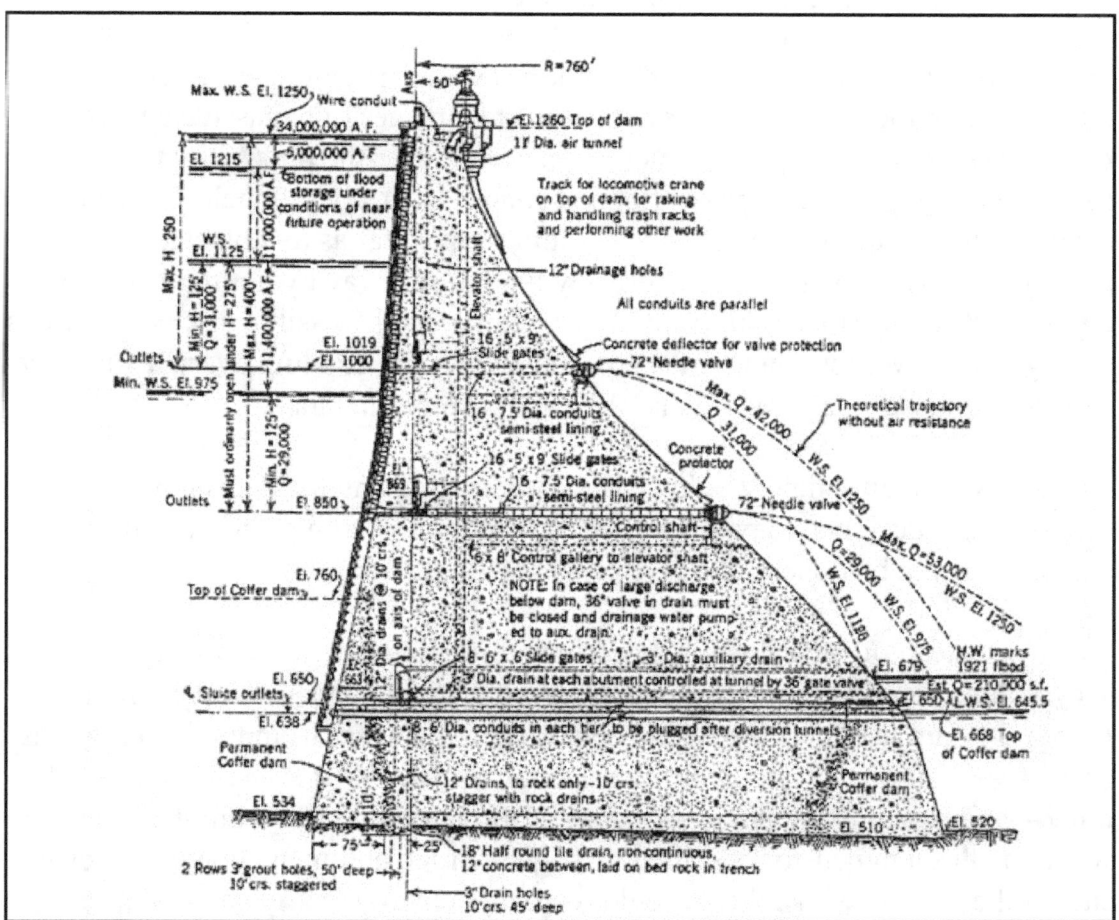

Figure 4-7: Frank E. Weymouth design for Hoover Dam at the Black Canyon Site showing outlets through the structure, circa 1924. Source: Bureau of Reclamation.

considered for the big storage dam. Instead, Weymouth reported only that "studies have been made of rock-fill and concrete dams of various types" and went on to explain:

> There is a grave question whether life and property below a dam of such unprecedented height and a reservoir of such enormous capacity should through the construction of a rock fill dam be subjected to a risk which could be removed by the adoption of a concrete dam. . . . With all possible safeguards taken in the construction of a rock-fill of the height proposed it must be admitted that its overtopping would result in its certain and sudden destruction with overwhelming disaster in the valley below. The dams adopted are believed to be the safest that can be built–concrete dams of the gravity type built on a curved plan–and estimates prepared indicate that the concrete dams could be built at less cost than rock-fills of the same height. [Note: The use of the plural 'dams' in this quotation refers to three designs of various heights—ranging from about 525 feet to over 700 feet—developed for the same site].[24]

In other words, Weymouth's report reveals that, although Reclamation estimated that there existed some economic advantages of a massive curved gravity design over a rock-fill structure, concerns over the possibility that a rock-fill design might someday be overtopped comprised a key rationale for selecting a curved gravity design. In fact, Weymouth went so far as to advocate a curved gravity design that would not feature any type of spillway by noting that overtopping could probably be prevented by opening up all possible discharge outlets through the powerhouse and the dam. But even if the flooding overwhelmed the capacity of these discharge outlets, Weymouth counseled that:

> Any overtopping would be of short duration and the dams have been designed to pass rare floods over the top with safety which can not be done in the case of a rock-fill dam.[25]

Instead of special spillways driven through the rock abutments, Weymouth proposed that outlet pipes (controlled by huge valves) be built directly into the dam itself. These would be able to draw water from the lower depths of the reservoir and discharge it from the downstream face of the structure. The other—and more advantageous—means of discharging water from the reservoir would be through penstocks drilled through the rock abutment along the Nevada side of the canyon walls; these would feed into a hydroelectric power house about a half mile downstream from the dam where they would deliver water to large-scale turbine/generator units. In formulating a basic plan for how best to construct the dam, Weymouth also proposed that the same tunnels used to carry water to the powerhouse provide vital service during the construction process. Specifically, they were to be used to divert the flow of the Colorado River around the dam site so that temporary rock-fill cofferdams could protect the site from flooding and allow excavation down to bedrock foundations in the middle of the stream bed.

Thus, by the beginning of 1924, the Weymouth Report laid out the basic features of what would become Hoover Dam. Over the next four years, Weymouth's proposal underwent careful consideration by Reclamation, and by 1928 it had undergone significant revision at the hands of John L. Savage. The most important of these revisions involved the drilling of diversion tunnels through both the Nevada and Arizona abutments (two tunnels on both sides of the river), the construction of two "glory-hole" spillways that would connect into the diversion tunnels and provide insurance that the dam would never be overtopped, and the construction of powerhouses in both Nevada and Arizona that would tap into the diversion tunnels (and to other tunnels connected to outlet towers built directly upstream from the dam).[26]

Clearly, these changes represent important alterations to the Weymouth design and are of importance in defining the form of the dam, powerhouse, penstock, and spillway system as it was actually built. But–beyond the driving of spillway discharge tunnels to feed into the diversion tunnels–they do not constitute anything that cannot be understood as an evolution of the Weymouth design. And even the addition of spillway tunnels represented an uncomplicated (yet no doubt imaginative) expansion of the diversion tunnel system.

During the mid-1920s, the specific character of the Hoover Dam design continued to evolve as more was learned about geological conditions and as Reclamation became interested in utilizing the "trial load" method of design to confirm the safety of the massive curved gravity design. While the "trial load" method of analysis (see chapter 3) undoubtedly figured into the final dimensioning of the dam's profile, it did not prompt any dramatic changes or modifications.[27] In fact, it is difficult to discern any radical differences between the preliminary profile that accompanied Weymouth's 1924 report and the design as built. Both represent curved gravity designs featuring extremely ample gravity sections and the use of "trial load" techniques of analysis did little in terms of altering the basic form of the design. In the same way, research on scale models of the dam also figured into Reclamation's analysis of structural safety and gave them greater confidence in its stability, but it is difficult to perceive how the basic form of the design was altered by such work.

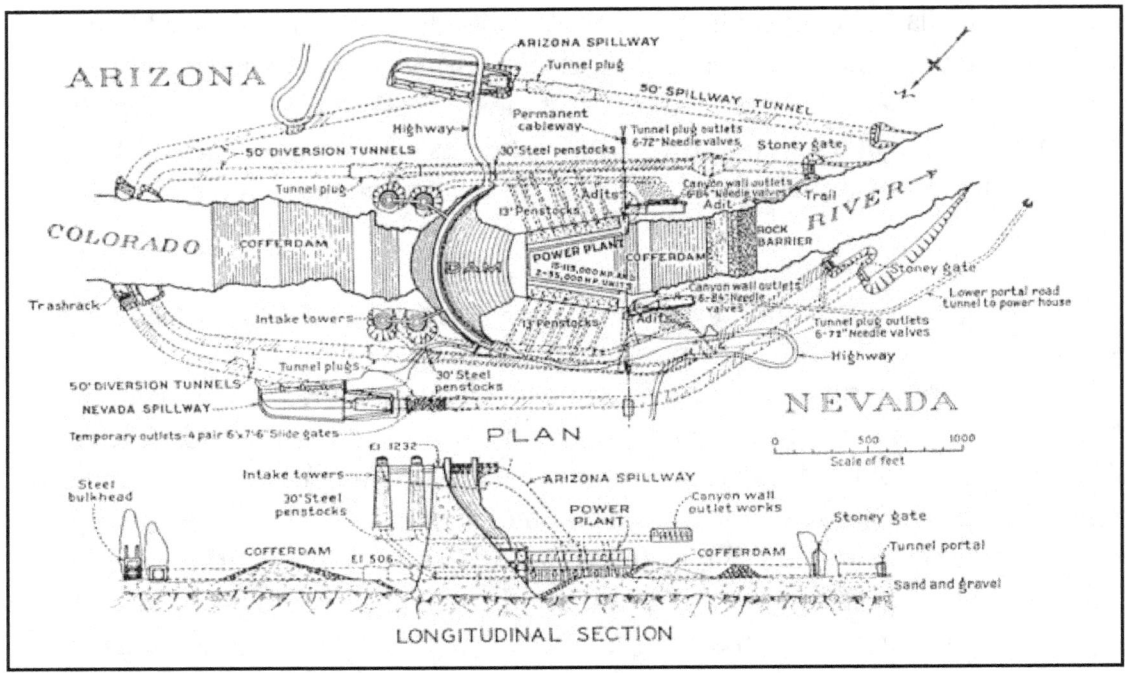

Figure 4-8: Plan and section of Hoover Dam as revised by John (Jack) L. Savage, circa 1928. Source: Bureau of Reclamation.

Near the end of the approval process for the Boulder Canyon Project, Congress authorized the formation of a special "Colorado River Board" that would be separate and distinct from Reclamation's own consulting engineers (and hence, would presumably provide an independent analysis of the proposed dam's safety and feasibility). This board approved the dam's basic design, but recommended that the maximum allowable stresses in the massive structure be reduced from 40 tons per square foot to 30 tons per square foot. Although this might have appeared to the layperson as a rather simple way to increase the strength of the design, to Reclamation it represented a problem in the sense that any real effort to strictly adhere to this requirement would have significantly added to both the bulk and cost of the dam.

Without directly resisting this directive, Reclamation, nonetheless, made no significant alteration to the design as proposed in Savage's November 1928 report. Instead, Reclamation opted to claim that more sophisticated mathematical analysis (in line with the "trial load" method) indicated that the proposed design in fact did not exceed a maximum allowable stress of about 33 tons per square foot, and this was considered adequate to meet the 30 tons psf criteria. In Mead's words: "It is not believed that the maximum stress as finally determined will appreciably exceed the 30-ton limit. It is believed that the general plan of the dam can be agreed upon without serious difficulties."[28] In the end, the Colorado River Board's recommendation had no substantive effect on the final design.

Figure 4-9: California's interest in development of the lower Colorado River was evidenced by William Mulholland's active support of the project. Pictured here about 1924 he gestures during an inspection tour near the Boulder Canyon damsite, indicated by the arrow. Source: Boulder Dam Association pamphlet, item in the private collection of Donald C. Jackson.

In his 1928 "Revised Plan," Savage took care not to criticize Weymouth's Report as being somehow faulty and in need of correction. Rather, he simply stated that "The Weymouth plan for the dam and power plant . . . constitutes a preliminary study on which to base an estimate of cost. This plan was not intended as a final design and should not be considered as such."[29] In this context, it is important to note that the design of Hoover Dam cannot be ascribed to any single individual, but instead represents a collaborative effort that extended over several years time. Davis, Weymouth, and Savage all played important roles in overseeing preparation of the basic design and, in concert with other Reclamation Service staff members, deserve credit as designers of Hoover Dam.[30]

The preceding discussion has focused on the technical and engineering aspects of the basic dam design. In contrast, the architectural treatment of the dam's surface features was handled in a very different manner and emanated from a source quite distinct from Reclamation's Denver office. During the 1920s, the architectural treatment of the dam was assumed to adhere to a neo-classic style featuring design motifs such as eagles with wide-spread wings. In 1931, long after all the major technical issues involving the design had been determined, Reclamation brought in Los Angeles architect Gordon Kaufmann to develop a more modern appearance for the dam. By simplifying the surface treatment of the design and utilizing a monumental art deco style, Kaufmann created an evocative, streamlined facade for the massive structure. Ironically, the prominence of the dam in American culture is, no doubt, tied in large part to its modernistic design motif, but the circumstance of hiring a non-government architect to carry out this work occurred very late in the design process and was very much separate and distinct from the rest of project.[31]

Selection of the Black Canyon Site

At the beginning of the Reclamation Service's work in developing plans for a storage dam on the lower Colorado, it was assumed that the structure would be built in the narrow gorge known as Boulder Canyon. During the early planning stages, attention did not focus so much on the precise location of where the "Boulder Dam" would be erected as it did on the notion that the dam should be built somewhere downstream from the Grand Canyon and in a location that would be relatively accessible to the electric power market of Southern California. Thus, in 1920-21, the advocacy of a dam at Boulder Canyon was undertaken by the Reclamation Service in the context that this represented a more desirable alternative than a dam upstream from the Grand Canyon at Glen Canyon or Lee's Ferry.[32] But as early as December 1921, Davis realized that it would be desirable to explore the possibility of using a site in Black Canyon as an alternative to Boulder Canyon. The two canyons were only about 20 miles apart (Black Canyon is further downstream) and both offered steep, narrow gorges topographically well suited for a dam. And in the larger context of the Colorado river basin, they provided essentially the same possibilities of service to southern California interests.

Because Black Canyon lay a bit further down the river, and at a somewhat lower elevation, it represented an opportunity to develop a small (yet not insignificant) additional amount of hydropower that would otherwise be difficult to develop. As Davis counseled Weymouth:

> I am inclined to think it best to make one or more borings at Black Canyon, because a dam at that point would utilize about thirty feet of fall which occurs between that point and our camp at Boulder Canyon, and this fall cannot be utilized in any other way.[33]

At the same time, Black Canyon was not so far downstream that it could not inundate the excellent reservoir site lying upstream from Boulder Canyon. By the beginning of 1922, the Reclamation Service was carrying out geological explorations at Black Canyon to discern the quality of bedrock at the site and the depth of excavation that would be required for dam foundations. In July, Weymouth reported to Davis that initial investigation of the upper end of Black Canyon (termed line "A") did not appear promising and he went so far as to state that:

> The foundation rock at line A in Black Canyon is not suitable for bearing pressures of 40 tons per square foot as used on the granite of Boulder Canyon, [and] the soft and porous structure of some of the rock may render this site entirely unsuitable for such a high dam. In this connection I will say that I am personally very doubtful of the feasibility of a dam 600 feet high in Black Canyon, unless the conditions at the lower site prove to be very much better....[34]

With this less than encouraging prognostication, studies soon focused on the lower end of Black Canyon (line "D"). As it turned out, conditions at this location proved better than at the upper end and, following a two day field visit in November 1922, Davis could report to Weymouth:

> No one doubts the entire feasibility of the Black Canyon site. The rock in the bottom of line D is much better than that secured at the head of the canyon last year ... I think we should make a choice between Black and Boulder Canyons as soon as possible so as to stop expenditures at the site rejected.[35]

With this endorsement and encouragement, attention soon shifted to Black Canyon, and, in early 1924, it was officially recommended as the site of the proposed dam. As the Weymouth Report explained the situation:

> An extensive geological examination has been made ... [and while] both dam sites [Boulder and Black Canyons] are excellently adapted to the construction of a very high dam ... the granite of Boulder Canyon is superior to the breccia of Black Canyon for carrying great loads ... [nonetheless] the investigations led to the adoption of the lower site in Black Canyon for the reason that it is more accessible [for construction equipment and materials]; the maximum depth to bedrock is less ... than at the upper site in Boulder Canyon and for the same height of dam the reservoir capacity is greater.[36]

Thus, the selection of Black Canyon was not made because it offered better geological conditions (in fact, by this criteria it was judged less desirable than Boulder Canyon), but because it could provide for a larger reservoir and would allow for a less costly structure based upon savings in material and logistical expenses. By the time this decision was made, however, so much effort had gone into the promotion of a "Boulder Canyon Project" that no effort was made to transform the nomenclature to the "Black Canyon Project" or "Black Dam."[37]

153

Nonetheless, from 1924 on, all work related to Hoover Dam revolved around the lower site (line "D") in Black Canyon.

Congressman Phil Swing

With the attention that attended submittal of the Fall/Davis Report, the possibility of federal involvement in building some type of storage dam across the lower Colorado River gained political credibility. The effort to keep this possibility alive and in the national public conscience was championed by Congressman Phil Swing, who represented the Imperial Valley as well as other parts of Southern California. Working in his official capacity as a member of the U.S. House of Representatives, throughout the 1920s, Swing assumed responsibility for keeping the All-American Canal and what quickly came to be known as the Boulder Canyon Project in the public eye. Working with California Senator Hiram Johnson, Swing kept abreast of all the political nuances related to the Boulder Canyon Project, and he made sure that, during the 1920s, Congress was presented with a series of Swing-Johnson Acts positing federal authorization of the project.

Herbert Hoover remains closely associated with the Boulder Canyon Project because of his work in negotiating what became known as the Colorado River Compact; and because he served as U.S. President during the time that construction work began on the high storage dam. His name was attached to the dam through the action of his Secretary of the Interior. Hiram Johnson is well remembered by historians for his early advocacy of progressive political reforms in the face of the Southern Pacific Railroad's "Octopus" (and for his intransigent "isolationism" in the realm of international politics during the 1930s). In contrast, Phil Swing never attained lasting fame as an advocate of the Boulder Canyon Project. But, in truth, Swing was the most important and persistent political proponent of the first high dam to be built across the Colorado River.

A native Southern Californian, Phil Swing was born near San Bernardino in 1881. After graduating from Stanford Law School in 1905, Swing moved to the Imperial Valley in 1907 where he experienced first hand the environmental and economic damage wrought by the floods of 1905-07. He quickly established a law practice in the valley and, in 1911, proved instrumental in the formation of the Imperial Irrigation District; in 1912, he formally entered the political arena and was elected District Attorney for Imperial County. Failing to win reelection in 1914, he reentered private practice and, after a new board of directors was elected for the irrigation district in 1916, served as legal counsel for the District. In this capacity, he took on the work of advocating federal support for the All-American Canal and for flood control along the lower Colorado. Specifically, in 1918 he journeyed to Washington D.C. where he negotiated a contract with Secretary of the Interior Franklin Lane whereby the district would pay two-thirds of the cost of a federal survey of possible right-of-ways for the All-American Canal.[38]

Later, in 1919, Swing again utilized his position as chief counsel for the Imperial Irrigation District to support legislation presented by California's 11th District Representative William Kettner that would have authorized federal guarantees for up to $30 million worth of district bonds to finance an All-American Canal. In July 1919 hearings before the House Committee, Swing first stepped before a national audience to make a plea that would be repeated and expanded upon in the decade ahead:

> Is the government to stand idly by and complacently watch foreign lands develop by sapping the life out of an American community when the remedy is easily within reach without cost to the government of a single dollar? We are here simply asking for a chance to live.[39]

The legislation proposed by Kettner never made it out of committee (although it served as an important precursor of the Kincaid Act passed in 1920). However, at the same hearings that Swing appeared before to advocate Kettner's bill, Arizona Congressman Carl Hayden raised questions as to the desirability of Congress acting too quickly or precipitously to support the All-American Canal. Specifically, Hayden brought to the forefront concerns over how such legislation would affect water rights along the Colorado River:

> But you [Kettner and supporters of the All-American Canal] are now coming to Congress asking that an extraordinary thing be done by the passage of this legislation, and Congress must look to the development not only of the Imperial Valley, which is your particular interest, but the Colorado River valley as a whole, and that can only be fully developed by storage.[40]

Hayden also raised concerns over how any major, federally supported irrigation work on the lower Colorado would relate to or affect the water rights of various states; in so doing, he helped accelerate events that, by early 1922, would result in Congress authorizing a conference among the states within the Colorado River watershed.

For Swing, it initially did not appear as though his appearance in Washington D.C., in 1919 would set the stage for a more dramatic and substantive move to the nation's capital. In August 1919, he was appointed a California Superior Court judge by Republican state governor William Stephens, and he held that prominent and prestigious position for the next year. But in the summer of 1920, he learned that Kettner would not run for reelection and Swing immediately announced that he would seek the now-open congressional seat. This district encompassed far more territory than the Imperial Valley—it stretched westward to the Pacific Ocean and northward to the Sierra Nevada and covered seven counties—but Swing proved adept in garnering support from San Diego and Anaheim and San Bernardino and myriad communities in between. With the Republican Party representing a majority of voters in the district, he won handily in the general election of November 1920.[41]

Upon arriving in Washington, D.C., as a U.S. Congressman, Swing took on the task of championing a Colorado River storage dam and an All-American Canal as his primary political responsibility. Acting in concert with California Senator Hiram Johnson, during his first term in office, Swing introduced legislation designed to accomplish this goal. Known publicly as the Swing-Johnson Act, this legislation called for:

> construction of the All-American Canal and of a dam at or near Boulder Canyon. It provided for the leasing of the power privileges by the Secretary of the Interior and stated that construction was not to begin until the lands to be irrigated were legally obligated to pay their proper proportion of the cost.[42]

The first Swing-Johnson bill remained in committee and never even came before Congress for a formal vote. Undeterred, the two legislators reintroduced their proposal three more times over the next six years and gradually fine tuned it in order to define more specifically the work to be financed. By 1928, the proposed Swing-Johnson Act called for a dam with a reservoir capacity of at least 26 million acre-feet and the construction of a powerplant by the federal government that could then be leased to other organizations (be they public or private) for actual operation and power generation.[43] But beyond discussion of the size and operation of the proposed dam and powerhouse complex, there existed a more basic issue that needed to be addressed before Congressional approval could become possible.

Debate over the Swing-Johnson Act continued in Congress until December 1928. In retrospect, the reasons it took so long for the legislation to win approval are not difficult to discern. First, the costs of the proposed work were estimated at several tens of millions of dollars (the final Swing-Johnson Act authorized a federal expenditure of $177 million); to many non-western congressmen as well as to proponents of privately financed electric power companies, this seemed a waste of taxpayer money and an ill-considered investment of public monies. Of course, repayment to the federal government was stipulated as part of the proposed legislation, with the dam proper to be paid for over a period of 50 years (including 4 percent interest) by proceeds from the sale of hydroelectric power generated at the dam.[44] Nonetheless, this plan to cover the cost of the project in a manner that hypothetically would relieve federal taxpayers from paying for the project prompted skepticism. But beyond concern over the cost of the project and the propriety of government power development, the most significant obstacle to the Swing-Johnson Act in its early history focused around widespread fears over California's desire to control the Colorado River. These concerns soon figured prominently into political negotiations related to the proposed project.

The Colorado River Compact

Before any serious political action could occur relative to the Boulder Canyon Project, important issues related to water rights needed to be addressed by the various states holding an interest in the water resources of the Colorado River. The river's watershed encompasses parts of seven states. Four of these (Wyoming, Colorado, Utah, and New Mexico) are termed the "Upper Basin" states; the other three (Nevada, Arizona, and California) are the "Lower Basin" states, which, essentially lie downstream from Lee's Ferry near the Utah/Arizona border. The river is one of the most important sources of water in the Southwest, and every state in both the Upper and Lower Basins wanted a share of the river's flow.

Except for water used by farmers in the Imperial Valley (and a few other locales such as the Palo Verde Valley near Blythe), most of the river's flow remained unused and unclaimed by the early 1920s. Under the doctrine of appropriation, rights to this water would accrue to whatever person or organization first diverted it for "beneficial use." As a result of a 1922 U.S. Supreme Court ruling focused on a dispute between Colorado and Wyoming over claims to the North Platte River, it became clear to students of western water law that a strict application of the appropriation doctrine ("first in time, first in right") would apply to competing claims no matter what state they originated in. With California poised to lay claim to vast quantities of flow stored at Boulder Canyon, the other states in the Colorado River watershed became concerned that California would eventually monopolize control over the entire river simply because water could be more quickly, easily, and profitably diverted along the lower reaches of the stream.

Congressman Phil Swing's most insightful biographer has perhaps best summarized why it was necessary for the southwestern states to come together and agree on how to allocate the river's annual flow. Acting under the authority of the federally sponsored Colorado River Commission (with Secretary of

Commerce Herbert Hoover serving as chairman) representatives from all seven southwestern states came together to forge a fractious document known as the Colorado River Compact:

> The upper states devoutly desired the water rights they would gain through the compact, rights which would be lost if allowed in the lower basin without it. The compact meant that the faster-developing lower states would have a limitation placed on their right to appropriate water.. . .. The upper states feared that their future growth, dependent upon water supply, would be forever stunted unless they restrained the ability of the lower states to use more water. A dam on the Colorado such as the Swing-Johnson bill called for could store the entire annual flow and regulate its release. . . . Fast-growing California was ready to put this water to beneficial use, thereby gaining a legal right to it. The upper states could not permit a dam without a compact which would assure them of their share of the river at some future date when they could utilize it.

Swing and other Californians appreciated that they needed the political support of most (although not necessarily all) western congressmen if they ever hoped to get federal authorization and financing for the dam and the All-American Canal. And to get this support they were willing to accommodate the Upper Basin states by agreeing to limit the amount of water that the Lower Basin states could legally claim.[45]

In the Fall of 1922, the seven states of the Colorado River Basin met near Santa Fe, New Mexico, for an extended conference in which they hammered out an agreement governing future water rights allocation.[46] The resulting Colorado River Compact divided the river into an Upper Basin and a Lower Basin. The annual flow of the river (generously estimated at 18 million acre-feet per year with two million reserved for delivery to Mexico) was to be divided equally between the two basins. In addition, California agreed to limit its consumption to 4.6 million acre-feet per year. The Upper Basin states were supportive of the compact's terms because they would be guaranteed future use of seven and one half million acre-feet per year without fear that California could preempt their claims. In the Lower Basin, Nevada was satisfied because the eventual construction of a dam at Boulder Canyon would foster general economic development in the state's southern region; California was generally pleased because, despite giving up potential rights to unclaimed water in the lower Colorado River, they garnered political support from Upper Basin states that would prove invaluable in the battle for congressional approval for the Boulder Canyon Project.

In contrast, Arizona remained bitterly opposed to the Compact and refused to ratify it.

Arizona's opposition stemmed from the fact that the Lower Basin, as a whole, and not individual states, was guaranteed rights to half of the river's flow. Because California's efforts to utilize the river were much more advanced than Arizona's, it was possible that California could monopolize control of the Lower Basin's allotment with Arizona permanently deprived of use of the river. The legal battle between Arizona and California over water rights came to the forefront with the Colorado River Compact and remained intense until a U.S. Supreme Court ruling in the early 1960s finally brought it to an end. In the short term, Arizona's opposition to the Compact was obviated by having the other states agree that it would be enforceable if ratified by six of the states in the basin. Although not ideal from California's perspective, this proved useful in furthering political support for a dam at Boulder Canyon.

After Arizona balked at approving the Colorado River Compact as drafted in November 1922, the states that did ratify the original compact were not legally bound to accept the revised Compact that called for the approval of only six states. Thus, when it came time to ratify the "six-state" Compact, the California state legislature voted that California's ratification would become effective only upon the passage of federal legislation that:

> authorized and directed the construction by the United States of a dam in the main stream of the Colorado River, at or below Boulder Dam, adequate to create a storage reservoir of a capacity of not less than twenty million acre-feet of water.[47]

Thus, the final and official ratification of the Colorado River Compact did not occur until March 1929, after federal legislation for construction of Hoover Dam had been approved by Congress and the President. Just as authorization of the Boulder Canyon Project had depended upon a political resolution of the Colorado River water rights issue, so too did official implementation of the Compact depend on approval of a federally sponsored storage dam at or below Boulder Canyon. The two were born out of the same desires of California political and business interests to gain access to the waters of the Colorado River.

CALIFORNIA AND POWER

Municipal Demand: Los Angeles and the Metropolitan Water District of Southern California

The origins of a large-scale storage dam on the lower Colorado are tied undisputedly to issues involving irrigation in—and flood control for—the Imperial Valley. But the political and economic leverage that Imperial Valley interests could bring to Congress in lobbying for the Boulder Canyon Project remained relatively limited so long as the initiative focused primarily on agricultural production. Since 1902, the track record of federally sponsored irrigation projects in meeting repayment schedules had proved dismal. This circumstance, more than any other, prompted Davis's dismissal as Director of the Reclamation Service in 1923 and brought about extensive political consideration of how the Service would carry on its future work.[48]

In official terms, Reclamation held to the ideals of promoting the family farm and of having Reclamation project beneficiaries pay back the federal government in a responsible, business-like manner. This was certainly the program that Elwood Mead promoted when he was appointed Commissioner of Reclamation in 1924.[49] But reconciling such a program with a huge undertaking designed to promote the interests of large-scale farmers in the Imperial Valley represented no simple task.

In this context, the notion that the Boulder Canyon Project would serve municipal water supply needs in greater Los Angeles also did not square easily with the small-farm agricultural orientation of Reclamation that Mead championed. Nonetheless, the alliance between Imperial Valley farmers and Los Angeles boosters quickly became central to the effort to win approval for the Boulder Canyon Project. It is impossible to know what would have transpired if Los Angeles had not come to embrace Hoover Dam as vital to its continued regional growth. But, as structured in the federal legislation actually passed in 1928, it is equally impossible to imagine how the fourth Swing-Johnson Bill could have become law without a strong financial commitment from the urbanized taxpayers (and voters) of greater Los Angeles.

The story of how late nineteenth and early twentieth-century Los Angeles grew as a result of its ability to control a regional and extra-regional water supply is well known. In the nineteenth century, city officials focused on gaining complete control over the Los Angeles River, based upon the idea that the city possessed a "pueblo right" to the entire stream as granted by the Spanish monarchy.[50] Working from the economic and political base afforded by control of the Los Angeles River, in the early twentieth century,

William Mulholland supervised construction of a remarkable, and controversial, 200-mile-long aqueduct to carry water from the Owens River to consumers in the City of Los Angeles.[51] Completed in 1913, the Los Angeles aqueduct from Owens Valley proved vital in supporting a huge regional economic boom that engulfed greater Los Angles during the teens and early 1920s. Specifically, the population of the region jumped from 668,000 to 1,085,000 between 1910 and 1920, and there existed every expectation and hope among Los Angeles boosters, businessmen, and the general population as a whole that this growth would continue for the foreseeable future. In fact, the population of the region jumped to 2,491,000 by 1930.[52] Such hopes, and expectations, of course, depended upon a reliable water supply.

In the early 1920s, the Owens River had not been tapped to its capacity by the City of Los Angeles, but–even as some farmers from the valley launched a dramatic (yet ultimately futile) bombing campaign to disrupt operation of the aqueduct–it was understood that the river could only partially meet the future needs of the region. Thus, with an electorate now attuned to the economic benefits that could accrue from massive public investments in large-scale water supply systems, city officials focused on potential new sources of water. The relatively small rivers within easy reach of Los Angeles offered little possibility of development because of existing claims to their stream flow. But the seemingly remote Colorado River presented a very different set of issues in terms of how it might serve as a source of supply.

With the Colorado, the issue was not so much the availability of water– in the early 1920s, most of the river's flow remained unused–but rather the difficulties of financing, building, and operating a reliable aqueduct across more than 200 miles of rugged, imposing escarpment encompassing the expansive Mohave Desert. With confidence that the physical and technological challenges of a Colorado River-to-Los Angeles aqueduct could be met, in June 1924, Los Angeles formally filed a claim to 1500 cubic feet per second of the Colorado's flow (amounting to about 550,000 acre-feet per year).[51] While significantly less than the 3 million+ acre-feet per year diverted into the Imperial Valley, this claim, nonetheless, represented an enormously significant event in the history of the Southwest. Los Angeles had announced its intention to utilize the Colorado for municipal development and, based on its track record in taking control of the Owens River, there existed little reason to think that the city would not act successfully upon this intention. In acknowledging the priority of claims made by the farmers in the Palo Verde and Imperial Valleys, Los Angeles did not threaten the legal status of water rights claimed by its California brethren (in contrast to the future claims of other states in the Colorado basin). As a result, the agricultural interests that Congressman Swing had originally perceived as the beneficiaries of the Swing-Johnson Bill could now look to urban areas along the Pacific Coast as partners in promoting the Boulder Canyon Project.

After Los Angeles formally proposed drawing water from the Colorado, lobbying for the project quickly expanded as urban boosters formed the Colorado River Aqueduct Association to complement and assist the Boulder Dam Association that had already been formed at the behest of Swing and the Imperial Irrigation District. Significantly, Swing himself addressed the initial organizational meeting of the Colorado River Aqueduct Association in Pasadena in September 1924.[54] Drawing upon a sophisticated sense of public relations, during the next five years the two associations churned out a succession of pamphlets designed to raise awareness of and appreciation for the importance of Hoover Dam. With titles such as "The Story of a Great Government Project for the Conquest of the Colorado River" and "The Federal Government's Colorado River Project," these promotional publications addressed both a national audience as well as Southern Californians who were, quite unabashedly, urged to contact their friends and relatives outside California to support federal approval of Hoover Dam.[55] One of the most overt of these promotional pamphlets (which appeared near the end of the approval process in late 1928) was actually printed by the Los Angeles Department of Water and Power and counseled:

> Write or telegraph today to your friends, relatives and former business associates in other parts of the United States. Give them the facts about Boulder Dam. Urge them, in turn, to communicate immediately with their Senators and Congressmen to the end that these members of Congress actively may support the pending Boulder Dam Bill. Help secure Boulder Dam legislation at this session of Congress![56]

In filing claims to the Colorado River, the City of Los Angeles served as the catalyst for what became known as the Colorado River Aqueduct. But the scale of the project represented something more than the city could (or wished to) develop on its own. At the September 1924 meeting of the Colorado River Aqueduct Association, city officials supported the formation of a region-wide committee to explore how the proposed aqueduct could benefit—and draw support from—communities lying outside the city's municipal boundaries. The formation of a powerful regional authority existing above the level of cities and municipalities but below the level of state government would have to be approved by the California State Legislature (as well as pass constitutional muster by the California Supreme Court) but, nonetheless, the Colorado River Aqueduct Association soon focused on the need for forming such an authority to finance, construct and operate the proposed aqueduct.

In early 1925, the state legislature considered legislation to authorize the formation of metropolitan water districts. The legislation passed the state senate but failed to win approval in the state assembly that spring.[57] In the wake of difficulties winning state approval for a regional metropolitan water district, Los Angeles voters plowed ahead in June 1925 and approved a $2 million bond issue to fund surveys, engineering, and other preliminary work on planning for the aqueduct. As Los Angeles Mayor George Cryer indicated in a speech before the Boulder Dam Association that same month:

> No circumstance has been too insignificant to be seized upon by them [opponents of the Boulder Canyon Project] and exaggerated into evidence of the desire of California, and particularly Los Angeles, to gain an undue advantage from the development of the [Colorado] river and ruthlessly to disregard the rights of less powerful communities and sections. As mayor of the City of Los Angeles, I hope I may, here and for all time, allay the distrust of the motives, purposes and objects of the great city for which I speak, and of the rich and populous area in Southern California having a common interest with Los Angeles. . . . The small city must be given an equal opportunity with the larger city to secure and enjoy the power benefits of the development. In bringing to the coast an additional water supply of domestic water, imperative to the growth of this section, all cities desiring to participate in the cost of the necessary works and the benefits to be derived therefrom must be given full and fair opportunity so to do—and this without any coerced annexation to or consolidation with Los Angeles.[58]

With this type of public reassurance, "small cities" such as Pasadena, Anaheim, Long Beach, Burbank, Glendale, and Santa Monica could take some comfort that the new aqueduct was not being promoted simply as a way for Los Angeles to extend formal control over them. In promoting an intermunicipal agency to administer the proposed aqueduct, Mayor Cryer further indicated that:

> Los Angeles desires partners in the benefits of the waters of the Colorado River, but it wants no unwilling partners. Neither does it wish to be an unwelcomed partner. Its great desire, both in respect to domestic water and in respect to power is to work in full harmony with its sister cities. . . .[59]

With this vision of cooperation in mind, the legislation that eventually fostered organization of the Metropolitan Water District of Southern California (MWDSC) allowed Los Angeles the right to appoint half of the board members charged with controlling the MWDSC; the remainder would be split between the other cities and water districts opting to join the district in accordance with the assessed value of land lying within their boundaries. Los Angeles could not outright dictate the actions of the MWDSC, but—given that it would generate the most financial support for the project based upon assessed land values—little could be done without its approval.

In early 1927, legislation was again proposed in Sacramento authorizing the formation of metropolitan water districts in California, and this time approval came over only minimal opposition (the State senate's approval was unanimous while the assembly passed it on a vote of 63 to 2).[60] To understand the importance of the legislation (and the consequent formation of the MWDSC) to Hoover Dam, it is necessary to appreciate how important it was for proponents of the Boulder Canyon Project to be able to assure skeptics that the federal government would actually be paid back the money used to build the Hoover Dam. In this regard, the Colorado River Aqueduct was not to be constructed using any federal funds; on the contrary, it was to be financed through bonds that would be guaranteed by the MWDSC's ability to levy real estate taxes against all property within the district. But even more importantly, the MWDSC was to be the most important customer for the power generated at Hoover Dam. As stated in the MWDSC's first annual report:

> It was early recognized that to secure favorable consideration [by the U.S. Congress], the [Boulder Canyon] project must be self-supporting and that the power to be generated from any development which was built must find a market which would eventually return all costs of the entire project to the Government. As additional engineering work for a Colorado River Aqueduct was done it became evident that any practicable diversion of the river must be made at an elevation lower than that of much of the area to be served, and would involve pumping. Such pumping was practicable only if a large amount of power could be obtained at a low price. This created, at once, a potential market for a substantial part of the power from any major Colorado River development. When these facts, as well as the need for an additional domestic water supply in Southern California were laid before Congress support for the Swing-Johnson measure became easier to obtain.[61]

In other words, the need to draw huge amounts of electric power to facilitate operation of the Colorado River Aqueduct provided a means of assuring hesitant Congressmen that Hoover Dam would not become some kind of white elephant, generating huge quantities of power that no one would want—or pay for. The MWDSC would be in a position to sign contracts guaranteeing power sales and, in turn, the federal government (and federal taxpayers) could rest assured that such contracts would be honored because California's Metropolitan Water District act granted to the MWDSC the right to directly tax land within its service area.

To make sure that nothing could impede the enforcement of this taxing authority—and hence the political foundation of Hoover Dam's financing—the legality of the Metropolitan Water District act was brought into question before the California Supreme Court in early 1928, and in August of that year, it was ruled constitutionally valid. Formal organization of the MWDSC quickly followed as the electorates of Los Angeles, Pasadena, Burbank, Anaheim, and several other cities in Southern California voted to include themselves in the district on November 6, 1928.[62] Thus, as consideration of the Boulder Canyon Project by Congress entered its final stage in December 1928, any questions as to the feasibility of power sales from Hoover Dam could be addressed and countered with great confidence. The Boulder Canyon Project was poised for final approval. All that remained was resolution of how public and private power interests would share control over the proposed dam's generating capacity.

The Politics of Hydroelectric Power and Approval of the Boulder Canyon Project

Of Hoover Dam's many "purposes," the most important in terms of both financing the project and heralding a new role for the federal government in the West involved the generation of hydroelectric power. But because of its importance, hydroelectric power was also a highly controversial aspect of the project. In the political context of the 1920s, when President Calvin Coolidge could arouse favorable support by averring "after all, the chief business of the American people is business" it was not at all apparent to a large constituency within the Republican Party that the federal government should take any active role in the generation or distribution of electric power.[61]

After the New Deal, it was easy to perceive the federal government as a primary—perhaps even central—participant in the West's electric power grid, but this was hardly the case at the start of the century. Prior to the 1920s, Reclamation had built a few hydroelectric powerplants with an aggregate generating capacity of less than 50,000 horsepower. In contrast, by the early 1930s, privately-financed electric power companies in the West owned and controlled plants with a combined generating capacity of more than 3.5 million horsepower.[63] Thus, a federally financed Hoover Dam, with an ultimate generating capacity of more than 1 million horsepower, represented something very different from anything previously proposed for the West.

165

Figure 4-10: Cover of pamphlet encouraging citizen support of Hoover (Boulder) Dam, 1928. Source: Boulder Dam Association pamphlet, item in the private collection of Donald C. Jackson.

The distinctive—if not exactly revolutionary—nature of what Reclamation proposed in terms of developing the hydropower potential at Black Canyon becomes apparent in its own description of the project as published in the 1927 edition of Edward Wegmann's *Design and Construction of Dams*. In describing the "primary objects" of the dam as (1) "permit[ting] the use of the normal flow of the Colorado River in the Upper Colorado Basin, without injury to the prior rights below the reservoir," (2) "[supporting] irrigation and domestic [use]," and (3) "provid[ing] flood protection," Reclamation very much downplays the importance of hydroelectric power. In fact, the proposed dam site was simply described as "presenting attractive possibilities of power development, [but these are] incidental to the use of the water for the primary objects of the reservoir." Unquestionably, the economic viability of the proposed dam was inextricably linked to hydropower revenues, but this did not mean that such a linkage was always trumpeted as a "primary objective" of the dam; especially while Calvin Coolidge resided in the White House.[65]

While the opposition of the private electric power industry to the Boulder Canyon Project was quite real, it manifested itself in ways that advocates often found frustrating (if not difficult) to counter. For example, Phil Swing believed that the reluctance of Utah Senator Reed Smoot to support the project derived from Smoot's close relationship with the Utah Power and Light Company, which was in turn controlled by the New York-based holding company Electric Bond and Share; but despite the logic underlying Swing's suspicions they were difficult to prove.[66] In opposing the power generating aspects of the Hoover Dam project, proponents of private power generally conceded that flood control constituted a useful and desirable objective. As such, they championed the construction of a dam about 300-feet high at a site near Topock, California, and hence referred to as Topock Dam, that would be big enough to hold back the river's annual floods but not big enough to foster large-scale, year-round electric power production.[67]

Like the Boulder Dam Association, private power interests sponsored the publication of pamphlets supporting their point of view. However, these often appeared without any direct sponsorship tying them to private industry. For example a handsome, 103-page booklet published under the seemingly benign title "Boulder Dam: Complete Bibliography, References, Engineers' Charts, Studies and Reports, the Swing-Johnson Bill, Minority Reports and General Comments" offers no information indicating who published it. But a reading of the text reveals a decided aversion to the power generating aspects of the project. For example, the booklet reprints an article entitled "Danger in the Boulder Canyon Project" written by Phillip Cabot, Lecturer on Public Utility Administration at the Harvard Business School. In this, Cabot expresses arguments that would have pleased private power executives:

> Considering the nature of the interests involved, the Colorado River flood control project is probably a legitimate enterprise on which the United States may embark. . . . But unfortunately the proposed development does not stop there. The scope of the enterprise has been greatly enlarged (if not inflated), so as to include the generation of electric power on a very large scale. . . . Clearly the production of electric power at this point is a highly speculative enterprise, and there is grave doubt whether the power generated in a desert hundreds of miles from any available markets can be sold at prices which will pay an adequate return upon its cost. . . . But whether the enterprise will pay or

ONE WAY TO HELP

With Congress once more in session, Southern California is presenting a solid front in the vitally important task of securing adoption of the Boulder Dam project Bill. The Bill already has been adopted by the House of Representatives. As Congress convened on December 3, the Bill was pending before the Senate as the first order of business.

Your help is urgently needed in securing the adoption of this Bill. The Los Angeles Chamber of Commerce has pointed out one way in which you may render aid. Here it is:

Write or telegraph today to your friends, relatives and former business associates in other parts of the United States. Give them the facts about Boulder Dam. Urge them, in turn, to communicate immediately with their Senators and Congressmen to the end that these members of Congress actively may support the pending Boulder Dam Bill.

Help secure Boulder Dam legislation at this session of Congress!

Figure 4-11: Los Angeles city officials distributed fliers in 1928 to encourage citizen support. Source: Boulder Dam Association, item in the private collection of Donald C. Jackson.

not is really beside the point, because there is a far more important objection to the development of electric power by the federal government as part of this flood control project.[68]

Cabot goes even further by characterizing California's advocacy of "Boulder Dam" as a bold assault on the federal treasury in order to benefit the state at the expense of the nation:

> It is easy to see why the Senators and Representatives from California are so keenly in favor of the Boulder Canyon Project as now proposed. They want cheap power in California.... But it is hard to believe that the Senators from the other parts of the country will develop equal enthusiasm. From the point of view of the nation as a whole there is no convincing argument in favor of the additional expenditure and additional risk involved in adding a 'tail' of electric power to the dog of flood control and reclamation. The result will probably be to burden the nation for the benefit of California, which is unwise, if not illegal.... It is disheartening to find some of our ablest Government leaders favoring such a scheme.[69]

Despite the efforts of private power advocates to convince Americans of the importance of keeping electric power production out of government control, the general public grew wary of what was frequently characterized as the "Power Trust." In the spring of 1928, this wariness blossomed into widespread skepticism following the release of a Federal Trade Commission study documenting

FACTS TO REMEMBER

Here are a few facts about Boulder Dam that may be of interest to your friends in other parts of the United States.

1. Construction of Boulder Dam is provided for in the Swing-Johnson Bill, adopted at the last session of Congress by the House of Representatives, and now pending before the Senate.

2. Boulder Dam will not cost the taxpayers of the Nation one cent. The Bill now pending before the Senate provides that it must be financed entirely through the sale of the vast quantities of hydro-electric power to be generated at the dam site.

3. Boulder Dam will control the flood waters of the Colorado River and forever protect Imperial Valley and other sections of California and Arizona against the growing Colorado flood menace.

4. Boulder Dam will create a great reservoir for the storage of flood waters now wasted into the sea.

5. A portion of the water stored in the Boulder Dam reservoir will be available for Los Angeles and other Southern California cities which urgently need additional domestic water. These cities stand ready to pay for this water when it is made available by the dam.

6. The high dam provided for in the Swing-Johnson Bill will be the site for the generation of 1,000,000 horsepower of hydro-electric energy to be used by factories, mines, ranches and homes in the Southwest.

7. A special engineering commission appointed by the Secretary of the Interior and headed by General William L. Sibert recently reported to President Coolidge that the Boulder dam project is entirely feasible and practicable from an engineering standpoint.

Figure 4-12: "Talking points" provided in 1928 to help citizens join the fight for approval of Hoover Dam. Source: Boulder Dam Association, item in the private collection of Donald C. Jackson.

how the private power industry worked to influence (if not outright manipulate) public opinion regarding the value of publicly-owned and administered power systems.[70] Evidence was brought to bear on how $400,000 was raised and allocated by the National Electric Light Association specifically to fight Hoover Dam and how politicians, academics, and other prominent figures were engaged to oppose the dam.[71]

Swing and Johnson were able to take these revelations and, during the final months of battle over the Boulder Canyon Project Act, use them to good advantage. Arguments denigrating the power generating aspects of the project were now much easier to characterize as propaganda from the "Power Trust" and the idea of building a dam only big enough for flood control faded from serious consideration. In December 1928, during the final "lame duck" session of the 70th Congress and after the Metropolitan Water District of Southern California had been formally organized to build the Colorado River Aqueduct, the fourth (and final) Swing-Johnson Bill was passed by Congress and signed by President Coolidge. In this Act, power generation remained as a keystone of the project, but some significant concessions were made to private industry.

First, while the power generating plant at Hoover Dam would be built and owned by the federal government, actual operation of the plant would be leased to non-federal organizations. Second, private power companies (most prominently Southern California Edison Company) would be allowed to compete in bidding for the purchase of power from the dam. The precise allocation of power privileges was not determined until 1930, when Secretary of the Interior Ray Lyman Wilbur finally authorized leases that governed use of Hoover Dam power for 50 years after the plant came on-line. As stipulated by Wilbur, over 64 percent of the dam's power was reserved for use in Southern California, 36 percent went to the Metropolitan Water District of Southern California to pump water through the Colorado River Aqueduct, a little more than 9 percent to the Southern California Edison Company (and other private power companies), and about 18 percent to the City of Los Angeles and other municipally-owned utilities in Southern California. In contrast, both Arizona and Nevada were allotted 18 percent of the dam's power, although it would be many years before these states were able to develop markets large enough to utilize their full allotments. With power allocations and contracts in place, financing of the Boulder Canyon Project became ensured, and in 1931 the focus shifted to actual construction.

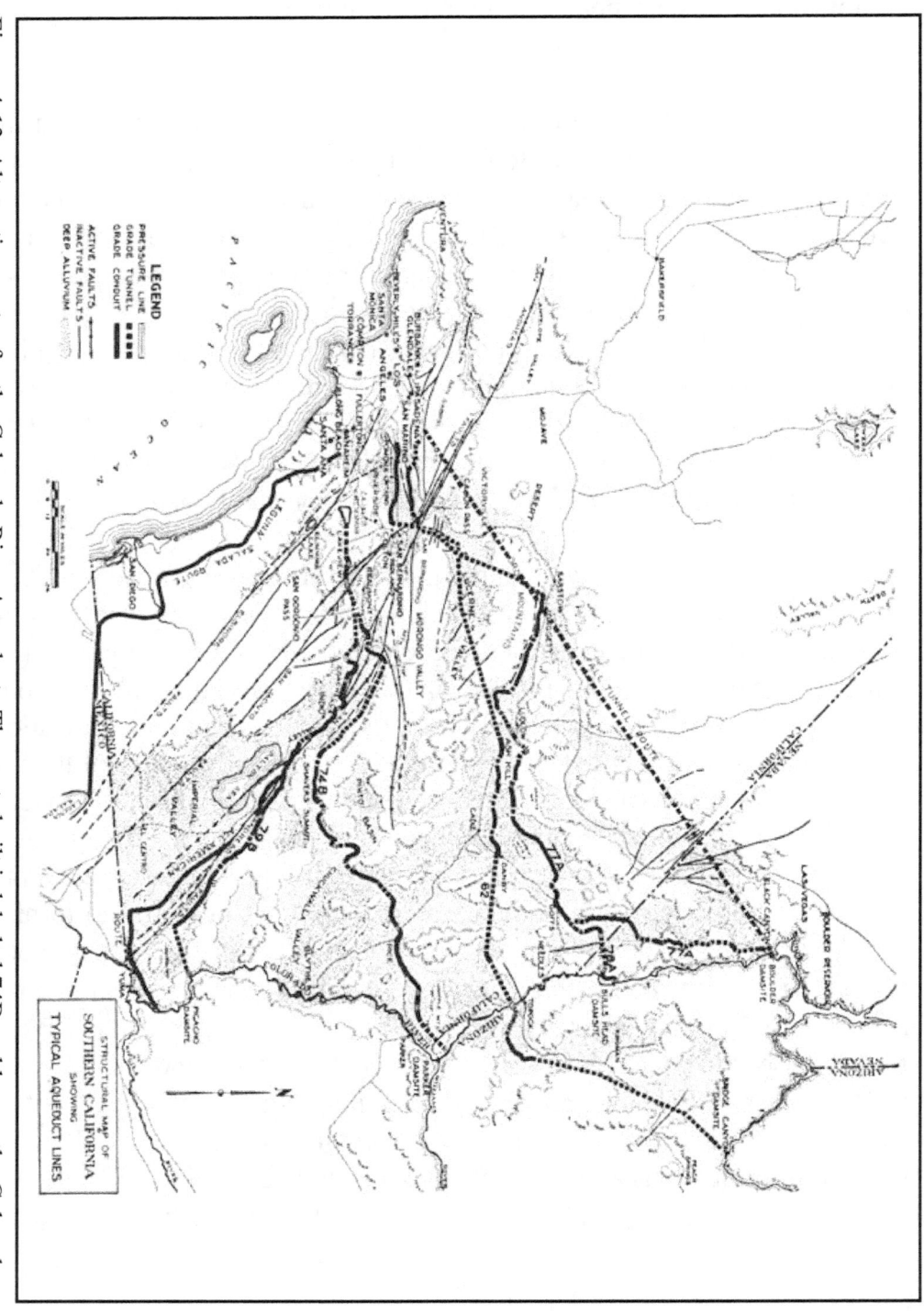

Figure 4-13. Alternative routes for the Colorado River Aqueduct. The route built is labeled 74B and leaves the Colorado River from Lake Havasu above the Parker damsite. Source: Charles A. Bissell, compiler and editor, *The Metropolitan Water District of Southern California: History and First Annual Report* (Los Angeles, 1939), 68.

HOOVER DAM

Construction of the Dam

Budgeted at $177 million, the Boulder Canyon Project would finance construction of a massive concrete storage dam and excavation of the All-American Canal feeding into the Imperial Valley. The Government billed it as a multipurpose project because it would foster irrigation in the Imperial Valley, supply municipal water for greater Los Angeles via the Metropolitan Water District's Colorado River Aqueduct, generate hydroelectric power for the Southwest, and provide flood control for the Lower Colorado Basin. But for all these goals to be met, it was first necessary to actually build a 726-foot-high barrier capable of impounding almost two year's average flow of the Colorado river. Throughout the 1920s, Bureau of Reclamation engineers planned how this might be accomplished. In 1931, the actual work commenced of building what now–thanks to the September 1930 pronouncement of Interior Secretary Ray Lyman Wilbur–was called Hoover Dam.

While work on the dam was to be closely supervised by the Bureau of Reclamation, long-time Reclamation engineer Walker Young took charge at the site, responsibility for construction was to be taken by a private contractor who had won the job by submitting a low bid in competition with other prospective bidders. The largest single federal contract ever let out for bids until that time, the building of Hoover Dam attracted the attention of some of America's largest civil engineering contractors when it was first advertised in January 1931. In a story that is well told by Joseph Stevens in his book *Building Hoover Dam: An American Adventure*, the scale of the project proved so great that several prominent entrepreneurs and companies—including the Utah Construction Company, Morrison-Knudsen, the Bechtel Company, and Henry Kaiser—pooled their talents into a joint corporate initiative called Six Companies Inc. and submitted a winning bid slightly less than $49 million that won them the right to build the dam.[72]

Although many people figured into Six Companies' successful effort in building Hoover Dam, in terms of the day-to-day work that actually transpired in the forbidding environs of Black Canyon, none was more important than Frank Crowe. For twenty years after joining the Reclamation Service as a young engineer in 1905, Crowe had worked on a variety of Reclamation Service projects, including construction of the concrete curved gravity Arrowrock Dam in Idaho. In 1925, he left government service and joined the dam-building outfit of Morrison-Knudsen, based in Boise, Idaho. Drawing upon his experiences in both government and private engineering work, Crowe took charge of the

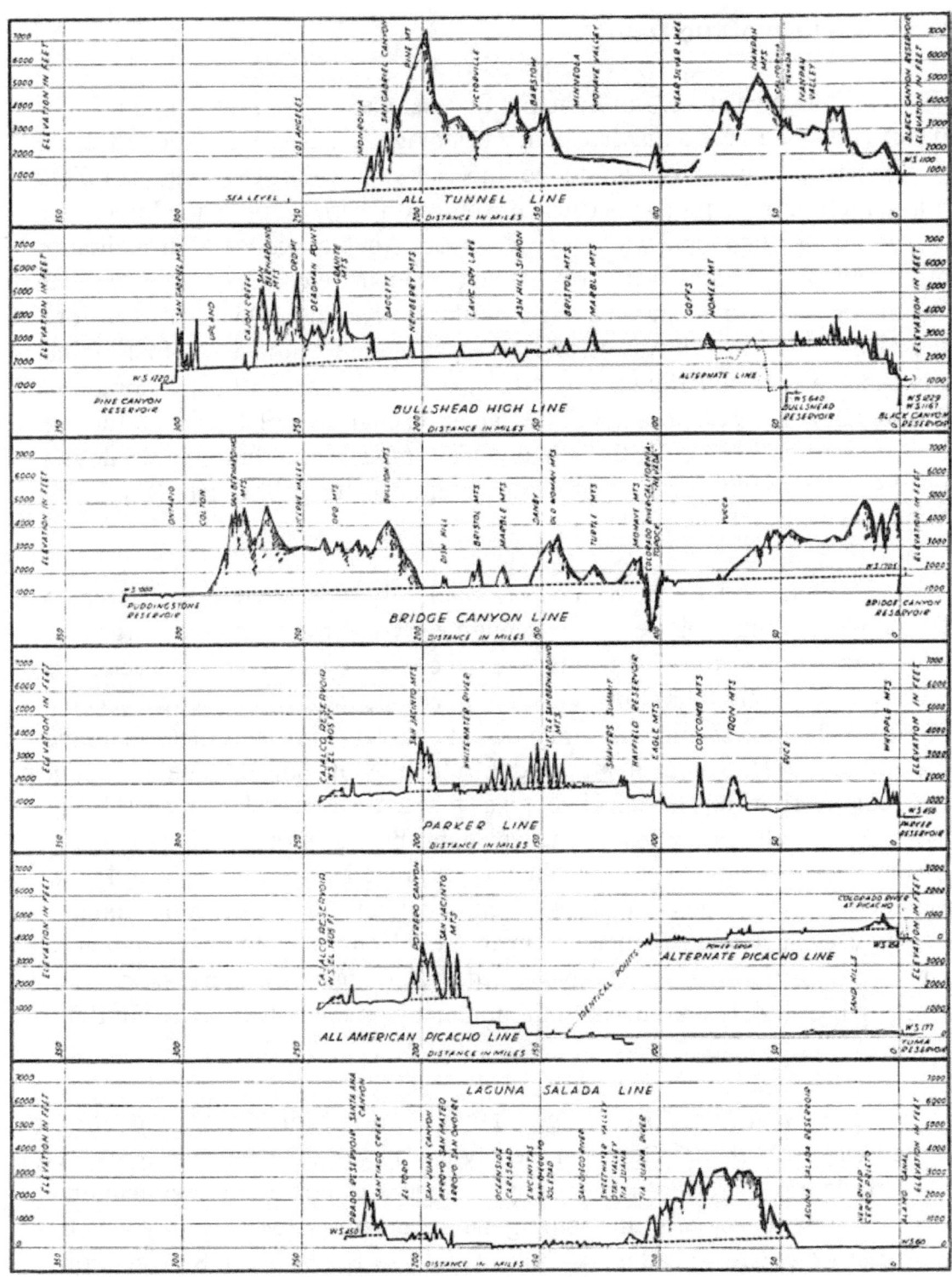

Figure 4-14. Profiles of preliminary Colorado River Aqueduct routes. Note the varying lengths studied. Source: Bissell, *The Metropolitan Water District of Southern California*, 81.

complicated task of coordinating work in Black Canyon so that Six Companies could meet its contractual obligation of completing the dam within seven years.[73]

In basic terms, building the dam relied on a few components that were key to its systematic and timely completion. Of course, construction of a rail and road transportation system connecting Black Canyon to the Union Pacific Railroad and to the outside world in general (via Las Vegas) was vital. And electric power needed to be brought in over a 220-mile transmission line extending out of San Bernardino, California. But for Crowe, the most critically important aspect of the project concerned the driving of four 56-foot diameter diversion tunnels (they would be reduced to 50-foot diameter after being lined with concrete) that would eventually carry the full flow of the Colorado River around the dam site and allow the foundations of the dam to be excavated down to solid bedrock. From a construction point-of-view, nothing of substance in terms of pouring concrete in the dam or powerhouses could be accomplished prior to diversion of the river through these lengthy tunnels (in aggregate they stretched for more than three miles through the rock abutments of the canyon). And in a financial context, Six Companies' contract stipulated that the full diversion would be accomplished by October 1, 1933. The company would incur a $3,000 per day fine for every day it was late in meeting this deadline.[74]

Thus, for both financial and practical reasons, the driving of the diversion tunnels became of paramount importance to the company and lent an air of urgency to the start of construction in the spring of 1931. This air of urgency was further exacerbated by the thousands of potential workers who migrated toward Las Vegas because they perceived the

Figure 4-15: Downstream view of Hoover Dam on the Colorado River. Source: Bureau of Reclamation.

project as a source of desperately needed jobs.[75] Although many arguments were made by Southern Californians during the 1920s as to why the Boulder Canyon Project should be built, the role of the project in providing employment in the midst of hard economic times was not one of them. But in the period following the dam's approval in December 1928, the nation's economy lay shattered in the wake of the stock market crash of October 1929. By the time actual construction commenced in Black Canyon, the project had spawned a new objective focused around what could be termed "work relief." In this context, the dam assumed a new role in the national consciousness, one that was not so much tied to the economic growth of Southern California, as it was to a larger national purpose focused around overcoming adversity in difficult times—and what could be more "American" than that?

Although the Hoover Administration remained generally opposed to federally financed relief projects designed to give an artificial boost to the economy, it was quite willing to promote and encourage the job opportunities that attended construction of Hoover Dam.[76] Thus, for a range of reasons, Reclamation, Six Companies, and Crowe were anxious to get work underway at Black Canyon as quickly and as fully as possible in the spring and summer of 1931. Unfortunately for prospective laborers and their families, housing conditions near the site were abysmal at this stage of the project (construction of accommodations at the federally controlled Boulder City would not be complete for several more months), and as the heat of the summer descended on the site, fourteen workers died from heat prostration. Working conditions in the diversion tunnels proved especially dangerous, but this did little to deter Crowe from pushing ahead as hard as possible on this critical phase of the project.[757]

As a result of tremendous physical hardship and apparent wage cuts invoked by Six Companies during the summer of 1931, labor unrest grew. In early August, discontent became manifest in a project-wide strike encouraged by radical labor leaders from the Industrial Workers of the World (the IWW or "Wobblies").[78] In fending off this strike, Six Companies and Reclamation shared a common interest in resisting any labor demands that threatened to impede completion of the dam. Elwood Mead succinctly expressed the federal government's position by characterizing the strikers as "impossible" and averring that "the present wage rate on Hoover Dam is considerably above that of the surrounding region."[79] Thus, once the strike was broken in mid-August by the importation of workers from Las Vegas willing to abide by the rules and wages established by Six Companies, particularly troublesome strikers found little support from Reclamation or other federal officials in their efforts to regain employment.

In even more substantive ways, Reclamation supported Six Companies in terms of working conditions in the tunnels and demonstrated how federal administration of the project proved advantageous to the contractor. Specifically, this involved Six Companies' reliance upon gasoline-powered trucks and internal combustion engines within the diversion tunnels in order to facilitate the rapid removal of rock that had been blasted from the tunnel facings. Of course, operation of internal combustion engines in poorly ventilated, confined spaces offers ideal conditions for carbon monoxide poisoning. For this reason, Nevada mining law specifically forbade use of such equipment underground. During the summer of 1931, Nevada officials took legal action to prevent Crowe from using internal combustion engines inside the diversion tunnels, something he strongly resisted because his plan for meeting the river diversion deadline required completing the tunnels as rapidly as possible; in turn, this depended upon large trucks that would drive directly into the tunnels and carry out debris on a "round the clock" basis—during January 1932 this reached a peak when as much as 16,000 cubic yards of rock was hauled away every day.[80]

Reclamation sided with Six Companies and argued that, because the dam was being built by the federal government on land that had been designated a "federal reservation," state law held no power over possible construction methods. After Nevada's state inspector of mines brought legal action to enforce state law and provide safe working conditions at the dam site, Six Companies and Reclamation obtained a restraining order allowing them to proceed until a panel of federal judges in San Francisco ruled upon the merits of the case. By the time the ruling came in April 1932, much of the tunnel excavation work had been finished; regardless, the court upheld Six Companies' and Reclamation's right to abrogate state law on this issue.[81] Later state court fights focused on civil law suits that sought to win financial judgments against Six Companies for workers who claimed to have been injured by the underground operation of internal combustion engines. A source of some embarrassment to Six Companies, these civil disputes dragged out until 1936 when they were ultimately resolved in out-of-court settlements for an undisclosed sum.[82] But in the larger scheme of things, the federal legal "umbrella" proved remarkably useful to Crowe and Six Companies in allowing them full control over the dam site without worrying about state regulations.

Work on the tunnels proceeded at a furious pace during 1931 and 1932.[83] Although some difficulties accompanied unforeseen floods in February 1932, the river ran relatively low during the spring and summer of that year and, with the coming of low water in the fall, conditions looked good for diversion of the Colorado River out of Black Canyon. Beginning in early November, Crowe's crews began dumping rock across the stream bed in order to erect a temporary cofferdam. As this mound of debris gradually rose upwards, it elevated the height of the river. On November 14, 1932, water began spilling into the diversion tunnels on the Arizona side of the construction site. With this, the mighty

Colorado River had been "tamed," and excavation of the site became possible. The building of the dam soon entered a new phase.

The diversion of the river allowed for site preparation to begin in earnest, but even before this occurred the abutments along the canyon walls had been subjected to an intensive effort to chip and drill away loose rock. Undertaken by daring "high scalers," this work was accomplished by men who rappelled down the slopes of the canyon walls carrying heavy jack hammers; their purpose was to remove any loose rock or potential debris that could impede a tight, solid connection between the dam's concrete and the rock abutments. Similarly, completion of the cofferdams (one on the upstream side of the site, one on the downstream side) allowed Crowe's men to remove all loose dirt, sediment, and rock lying in the riverbed that would prevent the concrete bottom of the dam from forming a tight, solid connection with the bedrock at the bottom of the dam. Excavation into the bottom of Black Canyon commenced in November 1932, and by early June 1933, Six Companies was ready to start the actual placement of concrete.[84]

Figure 4-16: Looking across Hoover Dam during construction from the Nevada side on June 2, 1934. Source: Bureau of Reclamation.

Figure 4-17: Hoover Dam and Lake Mead viewed from the Arizona side on April 20, 1954. Note, in the upper right corner, the Nevada side channel spillway which discharges into an inclined tunnel which connects to one of the four original tunnels built to divert water around the construction site. Source: Bureau of Reclamation.

Compared to the uncertainties that accompanied the efforts to drill the diversion tunnels, the pouring of concrete represented a much more predictable and controllable task—although it was, nonetheless, a complicated and potentially dangerous coordination of men and machines and huge batches of wet concrete. During the more than two years preceding the first placement of concrete on June 6, 1933, Crowe and Reclamation had overseen the erection of an elaborate concrete-making plant that could draw in materials (i.e., sand and gravel) from the local area and process them in a concrete mixing plant above the dam site on the Nevada side of the river. From here, large batches of freshly mixed concrete (carried by buckets with a capacity of eight cubic yards) could be delivered to various parts of the ever-rising dam via cableways strung across the width of the canyon.[83]

Recognizing that the dam could not be formed in one continuous pour, Reclamation's design called for the concrete to be placed in an assemblage of "blocks" that could be cast independently and allowed to harden before the pouring of adjoining blocks.[84] To allow the massive concrete structure to cool in a controlled (and relatively rapid) manner—something that was necessary because

177

of the heat released by the concrete as it hardened—an extensive system of one-inch diameter cooling pipes (measuring about 592 miles in total) was embedded in the dam. These pipes, which represented a technology that Reclamation had first experimented with during the construction of Owyhee Dam, in Oregon, a few years earlier, contained cooled water that served to draw off excess heat from the hardening concrete (resulting from what engineers term the "heat of hydration") and prevent the creation of potentially dangerous temperature cracks within the interior of the dam. Water passed through the cooling coil pipes at a rate of at least three gallons per minute and, at the completion of construction, Reclamation estimated that about 159 billion BTUs (British Thermal Units) of energy had been extracted out of the dam by this method.[85]

After the first few weeks of pouring concrete, the construction process became regularized and relatively routine (at least compared with the early stages of construction), although it never became simple. The growing skill of Crowe's workers and the efficiency of the concrete delivery system becomes apparent in reviewing the quantities of concrete placed on a monthly basis: In June 1933, 25,000 cubic yards of concrete were poured; two months later, in August, it reached 149,000 cubic yards; and in March 1934, it reached a peak of over 262,000 cubic yards—the equivalent of 1,100 buckets per day, or about one every 78 seconds. By December 1934, more than three million cubic yards of concrete had been poured into the dam; in early February 1935, the structure "topped out" with delivery of the last batch of concrete.[86] At this time, Reclamation dropped three of the massive bulkhead gates placed across the openings of the diversion tunnels, leaving the outer Nevada tunnel to release flows to meet the needs of downstream irrigators. Later, in 1935, the fourth gate fully blocked the Colorado River in its journey toward the Pacific. Gradually, the reservoir behind the dam began to rise and the hydraulic character of the once free-flowing stream experienced a dramatic transformation.

While work building the dam proper proceeded through 1934 and into 1935, other tasks such as building the powerhouses, the outlet towers, the spillways, and the tunnels that served them continued apace. A year later, in March 1936, the dam and powerhouses were officially declared to be complete and—as Six Companies officially left Black Canyon—Reclamation assumed responsibility for installing the initial set of hydroelectric turbine-generator units in the powerhouses.[87]

On October 7, 1936, water passed through one of the recently completed main turbine-generator units, and, two days later, hydroelectric power first surged out of Black Canyon into Southern California via a transmission line built by the City of Los Angeles. It would take until June 1937 before full-scale, continuous, commercial power transmission would occur, but, from that time on, the power generating capacity of the Boulder Canyon Project became a central feature of the Southwest's power grid. By 1940, ten distinct power lines emanated out of Black Canyon, including the Metropolitan Water District of Southern California's line to pump water through the Colorado River Aqueduct.[88] Although the installation of the 17 main turbine-generators (each with an initial generating capacity of 115,000 hp) in the dam's powerhouses extended over a period of years (the final unit would not come on line until 1961), the initial generation of power in 1936-37 was the beginning of the process whereby operation of the dam could (quite literally) generate the income that would pay for its construction.

The Legacy of Hoover Dam

Few actions better symbolize the political character and importance of the Boulder Canyon Project than Interior Secretary Ray Lyman Wilbur's decision in 1930 to change the name of Boulder Dam to Hoover Dam. President Herbert Hoover himself appears to have played no role in encouraging this name change, but he gladly accepted such a prominent and public association with the dam. In fact, perhaps the only other act that fully compares with Wilbur's move in symbolizing the political power projected by the dam's image involves the decision of President Franklin Roosevelt's Interior Secretary Harold Ickes to change the name back to Boulder Dam shortly after Hoover's departure from the White House.[89] Ickes perceived the dam as a major construct holding wide public appeal during the Depression, and, under his administrative eye, the Bureau of Reclamation continued to push hard to bring it to completion–during 1934, he made certain that $38 million from Public Works Administration funds were available to keep the project moving along as fast as possible.[90] At the same time, Ickes held no great political affection for Six Companies and—in contrast to Wilbur—was quite willing to confront the dam's main contractor over labor issues.[91]

The Roosevelt Administration's desire to affiliate itself with the dam became fully manifest in September 1935 when the President personally visited Black Canyon to dedicate the completed dam. In his remarks, Roosevelt celebrated the dam's "superlative" dimensions and avowed that "this morning I came, I saw, and I was conquered as everyone will be who sees for the first time this great feat of mankind." Paying homage to "the genius of their designers . . . the zeal of the builders . . . [and especially] the thousands of workers who gave brain and brawn to [the] work of construction" he characterized the dam as a "twentieth century marvel" and as "an engineering victory of the first order—another great achievement of American resourcefulness, skill and determination."[92] By portraying the dam in such broad terms, Roosevelt elevated the project above the political and economic aspirations that drove Southern Californians

179

to push for its authorization in the 1920s. In essence, Roosevelt raised the dam's importance to almost mythological proportions in portraying it as a great symbol of mankind's ability to tame nature through technology and human effort.

Although Roosevelt's speech in Black Canyon was hardly the first time that a politician had drawn upon a public works project to symbolize the advance of American civilization, the speech represented an important public event in which a large multipurpose dam was imbued with national values of seeming universal benefit. In such a context, the legacy of Hoover Dam continued to be felt for decades to come as multipurpose dams proliferated within river basins throughout America. Often viewed as a manifestation and extension of New Deal public works spending initiatives, the fact that the original, federally financed, multipurpose storage dam designed to be paid for by the commercial sale of hydroelectric power was not born out of the Great Depression (or out of efforts to alleviate the effects of the Depression) is something that should give pause to anyone wishing to paint large-scale federal dam-building in simple, broad-brush strokes.

In a broad context, Hoover Dam came to represent a major shift in three key aspects of western water policy: (1) municipal water supply now constituted a suitable purpose for Reclamation projects; (2) electric power generation was now considered acceptable as a major component of federally financed projects; and (3) because much of the privately held land in the Imperial Valley was in large tracts that stood in significant contrast to the "160-acre/small farm ideal" originally heralded in the 1902 Reclamation Act, implementation of the Boulder Canyon Project Act represented an expanded relationship between the federal government and private landowners in the West.

In a more focused context, the legacy of Hoover Dam was also expressed in how the Colorado River basin developed in the wake of the Boulder Canyon Project. First, construction of the river's first major storage dam as a federal project provided the rationale for preventing any private electric power company from ever erecting a hydroelectric powerplant along the mainstem of the river. Even before the completion of Hoover Dam, the Federal Power Commission had initiated wholesale rejection of any private power applications to develop the Colorado River on the grounds that they might interfere with interstate allocations of water under the terms of the Colorado River Compact (which, of course, did not become effective until authorization of the Boulder Canyon Project).[93]

Second, the Boulder Canyon Project marked a point at which individual states came to appreciate the implications of large, inter-basin transfers of water and the need to protect their interests through agreements (such as the Colorado

River Compact) or court action (such as Arizona's resistance to California's claims to the Colorado River). Looking more closely at the long-simmering dispute between Arizona and California, the completion of Hoover Dam in 1935 allowed Six Companies to immediately shift Frank Crowe and its attention to the construction of Parker Dam, and to commence work on a key component of the Metropolitan Water District of Southern California's Colorado River Aqueduct. Parker Dam would straddle the Colorado River about 150 miles below Hoover Dam. Significantly, this project–which was paid for by MWD–was officially built by Reclamation, a tactic made necessary by the fact that fully half of the dam was in the state of Arizona.

An important, yet oftentimes overlooked, legacy of the Boulder Canyon Project is that, in 1931, the U.S. Supreme Court ruled that Arizona could do nothing to prevent construction of Hoover Dam on constitutional grounds (derived essentially from the fact that the lower Colorado River was navigable and thus federal control over the dam could be justified in terms of the U.S. Constitution's interstate commerce clause).[94] Thus, despite Arizona's objections that Parker Dam was in truth simply a component of Southern California's water supply system, the fact that it was being officially built by Reclamation proved sufficient to counter such claims; as such, Arizona's attempt to launch an "Arizona Navy" to protect its side of the Parker Dam site foundered in the wake of the massive federal legal presence that had been established to control the Colorado River.[95]

Not surprisingly, construction of Parker Dam under the auspices of Reclamation did not bring an end to Arizona's struggle with California over the flow of the lower Colorado. Instead, it merely served to intensify the state's desire to push its legal claims in federal court. After almost three more decades of battle, the U.S. Supreme Court ruled in 1963 that California could be forced to limit its withdrawals from the Colorado in order to insure that Arizona would be able to take its fair share of the stream. On the surface it appeared as though this ruling did nothing to alter the basic allocations set forth in the Colorado River Compact. California was still assured of receiving 4.4 million acre-feet of Colorado flow per year; the difference came in California's ability to draw off half the "surplus flow" that might exceed the stipulated allocations.

In its 1963 ruling in *Arizona* v. *California*, the Supreme Court restrained California from using more than 4.4 million acre-feet per year and set the stage for Arizona to win congressional approval for a huge, federally financed aqueduct, known as the Central Arizona Project, to pump water out of the Colorado and deliver it to greater Phoenix and Tucson.[96] Without laboring upon the myriad details attending approval and construction of this project, what is important to stress is that resolution of the conflict between Arizona and California came within a federal forum and that federal financing of the Central Arizona Project was absolutely critical for it to be built. Whereas in the 1920s and 1930s the federal role in building Hoover and Parker Dams had been used to thwart Arizona efforts to block construction, by the 1960s, the federal government had become the means by which Arizona would implement its own technological system for tapping into the Colorado. The beneficiaries may have changed over the years, but the fact that the federal government had assumed legal prominence over the Colorado remained constant.

GLEN CANYON DAM

The Upper Basin

In the Upper Basin states, the Boulder Canyon Project represented the catalyst for the Colorado River Compact and—while it was recognized that construction of Hoover Dam would necessarily precede any work building large federal reservoirs in the Upper Colorado Basin—there developed a strong belief that federally financed dam projects would soon commence in the region upstream from Black Canyon. Public notice of such a program appeared in 1946 with the Bureau of Reclamation's publication of the large-format, 292-page *The Colorado River: A Natural Menace Becomes a National Resource* (it also carried the imposing subtitle, "A Comprehensive Departmental Report on the Development of the Water Resources of the Colorado River Basin for Review Prior to Submission to Congress").[98] In this report, scores of prospective projects were noted throughout the basin and justified in terms of the economic benefit that would accompany their completion. This plan represented the logical outgrowth of the water allocations stipulated in the Colorado River Compact. The time had come for the Upper Basin to get their share of federal support for developing the river basin and—under the name Colorado River Storage Project—this initiative received congressional approval in 1956. Like the Boulder Canyon Project, it was to be financed by revenue derived from the sale of hydroelectric power.[99]

As it turned out, the first major Reclamation project planned for the Upper Basin precipitated an environmental discussion on a scale comparable only to the dispute that raged over San Francisco's plans to build a dam in Yosemite National Park.[99] In the 1930s, Reclamation planners had targeted the large basin at the confluence of the Yampa and Green Rivers as an excellent site for a large storage dam in the Upper Colorado Basin. Located along the Utah/Colorado border about 50 miles south of Wyoming, the site of the proposed Echo Park Dam also happened to lie within the boundaries of Dinosaur National Monument (a part of the National Park System). In the late 1940s, Echo Park Dam was publicly proposed as one of the first components of the Colorado River Storage Project. Quickly, members of the Sierra Club, the Wilderness Society, and other advocates of the National Park Service rallied to protect Echo Park from inundation. They tenaciously lobbied Congress, and, by 1956, were successful in eliminating federal support for the dam. But in winning the environmental battle for Echo Park, they acquiesced in agreeing not to oppose another large dam that Reclamation planned to build in the Upper Colorado Basin.[100]

Figure 4-18: Glen Canyon Dam under construction on August 10, 1962. Source: Bureau of Reclamation

Of course, the Glen Canyon Dam site (located on the mainstem of the Colorado River only a few miles upstream from the spot marking the division between the Upper and Lower Basins) had long been familiar to Reclamation. In fact, it had figured as a possible alternative to Boulder/Black Canyon in the early 1920s. By the 1950s, Reclamation was eager to begin construction of a 700-foot high dam at Glen Canyon. Glen Canyon Dam would be another major step in the development of the Colorado as a source of hydroelectric power for the burgeoning Southwest. Whereas Echo Park lay within a part of the National Park System, and, thus, comprised a site well suited for wilderness advocates to defend, the Glen Canyon Dam and reservoir site simply encompassed federally owned land and thus was easier to justify in terms of inundating for the greater public good. Although the canyon lands upstream from Glen Canyon could certainly have been characterized as a natural (and national) treasure, they held no place in the national public consciousness and no great movement developed to protect them. Thus, when Congress agreed in 1956 to protect Echo Park, wilderness advocates offered little protest against approval of Glen Canyon Dam in what could later be understood as a de facto compromise regarding development of the two dam and reservoir sites.[101]

Design of Glen Canyon Dam and Beyond

In terms of design, the Glen Canyon Dam differed from Hoover in its use of an arch design featuring a profile insufficient to stand as a gravity dam. In this strictly technological context it diverged from the precedent set by the Boulder Canyon Project and, instead, drew from Reclamation's work in building thin arch dams that extended as far back as Pathfinder and Shoshone (Buffalo Bill) Dams prior to 1910.

The early version of the trial load analysis, improved by Noetzli in 1921 and further refined by Reclamation engineers in the late 1920s, laid a firm basis for its use on Glen Canyon Dam. This confidence helped lead design engineer Louis Puls to decide on a dam far thinner than Hoover and thus rely on arch action instead of only cantilever behavior. There were other reasons too. Concrete quality had improved since the 1920s, so the 415 psi stress limit at Hoover Dam could be increased to 1,000 psi for Glen Canyon.[102]

However, somewhat negating the advantages of improved concrete, the canyon walls at Glen Canyon were sandstone, a weaker material than the walls of Black Canyon. Therefore, the stress at the arches abutments was kept at 600 psi by thickening the arches as they approached the canyon walls. The weaker walls also required the injection of a grout curtain to strengthen the foundations and control seepage under and around the dam.

Glen Canyon Dam also differed from Hoover in that the storage capacity of the reservoir was not really necessary to provide for downstream irrigation and municipal use. But the two major dams shared a strong and common lineage in terms of electric power production. With a design capacity of more than a million horsepower, Glen Canyon Dam was intended first and foremost to be a "cash register" facilitating the sale of electric power to the greater Southwest,[103] and to deliver the Upper Basin's annual water commitment to the Lower Basin.

The approval of Glen Canyon Dam in the 1950s may have spurred little public protest, but by the time the huge concrete arch structure was completed in 1964, America's burgeoning community of wilderness advocates (a group often simply tagged as "environmentalists") had come to perceive the inundation of the canyon land above Glen Canyon as a terrible tragedy.[104] Subsequently, the outrage felt by environmentalists in the mid-1960s over Reclamation plans to build additional hydroelectric dams along the stretch of the Colorado River between Black Canyon and Glen Canyon—and flood portions of the lower Grand Canyon—proved of sufficient political potency that Congress refused to authorize their construction.[105] While other Reclamation dams were completed in the Upper Colorado Basin during the 1960s and early 1970s, the defeat of the so-called "Grand Canyon dams" heralded the beginning of a new era in western dam-building.[106]

Figure 4-19: Glen Canyon Dam looking upstream at the powerhouse on June 9, 1964. Source: Bureau of Reclamation.

In recent years, the place of Hoover Dam in the national consciousness has not escaped reassessment prompted by the environmental concerns over large-scale dams that flowered in the post-Glen Canyon era. To be sure, the image of Hoover Dam as a symbol of technological prowess and of the human spirit overcoming adversity still holds sway over many people. Not the least of these is Joseph Stevens, author of *Hoover Dam: An American Adventure*, who avows that "in the shadow of Hoover Dam one feels that the future is limitless, that no obstacle is insurmountable, that we have in our grasp the power to achieve anything if we can but summon the will."[107] In an essay with the less-than-subtle title "Hoover Dam: a Study in Domination," environmental historian Donald Worster offers a different view of the dam's larger cultural meaning. In ways that most likely would rankle, if not infuriate, everyone who supported the authorization—or cheered the construction and completion—of Hoover Dam during the 1920s and 1930s, Worster makes a point regarding the water storage along the Colorado River that "it is not 'man' who has achieved mastery over western American rivers, but some men."[108] Whether one agrees with Worster as to the social and economic benefits and costs associated with both Hoover Dam and dams built in its wake, it is, nonetheless, hard to deny that they derived from the interests of "some men" rather than "mankind" in general. Thus, we should not be so surprised that the dams have in the past and will likely in the future comprise sources of controversy. As part of that oftentimes controversial–yet often celebrated–process, the federal government (acting through the Bureau of Reclamation) came to play a critically important role in implementing the Boulder Canyon Project and all subsequent hydraulic engineering work of any significance within the Colorado River basin.

Endnotes

1. Hubbard, *Origins of the TVA*.
2. Jay Brigham, *Empowering the West: Electrical Politics Before FDR* (Lawrence: University Press of Kansas, 1998) offers good background on the debate over public vs. private power in the pre-New Deal Era.
3. U.S. Department of the Interior, Bureau of Reclamation, *The Colorado River: "A Natural Menace Becomes a Natural Resource"* (Washington, D.C., 1947), 31-41. This volume provides a good synopsis of the natural history and terrain of the river basin.
4. Paul L. Kleinsorge, *Boulder Canyon Project: Historical and Economic Aspects* (Palo Alto, California: Stanford University Press, 1941), 16-7.
5. Two early biographies of Powell remain useful. See William Culp Darrah, *Powell of the Colorado* (Princeton, New Jersey: Princeton University Press, 1951), and Wallace Stegner, *Beyond the 100th Meridian: John Wesley Powell and the Second Opening of the West* (Boston: Houghton Mifflin, 1954). However, both of these biographies have been superseded by Donald Worster, *A River Running Through West: The Life of John Wesley Powell* (Oxford: Oxford University Press, 2001).
6. Mildred de Stanley, *The Salton Sea: Yesterday and Today* (Los Angeles: Triumph Press, Inc., 1966), 17-8.
7. Hundley, *The Great Thirst*, 205.
8. De Stanley, *The Salton Sea*, 21-4.
9. Beverly Moeller, *Phil Swing and Boulder Dam* (Berkeley: University of California Press, 1971), 5.
10. Moeller, *Phil Swing and Boulder Dam*, 25-7.
11. Moeller, *Phil Swing and Boulder Dam*, 28-37.
12. Moeller, *Phil Swing and Boulder Dam*, 33-43.
13. Message from President Theodore Roosevelt to the Senate and House of Representatives, January 21, 1907, as quoted in de Stanley, *The Salton Sea*, 40-3.
14. De Stanley, *The Salton Sea*, 48.
15. Moeller, *Phil Swing and Boulder Dam*, 10-1.
16. Hundley, *The Great Thirst*, 207.
17. Hundley, *The Great Thirst*, 207.
18. Kleinsorge, *Boulder Canyon Project*, 76. The cost of planning studies for Hoover Dam eventually came to about $400,000 with approximately $170,000 of this coming from irrigation interests in Southern California.

19. See Lars Jorgensen to A. P. Davis, December 15, 1920; A. P. Davis to Lars Jorgensen, December 23, 1920; Lars Jorgensen to A. P. Davis, December 29, 1920; and A. P. Davis to Lars Jorgensen, January 10, 1921; all in National Archives, Denver, Records of the Bureau of Reclamation, RG115. General Administrative and Project Records, Project Files, 1919-45. Entry 7. Box 490. Folder 301.1, Colorado River Project, Correspondence re: Dams and Reservoirs, 1920 - November 1, 1921.

20. For evidence of Davis's early interest in Jorgensen's design work, see the file on "Yuba Dam" (later called Spaulding Dam) built by the Pacific Gas and Electric Company in northern California in 1912-13, in the John R. Freeman Papers, Massachusetts Institute of Technology Archives and Special Collections, Cambridge Massachusetts.

21. See index attached to F. E. Weymouth (Chief Engineer) to Director (A. P. Davis), December 2, 1921, National Archives, Denver. Records of the Bureau of Reclamation, RG115. General Administrative and Project Records, Project Files, 1919-45. Entry 7. Box 490. Folder 301.1, Colorado River Project, Correspondence re: Dams and Reservoirs, November 1, 1921 - December 31, 1923.

22. A. P. Davis to J. W. Reagan, October 5, 1920, National Archives, Denver. Records of the Bureau of Reclamation, RG115. General Administrative and Project Records, Project Files, 1919-1929. Entry 7. Box 490. Folder 301.1, Colorado River Project. Correspondence re: Dams and Reservoirs. To Dec. 31, 1923 thru 1929.

23. Bureau of Reclamation, "Report on the Problems of the Colorado River, Volume 5, Boulder Canyon: Investigations, Plans and Estimates," February 1924, 4. National Archives, Denver. RG115. Entry 7. General Administrative and Project Records 1919-1945, Project Files, 1919-39. Box 478, Colorado River 301, Weymouth Report, Parts 5-9.

24. Bureau of Reclamation, "Report on the Problems of the Colorado River, Volume 5, Boulder Canyon: Investigations, Plans and Estimates," February 1924, 6. National Archives, Denver. RG115. Entry 7. General Administrative and Project Records 1919-1945, Project Files, 1919-39. Box 478, Colorado River 301, Weymouth Report, Parts 5-9.

25. John L. Savage, "Revised Plan for Boulder Canyon Dam and Powerplant: Memorandum to Colorado River Board," November 24, 1928, National Archives, Denver. Records of the Bureau of Reclamation, RG115. General Administrative and Project Records, Project Files, 1919-45. Entry 7. Box 492. Folder 301.1, Colorado River Project.

26. For early confirmation on how the "Trial-Load" method was used to justify the final design, see Elwood Mead, "Memorandum to the Secretary re the Meeting of the Consulting Engineers to Approve Detail Plans of Boulder Dam," December 28, 1929, National Archives, Denver. Records of the Bureau of Reclamation, RG115. General Administrative and Project Records, Project Files, 1919-45. Entry 7. Box 490. Folder 301.1, Colorado River Project, Board & Engineering Reports on Construction Features, 1929.

27. The manner in which the design could be considered as meeting the 30 tons pounds per square foot criteria is described in Elwood Mead, "Memorandum to the Secretary re the Meeting of the Consulting Engineers to Approve Detail Plans of Boulder Dam," December 28, 1929. This same document also avers that: "Uplift is not an important consideration in the design of a curved dam where large proportions of the total load are carried by arch action." National Archives, Denver. Records of the Bureau of Reclamation, RG115. General Administrative and Project Records, Project Files, 1919-45. Project files, 1919-1929, Entry 7. Box 491. Folder 301.1, Colorado River Project, Board & Engineering Reports on Construction Features, 1929.

28. J. L. Savage, "Revised Plan for Boulder Canyon Dam and Power Plant: Memorandum to Colorado River Board," November 24, 1928, National Archives, Denver. RG115. General Administrative and Project Records 1919-1945. Project Files 1919-1929, Colorado River 301.1B, Entry 7. Box 492. "Revised Plan for Boulder Canyon Dam and Power Plant. Memorandum to Colorado River Board. By J. L. Savage, Chief Designing Engineer. Denver, Colorado. November 24, 1928."

29. In correspondence with Edward Wegmann concerning how design of the proposed Hoover Dam should be credited, it was reported that it "was first prepared in 1920 in the Denver Office of the Bureau of Reclamation by Mr. Walker R. Young, Member ASCE; under the supervision of Mr. J. L. Savage, Member ASCE (Designing Engineer, Bureau of Reclamation); and subsequently modified in minor particulars following review by Consulting Engineers A. J. Wiley, Member ASCE, of Boise Idaho; and L. C. Hill, Member ASCE, of Los Angeles, California." Significantly, both Davis and Weymouth—who were unquestionably involved in the evolution of the Hoover Dam design—were no longer employees of Reclamation or the federal government at the time (1927) when this letter was written and perhaps were excluded simply for this reason. See Chief Engineer (R. J. Walter) to Commissioner (Elwood Mead), February 21, 1927, and Elwood Mead to Edward Wegmann, February 25, 1927, both in the National Archives, Denver. RG115, Entry 7, General Administrative and Project Records 1919-1945, Project Files, 1919-29. Colorado River 301.03-301.1B, Box 490, Folder: 301.1 Correspondence re. Dams & Reservoirs, December 31, 1923 thru 1929.

30. The best discussion of Kaufmann's work in developing the architectural form of the dam is in Richard Guy Wilson, "Machine Age Iconography in the West," *Pacific Historical Review* 54 (1985) 463-93. Interestingly, after undertaking a considerable amount of research into the subject, Wilson is obliged to report that: "Why Kaufmann was selected for his post [as architect] remains unknown," see page 476.

31. The desirability of a Boulder Canyon reservoir over one located at Glen Canyon is prominently stressed in A. P. Davis to F. E. Weymouth, December 18, 1921, National Archives, Denver. Records of the Bureau of Reclamation, RG115. General Administrative and Project Records, Project Files, 1919-45. Entry 7. Box 490. Folder 301.1, Colorado River Project, Correspondence re: Dams and Reservoirs, November 1, 1921-December 31, 1923.

32. A. P. Davis to F. E. Weymouth, December 18, 1921, National Archives, Denver. Records of the Bureau of Reclamation, RG115. General Administrative and Project Records, Project Files, 1919-45. Entry 7. Box 491. Folder 301.1, Colorado River Project, Correspondence re: Dams and Reservoirs, November 1, 1921-December 31, 1923.

33. Chief Engineer (F. E. Weymouth) to Director (A. P. Davis), July 7, 1922, National Archives, Denver. Records of the Bureau of Reclamation, RG115. General Administrative and Project Records, Project Files, 1919-45. Entry 7. Box 490. Folder 301.1, Colorado River Project, Correspondence re: Dams and Reservoirs, November 1, 1921-December 31, 1923.

34. A. P. Davis to F. E. Weymouth, November 30, 1922, National Archives, Denver. Records of the Bureau of Reclamation, RG115. Entry 7. General Administrative and Project Records1919-1945, Project Files, 1919-45. Box 490. Folder 301.1, Colorado River Project, Correspondence re: Dams and Reservoirs, November 1, 1921- December 31, 1923.

35. Bureau of Reclamation, "Report on the Problems of the Colorado River, Volume 5, Boulder Canyon: Investigations, Plans and Estimates," February 1924, 3, National Archives, Denver. RG115. Entry 7. General Administrative and Project Records 1919-1945, Project Files, 1919-39. Box 478, Colorado River 301, Weymouth Report, Parts 5-9.

36. Moeller, *Phil Swing and Boulder Dam*, 15.

37. Phil Swing speech before House Committee on Irrigation of Arid Lands, July 1919; as quoted in Moeller, *Phil Swing and Boulder Dam*, 16.

38. Carl Hayden speech before House Committee on the Arid Lands, July 1919; as quoted in Moeller, *Phil Swing and Boulder Dam*, 16.

39. Moeller, *Phil Swing and Boulder Dam*, 19.

40. Kleinsorge, *Boulder Canyon Project*, 77-8.

41. Kleinsorge, *Boulder Canyon Project*, 78.

42. Kleinsorge, *Boulder Canyon Project*, 92-8. Also see Donald J. Pisani, *Water and American Government: The Reclamation Bureau, National Water Policy, and the West, 1902-1935* (Berkeley: University of California Press, 2002), 227-32.

43. Moeller, *Phil Swing and Boulder Dam*, 88.

44. A full history of this conference is presented in Hundley, *Water and the West*.

45. This legislation, known as the "Finney resolution" in recognition of its support by the Imperial Valley state legislator A. C. Finney, is quoted in Moeller, *Phil Swing and Boulder Dam*, 75. Utah also withheld its official ratification until these terms had been met.

46. The politically charged transition from the Reclamation Service to Bureau after Davis was fired is well described in: Cannon, "We Are Now Entering a New Era: Federal Reclamation and the Fact Finding Commission of 1923-24," 185-211.

47. See Kluger, *Turning on Water With a Shovel* for more on the ideals projected by Elwood Mead in promoting self-supporting irrigation colonies based around small-scale farms.

48. Hundley, *The Great Thirst*, 126-35.

49. William Kahrl, *Water and Power*.

50. Charles A. Bissell, compiler and editor, *The Metropolitan Water District of Southern California: History and First Annual Report* (Los Angeles, 1939), 5.

51. Bissell, *Metropolitan Water District*, 36.

52. Moeller, *Phil Swing and Boulder Dam*, 67.

53. "The Story of a Great Government Project for the Conquest of the Colorado River" (Los Angeles: Boulder Dam Association, 1927). See also "The Federal Government's Colorado River Project" (Los Angeles: Boulder Dam Association, September 1927).

54. "You Are Needed-Help Build Boulder Dam" (Los Angeles: Los Angeles Department of Water and Power, December 1928).

55. Bissell, *Metropolitan Water District*, 38.

56. "Colorado River Development: Statements by Congressman Addison T. Smith of Idaho and Mayor George E. Cryer of Los Angeles" (Los Angeles: Boulder Dam Association, 1925), 10-1.
57. "Colorado River Development: Statements," 11.
58. Bissell, *Metropolitan Water District*, 39.
59. Bissell, *Metropolitan Water District*, 38-9.
60. Bissell, *Metropolitan Water District*, 39-40.
61. Moeller, *Phil Swing and Boulder Dam*, 106, notes how Herbert Hoover favored the construction of a dam across the lower Colorado River to provide flood control but withheld overt support for power generation because "he would alienate a substantial portion of his 'regular Republican' business support if he became too vocal in his endorsement of the project."
62. The generating capacities of America's major hydroelectric stations as of 1930 are tabulated in "A Survey of Hydroelectric Developments." *Electrical Engineering* (July 1934), 1087-94.
63. Edward Wegmann, *Design and Construction of Dams*, 8th edition, 653.
64. Moeller, *Phil Swing and Boulder Dam*, 105.
65. The proposed Topock damsite is noted in Moeller, *Phil Swing and Boulder Dam*, 101.
66. Philip Cabot, "Danger in the Boulder Canyon Project," published in Frank Bohn, editor, "Boulder Dam: Complete Bibliography, References, Engineers' Charts, Studies and Reports, the Swing-Johnson Bill, Minority Reports and General Comments" (n.p. circa 1928), 101-2.
67. Cabot, "Boulder Canyon Project," 103.
68. Brigham, *Empowering the West*, 58-62, 77-8.
69. Moeller, *Phil Swing and Boulder Dam*, 111-2.
70. Stevens, *Hoover Dam*, 3-5, 35-46 provides detailed discussion of how Six Companies Inc. came to be formed and how it developed its winning bid for the Hoover Dam contract. This subject is also covered in Frank Wolf, *Big Dams and Other Dreams: The Story of Six Companies Inc.* (Norman: University of Oklahoma Press, 1996). The reason that Six Companies' contract fell far short of the $177 million authorized for the project is that much work necessary to build the dam and make it operational was carried out by Reclamation or let to other contractors; in addition, the cost of excavating the All-American Canal was included in the total cost of the Boulder Canyon Project.
71. Crowe's background is described in Stevens, *Hoover Dam*, 36-9.
72. Stevens, *Hoover Dam*, 59.
73. Some of the most moving and evocative parts of Stevens book focus on the description of labor conditions in the Black Canyon region during the early stages of construction. For example, see *Hoover Dam*, 50-4.
74. Stevens, *Hoover Dam*, 48, states: "President Hoover and Interior Secretary Wilbur had urged Elwood Mead . . . to begin work on the dam as soon as possible to provide jobs for . . . the nation's unemployed. Because of presidential pressure, the Boulder Canyon Project was to start six months earlier than originally planned, before the building of transportation and living facilities was finished."
75. The "hellish atmosphere" of Black Canyon and the numerous deaths of workers that occurred there during 1931 are described in Stevens, *Hoover Dam*, 56-65.
76. Stevens, *Hoover Dam*, 65-9.
77. As quoted in Stevens, *Hoover Dam*, 76.
78. Stevens, *Hoover Dam*, 102.
79. Stevens, *Hoover Dam*, 101, reports that "Six Companies lawyers, aided by government lawyers from the Bureau of Reclamation and the U.S. attorney's office, argued that Nevada did not have jurisdiction to enforce its laws on a federal reservation and that if the law prohibiting the use of gasoline-powered trucks was enforced, Six Companies would be deprived of its property without due process."
80. Stevens, *Hoover Dam*, 206-13.
81. The tunneling process is well described in Stevens, *Hoover Dam*, 81-100.
82. Excavation of the foundation and the abutments is described in U.S. Department of the Interior, Bureau of Reclamation, *Boulder Canyon Project Final Reports: Part IV—Design and Construction* (Denver, 1941), 77-94, the treatment of the foundations by grouting is described on pages 95-134. Also see Stevens, *Hoover Dam*, 185-90.
83. Bureau of Reclamation, *Boulder Canyon Project Final Reports: Part IV—Design and Construction*, 138-47 describe the techniques used to place the concrete while 148-66 describe how the concrete was manufactured. Also see Stevens, *Hoover Dam*, 179-85.
84. Bureau of Reclamation, *Boulder Canyon Project Final Reports: Part IV—Design and Construction*, 35-6, 136-7, 177-88 describes how the concrete was placed.
85. The cooling system is described in Bureau of Reclamation, *Boulder Canyon Project Final Reports: Part IV—Design and Construction*, 189-206. Also see Stevens, *Hoover Dam*, 190-9.

86. Stevens, *Hoover Dam*, 219, 230-1.
87. Stevens, *Hoover Dam*, 252.
88. Kleinsorge, *Boulder Canyon Project*, 214-8 provides a good overview of completion and initial operation of the powerplant.
89. Stevens, *Hoover Dam*, 174-5.
90. Stevens, *Hoover Dam*, 233.
91. Ickes' support for a federal investigation into Six Companies' adherence to the "eight hour day" provision of the Hoover Dam construction contract is documented in Stevens, *Hoover Dam*, 232-4. This investigation eventually resulted in Six Companies paying a $100,000 fine.
92. Roosevelt's remarks at the dam's dedication are quoted in Stevens, *Hoover Dam*, 246-8.
93. This issue is fully described in Kleinsorge, *Boulder Canyon Project*, 52-4.
94. The constitutional justification for the Boulder Canyon Act is explicated in Kleinsorge, *Boulder Canyon Project*, 123-36. As Kleinsorge notes, 134, "the navigability of the Colorado River, therefore, formed the basis for the constitutionality of the Boulder Canyon Project Act."
95. Hundley, *The Great Thirst*, 227-8.
96. Hundley, *The Great Thirst*, 303.
97. Bureau of Reclamation, "The Colorado River: A Natural Menace Becomes a National Resource."
98. An excellent, in-depth treatment of how the Colorado River Storage Project came to be authorized and funded in the 1950s and early 1960s is provided in Stephen Craig Sturgeon, "Wayne Aspinall and the Politics of Western Water" (Ph.D. dissertation, University of Colorado, Boulder, 1998), n.b., see also the subsequent book Sturgeon, *The Politics of Western Water: The Political Career of Wayne Aspinall* (Tucson: University of Arizona Press, 2002). U.S. Congressman Aspinall played a pivotal role in supporting federal water development of the Upper Colorado Basin that is comparable to that taken by Congressman Phil Swing in promoting the Boulder Canyon Project in the 1920s. The issue is also discussed in Marc Reisner, *Cadillac Desert*, 145-50.
99. Holway R. Jones, *John Muir and the Sierra Club: The Battle for Yosemite* (San Francisco: The Sierra Club, 1965) offers extensive documentation of how the early twentieth century environmental movement fought and lost the battle to keep a dam from being built at Hetch Hetchy.
100. The full story of Echo Park is documented in Mark Harvey, *A Symbol of Wilderness: Echo Park and the American Conservation Movement* (Albuquerque: University of New Mexico Press, 1994).
101. Harvey, *A Symbol of Wilderness*, 280-3, describes how wilderness advocates came to accept the need for Glen Canyon Dam in the mid-1950s. While there was some desire to push for the blocking Glen Canyon Dam it was much less vociferous than the effort to protect Echo Park.
102. *Glen Canyon Dam and Powerplant*, Bureau of Reclamation, Denver, 1970, 55-89.
103. Russell Martin, *The Story that Stands Like a Dam: Glen Canyon and the Struggle for the American West* (New York: Henry Holt and Company, 1989) describes the history of the dam in non specialist terms. Editor's note: It is true that Glen Canyon Dam was intended to be a "cash register dam," but Glen Canyon was also designed to insure that the Upper Basin States could reliably store and deliver their annual commitment of water to the Lower Basin states.
104. For example see Eliot Porter, *The Place No One Knew: Glen Canyon on the Colorado* (San Francisco: The Sierra Club, 1963). This book was published by the Sierra Club to bring to public attention to the loss of the canyon lands just as the dam was reaching completion. Also see Philip Fradkin, *A River No More: The Colorado and the West* (Tucson: University Press of Arizona, 1981). Martin, *The Story that Stands Like a Dam*.
105. Reisner, *Cadillac Desert*, 283-300.
106. Marc Reisner's book, *Cadillac Desert,* stands as one of the most popular publications championing the idea that dam-building in the West has been carried to the point that it has long since ceased to serve the greater public good. The political maneuvering that attended Congressman Wayne Aspinall's support for construction of the final group of large-scale Reclamation dams in the Colorado Basin is described in Sturgeon, *Wayne Aspinall and the Politics of Western Water*, 232-349.
107. Stevens, *Hoover Dam*, 266-7.
108. Donald Worster, "Hoover Dam: A Study in Domination," in *Under Western Skies: Nature and History in the American West* (New York: Oxford University Press, 1992), 73.

CHAPTER 5

BONNEVILLE AND GRAND COULEE DAMS AND THE COLUMBIA RIVER CONTROL PLAN

EXPLORING A CHANGING RIVER

By 1987, the Columbia River became "the largest hydroelectric energy producer in the world" and had realized its potential of supplying about 40 percent of the nation's hydropower.[1] This immense power has struck observers in two conflicting ways. The founder of the Sierra Club, John Muir, described the river as "gathering a glorious harvest of crystal water to be rolled through forest and plain in one majestic flood to the sea."[2] By contrast, President Franklin Roosevelt, in dedicating Bonneville Dam, portrayed a river providing "the widest possible use of electricity [to create] more wealth, better living and greater happiness for our children."[3]

These conflicting images, the pristine waters flowing to the sea versus the powerful current wired to the city, persist wherever a massive transformation of nature takes place. As modern civilization took shape with the continuing industrial revolution that began in the late eighteenth century, these two voices—one naturalist and one populist—have competed for political control. Perhaps nowhere has this competition been more intense than around the Columbia River, and also, perhaps nowhere is it less informed historically. In particular, the U.S. Army Corps of Engineers has received major criticism for activities which, in retrospect, seemed to be directed by the will of the people.

That will, defined here by the Congress of the United States, set in motion the federal damming of major American rivers for power by ordering a cost estimate for a survey through the Rivers and Harbor Act of March 3, 1925. The Corps responded on April 7, 1926, with its well known 308 Report that included the estimate of $7,300,400, of which the Columbia River estimate was $734,100.[4] The Congress then authorized the studies for some rivers, including the Columbia, on January 21, 1927, and on March 29, 1932, the Chief of Engineers submitted a 1,845 page report dealing with the Columbia River.[5]

This 1932 report laid out a system of ten dams along the mainstream up to the Canadian border, beginning with one at Warrendale (later moved to Bonneville) and ending with one at the head of Grand Coulee. But the engineers recommended that the federal government should pay only for the navigational aspects—locks and dredging—not for the dams and powerplants.[6] In early 1932, the guiding idea was local funding and hence local control. A summary of the

191

four mainstream dams between Bonneville and Pasco detailed the costs, the power capacity, and some of the technical problems for that 184 mile stretch having a fall of 309 feet.[7]

The other six dams, all above the mouth of the Snake, crossed the river as it flowed from the north to Pasco and included a dam at Rock Island Rapids already under construction by the Chelan County Public Utilities District (PUD), a private enterprise. But the main dam on the upper Columbia was Grand Coulee which local people had been pushing since before World War I. The new Reclamation Service in its first report, for 1902, discussed the Columbia Basin for irrigation, and, in 1908, the Corps began to survey the Grand Coulee region for navigation, considering also irrigation and power.[8]

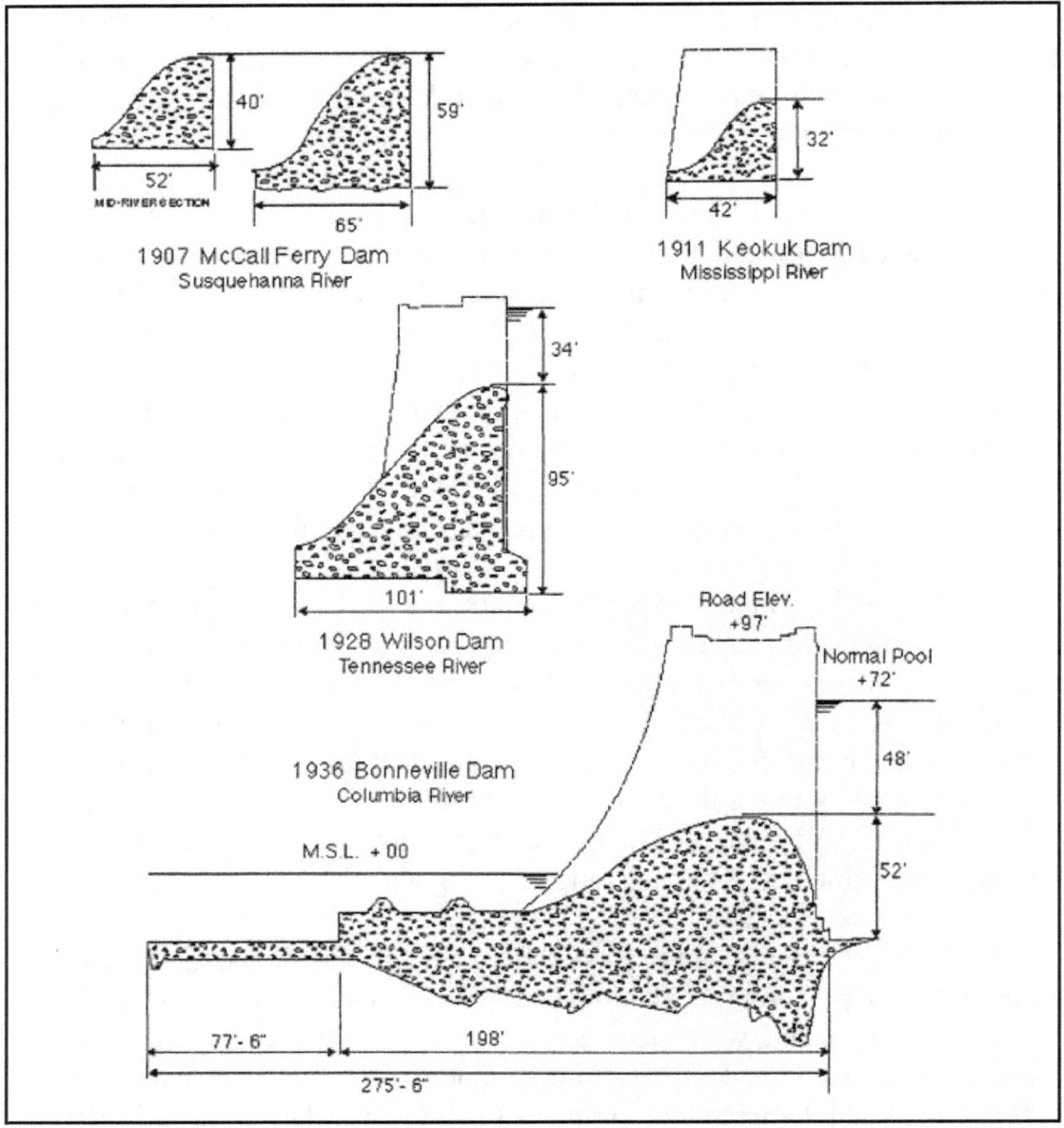

Figure 5-1: The cross sections of earlier concrete dams built by the U.S. Army Corps of Engineers influenced the design of the spillway dam at Bonneville. Source: David P. Billington.

The careful surveys that began with the 1927 act, and culminated in 1932, recommended activities that were suddenly politically feasible because of the depression and the election of Roosevelt later that year. He had made a campaign trip to the Northwest and promised development of the Columbia. Once elected, Roosevelt stimulated among many federal and local personnel an excitement in Washington over central planning which was, however, not shared by many people in Congress. The great water projects of the Tennessee Valley and the Columbia basin moved up high on the new administration's priority list. This enthusiasm focused ultimately on the construction of large multipurpose dams that would have been much more difficult had not the Corps and Reclamation, by 1932, already had experiences with such major works as the Panama Canal's Gatun Dam, Wilson Dam, Roosevelt Dam, and the greatest project of all, Hoover Dam. No large dam had yet been built on the Columbia; it was almost completely untamed, and the newly elected leaders were eager to put people to work. What happened next would be an acceleration in the major restructuring of the nation's largest river basins.

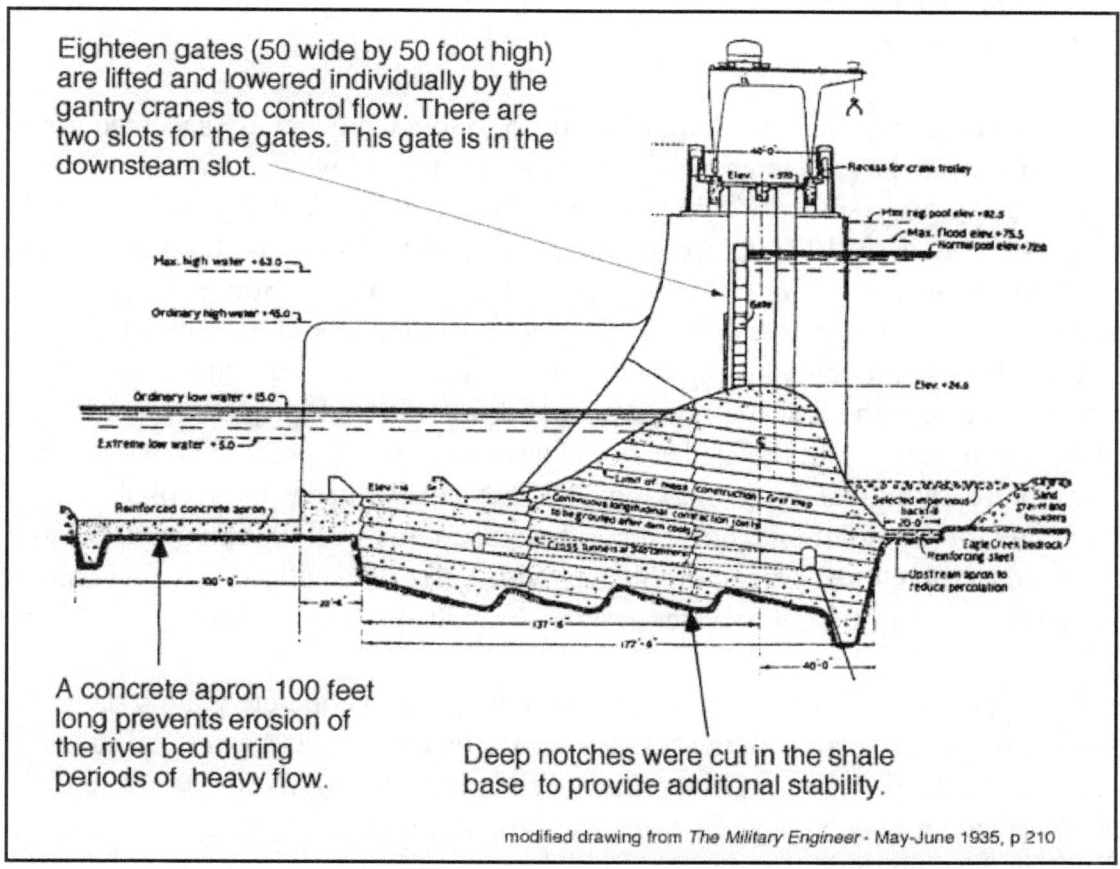

Figure 5-2: Bonneville Dam section. Source: *The Military Engineer*, May-June 1935, (printed with permission from the Society of American Military Engineers.), 210.

BONNEVILLE DAM: EMPLOYMENT VERSUS MARKETS

From Survey to Construction

The economic crisis of the early 1930s expressed itself dramatically in unemployment. From nearly full employment in 1929, the nation's idle had jumped to 25 percent by the time of Roosevelt's inauguration in March, 1933.[9] Dams meant jobs, and on September 30 the president authorized Bonneville Dam under the National Industrial Recovery Act. When work began on November 17, there had been little time for design.[10] The rush was from survey and preliminary planning to the letting of contracts for construction. The dam, therefore, resulted largely from precedent and not from innovation. Indeed, Bonneville, as a whole, would characterize a typically American approach to concrete design or, in a larger sense, an approach to industry. Concrete was to replace stone and express safety through mass just as industry was to replace craft and express economy through mass production. The federal government would be the new stimulus because private industry presumably had failed and local governments, like local crafts, were too small.

These ideas helped lead the Corps to follow the generally accepted practice as illustrated by the examples of the Gatun Dam spillway and Wilson Dam to design a gravity spillway dam at Bonneville and to do it quickly without much discussion. In early 1933, Corps engineers were considering alternatives in their preliminary planning. For example, a project upstream from Bonneville, at the Dalles, was to have an extensive multiple arch dam that could have resulted in considerable economy and exhibited major innovation in structural form; it would have been the first such structure to be part of a multipurpose project on a mainstream dam.[11] But the Dalles would have to wait and Bonneville would follow a more traditional design harking back to McCall's Ferry and Keokuk. The huge flow of the Columbia, however, did require some fresh thinking and led to the need to solve three crucial problems: wide variations in flow, soft foundation material, and high energy in water passing through spillway openings.[12]

The flow variation was met by designing a low concrete gravity dam flanked by high concrete buttresses and topped by eighteen large vertical steel gates above the dam to control the flow. When fully raised, these 50-foot-high, 50-foot-wide gates could pass a flow about one-third larger than any recorded flood on the Columbia River.

In addition, the Corps decided to use Kaplan turbines, which have blades that automatically adjust to changes in load and flow.[13] These turbines were larger than any built previously in the United States, and, furthermore, they were the most powerful Kaplan turbines put into service up to that time. Because of their unprecedented size, the Corps carried out a series of model tests at its Portland Hydraulic Laboratory, which led to numerous changes in the design of the concrete works surrounding these turbines.

The basic design was in the hands of the Corps, which was beginning to advance into the engineering of multipurpose dams in the face of this powerful river and the political controversies that it would call forth. The river demanded a new politics; politicians would have to react to its vast power potential. Not only was the water power a problem and a potential, but the bed upon which the river flowed presented new challenges to the engineers.

The soft foundations had led the engineers to design the low gravity dam 185 feet wide, only 75 feet high, and notched into the rock below. The wide base kept stresses low on the relatively soft volcanic rock, and the notches provided extra safety against sliding. The soft foundation was by far the central technical problem. The original plan to build the dam at Warrendale was abandoned because of poor foundation conditions, and President Roosevelt "refused to commit Federal funds for Bonneville unless he could be guaranteed that a suitable foundation existed."[14]

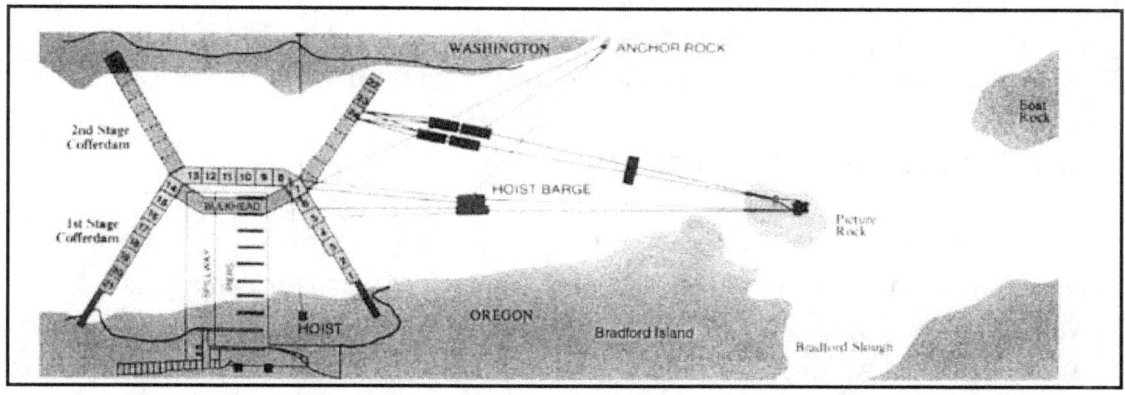

Figure 5-3: Construction at Bonneville Dam required huge coffer dams. The Corps built the south one during low water in the fall and winter of 1935-1936 and the north one during the low water season of 1936-1937. Source: U.S. Army Corps of Engineers. See discussion on page 199.

With the location settled and the president convinced, the engineers tackled the third problem of energy dissipation by shaping the dam profile to produce an upsurge of water after it passed over the dam crest and, in addition, by adding baffles to create turbulence. As a further means of protection against erosion and scour, the engineers designed a 100 foot concrete apron built onto the downstream face of the dam.

This upsurge of water, a strange effect, has the technical term of hydraulic jump, and it changes the flow from a relatively low depth to an increased depth. The result is a decrease in velocity and a corresponding loss of energy in the flowing water. With less energy to dissipate, the concrete spillway is, thereby, less susceptible to erosion or cavitation (creation of cavities in the concrete).[15]

To study the erosion problem, the Corps created an outdoor hydraulic laboratory in Portland, Oregon, which contained a 1:36 scale model of three openings for the spillway dam. These included three of the 50- by 50-foot crest gates (modeled at a size of about 1.4 by 1.4 feet) as well as two piers and two half-piers, thus, forming the three spillways. The engineers changed the configuration of the downstream spillway form for each of 140 separate tests and measured the resulting downstream scour. This highly empirical approach led them to the final design, which included baffles and a 100-foot-long apron or deck. The goal was to create a hydraulic jump equal to the normal downstream (tailwater) elevation.[16]

This laboratory was the Corps' first attempt to model the hydraulics in a major, multipurpose, mainstream dam spillway. Moreover, the Corps provided for similar measurements to be made on the dam itself to compare with the model results. Even before this work, the Corps had been making such studies for other mainstream dams.[17] The hydraulic laboratory at Bonneville eventually carried out about 100 model studies on other multipurpose dams and on navigation channel improvements before the Army transferred its functions to the Waterways Experiment Station at Vicksburg in March of 1982.[18]

From Construction to Dedication

The election of Roosevelt and the deepening Depression profoundly changed the Corps' mission at Bonneville. No longer was there any question of private construction of a power dam with the Corps responsible only for the navigation works. Now, only the federal government would rule, with the Corps as its agent. It was not simply the Corps seeking to enlarge its mission, but the country as a whole, through its newly elected president, seeking to alleviate the economic crisis through centrally funded public works.

There was no time to lose; jobs needed to be created and massive projects were an answer. But design takes thought and time while usually engaging few people compared to the large numbers needed for construction. The political imperative in such cases is clear: rush design and start construction. On October 12, 1933, the Public Works Administration allotted funds for design, on November 1, the Corps began to build a construction camp, and on February 6, 1934, the Corps issued the first main dam contract to excavate for the lock and powerhouse. In June the Corps awarded a contract for the main dam and the next month one for building the lock and powerhouse.

Figure 5-4: Turbulent flow between partially completed piers of Bonneville Dam during the flood season. Source: Records of the U.S. Army Corps of Engineers, Record Group 77, National Archives and Records Administration, College Park, Maryland.

In August, bowing to local political pressure, the Secretary of War ordered the Corps to build a ship lock instead of the planned, much smaller, barge lock. This new design—76 feet wide, 500 feet long, and 24 feet deep over the sill at low water—created the world's highest single-lift lock, raising sea-going ships up 60 feet.[19]

Not only did the dam shape require special study but so did the dam material. In December 1934 the Corps' consultants at the University of California, Berkeley, began a study of the cement to be recommended for the spillway. Three problems had to be solved: first, how to minimize the heat generated by the chemical reaction between cement and water (heat of hydration); second, how to use as little cement as possible; and third, how to produce a workable concrete (not too stiff when cast) without risking separation of gravel from the mortar (cement paste).[20]

The solution proposed by the Berkeley group and used at Bonneville was a mixture of Portland cement and puzzolan cement ground together in the process of making the cement itself. Here was a combination of something new, Portland cement developed in England in 1824 and puzzolan, or more properly pozzolana, a hydraulic cement discovered at least 2000 years earlier by Roman engineers. Pozzolana came originally from natural volcanic slag in deposits near Mt. Vesuvius at the ancient town of Pozzuoli near Naples.

Figure: 5-5: Two bays of the main spillway at Bonneville Dam with gates open on December 7, 1937. Source: Records of the U.S. Army Corps of Engineers, Record Group 77, National Archives and Records Administration, College Park, Maryland.

European engineers in the early twentieth century had used this composite of natural and artificial cements with success and it performed well at Bonneville. Subsequently, the Corps used it in the 1973 Dworshak Dam, and Reclamation put it into Davis Dam (1951), Hungry Horse Dam (1953), and Pueblo Dam (1975).[21] It served primarily as a means of reducing the heat of hydration, which had led to concrete cracking, and stimulated engineers at Reclamation to lay a pipe network within Owyhee and Hoover Dams for cooling.

Construction was unprecedented in scale because of the great flow, harsh weather, and deep water at Bonneville. George Gerdes, chief engineer for the main dam, designed huge timber-crib cofferdams that allowed concrete to be cast on dry land. At the time, it was the largest cofferdam installation on a river in the United States.[22] Gerdes designed one cofferdam to enclose the south half of the channel, where the contractor built the first sections of the concrete dam during the low water season between August 1935 and March 1936. The contractor concreted only part of the crest sections in the dam before removing the cofferdam; in this way the river could more easily flow between the concrete piers once a second cofferdam closed off the northern half of the channel.

Figure 5-6: The Bonneville spillway dam with thirteen gates open on October 30, 1939.
Source: Records of the U.S. Army Corps of Engineers, Record Group 77, National Archives and Records Administration, College Park, Maryland.

The contract for supplying the Portland puzzolan cement raised an issue that gives some insight into the strong competition existing during this depression era. On May 31, 1935, the Corps received bids for supplying 565,000 barrels of the cement for the spillway concrete. The low bidder, the Pacific Portland Cement Company, was a California concern, and the Governor of Oregon, Charles H. Martin, immediately wrote a letter to Major General Edward M. Markham, Chief of Engineers, protesting the acceptance of that bid over several Oregon companies whose bids were higher but still well below "the going price at that point."[23] Here was strong political pressure put upon the Army engineers, but General Markham replied quickly that:

> The bid of the Pacific Portland Cement Company was accompanied by the certificate of code compliance required under Executive Order No. 6646 of March 14, 1934. This company is a responsible concern and capable of performing satisfactorily in accordance with the terms of its bid. In view thereof, and of the above cited decision of the Comptroller General, the Department considered that it had no other proper course of action than to accept the bid of this company, which was the lowest one received in response to the advertised specifications.[24]

Figure 5-7: At Bonneville Dam the locks and powerhouse are on the left, Bradford Island is in the middle, and the main spillway dam is on the right. Source: Records of the U.S. Army Corps of Engineers, Record Group 77, National Archives and Records Administration, College Park, Maryland.

Martin had argued that the very low bid, which he claimed was well below the California company's costs, amounted to "dumping" and tended to run the smaller Oregon companies out of business. "As you are aware," Martin wrote, "the P.W.A. (The Public Works Administration) intended to distribute funds throughout, and for the benefit of the entire country and California certainly fared as well, if not better than, any other section."[25] As Markham replied, the U.S. Army Corps of Engineers was not permitted to engage in such judgments when the low bidder complied with the advertised specifications. There the matter stood, and the construction proceeded. In total, the project required about 1,000,000 cubic yards of concrete.

With the north section of the concrete dam fully cast by the spring of 1937 and the cofferdam removed, the contractor could complete the south part by casting concrete in prefabricated steel cofferdams placed between piers. By June 1938, all work on the spillway dam ended. Meanwhile, other builders completed the powerhouse, and the first two generators began to deliver power. In September 1937, President Roosevelt formally dedicated the dam and extolled its promise.[26]

Figure 5-8: Bonneville Dam powerhouse units 1-6 were complete by October 8, 1940.
Source: Records of the U.S. Army Corps of Engineers, Record Group 77, National Archives and Records Administration, College Park, Maryland.

Not everyone agreed with the president in 1937, and, 60 years later, some voices are even more critical of Roosevelt's acclaim. Probably the salmon industry remains the most controversial issue surrounding Bonneville and the other dams on the lower Columbia and lower Snake. Some writers claimed that the Corps had planned Bonneville Dam without any provisions for the fish, but that is clearly not so.[27] Already in 1929, Colonel G. R. Lukesh, Division Engineer in Seattle, had brought the issue to the attention of the Chief of Engineers, and the 1931 308 Report on the Columbia River included cost estimates for fish passage facilities on each proposed dam.[28] In early 1933, another Corps officer, writing about the proposed dams on the lower Columbia, stated that:

> The Columbia River is noted for its fishing industry, principally salmon. The average value of the annual catch is over $10,000,000. It is highly important that this industry be protected, and it involves a problem of no mean proportions. Before the actual construction of any dam is started, studies must be made to determine the best method of passing the salmon over the high structures required for power and navigation, or plans must be prepared to insure continuance of the fishing industry through other means, such as hatcheries below the dams.[29]

When the dam project was adopted in September 1933, the Corps began right away to design fishways by working closely with the U.S. Bureau of Fisheries and other groups.[30] In July 1934, the Interstate Fish Conservation Committee agreed to recommend fish ladders. The Director of the Department of Fish Culture agreed but also asked for funds to study the fish question during the coming year. Other experts urged the construction of fish lifts (or locks). On August 15, the Director sent to local Congressmen and others a plan for the proposed fishways (dated August 10, 1934) that included two fish ladders and three fish lifts. In the end, the Corps built three ladders and four locks; later studies showed that the ladders were more effective than the lifts.[31]

After Bonneville Dam was closed in January 1938, the fish passageways seemed to work successfully until the construction of other dams upstream and the increase in the number of turbines at the dams. Back in 1937, when President Roosevelt dedicated the Bonneville Dam, the fish problem was not the only controversial one. There were also the issues of a power oversupply and the suspicion of big government building monuments to itself.

From Dedication to Debate

In June of 1937, an article in *Collier's* magazine entitled "Dam of Doubt" reflected some perennial worries of big government critics that we might summarize as the building of useless monuments to the government itself.[32] The article used the voice of a down home sage, a prune farmer named Clark who observed that "Really it ain't nothin' new. Every civilization at some period—usually toward the end—goes in for monuments. Those old chaps over in Angor built the biggest temples on earth. The Aztecs did the same thing. The Egyptians had to build the biggest pyramids ever. The Greeks and Romans were no better." He concluded by observing that "it makes me plumb tired figuring out the human labor that's been wasted in building the biggest things on earth." In fact, this is no rural hick talking; this is the 1930s version of small is beautiful—Clark even showed the article's author the tiny dam and powerplant he built on his own farm to electrify his house.

The underlying argument was that Bonneville Dam was not needed, its power could not be used, industry would never settle in the Northwest, great labor was wasted, and all initial cost estimates are deviously low. As it turned out, Bonneville was to prove, along with its fellow monuments, Grand Coulee and Hoover, important in winning the war against Nazi Germany and Japan in the 1940s. This turn of events does not obscure some virtues in the homey arguments of Clark, but it does reveal how monuments in a democracy can serve society well and still be objects of critical reflection.

The first step in making Bonneville useful was an intensely political one. Models existed for the new challenge of river basin development in the Mississippi River Commission (1879), the Colorado River Compact of 1922 engineered by Herbert Hoover, and the Tennessee Valley Authority stimulated by Wilson Dam and Major Fiske's surveys (see chapter 3) and promoted by Nebraska Senator George W. Norris (1861-1944).

BONNEVILLE AND GRAND COULEE: CONFLICT OF POWER

The Columbia River 308 Report

Although the public descriptions of how Bonneville Dam came into being may seem rational and easily explicable in social terms, there is a personal story that illustrates the strong role of individuals in determining the way such history evolves. The case of Bonneville, as a prime example, reveals regional rivalries, strong personalities, and intergovernmental jealousies, especially as that dam became associated with Oregon just as the Grand Coulee Dam project became the project of Washington State. Here was one river with two quite different and competing visions, navigation in the south and irrigation in the north, and hydroelectric power was growing into the central issue. This twentieth-century conflict began with Theodore Roosevelt, who in 1908, wrote that "every stream should be used to the utmost [and] each river system . . . is a single unit and should be treated as such."[33] Only after World War I did the Congress follow up on Roosevelt's vision. First it began with the Colorado River Compact of 1922 and then came a more general goal with the 1927 directive ordering the Corps to produce documents on river surveys. By 1933, the Corps had completed the Columbia, Missouri, and Tennessee Basin reports. Ultimately, each would produce a different political solution to the way river basins would be organized. For the Columbia, the conflicts began in earnest with the release of the 308 report on the upper Columbia River.

Figure 5-9: Generators #1 and #2 in the Bonneville Dam powerhouse on August 29, 1939.
Source: Records of the U.S. Army Corps of Engineers, Record Group 77, National Archives and Records Administration, College Park, Maryland.

The investigations and surveys of the Columbia River were supervised by Colonel Gustave Lukesh, the Division Engineer for the North Pacific Division. These investigations and surveys were carried out by Major John S. Butler, District Engineer for the Seattle District, and Major Oscar O. Keuntz, District Engineer for the Portland District. The Seattle District investigated the river above the mouth of the Snake River, and the Portland District investigated the river below that point.³⁴ As described in a history of the Portland District, "Colonel Lukesh preferred to give the district engineer maximum latitude in arriving at opinions and conclusions."³⁵

Major Butler, district engineer in Seattle, had taken on responsibility for the 308 survey in 1927, and, by 1929, this work had gradually assumed a central significance for the future of that region. Butler was then 57 years old, a native of Tennessee, civil engineering graduate of Vanderbilt University, and a proud member of the Corps. His work related to the three significant controversies on the Columbia in the late 1920s.³⁶

First, there was rivalry between Washington and Oregon which shared the Columbia River once it left Canada. Senator Clarence Dill, a Democrat, represented Washington while Senator Charles McNary, a Republican, represented Oregon. Dill was for Grand Coulee while McNary pushed Bonneville. It appeared, in the early 30s, that there would not be funds for both, so the battle seemed to promise a winner and a loser.

Figure 5-10: The Bonneville Dam's Bradford Island fishladder intake in operation on February 24, 1938. Source: Records of the U.S. Army Corps of Engineers, Record Group 77, National Archives and Records Administration, College Park, Maryland.

Second, Butler and his staff worked within the context of a conflict between the eastern Washington city of Spokane and the much smaller central Washington city of Wenatchee. Spokane, the larger city, favored the use of upper Columbia water for irrigation on the Columbia Basin Project. This scheme would bring water by gravity from tributaries of the Columbia River to 1,500,000 acres of dry land in the south central region of the state. This "Gravity Plan" would require 130 miles of canals, tunnels, and siphons with a few minor dams, but this approach had little provision for the generation of power. Meanwhile, Wenatchee, the smaller city, had been pushing for a great dam at the bend in the Columbia just above a large dry basin called the Grand Coulee. Situated about 500 feet above the bed of the Columbia, the Grand Coulee (or Great Canyon) is 50 miles long and from one to six miles wide. At the Columbia's bend is an ideal dam site that is 4,300 feet wide with 600-foot-high cliffs on either side and nearly watertight canyon walls well suited for an upstream reservoir.[37] Thus, the Wenatchee group, led by newspaper publisher Rufus Woods, had clamored in 1918 for a dam to store water for irrigating the Columbia Basin Project. This counter proposal saw that some power from the dam could be used to pump reservoir water into the Grand Coulee from which it would then flow by gravity down to the project area for irrigation. Both sides focused on irrigating the potentially fertile region.

The third controversy facing Butler was one between the Corps and Reclamation. Frederick Newell, first chief engineer of the Reclamation

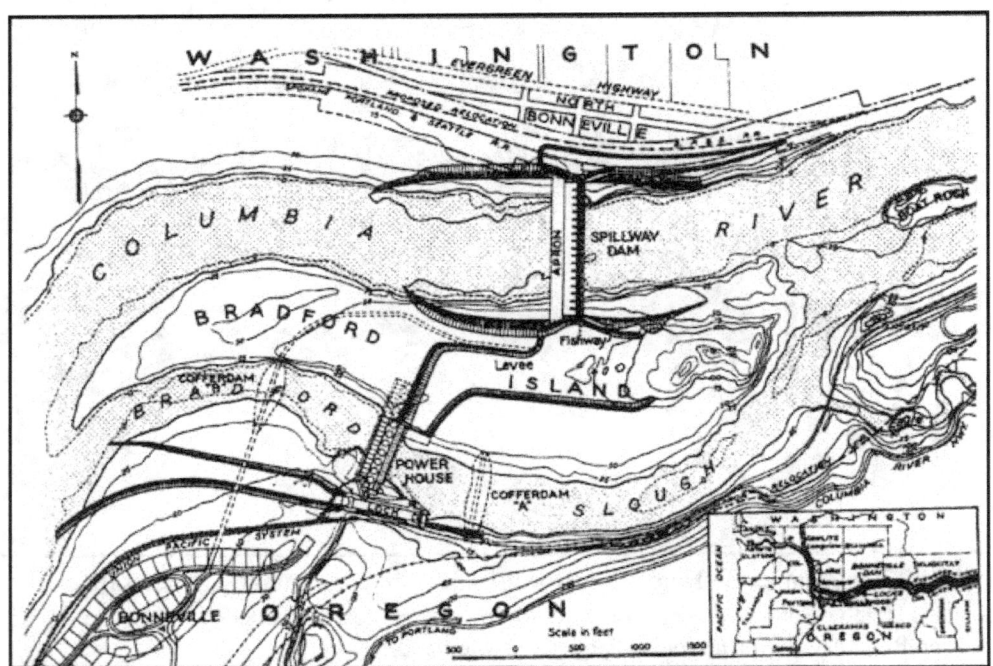

Figure 5-11: At the site chosen for Bonneville Dam, Bradford Island divides the Columbia River. The spillway structure is north of the island while the powerhouse structure is to the south. Source: *The Military Engineer* (May-June 1935), (Printed with permission from the Society of American Military Engineers.), 211.

Service, as it was then called, visited the Columbia Basin in 1903 and recognized the potential for irrigation, but the huge cost led his bureau to lose interest by 1906. At this time, the Army engineers began surveying the upper Columbia for dam sites to permit navigation, but they too gave up when it appeared that only power would result.[38] Interest built up in the late 1920s when Butler began his 308 survey. He had to produce a technically credible report, he had to choose

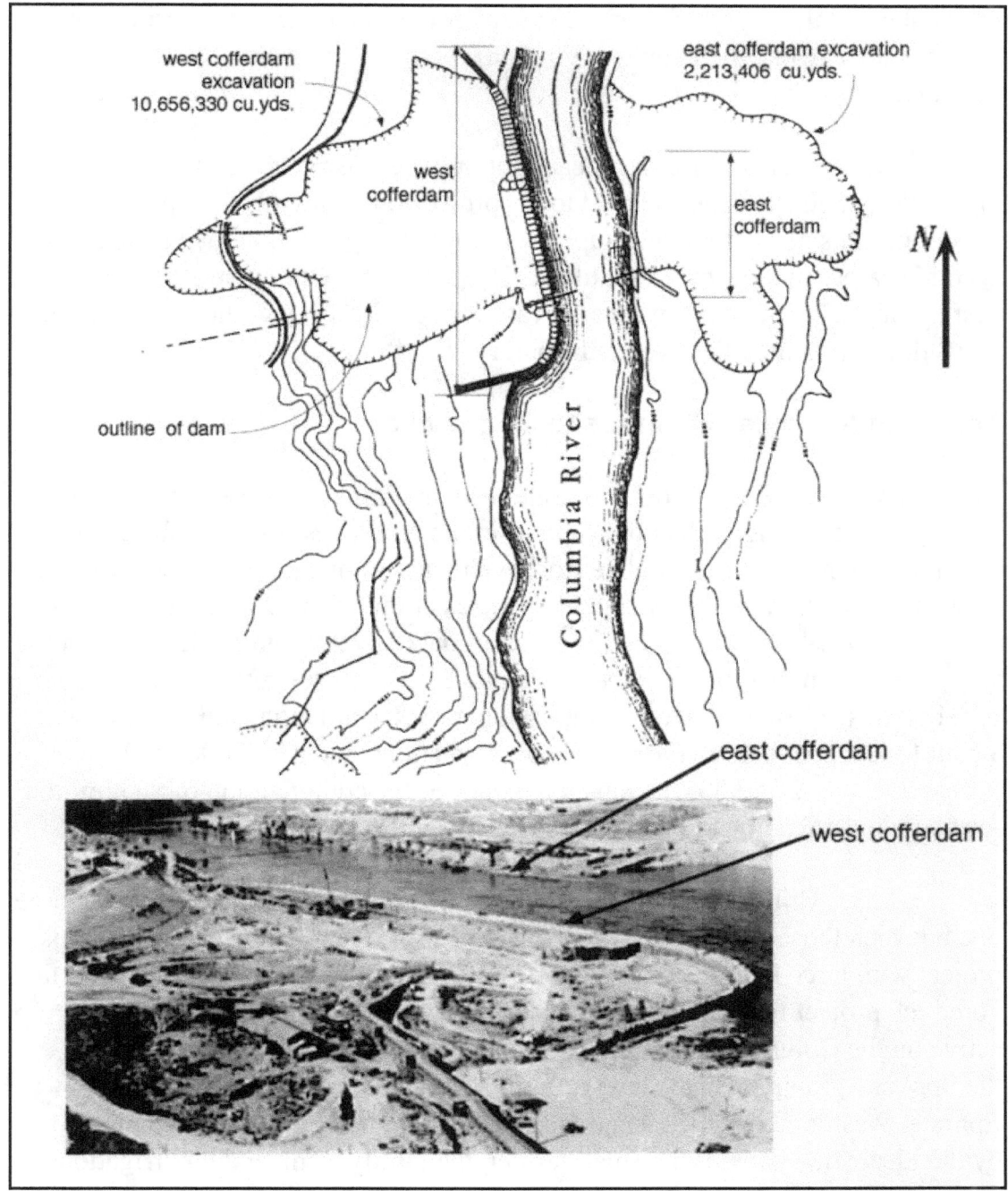

Figure 5-12: For Bonneville Dam the large and complex west cofferdam was key to the construction plan. Steel sheet piling was driven by multiple steam hammers. The main river flow of 350,000 cubic feet per second during the flood season in June 1935 was between the two cofferdams. Source: U.S. Army Corps of Engineers.

between the dam and the gravity canal, and he had to recommend economical plans to ensure that a project could be built.

In fact, Butler's final report on the upper Columbia was so convincing that it helped alert Reclamation to the possibility of losing the project to their dam-building rival.[39] Ironically, the very high quality of his report resulted in Reclamation's making a successful case for doing the work. Butler also decisively showed the superiority of the dam over the canal scheme. Furthermore, his work also awoke the people of Oregon who now saw the possibility of not getting a lower river dam.

Butler's efforts are significant not only because of his superior engineering work, but also because he held to his professional judgment in spite of early rejection of his report by Army superiors. This strong position may have damaged his career in the Army, but, ultimately, his study, with others, "shaped the destiny of the Pacific Northwest and was among the most significant influences on its development in the twentieth century."[40]

The Grande Coulee Stimulus to Bonneville

While Butler's report was under debate, Franklin Roosevelt won the presidency, receiving substantial support in the Northwest from Washington Senator Clarence Dill. Candidate Roosevelt had promised campaigner Dill that he would build Grand Coulee, but after the election, when Dill reported the estimated cost of $450 million, the New Yorker balked, reminding Dill that this price was greater than that for the Panama Canal. Roosevelt suggested a low dam, sent Dill to see Elwood Mead, commissioner of Reclamation, and the result was a plan for a 227 foot high dam with a powerplant having a 520,000 kilowatt capacity. On July 28, 1933, Roosevelt announced the authorization of $63 million for a low federal dam at Grand Coulee.[41]

Meanwhile, the rivalry between the Corps and Reclamation had become one between those like Rufus Woods, who wanted Grand Coulee to be a Corps project with large hydroelectric power, and those like James O'Sullivan who favored the project for irrigation and hence preferred Reclamation. O'Sullivan, active on the Columbia Basin Commission, had championed the dam against the Spokane gravity plan. A lawyer from Michigan, O'Sullivan had settled in Ephrata, Washington, in 1919 and took on the dam as an obsession dominated by the idealistic vision of the small farmer, the family farm, and the irrigation plans to make those farms possible. This was, of course, the early ideal of the Reclamation Service. O'Sullivan was, thus, partial to Reclamation and became, in the spring of 1933, a significant advocate for having Reclamation, not the Corps, build Grand Coulee Dam.[42]

All this activity over Grand Coulee distressed Oregon senators Charles McNary and Frederick Steiwer, who were pushing for dams on the lower Columbia. In late May, McNary and Oregon representative Charles Martin, a retired Army general, met with Roosevelt, who seemed willing to fund a dam project that served both navigation and power generation. He would consider setting aside $25 million from public works funds for the dam then planned near Warrenville. All dam politics focused on the president because of the enormous amounts of money that the Congress had allocated to him for national recovery from the Depression.[43]

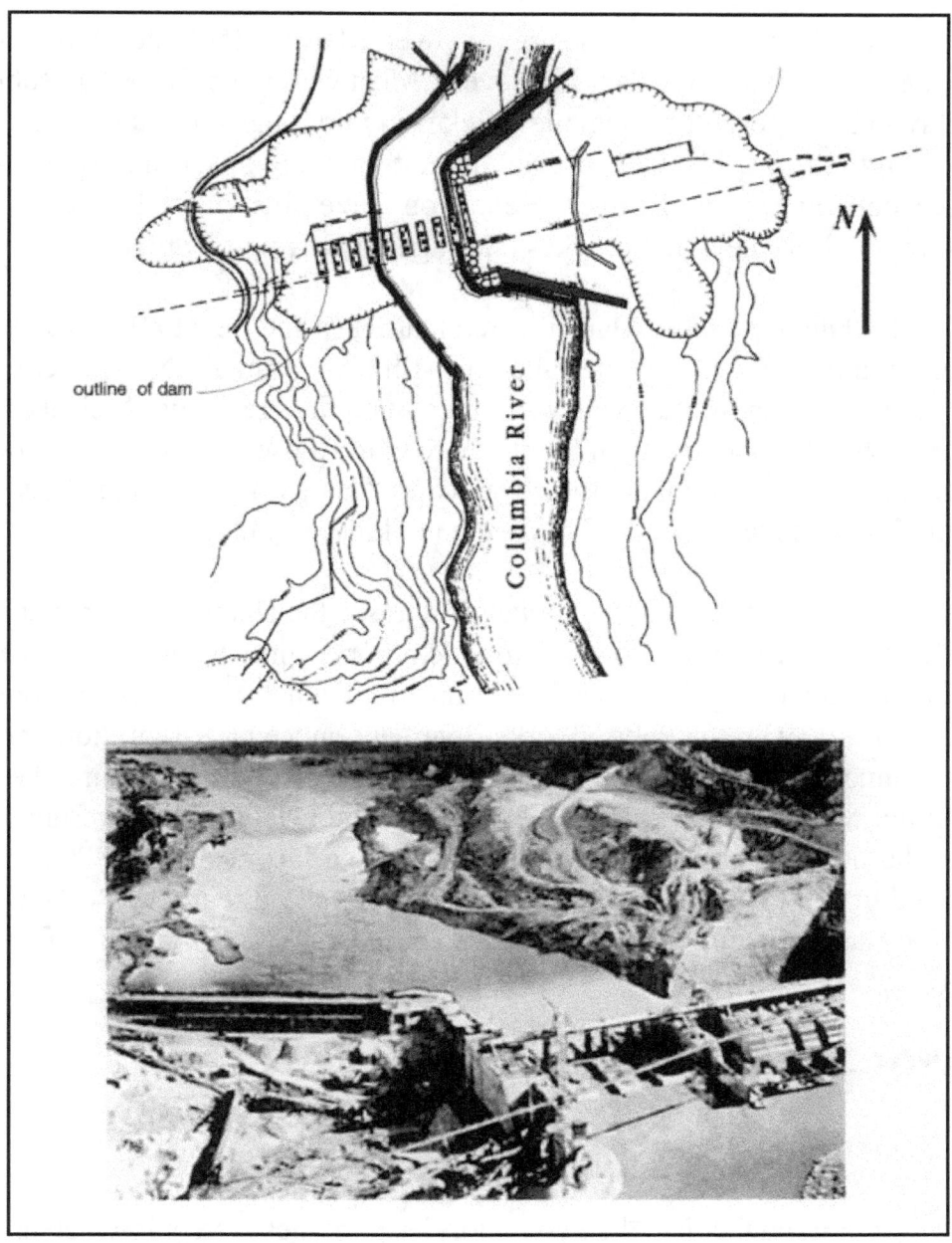

Figure 5-13: During April of 1937 upstream and downstream cross-river cofferdams directed the rising flow over the foundation concrete on the west side of the river. Excavation for the foundation then proceeded below the main river channel and on the east. Source: U.S. Army Corps of Engineers.

After Congress adjourned in June 1933, Martin returned to Oregon, but the two senators remained to talk again in early July with Roosevelt, who seemed still positive about their project. However, the next day (July 13), they met with the "ordinarily irascible" Secretary of the Interior Harold Ickes, who was not so positive, but seemed to be willing to follow the president's wishes.[44] Then came the announcement of the appropriation for Grand Coulee on July 28, with no word on the lower river dam, and that day the president left on vacation. McNary fell ill, and departed to recuperate in Oregon, and the Corps, in studying the dam site more closely, concluded that Bonneville was better than Warrendale.

By September 1, with no approval yet, Martin left Oregon for the capital to meet again with President Roosevelt. Meanwhile, the Corps had submitted their favorable report on Bonneville to Ickes in mid-August but the secretary did not send it to the president. Roosevelt did not see it until Martin gave the White House a copy in early September. He met on September 7 with Roosevelt, who still supported Bonneville, but Ickes once again threw up roadblocks.[45]

By late September Martin feared that Ickes and Senator Dill were intent on killing Bonneville, so he called the still recovering McNary, in Oregon, and urged him to cross the country and meet with the President. McNary arrived on September 24, but both men found the road to Roosevelt politically blocked. Unable to get an appointment, they decided to "camp on the [White House] doorstep until they kicked us out." They were not kicked out.[46]

The two Republican politicians got to see the Democratic president and after showing him the Corps' favorable report, which Ickes had withheld. Martin let McNary argue the case for Bonneville, which Roosevelt accepted, on September 29, 1933, the Public Works Board announced a $20 million grant to begin Bonneville.[47] Astounded that the two Oregonians had overcome the strong opposition within Roosevelt's cabinet (Ickes), a knowledgeable government official exclaimed to Martin, "You would have made a super salesman."[48] Although approved first, Grand Coulee did not get started nearly as early as Bonneville, partly because of its immense size but also because it was approved at first only as a low dam.

GRAND COULEE: COLUMBIAN COLOSSUS

The Design Questions

For Grand Coulee, the two major design issues were a low versus a high dam and a gravity versus a multiple-arch dam. Roosevelt had committed $63 million for the low dam with the understanding that it could be raised in the future without undue extra cost, compared to building the high dam all at once.

Reclamation studied the low multiple-arch dam and gave an estimate of its cost on May 24, 1933, the very day that Fred Noetzli, whose designs Reclamation engineers followed, died.[49] Comparative costs from estimates and bids invariably showed multiple arches to be less expensive than gravity designs. Reclamation, however, had followed a different route, becoming wedded to massive forms.

Nevertheless, Reclamation did make a detailed study of the lighter form during 1933 and 1934. When A. F. Darland, consulting engineer to the Columbia Basin Commission, visited Reclamation in Denver on December 2-4, 1933, the issue of cost comparisons arose, and the bureau's engineers were of "the unanimous opinion...that the hollow type dam should not be used...." By hollow, they meant multiple-arch, and those engineers were by then studying gravity designs. Darland incorrectly stated that Reclamation had "the largest engineering body in the United States" and was engaged in the design and construction of 40 dams. Darland met John L. (Jack) Savage, Chief Design Engineer, and a group of about ten people who had completed "an exhaustive study of the multiple arch, hollow dam...."[5] Darland had asked for comparative costs, and Raymond F. Walter, right after the meeting sent him a letter showing the low gravity dam to be less expensive than the low multiple arch design. Already, in November, Rufus Woods' paper announced that the dam would be "a gravity or solid section structure, instead of an Amberson-type (sic) multiple arch dam as originally intended."[51]

On February 1, 1934, Walter wrote Mead saying that "Altho most of the work in this office to date has been on the multiple arch type of low dam, there is some question whether, in the light of recent developments, this is a feasible plan for the initial development."[52] Savage informed Walter of his recommended plan on February 16, and then sent him a memorandum on March 8 giving the costs of alternate plans. He estimated the recommended plan to cost just under $63 million and the multiple-arch design to cost just over $65 million; however, since the power provisions were different for the two designs, no true comparison is possible because Savage gave no cost breakdown.[53]

But, the matter was settled. Reclamation was, of course, in the middle of building Hoover Dam as a massive curved gravity structure. Nevertheless, Savage had apparently taken seriously the new form, but, in the end, rejected it largely because of the perceived difficulty in raising it to the high dam level later. Reclamation published some results of the multiple-arch dam studies in September 1934, long after rejecting the form. The studies found that the multiple-arch dam would be "somewhat cheaper by about 3.5 percent than the gravity, for the low-head development."[54]

The question, however, remains: could the engineers have designed the multiple arch dam to have been safely raised the additional 191 feet to a total height of about 550 feet at its highest point. In 1939, Reclamation completed Bartlett Dam on the Verde River in central Arizona; it is the highest multiple arch dam in the United States, rising 287 feet at its highest point. It has performed structurally in a satisfactory way since its construction.[55] From a structural point of view, there is no reason why it could not have been designed to be 263 feet higher. André Coyne completed the Grandval multiple arch of 1959 to a height of 288 feet, and in 1968 his dam on the Manicouagan River in Canada rose to just over 700 feet.[56] Such a great jump of 245 percent in height is far greater than the increase for Grand Coulee of 92 percent (from 287 feet to 550 feet) which, however, represents a jump more consistent with Reclamation experience. For example, Reclamation went from Owyhee (417 feet high) to Hoover (726 feet high), or an increase of 74 percent, without undue danger.[57]

There seem to be no specific calculations or even technical discussion detailing why the multiple arch would not have been a "feasible plan." Indeed, Mead wrote on April 18, 1935, that for two years Reclamation had studied the Bartlett Dam and had included a multiple arch design with a maximum height of 347 feet. Reclamation abandoned this structurally feasible design because of cost (not of the dam but of the entire project) and went to a lower dam with a smaller reservoir.[58] The difference between Grand Coulee and the studies going on at the same time (1933-1935) for Bartlett represents an increase of 58 percent. Had Reclamation been focused on less massive forms, it seems likely that they could have succeeded in planning for a multiple arch dam at Grand Coulee.[59]

Raising the Dam

Meanwhile, in late 1933, work had already begun, with a contract for excavation let in November and with the building of roads and bridges in early 1934. On March 3, 1934, Reclamation called for bids on the low dam, 290 feet high with power houses but with no provision for irrigation. Senator Dill and others pushed for a start with enough money to build something substantial and with the promise of immediate jobs in the depressed society.

On June 18, four bidders submitted sealed envelopes. One was from a lawyer whose plan showed thought, but no experience or financial backing; another, from "Mae West," was a poem which promised diversion but no dam. The last two were serious bids, one from the six companies who were building Hoover Dam; they asked just over $34.5 million. The other bid came from a combine of three companies and totaled just over $29.3 million. The low bidders, led by Silas Mason of New York, were declared the winners and Ickes formally awarded them the project on July 13. They were required to complete the low dam in four and one-half years.[60]

Even before the bid, local people had begun to call for the high dam, but Senator Dill cautioned them to lay low until work got well underway; then, he promised, "You will find me just as aggressive for the high dam as you have ever been."[61] Indeed, so was Reclamation, but they too held their fire. The combine, taking the acronym MWAK (for Mason, Walsh Construction Company of Davenport, Iowa, and Atkinson-Kier Company of San Francisco), immediately began work. The site was, therefore, busy on August 4, 1934, when, at 11:05 A.M., President Roosevelt's car drove up and he gave a 20 minute talk to the assembled 20,000 visitors and workers.[62]

The major job for MWAK was to design and build huge cofferdams so that the concrete could be cast on a dry river bed while the raging Columbia was directed into only part of its normal width. There would be no tunnels drilled through mountains as at Hoover Dam, but controlling the river would still be risky. Even before construction on the coffers began, Reclamation had made its arguments explicit for a high dam.

On December 4, 1934, Commissioner Mead held a conference in Denver with his key engineers, led by Chief Engineer Raymond F. Walter, and Chief Design Engineer John L. (Jack) Savage. Two recommendations resulted: first, that the high dam be approved and, second, that construction for the Columbia Basin Irrigation Project be started immediately. Mead's report estimated the total cost of the high dam to be about $114 million, the powerplant to be about $67 million, and the interest an additional $15 million. In addition to this total of $196 million, the cost of irrigating 1.2 million acres added $209 million to give a total project cost of $405 million.[63]

Reclamation's report reveals the interplay between politics and engineering. It states that "a number of factors have developed since construction...was undertaken which have greatly changed the economic situation." First, the report notes the excess of power in the region due to Bonneville Dam, the Skagit River power development, and Grand Coulee. This would seem to argue against the high dam with its immense power potential, but the report goes on to take the opposite tack, "the damsite is not an economical site for a low-head power development...." Therefore, the engineers conclude, it will be more economically sensible to go right to the high dam. Surely, this is not a factor "developed since construction" began.

A second major reason for the high dam was the difficulty of making a safe construction joint between low and high dam, if the low one were completed first. Again, the problems identified did not just arise, but were clearly there before the low dam was decided upon. The third crucial objection was to the turbines; efficient low head turbines would not be efficient for high heads and vice versa, again an issue well known before the construction began. Thus, all three of these reasons are really back-rationalizations for building a high dam, or, more properly, they are convincing arguments against ever starting the low dam at all.

One argument that does seem topical was related to the unprecedented drought throughout the Midwest and the need for irrigation lands on which to settle displaced farmers. Still, the overriding impression one gets is that the high dam could only be gotten politically if Reclamation began the technically indefensible low dam first and saved all its rationale for a time when substantial funds had been committed. Roosevelt surely understood this procedure and the engineering report of December 4, 1934, reflects the process which culminated in Mead's letter of December 27, 1934, to Ickes, formally requesting that Reclamation change plans and begin the high dam construction as soon as possible.[64]

By January 1935, the contractor already had 2,500 men on the site working on the plan to build the low dam. Frank A. Banks, Reclamation engineer in charge of construction, wrote about the dam in late November 1934, still describing the two stage procedure without raising any of the objections that would surface during the December 4 meeting (which Banks did not attend).[65] Then, in early January, *The Engineering News-Record* lowered the boom on the low dam. In a brief unsigned article (most likely by the editor) entitled "A Mistake That Should Be Corrected," the author stated that "doubts arose in the public mind... when it became known that the project had not been studied either as to design or as to ultimate service value."

The article stated bluntly that the low dam plan, with the eventual addition of 200 feet to make the high dam, was "an economic error of first magnitude" and a risky technical project as well. "Before more money is sunk in this wildcat investment the blunder should be corrected by abandoning the low dam power scheme and building instead the high dam." Otherwise, the *News-Record* concluded that, "it is better to abandon the work entirely than to continue useless expenditure."[66]

While noting the same disadvantages as recorded on December 4 by Reclamation, the article implies that the technical basis for work under contract was faulty. The debate was clouded by competing visions for the dam just because it was so clearly a multipurpose dam having benefits in irrigation (its original goal), in power, and in construction jobs. Nearly everyone in early 1935 agreed that the power market was too small to use Grand Coulee's capacity, especially with Bonneville soon to come on line. Most believed that the power needed for irrigation would be the greatest benefit, one which the low dam could not provide.

Meanwhile, MWAK continued work to prepare for the great cofferdams. A consulting report in November 1934 confirmed the contractor's design, and the foundation consultants saw that "no serious doubt can be cast on the safety of the project."[67]

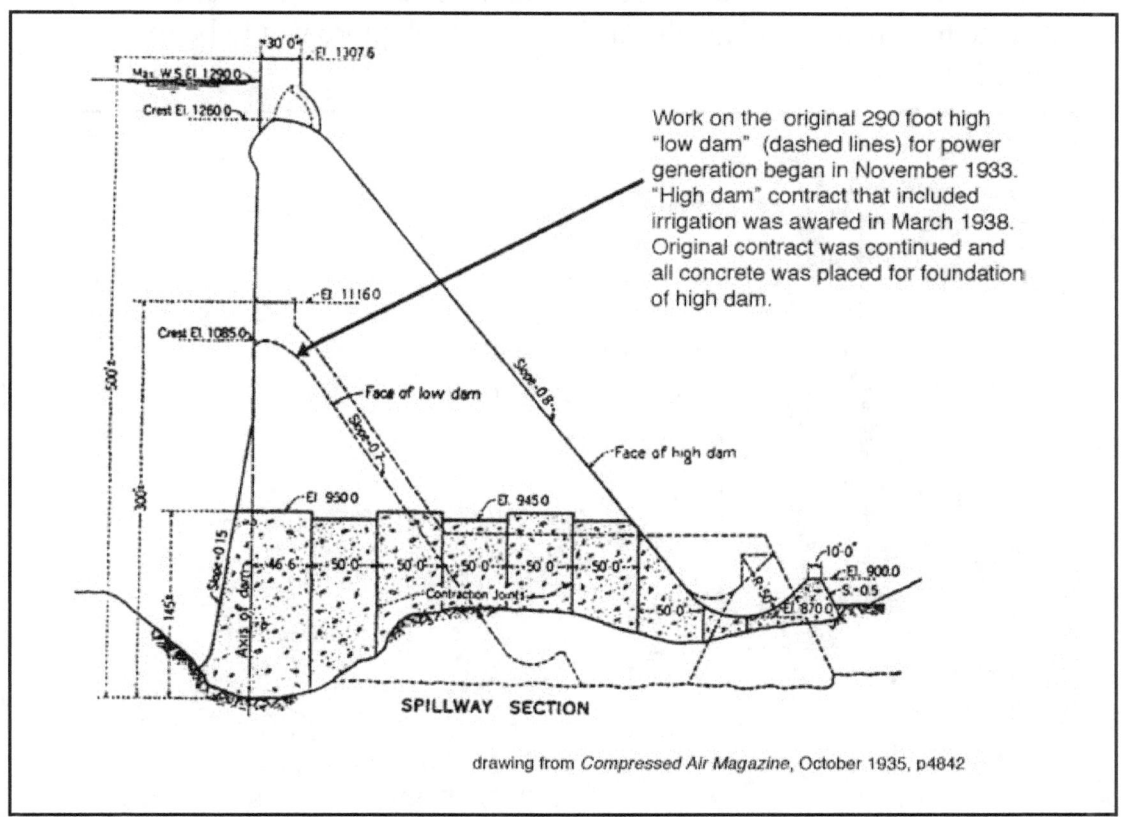

Figure 5-14: This spillway section of Grand Coulee Dam shows both the original low dam plan and the final high dam as built. Source: With permission from *Compressed Air Magazine* (October 1935), 4842, a publication of Ingersoll-Rand Company.

In early 1935, another voice cried out for the high dam, Carl E. Magnusson, professor of electrical engineering at the University of Washington. In a Bulletin from the Engineering Experiment Station, Magnusson presented eleven advantages for the high over the low dam. While power remained central, he answered the criticism of no power market by noting that the irrigation project would increase the state's population and hence provide new markets for electric power.[68]

MWAK had started the cofferdam as 1935 began and, by early April, had driven the last sheet piling. The cofferdam, parallel to the river flow, withstood a high spring flood and served to keep dry the western third of the main damsite so that the foundation could be cast relatively easily. Meanwhile, Reclamation engineers kept at work on designing the high dam. In March, the Senate passed a huge $4.8 billion bill for work relief which included more money for Grand Coulee, but in a series of rulings, the Supreme Court found unconstitutional many aspects of Roosevelt's public works program. Finally, on June 7, Ickes signed a change order to go to the high dam, but that meant only that MWAK would now build the full high dam foundation instead of that part of the foundation necessary for the low dam plus the low dam itself. Thus, on June 17, the Public Works Administration allotted $23 million to Grand Coulee Dam so the high dam could go ahead although much more would be needed to complete it.[69]

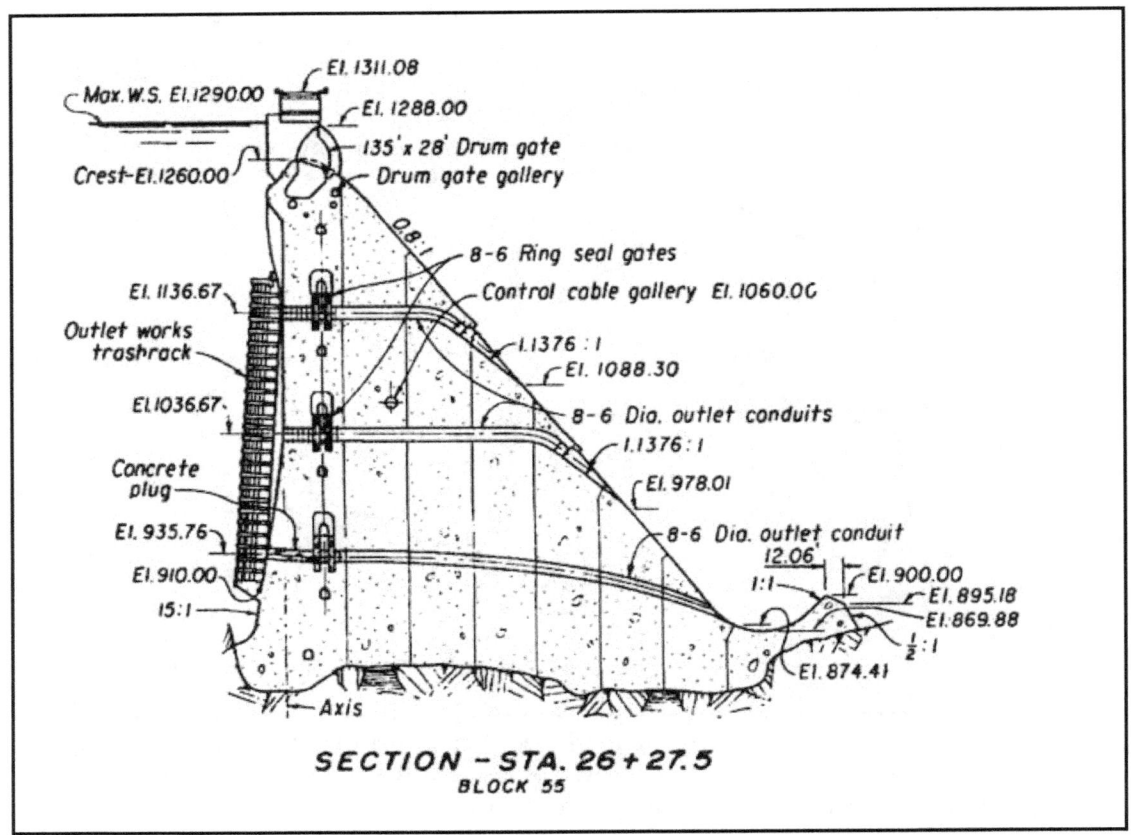

Figure 5-15: At Grand Coulee Dam, outlet tubes were built through the spillway section to facilitate construction. Source: Bureau of Reclamation.

The *Engineering News-Record*, feeling fully vindicated in its earlier harsh criticism of the low dam, exulted, "What heretofore was an engineering and economic error is transformed into a sound and justifiable work-creating undertaking."[70] The *Saturday Evening Post*, recalling Montgomery Schuyler's 1883 forecast on the Brooklyn Bridge, predicted that "Five thousand years from now . . . archaeologists . . . intent upon seeking new clues to a vanished race . . . will come upon a vast waterfall . . . [while all else] will probably have vanished long since . . . the dam . . . will still stand."[71] Even before any concrete had been cast, people began to visualize Grand Coulee, along with Hoover, as somehow defining the civilization of twentieth-century America.

The *Engineering News-Record* devoted 23 pages to the construction up to August 1, 1935, and its concluding editorial makes three claims that characterize the ambiguity of the mid-thirties in America. First, it puts the construction skill on the same level as the "great heights . . . of the designing engineer when he originates and develops his conception." Yet Grand Coulee and Hoover Dams both show how innovative was the construction and how traditional was the design. Second, the editorial justifies the high dam because it will not generate power except to pump water for irrigation. It "does away with the economic paradox of building a huge power project in a market already greatly oversupplied."[72]

The third claim was for state planning which, according to the editorial, "has much to its credit even though its tangible achievements are few." The depression years, which stimulated the New Deal, emphasized how state planning would bring rational decisions into politics and allow development problems to be solved by central planning boards. The great exemplar here was the Tennessee Valley Authority, which had already spurred planners to imagine a Columbia Valley Authority along with many other such organizations. Early in January 1935, Congress began to consider such an authority for the Columbia River Basin and when the *Engineering News-Record's* editorial appeared, the issue was still under consideration. Soon, however, that idea would put both the Corps and Reclamation on the same side, firmly opposed to such plans.[73]

In August 1935, therefore, the situation of the dam was this: out of the $60 million allotted, $15 million was immediately available and had been spent. The new allocation of $23 million, now being used on the modified plan, left $22 million to be made available in 1937. Mead's estimate to Senator McNary for the additional funds needed to complete the high dam was $106 million, with $209 million more required to carry out the irrigation of the Columbia Basin Project.[74] This figure of $166 million for the dam is substantially lower than what he gave in late 1934, $196 million.

Then, in late August 1935, Congress finally passed a Rivers and Harbors Act that specifically authorized construction of Grand Coulee Dam as a full multipurpose structure. On December 6, Washington Governor Clarence Martin pulled the release handle on a four-cubic-yard bucket of concrete to inaugurate the casting of concrete.[75] Before completion, almost 12 million cubic yards would be cast; it would be the largest concrete structure in the world. Superlatives came easily to this wilderness project now, in late 1935, on its way to a structural height of 550 feet and a crest length of 5,673 feet.

Completing the Dam

MWAK had completed the west cofferdam through the winter of 1934–35, and, in the fall of 1935, they completed the east coffer. The contractor began the foundations for the high dam, approved in July, by casting concrete within the west cofferdam. Although halted by freezing weather in January and February 1936, work progressed rapidly in the spring and summer until, by August 17, with all concrete cast, the contractor had dismantled part of the west cofferdam and the river started to flow behind and then over the concrete foundation. Meanwhile, beginning in February, MWAK built cofferdams across the channel both above and below the damsite so that, in December, the engineers could divert the entire river through the foundations cast behind the old west cofferdam. On December 15, 1936, the *Wenatchee Daily World* announced that the river was diverted, and by January 11, 1937, people began flocking to see the exposed bed of the great river.[76]

But two of the principals in that great achievement did not live to see the river bed; Elwood Mead died on January 27, 1936, aged 78 after 12 years as commissioner, and Silas Mason, at only 56, died in April. Mead would be permanently memorialized by the lake behind Hoover Dam, which bears his name, while the contractor had only a temporary fame through the workers' town for Grand Coulee named Mason City.

Concrete began to flow into the east side foundation on November 28, 1936, stopped again in January, and picked up in earnest in March. Despite a leaky downstream coffer, MWAK completed the concrete foundation at the end of 1937. In late November, the Columbia had returned to its familiar channel, except that there was now a dam about 60 feet high which created a new waterfall. Work by MWAK ended in early 1938 with a total cost reaching just over $39 million–about $10 million above its original bid, but 14 months ahead of schedule. Changes from the low to high dam accounted for $6 million of that increase.[77]

As the contractor rushed the foundation to completion through 1937, Reclamation pressed forward with the high dam, for which final plans appeared on November 2 when Ickes set the bid opening for December 10, 1937. The day before that opening, Henry Kaiser met with Guy Atkinson, head of MWAK, and the two agreed to join forces in the bid. Kaiser and his Six Companies teams had lost out to Atkinson's group in the earlier bid, but now they realized the desirability of working together. Kaiser's group had finished Hoover Dam and would soon complete work on Parker Dam, so the timing fit. They bid $34.4 million and won the contract over the only other bidder, another conglomerate, whose price was just over $42 million. Thus, the two former rivals came together, under the new name of Consolidated Builders Incorporated (CBI), to begin this second phase officially on March 18, 1938.[78]

By mid-May, nearly 1,700 men were at work on the dam, and the casting of concrete began in late July. Three shifts daily was the routine by late August. Because the Columbia was free to run over the previously cast concrete, CBI had to erect movable steel gates fifty-two feet wide and thirty-five feet high within which to cast concrete in increments of five feet high until it exceeded the heights of adjacent blocks. The river could then flow over the lower blocks. In this way, the dam progressed upward in a crenulated form while the rising river spilled over the lowest parts as well as through outlet tubes cast with the spillway. The entire scene of rising walls, shooting water, and steel trestle above created a spectacle that drew thousands of visitors to the wilderness site of the world's largest concrete work. The engineers mechanized the entire project such that, for example, the mixing of a four cubic yard batch of concrete took only thirteen seconds. A train on standard-gauge tracks carried the mix in buckets over the trestle where it was then lowered to the dam for placement.[79]

During late 1938, the contractor developed proposals to speed up the concrete work. After careful study, Reclamation engineers responded on December 1 through a memorandum from John Savage to Raymond Walter. This crucial document gives insight into the thinking of Reclamation engineers as they were in the middle of the largest dam building program ever contemplated. Savage warned his chief engineer that Grand Coulee "is the boldest that the Bureau of Reclamation has approved to date for one of its major structures."

Only Shasta (not yet under construction) was comparable, but Hoover was far less bold, as he explained: "the arched design of Boulder Dam makes it a far safer structure than is economically possible for a long, straight-gravity dam [like Grand Coulee]." Savage continued with the dramatic statement that:

> Two-thirds of the total water load against Boulder Dam is carried to the "states of Arizona and Nevada"... to the abutments of the dam, and only the remaining one-third ... goes to the foundation. In the case of a long gravity dam <u>all</u> of the water load is carried to the base of the structure.

As Savage surely knew, this was not a correct comparison; Hoover Dam was more conservative because he designed it to carry all the load to the foundation by neglecting the arch effect, and, thus, it was the neglected arching that made it less bold. Had Hoover been designed by strictly applying the trial-load method and fully utilizing the arch strength, the safety would have been comparable to Grand Coulee.

Savage clearly wished to stress the need for the contractor not to compromise dam safety for construction speed, and he went on to emphasize the boldness of Grand Coulee by telling how:

> It was only as a result of nearly six years of intensive research . . . [on] Boulder Dam that Bureau engineers acquired the necessary personnel and technical knowledge . . . [to design] a structure like Grand Coulee Dam.

Figure 5-16: Grand Coulee Dam and Lake Roosevelt on July 3, 1968. Source: Bureau of Reclamation.

He proceeded to identify the problems inherent in Grand Coulee: there was essentially one major problem (he gave four but all were closely related)– that of ensuring that a series of discrete parts, cast separately as blocks, will create a fully monolithic structure free of dangerous cracks. Savage objected strongly to the contractor's proposal to cast concrete during the freezing months of December, January, and February. He was otherwise sympathetic to the main changes in the construction plan.[80]

As directed by Savage's memorandum, winter work was stopped until the spring of 1939. By April, 5,500 men were at work around the clock, seven days a week, much as at Hoover Dam. The contractors pushed relentlessly onward as it became clear that they would finish a year ahead of schedule. While mechanized in a way similar to Hoover Dam, the Grand Coulee Dam was far more a horizontal than a vertical challenge. No great airy high wire cables carried concrete, but, rather, now it was carried over a dense but spindly steel trestle built right on top of the already cast concrete.

Also, the Columbia River was always present at the site, unlike the Colorado's total disappearance from Black Canyon. The two dams illustrate clearly the two primary components of hydroelectric power: the height (H), in feet, of the water fall and its flow through the powerhouse (Q), in cubic feet per second. Power is equal to height multiplied by flow and by the density of water (w = 62.4 pounds per cubic foot), or HwQ foot-pounds per second. One horsepower (P) is defined as 550 foot-pounds per second, or P = HwQ/550, which, for the estimates of 1934, result in H = 351 feet, a mean flow estimate of 109,000 cubic feet per second and P = 351 (62.4) 109,000/550 = 4.35 million horsepower. Since one horsepower equals 0.746 kilowatts, this power potential would be 4.35 (0.746) = 3.23 million kilowatts of electrical power. The design being constructed in 1940 provided for nearly two million kilowatts, so that when the flow exceeded about 68,000 cubic feet per second, the overflow would go through outlet conduits on the dam or over the spillway. Once the lake behind Grand Coulee filled, then a steady flow higher than the mean was possible, so eventually the dam could supply much more power.[81]

Once again in late 1939, as war began in Europe, construction stopped until late February of 1940. Thereafter began a rush to install the electrical equipment. Concrete casting being nearly completed by September of 1940, many workers headed south to work on the Friant Dam for the Central Valley Project in California. In early 1941, Frank Banks turned on the first small service generators, and the first large-scale turbines began their journey from the east coast, requiring strengthening of rail beds along the way. The first generator arrived in pieces that filled 38 flat cars. Assembly began on February 3, 1941, and the first completed turbine-generator unit officially began producing electricity in early October.

There was a frantic push to complete the project as 1941 came to a close. The contractor stated that all concrete work was completed. The announcement came on December 12, 1941, five days after the Japanese bombed Pearl Harbor and United States isolation from the war was finally ended. By June 1, 1942, water in the lake behind Grand Coulee Dam had risen to the top of the spillway and began its smooth descent to the river below. Early the next year, CBI turned the project over to Reclamation. The final bill totaled $41,400,000 compared to the initial bid of $34,400,000. In spite of such overruns, the total construction cost of Grand Coulee Dam through June 30, 1943, was only

$162,600,000 compared to the government allotment of $179,500,000, a record that would be difficult to match after the war.[82]

THE BONNEVILLE POWER ADMINISTRATION AND THE CONTROL PLAN

Power and the Restudy of the River Basin

With Grand Coulee under construction and Bonneville nearing completion, political pressure built up for a Columbia Valley Authority modeled after the much heralded but still controversial Tennessee Valley Authority. On August 20, 1937, Roosevelt signed the Bonneville Power Act, a compromise that left control of the dams in the hands of the Corps and Reclamation, but created an independent administration within the Interior Department to sell and distribute the power from Bonneville Dam and Grand Coulee. On August 26, 1940, Roosevelt ordered the Bonneville Power Administration (BPA) to take charge of the power from Grand Coulee, thus creating a complex set of three separate agencies, each responsible for separate pieces of the total Federal Columbia Valley System.

Item	Reservoir (see Fig. 2)	Status as of July, 1950	Construction cost (thousand dollars)	Usable storage* (acre-feet)	Normal head of dam (feet)	Installed capacity (kilowatts)
(1)	(2)	(3)	(4)	(5)	(6)	(7)
1	Hungry Horse	Under construction (USBR)	107,000	2,100,000 (J) / 880,000 (P)	485	300,000
2	Glacier View	Alternate to be recommended	94,962	3,160,000 (J)	402	210,000
3	Libby	Authorized (USED)	239,077	4,250,000 (J)	353	588,000
4	Albeni Falls	Authorized (USED)	31,070	1,140,000 (P)	25.5	42,600
5	Grand Coulee	In operation (USBR)	273,000	5,120,000 (J)	348	1,944,000
6	Chief Joseph	Under construction (USED)	225,000	Pondage	170	1,280,000
7	Priest Rapids	Authorized (USED)	326,124	2,100,000 (FC)	146	1,219,000
8	Hell's Canyon	Recommended	342,076	3,280,000 (J)	575	980,000
9	Lower Snake River dams	Authorized (USED)	359,000	Pondage	80 to 100	980,000
10	McNary	Under construction (USED)	227,000	Pondage	90	980,000
11	John Day	Authorized (USED)	379,826	2,000,000 (FC)	95	1,105,000
12	The Dalles	Authorized (USED)	286,286	Pondage	88	980,000
13	Bonneville	In operation (USED)	85,000	Pondage	65	518,000

* Abbreviations: "J" denotes joint use of storage; "P" denotes power storage; and "FC" denotes flood control storage. † Flood control use recommended.

Figure 5-17: The reservoirs of the Main Control Plan, Columbia River Basin. Source: U.S. Army Corps of Engineers.

There followed substantial wrangling among the new power administration, Reclamation, and private power companies. BPA wanted to sell the power as cheaply as possible to reduce the influence of private power; Reclamation wanted higher prices to help subsidize irrigation for the Columbia Basin Project; and private power wanted even higher prices to allow it to remain profitable.[83] BPA won, and throughout the next 60-plus years Pacific Northwest power has remained some of the least costly in the nation.[84]

But the two great dams were only the beginning of the Columbia restructuring. So, with the war still raging in 1943, the Congress directed the Corps to review the original 1932 report in the light of the completed dams and the newly formed Bonneville Power Administration. Just as the earlier report resulted, largely, from the dedicated effort of one Corps officer, so the new report would be primarily the work of a 39-year-old colonel, William Whipple, Jr.

Sent to Portland in 1947, Whipple reported to Colonel Theron Weaver, the Division Engineer, Northwest Division, U.S. Army Corps of Engineers, for whom he was to be the executive officer. Whipple began to study the new report, already four years in progress, on which the government had spent $4 million while surveying over 1,000 dam sites. He quickly recognized that the study was essentially leaderless and without clear direction.[85] The missing focus resulted partly from the complexity of issues on the Columbia and partly from the lack of leadership. The Corps was still unsure, as a bureau, how to deal with questions of irrigation, flood control, power, and fisheries along with its historic mission of navigation.

The division engineer put Whipple in charge and he quickly focused the study on a few large power projects plus a navigation plan permitting ships to get to Idaho. Flood control, he believed, was not a major problem and could be handled by a few levees and some channel improvements.[86] In 1946, Reclamation had published its own study of the Columbia River Basin, focusing on irrigation and dams.

The Flood and the Control Plan

Meanwhile, Whipple worked on his study, and, by early June 1948, he was far enough along to justify a vacation, so he went with his wife and their three young children north into British Columbia. Once there, he saw the valleys in flood, "the greatest flood the country had seen since Indian days." It was descending south into the Columbia valley and Whipple realized that he needed to be back in Portland immediately.

The public alarm mounted, water at Bonneville Dam was so high that power production stopped; the Portland Airport was under 15 feet of water; and Vanport, a community of 20,000, was inundated beyond repair. Plans to minimize flood control were doomed as the Corps struggled to keep losses down. The Chief of Engineers, Lieutenant General Raymond A. Wheeler (1885–1974), flew out from Washington, D.C., and Colonel Whipple met him.[87] The next morning, Wheeler held a 25 minute meeting with his local engineers in preparation for a press conference at 8:30 A.M. Thirty years later, Whipple reported Wheeler's comments at that meeting:

"The news people have been after me since 6:30 this morning; and I announced a press conference here at 8:30. That gives us about 25 minutes. Now I intend to say that the situation remains in the hands of the district and division engineers, in whom I have implicit confidence. I will also say that my office has already promised to give complete support in terms of men, equipment, and money. That's right isn't it?" It certainly was, and we all nodded. "Now what is a rough estimate of the damage caused by the flood?" Weaver and Walsh had a moment's consultation and told him about $100 million.[88] "OK," he said "And of course I'll refer to the assistance of the troops and other agencies. Now what about the big 308 report? As I recall, it was due in our office on 1 October; but you requested a year's extension of time. Of course, under the circumstances, there can be no extension of time. You will have the report in by 1 October, won't you?" Colonel Weaver turned to me, and I said "Yes, Sir, we will." "Now, in view of this flood and all of the interest nationwide, we practically have to have a flood control plan in it. The report does have a flood control plan in it, doesn't it?" Again Colonel Weaver looked at me for the answer. I felt strength mounting in me. "The plan we are working on does not have flood control in it now" I said, "But when we submit it, it will." Colonel Weaver looked as relieved as his normally impassive face permitted. General Wheeler gave me a quick look of understanding and appreciation.

"Well, gentlemen" he said, "I have all the information I need for the press conference. I see that my confidence in the district and division offices was not misplaced." This time he looked also at me. "Now, since we have lots of time before the press arrives, let's all have a cup of coffee."

And so the Comprehensive Plan proceeded at a furious rate to include flood control and to be ready by the end of September. Before the war, many people doubted the need for much new power. But the wartime growth of industries such as aluminum had made power the primary objective, while the flood of 1948 highlighted the competing objective of flood control. For maximum power, the reservoirs must be filled, but for maximum flood protection they must be drawn down to prevent flood crests from moving downstream. Thus, the more power the less flood control and vice versa. Whipple organized the study around a Main Control Plan focused on a few major dams and a system-wide coordination of flood predictions with the necessary draw down of reservoirs. That plan was developed by 1948 and published in 1950.[89]

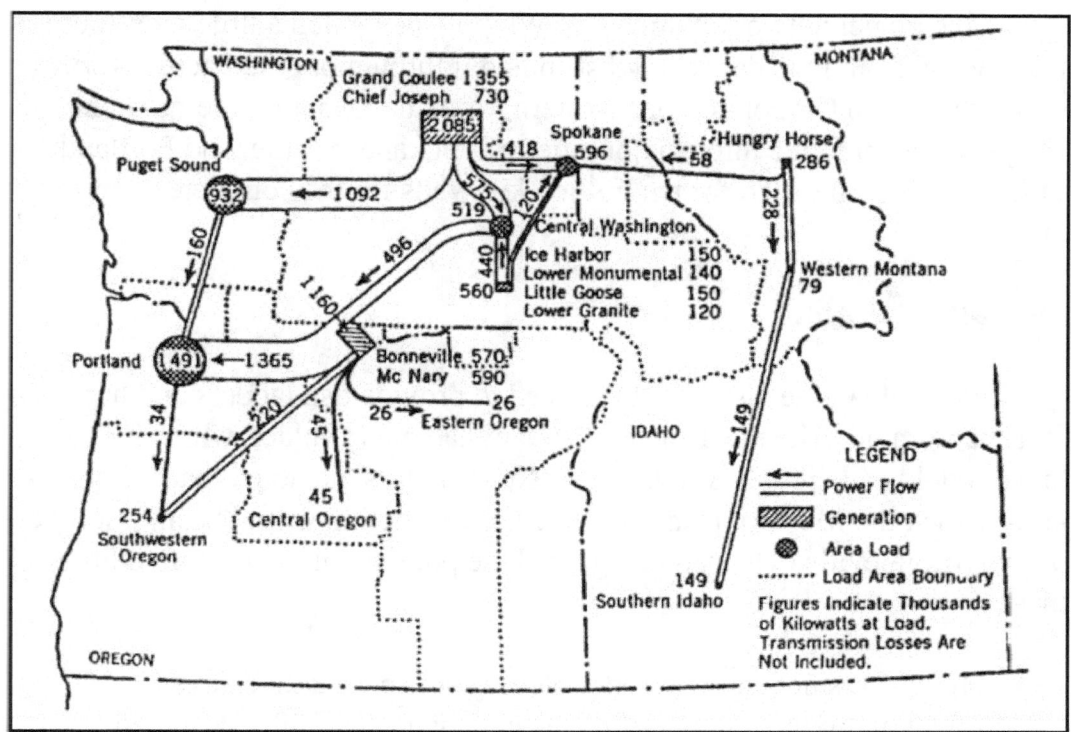

Figure 5-18: The power network planned for Washington state by Colonel William Whipple, Jr. Source: U.S. Army Corps of Engineers.

Of the 16 dams with reservoirs, only Bonneville and Grand Coulee were then operating, but three others were under construction and Congress had authorized nine others. The system was to provide enough emergency storage in seven reservoirs to hold the flow down to a level that did not exceed the capacity of the downstream levees.[90] At the same time, this increased storage capacity leads to a decrease in power because the reservoirs must be partially emptied in preparation for the flood crest.

Dam	Owner	Date in service	Power kw	Discharge cu.feet/sec	Reservoir acre-feet	River
Bonneville	Corps	1938	1,078,000	183,300		Columbia
Grand Coulee	Bureau	1941	7,003,000	107,700	5,190,000	Columbia
Hungry Horse	Bureau	1952	328,000	3,517	3,160,000	FlatHead
McNary	Corps	1953	1,133,000	169,800		Columbia
Albeni Falls	Corps	1955	43,000	25,340	1,160,000	Pend Oreille
Chief Joseph	Corps	1955	2,075,000	108,000		Columbia
The Dalles	Corps	1957	2,052,000	177,900		Columbia
Ice Harbor	Corps	1961	643,000	47,680		Snake
John Day	Corps	1968	2,484,000	172,400		Columbia
Lower Monumental	Corps	1969	930,000	47,670		Snake
Little Goose	Corps	1970	932,000	47,230		Snake
Dworshak	Corps	1973	400,000	5,280	2,020,000	Clearwater
Lower Granite	Corps	1975	932,000	49,680		Snake
Libby	Corps	1975	604,000	11,350	4,980,000	Kootenai

Figure 5-19: Dams on the Columbia River system. Source: U.S. Army Corps of Engineers.

The power network planned by Whipple generated a little over four million kilowatts from nine dams and distributed it throughout the Pacific Northwest from southwestern Oregon to northwestern Montana. About three-quarters of the power went to the three major cities of Spokane, Seattle, and Portland. Eventually, the plan called for ten million kilowatts for the completed control plan.

Power and Salmon

Whipple wrote also about the need to provide fish ladders and hatcheries for the salmon. Already, Bonneville Dam had a fish ladder, which by 1950, Whipple could call "highly successful." Nevertheless, he noted that the provision for fish at the McNary Dam then under construction would be an improvement over Bonneville. He also recognized the potential influence of additional dams:

> However, since there is undoubtedly some cumulative adverse effect of a large number of dams, even when carefully designed to aid the passage of fish, the comprehensive plan calls for a compensatory program of $20,000,000, known as the Lower Columbia Fisheries Plan, to clear natural obstacles from the lower tributary streams and to build hatcheries. This plan was developed by the Fish and Wildlife Service. Although the plan as a whole has not yet been authorized by Congress, some money has already been appropriated, and the State of Washington has taken legislative action to prevent encroachment upon the streams concerned in that state.[91]

Clearly, the fish were on the minds of the Corps, and five years later it would establish a laboratory at Bonneville Dam for research on anadromous fish (fish which go from fresh water to salt water and return to spawn). But as late as 1994, a former Director of the National Marine Fisheries Service could say "We still don't know the best way to protect the fish."[92] A more detailed discussion of the fish and other environmental issues is reserved for chapter 9.

The Main Control Plan had become essentially a power plan lighting up the Pacific Northwest and bringing it industry and wealth. Although the Corps took seriously the fish problem, it was not central to its planning. Even more crucial at the time was the Corps' lack of emphasis on irrigation. Whipple did refer to Reclamation's studies of the Columbia Basin, but he considered it a secondary problem in light of the power and flood control issues. The origins of the Grand Coulee project gave way to the development of its power, which by the 1980s had reached the phenomenal total of over seven million kilowatts.[93]

In the conclusion to his report on the comprehensive plan, Whipple emphasized strongly the engineering, rather than the scientific, nature of the Control Plan. As he put it, the "general basin-wide plans and criteria . . . are not general ideas or principles but highly specific engineering conclusions and relationships which vary in accordance with the stage of development." In other words, scientific and abstract planning principles are of little value in developing such comprehensive projects. Rather, they are "empirically derived from engineering relationships and physical and economic conditions."[94]

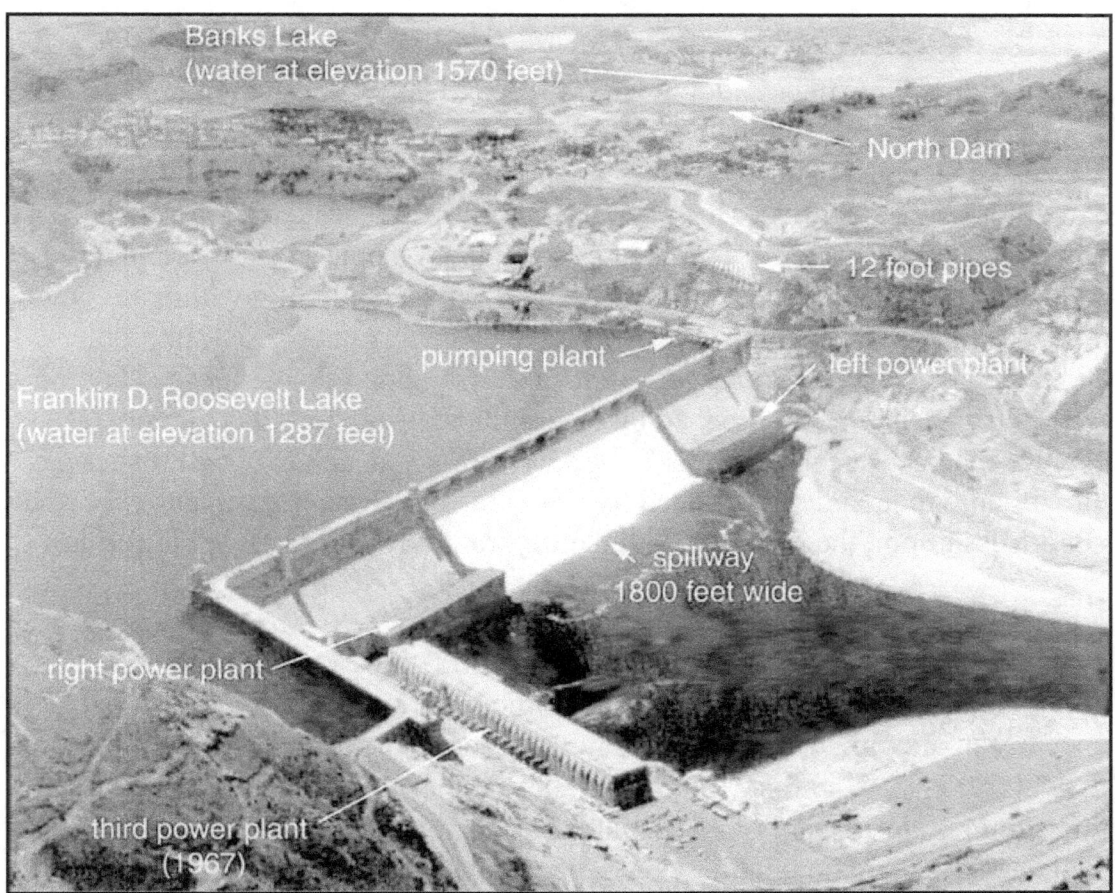

Figure 5-20: An aerial view of Grand Coulee Dam with the new Third Powerhouse at the bottom center. Source: Bureau of Reclamation.

Whipple was announcing here his ideas behind multipurpose dams, and he was doing that in the light of both helping the public understand the issues and responding to the public's cry for flood protection and power, which by 1948 had become a major goal in the Northwest.

Following publication of the Control Plan, Whipple became the first district engineer of the newly formed Walla Walla District, where he oversaw completion of McNary Dam, named for the senator most responsible for the lower Columbia development and the Republican running mate of Wendell Wilkie in the 1940 presidential election.

Subsequently, the dams were built as shown in Table 2, and, by 1975, the Corps and Reclamation had essentially completed the 1948 plan in which the Bonneville Power Administration distributed the power.[95] In 1966, the Congress authorized construction of a third powerplant for Grand Coulee, and construction began in July of 1967. The contractor completed the necessary dam alterations in 1978, and the full powerplant was in operation by mid-1980. In 1955, Grand Coulee had lost its position as the world's largest powerplant, but with the Third Powerhouse in operation, that title returned to the upper Columbia River for a time.[96]

That powerhouse, along with the second powerhouse at Bonneville, ended major dam modifications in the Northwest. As Whipple's report shows, he was also concerned about the environment, but he knew that he could not predict all the consequences of his plan. He knew that he needed a plan, yet also a flexibility to respond to new problems. In 1948, Whipple put power and flood control above the salmon, and for that time we must credit him with reasonable priorities. Even more to the point, he had relegated irrigation to a subproblem, and there he had considerable foresight. The irrigation scheme for the Columbia Basin Project has been criticized by environmentalists. But the Columbia River Control Plan is a special case, as Whipple emphasized, and to get a broader picture of large federal dams we need to turn to the other major basin projects: the Ohio-Mississippi, the Missouri, and the Central Valley Projects of California.

The year 1948 also saw the victory of Harry S. Truman in the presidential election. "The president was certain that the west had given him a mandate. . . . His White House advisors and officials at the Interior Department at once began drawing up plans for the reintroduction of regional valley authority legislation."[97] The first Columbia Valley Authority bill had appeared before Congress in January of 1935 but had strong opposition from the Army engineers, Reclamation, public power people, irrigation people, and Senator Charles McNary. The Bill died in Committee. Other efforts failed up to 1937, when Congress created the Bonneville Power Administration.[98] Thus, when the Truman advisors began to rework a valley authority, there already existed a complex of three major agencies in the Columbia Basin. Also significant in 1948 were two other factors. First, there was the image of the TVA and the proposal for similar authorities, all of which were examples of centralized planning and therefore considered by some as threats to American free enterprise. Second, and more fundamental, was the issue of centralized federal control over local authority. Ultimately, these two factors doomed the Columbia Valley Authority, which lost substantial support after the election of 1950 and never surfaced again.[99] A similar fate awaited President Roosevelt's hopes for an authority for the Missouri Valley, a story to which we now turn.

Figure 5-21: Grand Coulee Dam on the Columbia River, the feeder canal from the Grand Coulee Pump-Generating Plant outlet into Banks Lake (lower right). Banks Lake provides initial storage for water to be delivered to the Columbia Basin Project for irrigation of over 670,000 acres. Source: Bureau of Reclamation.

Figure 5-22: The west cofferdam at Grand Coulee in 1935. Source: Bureau of Reclamation.

Figure 5-23: Excavating to bedrock behind the west cofferdam at Grand Coulee in 1936. Source: Bureau of Reclamation.

Endnotes

1. William F. Willingham, *Water Power in the "Wilderness": The History of Bonneville Lock and Dam* (Portland, Oregon: U.S. Army Corps of Engineers, Portland District, 1988) [undated but certainly 1988 because it commemorates the 50th anniversary of the dam dedication of July 9, 1938], 3, 55.
2. Quoted in G. R. Lukesh, "The Columbia River System," *The Military Engineer* 22, No. 124 (July-August 1930), 328.
3. Willingham, *Water Power*, 27.
4. "Estimate of Cost of Examinations etc. of Streams Where Power Development Appears Feasible," *House Document No. 308*, April 12, 1926, 6 pages.
5. "Columbia River and Minor Tributaries," *House Document No. 103*, March 29, 1932, 1845 pages plus appendices.
6. "Columbia River and Minor Tributaries," 13.
7. Oscar O. Kuentz, "The Lower Columbia River Project," *The Military Engineer* 25, No. 139 (January-February 1933), 38.
8. Paul C. Pitzer, *Grand Coulee: Harnessing A Dream* (Washington State University Press, 1994), 11-3.
9. "Great Depression," *Encyclopedia Britannica*, 1997, 5, 443.

10. Willingham, *Water Power*, 4, describes the process of authorization as follows: The Federal Emergency Administration of Public Works authorized Bonneville Dam on September 30, 1933 as Federal Works Project No. 28, under provisions of the National Industrial Recovery Act. When work began on November 17, 1933, the plans called for locating a dam, a powerplant with two units, and a navigation lock in the vicinity of Bonneville, Oregon. The initial allotment contained $20,000,000 for construction and $250,000 for preliminary study and design. Before Congress formally adopted the project on August 30, 1935, putting it under the regular appropriations process, $32,400,000 in public works funds had been spent. It cost another $7,500,000 to complete the undertaking as originally planned. Subsequently, the Corps installed eight additional power units to complete the project at a total cost by 1943 of $75,000,000.

11. Kuentz, "Lower Columbia River Project," 41.

12. J. S. Gorlinski, "The Bonneville Dam," *The Military Engineer* 27, No. 153 (May-June 1935), 210-2.

13. C. G. Galbraith, "Kaplan Turbines for Bonneville," *Engineering News-Record* (May 27, 1937), 765-9. For the history of Kaplan turbines, see Norman Smith, *Man and Water: A History of Hydro-Technology* (New York: Scribner's Sons, 1975), 189-99.

14. Willingham, *Water Power*, 4.

15. R. C. Binder, *Fluid Mechanics* (New York: Prentice-Hall Inc., 1943), 211-4.

16. J. C. Stevens, "Models Cut Costs and Speed Construction," *Civil Engineering* 6, No. 10 (October 1936), 674-7.

17. J. C. Stevens and R. B. Cochrane, "Pressure Heads on Bonneville Dam," *Transactions American Society of Civil Engineers*, 109 (1944), 77-85.

18. William F. Willingham and Donald Jackson, "Bonneville Dam," HAER OR-11 (April 1989), 10. Prints and Photographs Division, Library of Congress.

19. Willingham, *Water Power*, 14.

20. R. R. Clarke and H. E. Brown Jr., "Portland-Puzzolan Cement as used in the Bonneville Spillway Dam," *Journal of the American Concrete Institute* (January - February 1937), 183-221.

21. Kollgaard and Chadwick, editors, *Development of Dam Engineering in the United States*, 34-5, 38, 285, 295, 450, 552.

22. Willingham, *Water Power*, 1988, 16-7.

23. Charles H. Martin to Major General Edward M. Markham, June 8, 1935, 3 pages, National Archives and Records Administration, College Park, Maryland, Records of the U.S. Army Corps of Engineers, RG77, Entry 111, Box 130, Folder 3362.

24. Major General E. M. Markham to Charles H. Martin, June 13, 1935, National Archives, College Park, RG77, Entry 111, Box 130, Folder 3362.

25. Martin to Markham, National Archives, College Park, RG77, Entry 111, Box 130, Folder 3362.

26. Willingham, *Water Power*, 27.

27. Richard White, *The Organic Machine* (New York, Hill and Wang, 1995), 94-5 states that "A folklore has grown up about the Bonneville that, as originally designed, the dam contained no fishways. But except for artists' sketches, there were always provisions for fish passage at the Bonneville. . . ."

28. Willingham, *Water Power*, 47. For Bonneville, see Lisa Mighetto and Wesley J. Ebel, *Saving the Salmon: A History of the U.S. Army Corps of Engineers' Efforts to Protect Anadromous Fish on the Columbia and Snake Rivers* (Seattle: Historical Research Associates, Inc., 1994), 53-8.

29. Kuentz, "The Lower Columbia River Project," 44.

30. Willingham, *Water Power*, 47-9.

31. Letter from the Director of the Department of Fish Culture to the Honorable John C. Veatch et al., August 15, 1934, Thomas M. Robins Papers, Office of History, Headquarters, U.S. Army Corps of Engineers, Alexandria, Virginia. Attached to this letter was a plan dated August 10, 1934 and an eight page memorandum to the Commissioner of Fisheries (unsigned and undated) explaining the August 10 plan. Copy. Original in Special Collections, Knight Library, University of Oregon.

32. Jim Marshall, "Dam of Doubt," *Colliers* (June 19, 1937). Reprinted in Willingham, *Water Power*, 44-5.

33. John R. Jameson, "Bonneville and Grand Coulee: The Politics of Multipurpose Development on the Columbia," Office of History, U.S. Army Corps of Engineers, Fort Belvoir, Virginia, 64 pages. The Roosevelt quote is from U.S. Congress, Senate, *Preliminary Report of the Inland Waterways Commission* (Senate Document No. 325), 60th Congress, 1st Session, 1908, iv-v.

34. Thomas M. Robins, "Improvement of the Columbia River," *Civil Engineering* 2, No. 9 (September 1932), 567.

35. William F. Willingham, *Army Engineers and the Development of Oregon: A History of the Portland District, U.S. Army Corps of Engineers* (Washington, D.C.: U.S. Army Corps of Engineers, 1983), 93.

36. Pitzer, *Grand Coulee*, 48.

37. Pitzer, *Grand Coulee*, 1-6.
38. Pitzer, *Grand Coulee*, 12-3.
39. Pitzer, *Grand Coulee*, 50.
40. Pitzer, *Grand Coulee*, 54. For a brief discussion of the report, see John S. Butler, "Comprehensive Study by Army Engineers," *Civil Engineering* 1, No. 12 (September 1931), 1075-80.
41. Clarence C. Dill, *Where Water Falls* (Spokane, Washington: C. W. Hill Printers, 1970), 167-70. Jameson, "Bonneville and Grand Coulee," 27-30, describes the meetings between Dill and Roosevelt. The formal approval came from Harold L. Ickes to the Secretary of the Interior, August 10, 1933, National Archives, Denver, RG115, Entry 7, Grand Coulee Dam Project files. Ickes transferred $63 million in funds under the National Industrial Recovery Act to Reclamation for the Grand Coulee project. See also *Engineering News-Record* (August 3, 1933), 145.
42. Pitzer, *Grand Coulee*, 25-7, 72.
43. Jameson, "Bonneville and Grand Coulee," 32-3.
44. Pitzer, *Grand Coulee*, 74-7.
45. Pitzer, *Grand Coulee*, 41-4.
46. Pitzer, *Grand Coulee*, 49.
47. Pitzer, *Grand Coulee*, 49-50. *Engineering News-Record* (October 5, 1933), 420, announced the grants to both Bonneville and Grand Coulee.
48. Willingham, *Water Power*, 4.
49. R. F. Walter to A. F. Darland, December 7, 1933, National Archives, Denver, RG115, Entry 7, Grand Coulee Dam Project files.
50. A. F. Darland, report of a conference held at the office of Reclamation, December 2-4, 1933, received February 14, 1934, National Archives, Denver, RG115, Entry 7, Bureau of Reclamation Project correspondence File, 1930-1945. Columbia Basin Project. 301. Box 527. File folder: 301. Columbia Basin Project. Board and Engineering Reports on Construction Features. Jan. 1, 1933-June 30, 1934. Darland was an electrical engineer who became an assistant construction engineer for Reclamation at Grand Coulee on April 17, 1934. See L. Vaughn Downs, *The Mightiest of Them All: Memories of Grand Coulee Dam* (Fairfield, Washington: Ye Galleon Press, 1986), 24. Walter's claim to be head of the largest engineering body in the United States is not true if one considers the total number of engineers (civilian and military) employed by the U.S. Army Corps of Engineers.
51. "Plans Changed—Foundation Going in for High Coulee Dam," *Wenatchee Daily World* (November 27, 1933), 1, 5. Savage had directed the study of the multiple-arch dam which "Bureau engineers . . . speculated . . . could be raised later," Pitzer, *Grand Coulee*, 71.
52. R. F. Walter to Elwood Mead, February 1, 1934, National Archives, Denver, RG115, Entry 7, Grand Coulee Dam Project files.
53. J. L. Savage, "Memorandum to Chief Engineer," Denver, March 8, 1934, 6 pages. An earlier letter from Acting Chief Engineer S. O. Harper to the Columbia Basin Commission, February 21, 1934, stated that the dam would be a gravity section. The relevant portion of this letter is quoted by B. E. Stoutmeyer to Chief Engineer, Bureau of Reclamation, March 26, 1934, National Archives, Denver, RG115, Entry 7, Bureau of Reclamation Project Correspondence File, 1930-1945. Box 535. Columbia Basin Project 301.1. Folder: 301.1 Columbia River Basin. Dams & Reservoirs. Grand Coulee Dam. July1, 1933-June 30, 1934.
54. W. O. McMeen, "Memorandum to Chief Designing Engineer: Summary of Multiple-Arch Dam Studies for Grand Coulee Project," Denver, September 17, 1934, 86 pages, National Archives, Denver, RG115, Entry 7, Grand Coulee Dam Project files.
55. James Legos, "Concrete Buttress Dams," in Kollgaard and Chadwick, editors, *Development of Dam Engineering in the United States* 550-1, 655-62.
56. Editor's note: This is the Daniel Johnson Dam which was completed in 1968 as a hydroelectric dam in the Province of Québec by Hydro-Québec.
57. Schnitter, *A History of Dams*, 186-7.
58. Elwood Mead to Fred Schnepfe, Director of Projects Division, Federal Emergency Administration of Public Works in Washington, D.C., April 18, 1935. The Reclamation report is in a "Memorandum to Chief Designing Engineer," by E. R. Dexter, Denver, June 29, 1936, National Archives, Denver, RG115, Entry 7, Grand Coulee Dam Project files.
59. The problems of raising a dam once built would have been just as difficult for the massive as for the light design. At the hydraulics laboratory at Colorado State College, there were model studies made of the spillway section for the multiple arch design for Grand Coulee. See "Report" by A. F. Darland, National Archives, Denver, RG115, Grand Coulee Dam Project files. It is not stated in the correspondence noted here what influence these model studies had on the decision to abandon multiple arch design.

60. Pitzer, *Grand Coulee*, 97-101.
61. C. C. Dill to James O'Sullivan, February 4, 1934, quoted in Pitzer, *Grand Coulee*, 117. See also letters from Dill to S. H. Hedges, July 24, 1934, and B. E. Stoutmeyer to Elwood Mead, July 18, 1934, National Archives, Denver, RG115, Entry 7, Grand Coulee Dam Project files.
62. Pitzer, *Grand Coulee*, 101 and see Downs, *Mightiest of Them All*, 8, for a photo of the visit.
63. Elwood Mead *et al.*, to the Secretary of the Interior, December 21, 1934, National Archives, Denver, RG115, Entry 7, Grand Coulee Dam Project files. Costs are on 6.
64. Elwood Mead to Secretary Ickes, December 27, 1934, National Archives, Denver, RG115, Entry 7, Grand Coulee Dam Project files.
65. F. A. Banks, "Columbia Basin Project is Described by Construction Engineer," *Southwest Builder and Contractor* (November 23, 1934), 8-9.
66. "A Mistake That Should Be Corrected," *Engineering News-Record* (January 3, 1935), 23.
67. F. A. Banks to R. F. Walter, January 22, 1935, enclosing S. H. Woodard, "Cofferdam for Construction of Grand Coulee Dam," November 13, 1934, National Archives, Denver, RG115, Entry 7, Grand Coulee Dam Project files. See also Charles P. Berkey, "Foundation Conditions for Grand Coulee and Bonneville Projects," *Civil Engineering* 5, No. 2 (February 1935), 68.
68. Carl Edward Magnusson, "Hydroelectric Power in Washington," *Bulletin No. 78*, Engineering Experiment Station Series, Seattle, February 1934, 28 pages.
69. Pitzer, *Grand Coulee*, 120-3. See also "Constructing the First Cofferdam," *Engineering News-Record* (August 1, 1935), 148-9.
70. "Grand Coulee Revised," *Engineering News-Record* (June 20, 1935), 888. A Department of the Interior press release announced Ickes' action, June 8, 1935, National Archives, Denver, RG115, Entry 7, Grand Coulee Dam Project files.
71. Robert Ormond Case, "The Eighth World Wonder," *The Saturday Evening Post*, (July 13, 1935), 36. Also see Montgomery Schuyler, edited by William H. Jordy and Ralph Coe *American Architecture and Other Writings* (New York: Atheneum, 1964), 164.
72. *Engineering News-Record* (August 1, 1935), 164.
73. Pitzer, *Grand Coulee*, 234-6.
74. Elwood Mead to Senator Charles L. McNary, August 17, 1935, National Archives, Denver, RG115, Entry 7, Grand Coulee Dam Project files. It is unclear why Mead referred to the initial allotment as $60 million when Roosevelt publicly identified it as $63 million. See also Charles H. Carter, "Change in Plan for Grand Coulee Dam Explained by Engineer," *Southwest Builder and Contractor* (August 23, 1935).
75. Pitzer, *Grand Coulee*, 129-30.
76. Pitzer, *Grand Coulee*, 140-1. See also Alvin F. Darland, "The Columbia Basin Project," *Electrical Engineering* 56, No. 11 (November 1937), 1344-5.
77. Pitzer, *Grand Coulee*, 150-4.
78. Pitzer, *Grand Coulee*, 196-7.
79. Pitzer, *Grand Coulee*, 198-204. For a more complete description see "Concrete Placing at Grand Coulee: from Gravel Pit to Forms," *Western Construction News* (June 1939), 198-203.
80. John L. Savage, "Memorandum to Chief Engineer," Denver, December 1, 1938, 3 pages. Transmitted to Commissioner John C. Page as an attachment to a more report dated December 12, and signed by Walter, Savage, Banks, McClellan, and Hammond, National Archives, Denver, RG115, Entry 7, Bureau of Reclamation Project Correspondence File, 1930-1945. Box 534. Columbia Basin Project. 301.1. Folder: 301.1 Columbia Basin. Dams & Reservoirs, Grand Coulee Dam, 1938. Grand Coulee Dam Project files.
81. "Undertakings Without Precedent," *Engineering News Record* (November 29, 1934), 679. For the power capacity of the original 1943 dam, see Kollgaard and Chadwick, editors, *Development of Dam Engineering in the United States*, 136. The initial power design was for 18 units each generating 108,000 kW plus three smaller service units of 12,500 kW each so that the total was 1,981,500 kW.
82. Pitzer, *Grand Coulee*, 210-3.
83. Pitzer, *Grand Coulee*, 236-8.
84. Timothy Egan, "A G.O.P. Attack Hits Bit Too Close to Home," *New York Times* (Friday, March 3, 1995), A14. The article states that "Electricity in the Northwest, most of it generated by Federal Dams along the Columbia River, costs residents about $1 a day, less than half the national rate, according to Federal figures." The table shows that Con Edison of New York charged 15.97 cents per kilowatt-hour compared to Seattle City Light whose price was 3.77 cents per kilowatt-hour.
85. William Whipple, Jr., autobiography, unpublished manuscript, 1978, 224-7. Copy in possession of David P. Billington.
85. Whipple Jr., autobiography, unpublished manuscript, 228-9.

87. Whipple Jr., autobiography, unpublished manuscript, 155.
88. Whipple Jr., autobiography, unpublished manuscript, 156-7. As Whipple stated in a footnote, "The final estimate of damages made after intensive data gathering for seven months afterwards, was $100 million dollars. Whether this exact correspondence represented a lucky guess the first time, or some other fudging afterwards, could not be determined. The staff maintained that the final estimate was honestly made."
89. William Whipple Jr., "Comprehensive Plan for the Columbia Basin," *Proceedings of the American Society of Civil Engineers* 76 (1950), paper No. 2473 (published in November 1950 as *Proceedings-Separate No. 45*), 1-73.
90. Whipple Jr., "Comprehensive Plan," 1430. The seven reservoirs were those behind John Day, Hells Canyon, Priest Rapids, Grand Coulee, Libby, Glacier, and Hungry Horse Dams, all totaling 22,010,000 acre-feet or 965×10^9 cubic feet of water (one acre-foot is 43,560 cubic feet). Filling this capacity at the rate of 440,000 cubic feet per second would take about 25 days, where that flow is the difference between the maximum recorded peak flow of 1,240,000 cubic feet per second and 800,000 cubic feet per second that the levees could hold. See 1432-3.
91. Whipple Jr., "Comprehensive Plan," 1-23. Quotation on 20-1.
92. Mighetto and Ebel, *Saving the Salmon*, 195-8. Until other dams were built the fishways of Bonneville were successful as were the efforts to divert salmon to tributaries downstream from Grand Coulee, see Richard Lowitt, *The New Deal and the West* (Bloomington: Indiana University Press, 1984), 158-9.
93. U.S. Department of Energy, Bonneville Power Administration; U.S. Army Corps of Engineers; and U.S Department of the Interior, Bureau of Reclamation, *The Columbia River System: The Inside Story* (Portland, Oregon, 1991), 15.
94. Whipple Jr., "Comprehensive Plan," 22-3.
95. Bonneville Power Administration, "Columbia River System . . . ," 14-5.
96. Pitzer, *Grand Coulee*, 341. Today the largest powerplant in the world is Itaipu on the Paraná River between Brazil and Paraguay. Construction of the dam began in 1975. The first generating unit went into operation in 1983 and the last in 1991.
97. Elmo Richardson, *Dams, Parks and Politics: Resource Development and Preservation in the Truman-Eisenhower Era* (Lexington: University Press of Kentucky, 1973), 30.
98. Pitzer, *Grand Coulee*, 236.
99. Richardson, *Dams, Parks and Politics*, 25, 31-8.

CHAPTER 6‡

EARTH DAMS ON THE MISSOURI RIVER: FORT PECK AND GARRISON DAMS AND THE PICK-SLOAN PLAN[1]

THE BASIN PLAN

From Navigation to Power

The Missouri River begins at the juncture of Three Forks (the Jefferson, Madison, and Gallatin Rivers) in southwestern Montana and flows east and south 2,470 miles to its mouth where it joins the Mississippi River about 15 miles north of St. Louis. The watershed, or drainage area, is 529,000 square miles, or about the same as that of the Volga River, making it, alone, the sixteenth greatest watershed in the world, and after the Mississippi, the largest in the United States. It is the fifteenth longest river in the world. The Missouri watershed is greater than the combined land area of France, Germany, Spain, and the Netherlands; its river length is slightly greater than the flight mileage between Los Angeles and Newark, New Jersey. If the river were taken from its source to its delta, below New Orleans, it would be the fourth longest river in the world (after the Nile, the Amazon, and the Yangtze), and its drainage area would be fifth (after the Amazon, Parana, Congo, and Nile).

Although the drainage basin is extensive, the amount of water making its way into the Missouri River is not remarkable—at least not usually. Its discharge can be as low as 12,500 cubic feet per second, which is about the mean flow of the Merrimack River in New England. Yet, at full flood, the discharge at the mouth can rise to 900,000 cubic feet per second, which is well above

---

‡ (Editor's note) The Bureau of Reclamation's embankment dams are not a focus of this study, but it should be noted that from Reclamation's beginnings it contributed to embankment dam engineering through a research and construction program that eventually encompassed 240 embankment dams. These dams include such significant projects as Minidoka, Belle Fourche, Cold Springs, Tieton, Echo, Taylor Park, Alcova, Green Mountain, Anderson Ranch, Trinity, Navajo, Fontenelle, and San Luis Dams. For additional information see John Lowe III, "Earthfill Dams," in *Development of Dam Engineering in the United States,* Eric B. Kollgaard and Wallace L. Chadwick, Editors (New York: Pergamon Press, 1988), 671-884. Or access [www.usbr.gov/history] Richard L. Wiltshire, P.E., "100 Years of Embankment Dam Design and Construction in the U.S. Bureau of Reclamation" (September 2002), a paper prepared for Reclamation's centennial Symposium on the History of the Bureau of Reclamation at the University of Nevada, Las Vegas, June 18-19, 2002. You may request a copy of the paper at: History Program, D-5300, Bureau of Reclamation, P. O. Box 25007, Lakewood, Colorado 80225-0007.

the mean discharge of all rivers in the world save the Amazon, the Congo, the Yangtze, and the Ganges. The Missouri is big, unique, and temperamental.[2]

The river rises in the Rocky Mountains, runs in gorges for about 200 miles, then the terrain opens out to sloping terraces with some "badlands" until it reaches Yankton, South Dakota, where it becomes navigable and goes through rolling hills, skirts the Ozarks, and finally empties into the Mississippi. But these overall features tend to obscure the huge task of a detailed, localized survey aimed at imagining a gigantic compromise between nature and society.

The first American survey of the Missouri followed the Louisiana Purchase, a major event in America history, which is the story of two rivers: the Missouri and the Columbia.[3] Both rivers would become major sites for navigation dams designed and built by the U.S. Army Corps of Engineers in the twentieth century. The former river became famous thanks to Lewis and Clark, who began and ended their famous trek on the Missouri between May 14, 1804,

Figure 6-1: The Missouri River Basin. Source: *Big Dam Era*, Missouri River Division, U.S. Army Corps of Engineers, xvi.

and September 23, 1806. The two Army officers were pioneers and explorers who studied the natural environment and focused on the river as a trading route. Technically, their vision was more pre-industrial; they accepted nature and sought to catalogue and explain it rather than exploring nature to change and utilize it. It took seventy-two years after their return to St. Louis for the United States Congress to authorize the Army to make accurate large-scale maps of the Missouri River.[4] Then, in 1926, an imperative for action once again came from the Congress.

If the 1804 expedition was a monument by the Army officers to the national expansion, then the House Document 308 of 1926 has been called a monument to "the most extensive and comprehensive engineering study of all times."[5] The Missouri River 308 report of 1932 was one of the major results of this study. It is an immense document, somehow commensurate in size to the river basin that it describes. Unlike the 1804-1806 description, the 1932 report is an engineering one; that is to say, it focuses on changing the river environment and on restructuring the entire basin. That meant, primarily, the building of dams that create reservoirs.

Congress had expressly ordered the U.S. Army Corps of Engineers, in the 308 reports, to investigate "those navigable streams in the United States, and their tributaries, whereon power development appears feasible and practicable . . . ." The charge, including navigation, irrigation, control of floods as well as water power, clearly gave the Corps the direction to think in terms of entire river basins rather than isolated projects. Congress put the investigations into law by act in 1927, and the Kansas City District office of the Corps began the Missouri Basin study right away. District Engineer Captain Theodore Wyman submitted his 555 page report (with 243 charts and a 634 page appendix) in late September 1932 to his Division Engineer in St. Louis.[6] Long before that, the Corps began work at the site of the major work recommended by Wyman, the massive Fort Peck Dam.[7]

The Corps made preliminary surveys for the dam in 1928 and then detailed ones over the next three and one-half years. These were largely made to discover the natural processes of basin behavior, the most important of which was flow, or discharge, in the main stem of the Missouri River. The engineers did separate studies of each of the 23 major tributaries, and Wyman summarized them in the third part of his report. The first part was devoted to the main stem, the second to the minor tributaries, and the third part summarized all work and plans for the entire basin.

Wyman ended that third and final part with three sections: Conclusions, Plan of Development, and Recommendations.[8] Conclusions centered on the six engineering issues of levees, reservoirs, irrigation, hydroelectric power, navigation, and bank erosion and silt. The plan addressed each issue with detailed t ables giving proposed engineering projects: flood protection for cities, irrigation, power sites, navigation channels, and reservoirs. The recommendations essentially included all the plans but put the central emphasis on one major project, the construction of a huge dam at Fort Peck, Montana.

Wyman made the case for a huge reservoir of 17 million acre-feet (second at that time in volume only to Lake Mead, which was to be created by Hoover Dam, then under construction) to supply water for the lower Missouri when low discharge would make navigation hazardous, if not impossible, in parts of the river between Yankton, South Dakota, and the mouth.[9]

Figure 6-2: General Lytle Brown, U.S. Army Corps of Engineers. Source: U.S. Army Corps of Engineers.

In his covering letter submitting Wyman's report to the Secretary of War, the Chief of Engineers, General Lytle Brown, recommended the entire plan, but added two emphases:

> I further recommend that the project for navigation on the main stem as heretofore authorized, namely from the mouth to Sioux City, Iowa, be vigorously pressed to completion, and that, in addition, the reservoir at the site of Fort Peck be built to the maximum practicable capacity; and be operated primarily for navigation, with such arrangement for future installation of power as will permit the maximum production of hydroelectric power consistent with the primary demands of navigation. . . .[10]

The complex document then passed through the government bureaucracy, which included the Mississippi River Commission; the Board of Engineers for Rivers and Harbors; and the Chief of Engineers, who one year later on September 30, 1933, sent it to the Secretary of War, who promptly passed it on to the Speaker of the House of Representatives. It was referred to the House Committee on Rivers and Harbors on February 5, 1934.

Between the time that Wyman completed his report and General Brown sent it to the Secretary of War, the creation of jobs had become a high priority of the Roosevelt Administration. Just as with the Columbia Basin, construction actually became a high priority, so that 15 days after the Secretary received Brown's recommendations, the Fort Peck Dam was approved as Public-Works Project Number 30 (October 15, 1933). Funds immediately flowed to the Corps, and on October 23, seventy men began work at the dam site.[11] Fort Peck had joined the ranks of the world's largest structures in preparation, all dams in the West and all destined to radically change their river basins. But now there was one major difference. Unlike Wilson, Hoover, Bonneville, and Grand Coulee, this Montana dam was to be made of earth, not concrete, and 126 million cubic yards of earth instead of the mere 12 million cubic yards of concrete in the record setting mass for Grand Coulee.

EARTHFILL DAMS AND FORT PECK

The Birth of Soil Mechanics

Three conditions usually make earthfill dams practicable: they can be built on earthen foundations more easily than can masonry dams, they can be made of a variety of locally available materials, and they are usually competitive in cost. Of the nearly 70,000 dams in the United States, about 85 percent are earthfill embankment dams, and in the twentieth century, even with the advent of concrete, still about 65 percent of the dams built have been earthfill.[12] There are two types of earthfill embankments: levees and dams. The difference arises primarily from the fact that levees usually (but not always) hold back water only in floods, whereas dams create permanent reservoirs. The earliest documented dams, found in present day Jordan, date from about 3,000 B.C. and consist of masonry walls enclosing an earth core, an impervious upstream blanket over more earthfill, and a wide downstream earthen embankment. Thus, the primary elements, if somewhat complex, were in place in antiquity: a core, an earthen embankment on both sides, and some protection on the upstream face.[13] Since the time of the earliest dams, and especially in Roman times, masonry played a major role in dams. The height of embankment dams gradually increased, reaching seventy-nine feet as opposed to a maximum of only forty-three feet for masonry works. The embankments had masonry protecting the upstream face. Many medieval dams were embankment, as were the power dams in renaissance Europe and the pre-industrial dams of Central Europe.[14]

The industrial era influenced earth dams first theoretically in France, then empirically in Britain. The most famous name in earthworks of the eighteenth century is certainly Charles A. Coulomb (1736-1806), who established the basic concept that sliding resistance of soils depends upon their cohesion. Other French theorists developed his ideas further, but the actual construction of embankments shifted to Britain. That country's most famous structural engineer, Thomas Telford (1757-1834), built Britain's first major water supply structure, the Glencorse earth dam, between 1819 and 1824, near Edinburgh. The dam had a clay core with wide earth embankments (shells) on either side and a stone surface on the upstream side.[15]

Just as with bridges (a major focus for Telford), the British works tended to be conservative and mostly safe compared to their American counterparts which were more quickly built and experienced many more failures. Up to 1930, there had been four failures in Britain compared to thirty-four in the United States. Of course, the South Fork Dam failure near Johnstown, Pennsylvania, in 1889 became the most notorious by killing 2,209 people following the bursting of its overtopped earthen structure. Nevertheless, earth dams persevered, and in the 1930s they received considerable theoretical help from abroad.[16]

The French provided the seminal idea. Henri-Philibert-Gaspard Darcy (1803-1858), while designing and building a water supply system for his home town of Dijon, developed a theory for explaining the flow of water through granular media. He published his seminal idea in a delightfully French manner embedded in an 1856 writing entitled "The Public Fountains for the City of Dijon."[17]

Eventually, the Anglo-Saxon world learned the applications. First an Austrian professor, Philipp Forchheimer (1852-1933), discovered that Darcy's idea could be expressed more mathematically by using the ideas of another Frenchman, Pierre S. de Laplace (1749-1826). Forchheimer, an engineer well-schooled in mathematical physics, recognized the similarity of ground water flow to heat flow and from those ideas he proceeded to develop a graphical method of analysis, which appealed to hydraulic practitioners in the same way as graphic statics had appealed to structural engineers.[18] An English physicist had earlier developed the same graphical approach, but it did not become widespread until the 1930s after two Austrians had laid a more systematic foundation for soil mechanics.

Karl Terzaghi (1883–1963) was the principal figure after publication of his ground-breaking book in 1925. He took the earlier ideas and put them systematically together in a theoretically consistent and practically oriented way, combining the properties of soil under stresses with the action of water flowing through it.[19] All these developments began to influence earth dam design, and they were directly aided, as well, by the publication of Arthur Casagrande's 1937 paper on seepage through dams.[20]

A further major development in the 1930s was the acceptance of stability analysis for explaining and, hence, preventing the many earth dam failures by sliding. Pioneered by Swedish engineers in 1916 and formalized in 1926 by the Swedish professor Wolmar Fellenius (1876-1957), the sliding circle stability analysis became a standard method by the late 1930s and was used to analyze the great slide at Fort Peck.[21] All of this theoretical work encouraged engineers to design earth dams and led to the "vast proliferation of embankment dams around the world . . . (even for) large dams, especially after World War II."[22]

But when the U.S. Army Corps of Engineers began studies of Fort Peck Dam, in preparation for the 308 Report, Terzaghi and Casagrande were not widely known even in the engineering community, even though they had already been active in the United States. The design criteria used in planning Fort Peck in the early 1930s were those presented most completely by Joel D. Justin in 1924, before the beginning of soil mechanics as a recognized discipline. These criteria involved, primarily, the geometry of the dam cross section and the materials in the dam, particularly as they relate to seepage. Missing is a method of evaluating the stability against slides, for as Justin stated, "it is not generally necessary to investigate closely the safety of earth dams against sliding and overturning."[23] Justin cautioned about slides in his paper when the slope of the dam section is too steep or when an unstable foundation becomes saturated before the reservoir is only partly filled. Those cautions would become reality at Fort Peck fourteen years later as Justin himself would describe in his discussion of a 1942 paper.[24]

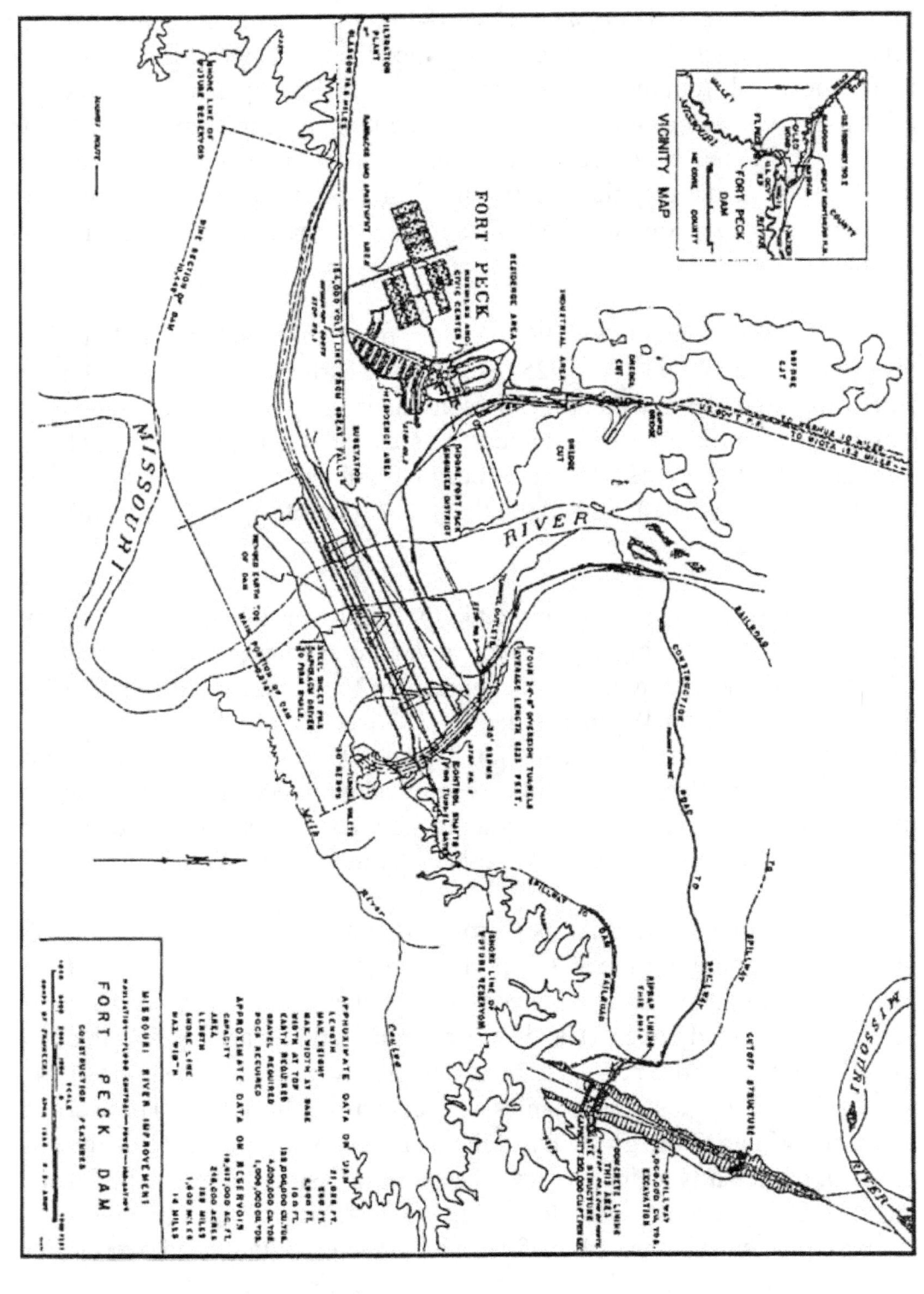

Figure 6-3: Section and plan for Fort Peck Dam. Source: U.S. Army Corps of Engineers.

Thus, the period between 1924 and late 1933, when construction began was just when the modern engineering study of earth dams began, but nothing on the scale of Fort Peck had ever before been attempted. From the dam section of 1934, it appears that the design took into account Justin's criteria and thus would be judged acceptable state-of-the-art. The major issue foreseen in 1934 was not the design but the construction. The project could not be justified economically, and, so, there was considerable pressure to build it as inexpensively as possible. The biggest cost would be moving 126 million cubic yards of earth. Therefore, the Corps decided to use the cheapest transportation method available—hydraulic fill.

Construction of Hydraulic-Fill Dams

California mining of the nineteenth century evolved the practice of "hydraulicking" by excavating surface materials with a moving stream of high-pressure water, ejected from a nozzle, over a mining face or hillside. This powerful force loosens the solids, which flow with the water into a sluice or pipe. This process transports the slurry and separates out heavier particles on the way. Proven effective for moving large quantities of earth, hydraulicking seemed a good way to transport earth for dams in the 1890s as reservoirs for power became essential in the West.

This dam practice reached the profession convincingly in 1907, when James D. Schuyler, a noted dam engineer, responded to urging of colleagues and "cheerfully...prepared (a paper) in the few and widely separated leisure moments of a busy life, mostly on railroad trains, or trolley cars."[25] This leisure time paper is fifty-seven pages with detailed descriptions of early hydraulic-fill dams, beginning with one built in Tyler, Texas, in 1894 and focusing on two California dams completed just prior to the paper's publication. Schuyler succinctly explained the theory of such dams, wherein muddy material is deposited in the dam such that the clay with water forms the core (clay puddle), which becomes impervious once the water drains out through the adjacent porous layers. The semiporous layer allows core water to escape, but prevents it from carrying out the fine particles in the core. It is possible to control the type of solid material that will settle out so that the clay will concentrate in the core and the sand and gravel in the outer parts of the dam. Not only is the method of transport cheap but the segregation of solid material can be controlled easily, according to Schuyler.[26]

The Reclamation Service built the Conconully Dam in 1910 on its Okanogan Project in North Central Washington. This was one of many hydraulic fill dams built between Schuyler's paper and 1920. In alarming numbers they also failed, prompting the distinguished hydraulic engineer, Allen Hazen

243

(1869–1930), to emphasize in 1921 the importance of keeping particles of small sizes out of the core. He noted that water retained in the hydraulically-placed core could not then easily drain out, and the resulting outward water pressure could cause failures, especially during construction.[27] He concluded that more careful attention to particle size and more careful control of materials placed would avoid hydraulic-fill dam failures, although the additional care would tend to make them more costly.

Two more papers appeared right after Hazen's, and both demonstrated how the control of particle sizes led to successful hydraulic-fill dams. The success of hydraulic fill dams allowed the second of these papers to state that "many of the doubts that have existed heretofore regarding the stability of hydraulic-fill cores, have been dispelled."[28] The author of the earlier paper noted that:

> During its construction period, a hydraulic-fill dam may be likened to a cantilever bridge in process of erection. Such a bridge will not again be subjected to the stresses that must be withstood during erection, nor will the dam, with its central pool, again be in so unstable a condition as it is in during construction. In designing an earth dam, this construction condition must be considered, just as it is necessary to provide for erection conditions in the case of a cantilever bridge.[29]

This apt analogy would have made a strong impression on engineers in 1921 because of the two dramatic collapses of the longest spanning cantilever bridge in the world, the 1917 Quebec Bridge. It failed twice during construction.[30]

Thus, when Justin came to summarize earth dam design in his prize winning 1924 paper, he could refer to the recent literature and communicate with confidence that such a construction procedure could be relied upon if properly controlled. He gave a table of 118 dams listed as "Statistics of Successful Earth Dams." Twenty-six of these were hydraulic-fill. In the last edition (1927) of his noted book on dams, Edward Wegmann summarized the hydraulic-fill method for building dams, lending additional prestige to the technique.[31] Finally, as the U.S. Army Corps of Engineers completed designs for the Fort Peck Dam, a paper on hydraulic-fill dams appeared in the *Transactions* of the ASCE, which stated on the front page that "Hydraulic-fill dams are almost new compared with other types, and they are becoming more popular because of their many advantages." The paper called for standardizing the testing of materials in such dams, using a dam in Massachusetts for reference, and most significantly, based its theoretical section on the emerging field of soil mechanics, with special reference to Karl Terzaghi.[32]

Also in 1934, a professor at the Massachusetts Institute of Technology, Glennon Gilboy, wrote a paper on hydraulic-fill dams that lent further credibility to the method.[33] He noted that his theoretical approach used "simplifications and idealizations. . .no more radical in nature and extent than those associated with other types of engineering theory, such as the concepts of frictionless joints in trusses, simple bending of beams, and so on." Academia and practice seemed to agree on the validity of hydraulicking for earth dams. In a discussion, Arthur Casagrande commended Gilboy and lent even further support to his conclusions.

Just as earth dam design seemed well developed by 1934, so did the construction method of hydraulic-fill. The Corps had reason to be confident that its design for Fort Peck was sound and that its plan for construction would be both economical and safe, but they also had reason to be wary of construction slides. The unknown factor in all of this planning was the scale, and that, for engineering works, is always a crucial concern. Fort Peck Dam would be five times larger than the then largest earth dam in the world, the Corps designed Gatun Dam for the Panama Canal, also built using hydraulic fill.

Fort Peck: Skilled Generalship and Ultra Modern Mechanization

On May 9, 1935, the *Engineering-News Record,* in an editorial, expressed the wonder that most knowledgeable people felt about the great dam constructions in the West. The size and quantities of materials "strain ordinary comprehension," and the work is "made possible only by bringing into action all the resources of invention, of business organization, and of technical skill that engineering and contracting and the equipment manufacturers have created. . . ." The editorial characterized this momentous effort by calling:

> Fort Peck Dam progress . . . the child of skilled generalship plus ultra modern mechanization . . . without this parentage the great Missouri River dam construction . . . would not be giving employment to so many men needing work in so many fields of industrial labor.

But that fast deployment meant that field work began before engineers had completed design work.[34]

The dam's immediate justification became a reality quickly. By December 1933, more than 700 men were at work, increasing to 1,000 as the new year began, and growing rapidly even in the harsh Montana winter so that by July 15, 1934, 7,000 men were productively on the site.[35] What were they doing? During 1934, they built 18 miles of railroad, buildings for barracks, a

245

Figure 6-4: The spillway gate structure and concrete lined canal at Fort Peck. Source: Records of the U.S. Army Corps of Engineers, Record Group 77, National Archives and Records Administration, College Park, Maryland.

Figure 6-5: Fort Peck spillway and dam. Source: Records of the U.S. Army Corps of Engineers, Record Group 77, National Archives and Records Administration, College Park, Maryland.

mess hall, construction equipment, a bridge across the Missouri, wooden railroad trestles around the dam site, barges and dredges, and the new town of Fort Peck. They also were stripping loose overburden from the dam site, beginning construction of the diversion tunnels, and starting work on the spillway.[36] By February 1935, everything stood ready for the biggest operation of all, the hydraulic filling of the dam.

Many features of the dam were without precedent. One was the steel sheet pile cutoff wall that was driven, mostly in 1934, deep down and into the firm shale all across the valley and close to the center of the dam. This immense project, by itself, over 10,000 feet long and up to 190 feet deep, sought to control the seepage (percolation) underneath the dam.[37] While the sheet piles were being driven 40 feet deeper than any sheet piles had previously been driven, the hydraulic-fill started its journey from the upstream borrow pits in the river.[38]

Before winter closed down operations in late 1934, 843,000 cubic yards of material came to the dam site, less than 1 percent of the total. Operations resumed in the spring with the four dredges named *Gallatin*, *Jefferson*, *Madison* (the rivers that form the Missouri at Three Forks), and the *Missouri* sucking up the alluvial mix of sand, silt, clay, and gravel, all floating in water sent to the dam site through pipes often greater than two and one-half miles long. At the dam site the coarser sand and gravel settled out along the flat sloping beaches, while the water carried the finer silt and clay particles into the central or core section

Figure 6-6: This September 12, 1935, aerial shows the entire hydraulic fill area of Fort Peck Dam, including trestles for dumping gravel and rock in the toes of the dam. Source: Records of the U.S. Army Corps of Engineers, Record Group 77, National Archives and Records Administration, College Park, Maryland.

Figure 6-7: Laying 28-inch ball and cone pipeline in the northwest corner of the hydraulic fill area at Fort Peck Dam on May 15, 1936. Source: Records of the U.S. Army Corps of Engineers, Record Group 77, National Archives and Records Administration, College Park, Maryland.

Figure 6-8: The completed east section of Fort Peck Dam before the slide. Hydraulic fill material was discharged through traps in the bottom of the dredge pipe. As the fill rose, the rail lines at the upstream edge (right) were relocated at higher levels to provide for placing the blanket of gravel and protective stone. June 29, 1938. Source: Records of the U.S. Army Corps of Engineers, Record Group 77, National Archives and Records Administration, College Park, Maryland.

of the dam. The solids comprised only about 13 percent of the volume, the rest was water that had to be drained off back into the river. Some solid materials were lost through drainage, so that by the end of June 1936, the dam had retained slightly more than 30 million of the 37.5 million cubic yards pumped through the pipes.[39]

A visitor to the site at that time would have seen not a solid dam taking shape but rather a swamp of water and mud. It seemed to be a perverse way to seal off the great river, but the method was working ahead of schedule. At the same time, three other major components of the Fort Peck plan were also taking shape, and, as with the hydraulic fill, the shape they had in the summer of 1936 was quite different from that which a visitor would see today.

Since the earth dam was to block completely the flow of the muddy Missouri, the Corps designed four diversion tunnels, each more than a mile long, driven into the firm shale bluff on the right bank, and each one sized to carry the normal flow of the river. Each could carry 84,000 cubic feet per second. With a

Figure 6-9: At Fort Peck Dam the Corps used timbers at the end of the 28-inch pipe to help spread the hydraulic fill as it discharged. Shear board walls fanned the flow out and reduced velocity. May 13, 1936. Source: Records of the U.S. Army Corps of Engineers, Record Group 77, National Archives and Records Administration, College Park, Maryland.

three-foot thick lining of reinforced concrete, these tunnels have an inside diameter of just over 24 feet. Invisible today, the driving of them through temperamental shale, the masses of muck removed, and the huge jumbo traveling crane could all be seen by the adventurous visitor in 1936. Considered at the time to be "the most interesting and difficult tunneling work under way in the country," these conduits were designed to carry the entire river.[40]

In late 1935, the Corps had simplified the design to permit quicker tunnel driving because the hydraulic-fill operation was ahead of schedule. The engineers now saw that the dam could be closed during the summer of 1937, a year earlier, but only if the diversion tunnels were ready to take the entire river. The tunnel design then became the bottleneck because its complex three-layer lining, as specified in the original design, could not be finished until 1938. Therefore, success in one part of the project forced a rethinking of another part, and the lining now became only a single layer. It was this major change in tunnel design that led to a dispute between the Corps and the contractor. Because they failed to agree on a price for the changed design, the government decided to take over the contract itself. Also, the Corps decided to keep a steel liner in one tunnel to serve as a penstock and thus permit some power generation.[41]

With the government now in full charge, the tunnels moved along quickly, and the tunnel work force alone increased from 1,000 to 3,500 men

Figure 6-10: This June 3, 1936, view of the *Jefferson* (right) show the 28-inch pipeline from the dredge and a derrick towboat next to it for lifting heavy items. The cutter head at the far end of the *Jefferson* was in a raised position. Hydraulic fill was pumped up 210 feet to the dike at the top of the dam. Source: Records of the U.S. Army Corps of Engineers, Record Group 77, National Archives and Records Administration, College Park, Maryland.

by July 1, 1936.[42] One year later, on June 7, 1937, the government-hired labor force completed the tunnels and two weeks later, when the river channel was definitely closed, the entire Missouri flowed through the tunnels. The hydraulic-fill proceeded rapidly that summer and fall, picked up again in the spring and summer of 1938, and gave every indication that the fill would be completed by November, before the winter set in.[43]

Figure 6-11: By August 8, 1936, stone placement on the upstream face of Fort Peck Dam was nearly complete. Source: Records of the U.S. Army Corps of Engineers, Record Group 77, National Archives and Records Administration, College Park, Maryland.

Figure 6-12: This September 20, 1938, image show the corepool near the top of Fort Peck Dam. Source: Records of the U.S. Army Corps of Engineers, Record Group 77, National Archives and Records Administration, College Park, Maryland.

The Spillway and the Castle

In his 308 Report, Captain Wyman had given a brief description of the dam, the tunnels, and the spillway—the three primary features of the project. This last part, he had described as a series of twenty-four openings thirty feet wide with sills set twenty feet below the maximum design elevation of the reservoir. He planned twenty-four tainter gates twenty feet high (to retain the reservoir height) and thirty feet wide. When fully opened, these could discharge 200,000 cubic feet per second, which just exceeded the maximum flood of 173,000 cubic feet per second that he calculated. This flood, he assumed, would occur once every one hundred years, and, according to standard estimates of the time, that would imply an average annual twenty-four hour flood of 66,500 cubic feet per second.[44]

Between the time of Wyman's report and the final design, the U.S. Army Corps of Engineers had revised the design to sixteen stoney gates, twenty-five feet high and forty feet wide, designed to carry 255,000 cubic feet per second with the sills at twenty-eight and one-half feet below the reservoir surface (stoney gates are plate steel gates that move vertically.). The Corps calculated the maximum probable flood to produce a flow into the reservoir of 380,000 cubic feet per second with the improbable frequency of once in 8,000 years, or about two and one-half times the maximum flood ever recorded (about 150,000 cubic feet per second).[45] This conservative approach is essential for earth dams because if the spillway cannot carry off the flood it may overtop the dam. Overtopping earthworks is fatal and dam failure is highly likely.[46]

Carrying away such a huge volume of water required a monumental structure to be built on solid foundations and to be itself impervious to erosion. The contractor had to strip away six million cubic yards of earth and seven million cubic yards of shale to make safe room for 400,000 cubic yards of concrete reinforced with about twenty-two million pounds of steel bars.[47] These statistics dominated the engineering literature along with drawings of the gate structure—elevations and sections, those carefully drafted documents from which builders convert the engineers' calculations into constructed objects. But there was not a word about how the huge concrete structure would look in three dimensions or even in an isometric drawing. It would take someone with an artistic eye and the technique to record the sight to turn the engineers' spillway into the artist's castle.

Since early in the nineteenth century, some painters had been influenced deeply by industrial development. For example, Joseph Turner, in "Rain, Steam and Speed" (1844), painted the Great Western Railway; and a great number of American landscape painters also were stimulated by the railroad.[48] In the early twentieth century, the Eiffel Tower and the Brooklyn Bridge became icons for painters at the same time that photography began to be recognized as an art form in its own right. Charles Sheeler, an American painter and photographer, photographed the industrial landscape, including New England mill towns; Henry Ford's immense River Rouge car factory (1927); and, later, Hoover Dam (1939).[49] So, in the 1930s, attention moved from industrial plants to public works. Another of the great photographic artists was Margaret Bourke-White who had photographed machines, factories, and industrial titans for Henry Luce's *Fortune*.

In 1936, Luce was going to inaugurate a new magazine, called *Life*, which would rely heavily on photography. To start their depression-era journal, his staff determined to focus the first issue on the greatest public works project of the day, Bonneville Dam and the domestication of the Columbia River. Bourke-White was the ideal candidate to go west and visually capture the spirit of the new age of progress and the happily employed men who were rebuilding nature.[50]

As September plans for the new magazine reached the final stage, the well-known journalist, Ernie Pyle, wrote a front page feature for the now defunct *New York World-Telegram* devoted to "Roosevelt's New Deal Dam in Montana" and the Wild West images it created. As a consequence, *Life's* editor directed Bourke-White to stop there on her way to Bonneville.

He did not realize what could result. As the editors wrote in their introduction to the magazine's first issue:

> If any Charter Subscriber is surprised by what turned out to be the first story in this first issue of *LIFE*, he is not nearly so surprised as the Editors were. Photographer Margaret Bourke-White had been dispatched to the Northwest to photograph the multi-million dollar projects of the Columbia River Basin. What the Editors expected—for use in some later issue—were construction pictures as only Bourke-White can take them. What the Editors got was a human document of American frontier life which, to them at least, was a revelation.[51]

The revelation took two distinct and utterly different forms. First, they got a cover, the great Montana Castle with crenulated turrets and two minuscule people set just above the dateline, November 23, 1936. Often misidentified as Fort Peck Dam, it is, of course, the spillway gate structure as it appeared that fall, incomplete but splendidly majestic in bare concrete. It is one of those concrete sculptures that is designed to be fully functional when complete but is

Figure 6-13: Concrete placement on the reinforced spillway floor at Fort Peck. Source: Records of the U.S. Army Corps of Engineers, Record Group 77, National Archives and Records Administration, College Park, Maryland.

Figure 6-14: The Fort Peck Dam spillway gate structure on August 5, 1943, with the water level near the maximum. Source: Records of the U.S. Army Corps of Engineers, Record Group 77, National Archives and Records Administration, College Park, Maryland.

totally without function until then. The turrets carry the light railroad, and the lower battlements are a pedestrian walk, all of which looks quite appropriate now but not nearly so exotic as when Miss Bourke-White saw it in 1936.

The second visual surprise *Life's* charter subscribers got was a close look at the workers, their shanty towns, and the night life of raw frontier communities in the otherwise desolate landscape of eastern Montana. *Life* made sure that the carefully posed portraits of the Corps' two men in charge, Lt. Col. T. B. Larkin and Major Clark Kittrell, were set in facing pages that juxtaposed photos of two bars and two pictures of the best known brothel in town. A huge steel liner section for one diversion tunnel unified the composition.[52] The spillway remains today; the shanty towns disappeared years ago.

The Corps had made a serious effort to provide decent living quarters for the workers by building a model community called the town of Fort Peck, but it made one serious miscalculation. The Corps assumed that most workers wo uld be single men, so barracks were built for them, but Montana state law gave hiring preference to married men. So families lived in the surrounding new towns of Wheeler, Delano Heights, New Deal, Square Deal, Park Grove, and Wilson. Whereas the town of Fort Peck had all the amenities, the others were short on indoor plumbing, clean water, and electricity. When engineer

Figure 6-15: President Franklin D. Roosevelt inspects the Fort Peck Dam project on August 6, 1934. Source: Records of the U.S. Army Corps of Engineers, record Group 77, National Archives and Records Administration, College Park, Maryland.

Captain Ewart G. Plank, who served as town manager, boasted that the town "stands in striking contrast to construction camps of even the recent past" he conveniently ignored the nearby shanty towns.[53]

Captain Plank's article appeared in the *Military Engineer* during the time of Margaret Bourke-White's visit to the towns that he had ignored. Further national publicity came to the dam a year after Bourke-White's visit when, on October 3, 1937, President Roosevelt traveled to a location just below the dam to speak of the great work. Excursion trains and charter buses brought people from all over Montana.[54]

Slide and Recovery

After closure, the work continued where "phenomenal performance records are being established by the four hydraulic dredge units...(which now operate at) double the designed capacity of the plant."[55] Such performance continued into 1938, and it seemed clear that the placing of hydraulic-fill would be finished in November. On February 1, 1938, the Fort Peck District office issued a statement that "at the present time, nothing is foreseen which will prevent the orderly completion of the project by the fall of 1939." The continuing good performance of the dredges during 1938 "gave employees 'itchy feet' and many started looking for the next job." The Corps had already begun demobilization and disposal of workers' buildings during the summer.[56] One person who did not look for another job was Jerold Van Faasen, who had been working on the dam since October 1935 and who, on September 15, 1938, took over, from his departing boss, the job of supervision in the Soil Engineering and Fill Control Section. He had to decide where the dredging would be done based on results from core samples, and he had to supervise survey crews on the dam to ensure proper elevations of the growing dam.

On the morning of September 22, at about 9:30 A.M., one of his survey party chiefs reported that at the east end of the dam the upstream embankment or shell was only one foot higher than the core pool's water. Normally, it was kept from four to five feet higher. Something was very wrong, so Van Faasen ordered a second survey to confirm the hazardous situation, and he immediately contacted his superior.[57] While Van Faasen waited, the survey parties were struggling and were finding more disturbing distortions. The water level of the core pool had not risen or dropped and the downstream freeboard remained safely where it had been the day before. The upstream embankment was moving down. As one surveyor, Ray Kendall, later reported, "[I] was having trouble setting up the level when I noticed a small crack in the ground underneath my feet.... In a few seconds it got considerably deeper and longer." He saw two fellow workers Douglas Moore and Nelson Van Stone, down on the upstream road "waving and yelling." Then suddenly they disappeared. "I looked at my watch. . . . It was 1:21 P.M." Kendall climbed to safety from where "I searched the terrible scene for bodies but all was in vain."[58]

Figure 6-16: On September 22, 1938, a 5,000,000 cubic yard slide occurred on the upstream embankment of Fort Peck Dam. Source: Records of the U.S. Army Corps of Engineers, Record Group 77, National Archives and Records Administration, College Park, Maryland.

Moments before, a maintenance man, Ralph Anglen, on one of the huge drag lines (cranes), saw the same scene and was close enough to Van Stone so that "as I dropped he [Van Stone] went out of sight and I heard him say, 'Goodbye, boys.'" Anglen came up in the lake, covered with mud but alive; Van Stone never did. He was one of eight who died. Anglen was one of 26 who went under but lived.[59]

Others were luckier and could run to safety, among them the project leader, Clark Kittrell, who had been driven out to the troubled area in the early afternoon. His car, driven by Eugene Tourlotte, approached from the west and arrived at the site at about 1:15 P.M. Tourlotte saw the upstream shell begin to move out beneath the car so he slammed on the brakes and went in reverse at its highest speed. The dam break seemed to follow them as the car just made it to beyond the point where the sliding section broke away and swung out into the partly filled reservoir.[60]

Figure 6-17: Gravel and stone facing can be seen in the material that slid away from the upstream embankment at Fort Peck Dam on September 22, 1938. Source: Records of the U.S. Army Corps of Engineers, Record Group 77, National Archives and Records Administration, College Park, Maryland.

Army discipline then prevailed, and after searching for survivors, work began immediately to repair the damage, which included about 5 million cubic yards of slide material, a broken rail line, and a disruption of dredging. Stabilizing the dam by placing quarry stone along the upstream face of the damaged area which could be begun after the workers rebuilt the rail line in just five days. By October 13, dredges began sending hydraulic fill into the upstream region behind the quarry stone. This protective dike work proceeded rapidly until the winter shutdown.[61]

Meanwhile, the Corps assembled a Board of Consultants to study the slide. Three had been consultants on the dam design and six more were added after the slide. Prominently among those added were Arthur Casagrande and Joel Justin. Their final report was to address two major questions: How to rebuild the dam and what was the cause of the slide?[62]

The Board of Consultants met at the site between October 3 and 6, 1938, to inspect the slide and discuss temporary measures and tests. The Corps had already begun discussing flatter slopes in the upstream areas, and, following the

Board's meeting, Division Engineer C. L. Sturdevant; Thomas Middlebrooks, the Corps' expert on soil mechanics; and Kittrell all recommended flatter slopes in the rebuilding programs.[63] Strong differences of opinion began to surface among the members.[64] A final meeting in Washington, D.C., on March 1 and 2 led to the formal Report of the Board of Consultants, which was published in July.

The report, with three of the ten members dissenting, reached the major conclusion "that the shearing resistance of the weathered shale and bentonite seams in the foundation was insufficient to withstand the shearing force to which the foundation was subjected." It then recommended that the dam section be increased such that the upstream slope be flattened from the average of one in four (a slope angle of 14°) to an average of one in eight (an angle of about 7°). Indeed, the final cross section that they recommended is even flatter, being more than one in 20 in the area of the slide and an average of over one in ten in the rest of the dam.[65]

On the sensitive issue of the hydraulic fill, the Board recommended that the shell could be completed up to within about twenty-five feet of the crest by that method, but above that only by dumped fill compacted by rollers. In addition, it stipulated that the core be completed only by rolled fill, and here, Mr. William Gerig dissented. As one of the original members of the board of consultants and senior engineer in the employ of the Army Corps, Gerig still believed that hydraulic fill could complete the core.[66]

Since adequate stability against sliding depends heavily on the ratio of the slope derived from the material's internal friction ($\tan f$) to the slope of the embankment ($\tan a$), i.e., the factor of safety, F.S. ë $\tan f / \tan a$; thus, a flatter slope to the embankment (smaller $\tan a$ = rise / length) will make a safer dam even if the cohesion is zero.[67] Thus, the primary fact of the Fort Peck slide was the steep upstream slope, and in future dams, Garrison and Oahe for example, the upstream slopes would be one in eight or flatter.[68] This was the great lesson from Fort Peck, but there were additional ones that arose from the other two dissenters and from persons in high places.

On November 2, the Chief of Engineers received a confidential memorandum from President Roosevelt. It said in toto:

> Various reports are coming to my office relating to the Fort
> Peck Dam such as serious damage of various kinds, danger of
> the whole thing going out, and similar alarmist stories.
> Would you be good enough to let me have a memorandum.[69]

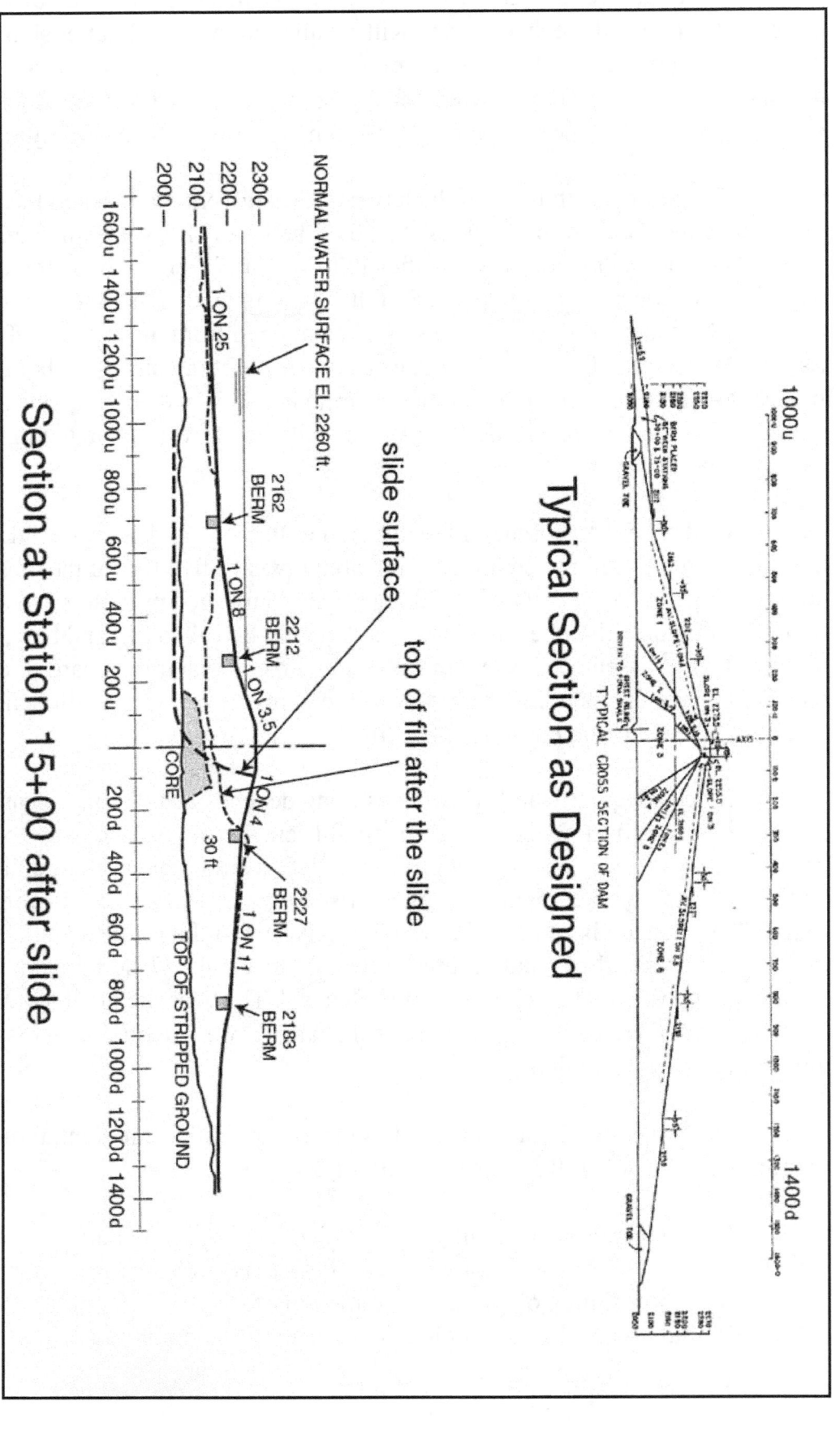

Figure 6-18: The original cross section of hydraulic fill at Fort Peck Dam was designed at 1:4. After the slide, during redesign the Corps decreased the slope to 1:8. Source: Records of the U.S. Army Corps of Engineers, Record Group 77, National Archives and Records Administration, College Park, Maryland.

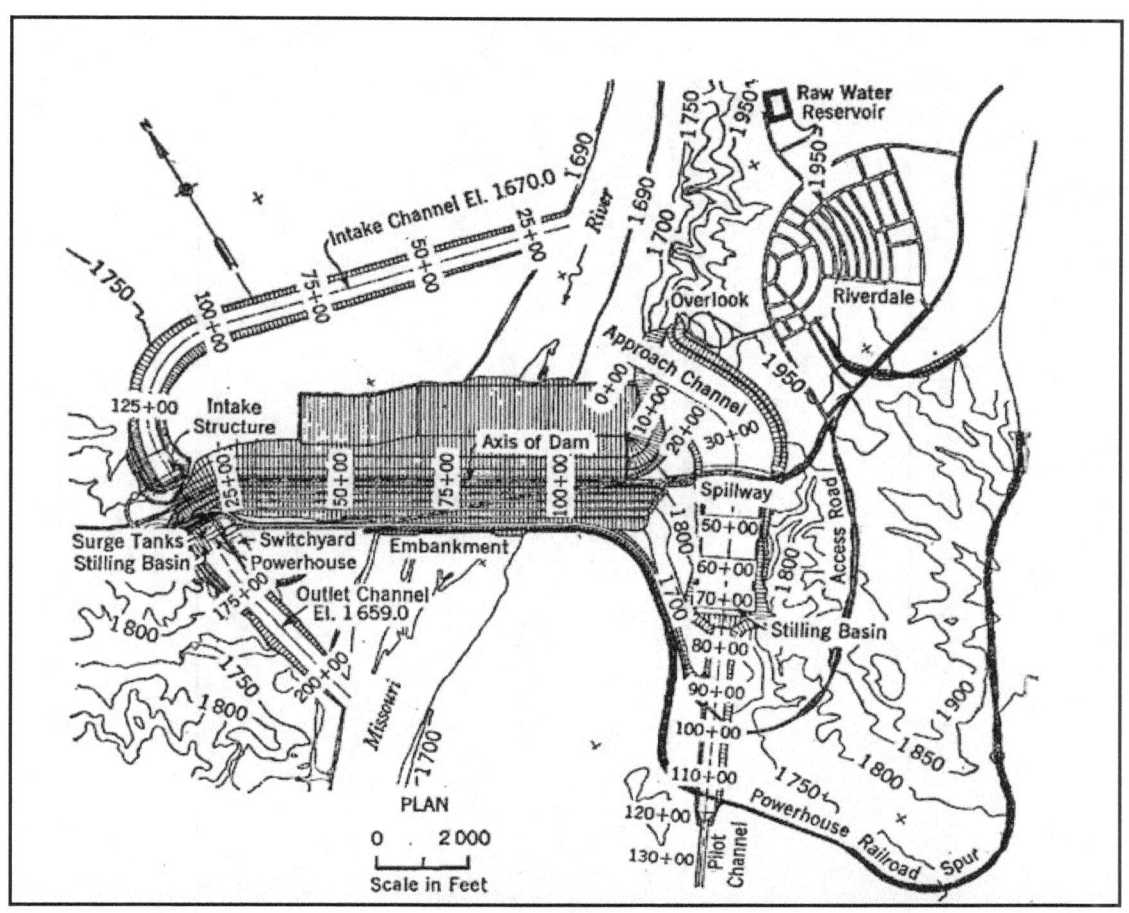

Figure 6-19: The plan for Garrison Dam, North Dakota. Source: U.S. Army Corps of Engineers.

Needless to say the Chief of Engineers did not dally; a full memorandum appeared the next day on the desk of the Secretary of War from General M. C. Tyler, acting Chief of Engineers, detailing the situation. It is a concise, but accurate, assessment of the dam's design and construction, a description of the slide, and an outline of steps taken to plan reconstruction and determine the cause. The last sentence stated that "the Chief of Engineers knows of no grounds for the reports and alarmist stories to which the President refers in his memorandum." A major feature of the Corps' response was to identify the new members of the Board with two of them, Casagrande and Gilboy, specially noted as coming from Harvard and MIT respectively.[70]

While that memorandum satisfied the commander-in-chief, two members of the Board were not to be so placated. The most unhappy was Thaddeus Merriman (1876-1939), who had been suspicious of the Board's direction even after the first meeting.[71] On November 22, he had sent Kittrell a detailed, four-page letter expressing grave concern about the plan of action that the Board and the Corps were following. His primary worry was over the weathered shale foundation, which he believed to be unable to sustain any dam and which,

261

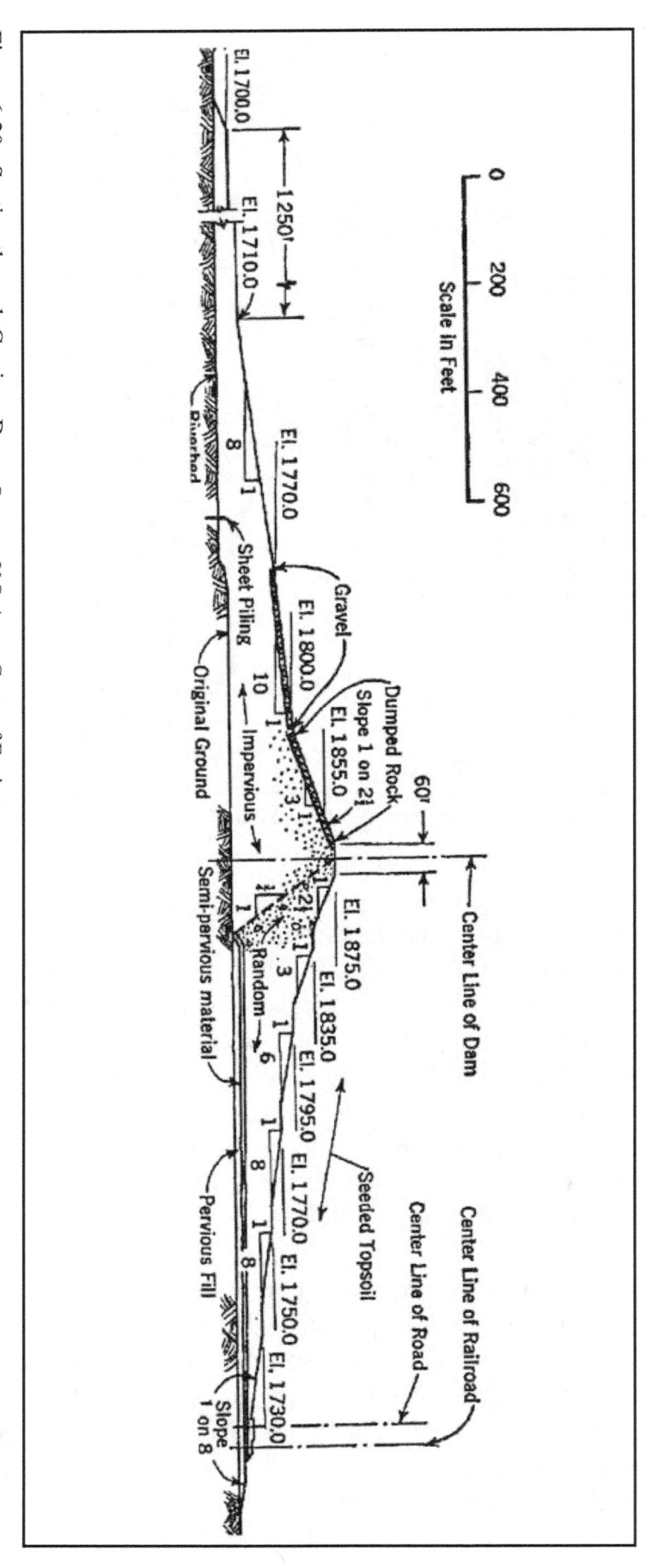

Figure 6-20: Section through Garrison Dam. Source: U.S. Army Corps of Engineers.

therefore, needed to be excavated before rebuilding the dam. His conclusion was that the problems were "both difficult and far flung," that "neither time nor cost nor effort can be permitted to interfere in even the slightest degree with their correct solution," and that "no error of design and no lapse in construction procedure on the repair of this slide will ever be explainable."[72] This is the position of absolute safety with no thought to economy. His proposals were so complex that they would have made the cost of the dam prohibitively high. Meanwhile, the Corps proceeded on the course of action developed in conjunction with the majority of the Board members in October and November.

Merriman was a distinguished engineer, having been chief engineer of the Board of Water Supply of the city of New York from 1927 until his retirement in 1933. He became associated with the Fort Peck Project when his research on cement led to a sulfate resistant Portland cement, which the Corps adopted for use in the spillway. He had experience with dams in the New York area but little detailed background in large-scale embankments. He was still a leading expert on water supply, having served as a consultant to the Metropolitan Water District of Southern California on its project to get water from the Colorado River. He also was a consultant to the Tennessee Valley Authority on Norris Dam and other large dams as well as to major agencies in Pennsylvania and Massachusetts.[73]

Meanwhile, the Corps had hired one of the foremost theorists in Civil Engineering, the Danish-American Harvard professor, H. M. Westergaard, to prepare a report on the dam's stability, which he submitted to Kittrell in time for the Board's meeting in late January 1939. The civilian engineer at the site, W. F. Cummins Jr., reviewed Westergaard's report on the dynamics of the slide and concluded that his results are close to those obtained from a static analysis (which uses the safety factor approach referred to above).[74] After the draft report of February 4, Merriman put his objections into a ten page report of his own. He noted that he had been unable to participate in the preparation of the Board's report, and he reiterated his objections from November. In addition, Warren J. Mead, a professor of geology at MIT, wrote later in February stating his agreement with the engineering principles of the Board's report and even with Merriman's analysis, but he announced that he could not sign the report because there was no program analysis of the risk versus the benefits. As he put it, "It is my belief that the purposes of building the Dam do not justify the existence of this admittedly remote threat to the Missouri River Valley."[75] Here he is getting at the issue of benefit, which was bypassed in 1933 when the President chose to provide immediate employment for depression-stricken men and ignored issues of project benefits.

The deeper issue raised by both formal dissenters (Gerig dissented only with respect to the core fill and, unlike Merriman and Mead, did sign the majority report) relates to the safety of the dam or of any public works project whose failure would risk life and property. As any engineer knows, no construction can be made free of risk, especially one that seeks to control the force of a huge river. Mead was, perhaps, correct in implying that no cost-benefit analysis would justify such a dam, and Merriman was undoubtedly correct that it would have been safer to have excavated the unreliable foundation material. But the decision to build was politically irrevocable; Roosevelt's memo never implied that the dam be abandoned or even put on hold; but only that it be repaired based on the best engineering judgment in the nation.

But the dissenters' arguments required a careful response by the U.S. Army Corps of Engineers—more detailed than its November answer to Roosevelt. Therefore, Major General Julian Schley, Chief of Engineers, directed the Board to report to him on the views of Merriman and Mead. The Board replied on March 3, 1939, having known the objections since their February meeting. Of Merriman's many recommendations, the Board rejected the central one of excavating the shale by stating that "an ample factor of safety is provided by the extremely flat slopes adopted for the new design." The report went on to reject Merriman's other recommendations and then turned to

Figure 6-21: General Julian L. Schley. Source: U.S. Army Corps of Engineers.

Mead's letter, which it dismissed with the observation that "the Board does not consider that a discussion of the economic and social values of the Fort Peck Project is within its province." Mead had agreed with the Board on the "engineering principles involved."[76] General Tyler wrote a memorandum to General Schley summarizing the project and recommending that the Board's report and response be approved, which Schley did on October 7.[77] With that, the Board's arguments ended and the project continued as planned.

Completion and Controversy

The last earthfill rolled onto the dam on October 11, 1940, instead of late 1938 as originally planned.[78] The slide delay, together with the substantial redesign, added about 25 percent more earthfill than originally planned. But while the dam continued toward completion, the controversies within the Board became more public. The Corps recognized that the great importance of the project demanded a public accounting, not only in the official report but also in a journal paper. Therefore, in the spring of 1939, Thomas Middlebrooks, the Corps' Chief Engineer for Soil Mechanics began to prepare a paper for the *Proceedings of the American Society of Civil Engineers*. At the same time, the ASCE organized a session at its San Francisco meeting to discuss the paper.[79]

Already, Karl Terzaghi, writing in Britain's *Journal of the Institute of Civil Engineers* in June 1939, stated that "This [a flow failure] was demonstrated . . . by the failure of a dam . . . in the U.S.A. Without warning, 10 million cubic yards of earth were lost within a few minutes."[80] Flow failure means that the sand shell loses all its shear strength—it becomes quick sand—and flows like a liquid (the condition engineers call liquefaction). The Board's conclusion had been quite different; it had identified the failure in the foundation, not in the shell, and due to weak seams of weathered shale and bentonite.

After those comments, Middlebrooks wrote Terzaghi a letter expressing strong disagreement with the distinguished professor.[81] In early September, Terzaghi replied in some detail, even drawing diagrams to illustrate his view of how the slide had occurred. His letter contains the outline of a theoretical analysis of the stresses in the foundation, but his principal arguments have more to do with human nature than with natural laws.

First, Terzaghi used his experience:

I have investigated at least twenty foundation failures due to . . . weak layers of clay beneath the base of embankments . . . characteristic for all these slides was an energetic compression of the lowest part of the downstream slope, which is conspicuously absent in the Fort Peck case.

So he claimed it must have been a flow failure. "For twenty years I have collected whatever data I could secure concerning flow failures and, as a consequence, I am fairly familiar with these phenomena."

Second, he questioned the Board's judgment, being careful not to impugn any individual. He admitted that a foundation failure had occurred but it was only an agent "which pulled the trigger" to set in motion the inherently unstable sand embankment saturated by the method of hydraulicking. He implied that the Board, partially made up of people who had, in Terzaghi's words, "fatally overestimated the safety of the foundation (in 1935)," had not clearly explained the problems. As support, he quoted an editorial in the May 11, 1939, *Engineering News-Record* which stated:

> Two or three years ago it was discovered that a loosely deposited mass of sand, if later saturated, might undergo sudden liquefaction when disturbed by a shock or otherwise, and the Fort Peck slide was believed by many to be due to such liquefaction. It was therefore to be expected that the report would place the question beyond any doubt. The two sentences which the board devotes to the question of cause are quite insufficient for this purpose.[82]

Middlebrooks' 1942 paper devoted far more than two sentences to liquefaction. In nearly two pages of the twenty page paper, he discussed a major 1936 paper of Casagrande's, a series of tests on materials taken from the upstream shell at Fort Peck, and ended with a conclusion "that the shell material was of sufficient density to prevent it from becoming liquefied even under the severe stress conditions that occurred during the slide."[83] Here was his answer to Terzaghi. However, Glennon Gilboy, in discussing Middlebrooks' paper, stated that "the critical density tests (crucial to Middlebrooks' argument) do not prove the absence of liquefaction; and that the balance of other evidence is strongly in favor of the view that some liquefaction took place. . . ." Gilboy was a member of the Board but also a Cambridge ally of Terzaghi.[84] In his closure, Middlebrooks downplayed that viewpoint and stuck to his two major conclusions: "that the hydraulic fill was not at fault" and that the failure was not by liquefaction.

Behind this disagreement was a more typical twentieth-century debate, the one between academics and practitioners or more precisely in this case, between those who consult and those who construct. Terzaghi's experience, like Casagrande's and Gilboy's, was substantial but it was advisory, whereas Middlebrooks and Gerig worked for the bureau that was responsible for building. Indeed, after Middlebrooks' article appeared in December 1940, the ASCE wrote Gerig asking him to contribute a discussion. He wrote right away to Middlebrooks, saying that he would do it even though he did not agree with everything in the paper, but he stressed, "I am not prepared to accept everything that Casagrande, Terzaghi, and Gilboy accept."[85]

In his reply, Middlebrooks took up the same theme:

> There is no doubt that his [Terzaghi's] statement at that time was prompted by Casagrande's conclusions. Incidentally Casagrande has cooled off considerably recently in regards to his 'critical density' theory. You know, they have "fads" like the ladies, which come and go with the seasons.

Middlebrooks was not so certain about everything at the dam; later in the letter he noted that he had not been to the site recently and that there was so far very little seepage through the dam. However, the reservoir was still low. "I expected some seepage out-crops" he wrote, "on high stages, but nothing serious (I hope)."[86] His hopes were not fulfilled, seepage was too great, and this led the Corps to develop a new device, the relief well, to control seepage.[87]

Years later, Middlebrooks, in a review paper presented at the ASCE Centennial Convention in Chicago, noted the need for the relief wells at Fort Peck, but he also returned to the slide and reiterated his view that it was not a flow failure. He indirectly defended hydraulic fill by explaining that few such dams followed Fort Peck because of "the improvement in earth excavating, hauling, and compaction equipment [that] has reduced the cost of rolled fills to such a degree that hydraulic fills cannot compete unless circumstances are exceptionally favorable."[88]

Nevertheless, the standard book on earth dams pronounced that the Fort Peck slide and:

> at least a dozen similar failures of major dams . . . occurred because the hydraulic fill method of construction resulted in an embankment which was, at least in part, composed of very loosely compacted saturated sand which 'liquefied.' There are no records of similar trouble during construction of compacted earth dams."[89]

What, finally, was the cause of the slide?

A review of this controversy more than 60 years later reveals that there was not one cause but, as is usually the case in such major events, several. We can identify at least five: the foundation slip, the relatively steep upstream slope, the saturated sand shell, the high speed of construction, and the redirection of the river. Terzaghi, Middlebrooks, and the Board all agreed that the foundation slip instigated the failure. Of all the experts, only Merriman seems to have raised the foundation issue before the slide, and he did it in 1935 during discussions of design. In 1933, the Corps had collected a respected group of consultants, and, in that sense, the design should be seen as the state-of-the-art.

No one seems to have objected to the steep upstream slope, and, yet, after the slide, no one would defend it. Therefore, it appears that the profession fully accepted the two primary causes: foundation slip and steep slope. The biggest controversy arose over the flow failure and its connection to the hydraulic fill. This was a seductive construction method for the Corps in the depression years because it promised relatively easy access to fill materials; it involved dredging, at which the Corps was expert; and it had the approval of leading engineers, especially Wegmann, Justin, and even Gilboy. It seems clear that had there been no slip and had the slope been flatter, there would have been no slide even with the hydraulic fill. Surely with those two primary features, the effects of hydraulicking, as described by Sherard, *et al.*, did cause the slide to move more quickly and more extensively. Also, the speed of construction contributed to the flow, and the redirected river–partly undermining the region beyond the toe–helped to extend the slide.

This situation of multiple causes was to be repeated even more dramatically two years later near Tacoma, Washington, on another depression era project, the doomed Tacoma Narrows Bridge. The third longest spanning bridge in the world completely failed in a moderate wind storm while movie cameras recorded it tearing apart. The bridge had too little deck vertical stiffness, almost no torsional stiffness, and had a solid plate girder facing the wind.[90] Just as at Fort Peck, the rebuilt work corrected those major faults and changed the profession's approach to large-scale construction. In both cases, the profession benefitted from disaster.

One final word on the Terzaghi-Middlebrooks controversy. The profession has long since recognized these two pioneers as exemplary engineers. Terzaghi won many ASCE awards for this research, and Middlebrooks twice won the Laurie Award for highly meritorious papers (including the one on the slide). After his death in 1955, the American Society of Civil Engineers established the Thomas A. Middlebrooks Award for a paper on geotechnical engineering (soil mechanics), the first such award established by the Society. In 1960, the Society established the Karl Terzaghi Award for contributions to the same field.[91] The debate between these two engineers is today of less significance than the Fort Peck Dam, which was, in the end, a vast and successful experiment. Its construction period coincided with the time when soil mechanics entered the world of earthfill dams. Its huge scale made it a daring design, and its partial failure revealed defects in that design. But it, nevertheless, served to confirm the validity of using embankments for mainstream dams on the great Missouri. This success was crucial to the next major event on that river, an event which would completely restructure that waterway after World War II.

THE PICK-SLOAN PLAN

The Pick Plan

On March 4, 1939, Joel D. Justin, chairman of the Board of Consultants for the Fort Peck slide, wrote to General Tyler, Chief of Engineers, telling of a March 13, meeting that he was to have with Reclamation on an earthfill dam in Nebraska. Because of the slide, some engineers thought the Reclamation dam was unsafe and Justin wanted to show data from Fort Peck to convince them that the Reclamation dam was very different and, thus, safe. He was on the Board of Consultants for that dam. The general gave him permission by phone to show them data but not to give them anything.[92] It was one small example of the suspicions that kept the two dam building bureaus apart. The Missouri Basin planning would bring those competitive strains to the surface in the most public political arena, the Congress of the United States.

Later that year, the Congress passed the Reclamation Project Act of 1939 (August 4, 1939). Section 9 directed Reclamation to broaden its irrigation studies to include hydroelectric power, municipal water supply, and other features which could include navigation and flood control. In these latter two cases, Reclamation was to consult the Chief of Engineers. To head the study for the Missouri Basin, Reclamation picked an Assistant Regional Director and seasoned engineer, William Glenn Sloan, who had worked for the Corps, for the Department of Agriculture, and in private practice. Sloan had been with Reclamation since 1936.[93] Sloan's studies had been progressing slowly and carefully for almost four years when Missouri River Division Engineer Colonel Lewis A. Pick entered the planning scene in a quick and decisive way. On May 13, 1943, Pick appeared before the House Committee on Flood Control. The immediate issue for his division centered on the recent flooding of the Missouri, but a more general worry was the fear of a post war depression and the need, therefore, of public works to counteract veterans' unemployment.[94] Lack of work right after World War I combined with the New Deal efforts in the 1930s led the Congress and the president to think about preparing for large-scale river engineering once the war ended.

The result of Pick's testimony was a committee resolution directing the Board of Engineers for Rivers and Harbors to review earlier reports on the Missouri River and to determine "whether any modification should be made therein . . . with respect to flood control along the main stem. . .from Sioux City, Iowa, to its mouth."[95] That rather innocent sounding order set Pick off onto the preparation of a plan for the Missouri River Basin.

Figure 6-22: Lewis A. Pick after his promotion to general. Source: U.S. Army Corps of Engineers.

Figure 6-23: (William) Glenn Sloan. Source: Bureau of Reclamation

Moreover, Pick's report was instigated by the Committee on Flood Control, and two major floods had just occurred when Pick was testifying. The first one, in March, breached levees all the way from Sioux City to Kansas City, and the second, in May, breached or overtopped most of the levees between Jefferson City and the mouth of the Missouri.[96] This was a graphic illustration of the need for better flood control and, as Pick emphasized in his report of August 10, 1943, "complete protection against all floods of record by levees alone is impracticable."[97] This fact, already stressed in the 308 Wyman Report, led him to propose a series of huge mainstream dams and reservoirs between Fort Peck and Sioux City and to stress that "for the maximum utilization of the waters of the basin, the reservoirs proposed above Sioux City should be multiple-purpose projects." He identified five such projects on the Missouri: Garrison, Oak Creek, Oahe, Fort Randall, and Gavins Point.[98]

Colonel Pick held a few hearings on his tentative plans in Montana, Iowa, and Nebraska between June 8 and 10, while the district offices of Kansas City and Omaha prepared the data that the division office would correlate.

Pick submitted his plan, described in a mere 12 pages, to the Chief of Engineers on August 10, 1943, less than three months after the House Committee resolution. The plan was exceedingly simple; it included strengthening levees below Sioux City for $80 million, building the five mainstream dams for $330 million, constructing two other multipurpose dams for $55 million, and constructing five smaller dams for $25 million.[99] Although the total of $490 million has little meaning in post World War II money, the breakdown shows clearly Pick's priorities. And these priorities would not go uncontested.

Pick sent his report to the Chief of Engineers and to the Board of Engineers for Rivers and Harbors (a congressionally authorized review board consisting of Corps officers). The Board sent approval to the Chief of Engineers on August 23, and five days later the Chief of Engineers sent both the Board's approval letter and Pick's report to Commissioner Harry Bashore of the Bureau of Reclamation and to Leland Olds, Chairman, Federal Power Commission. The Acting Chief of Engineers, Major General Thomas M. Robins sent the two documents to the Department of Agriculture on November 10. The responses from Olds and the Agriculture Department were favorable, yet cautious; but Bashore's response on December 17, 1943, was more detailed and with substantial negative implications. Bashore's primary worry was that Congress would ignore Reclamation's detailed, but unfinished, plan and move too quickly to adopt the Corps' proposal. He urged the Chief of Engineers to recognize "that a truly comprehensive plan can be developed best through integration of these two approaches." He was hampered, of course, because Reclamation's approach would not be ready until the spring of 1944. Therefore, he identified a set of principles that any coordinated plan should follow. Bashore's primary goal, after the need to work with the Corps, was to insure that the region above Sioux City would provide irrigation and that, where irrigation and power predominated, Reclamation be the bureau of design, construction, and operation. He did, however, agree that the five big dams between Fort Peck and Sioux City be constructed, operated, and maintained by the U.S. Army Corps of Engineers. Since this had been Pick's central concern and the greatest priority of his plan, Bashore was laying the basis for a single coordinated plan.[100]

Figure 6-24: General Eugene Reybold.
Source: U.S. Army Corps of Engineers.

In his summary of the Pick Plan and of responses to it from other agencies, the Chief of Engineers, Major General Eugene Reybold, strongly urged that the Corps' plan be approved right away and that its first phase be authorized. He stressed the need to work with Reclamation. Reybold's summary of December 31, 1943, went to Henry L. Stimson, Secretary of War, who sent it on January 7, 1944, to Harold D. Smith, the Director of the Bureau of the Budget. Smith responded a month later by agreeing that Stimson could submit the Pick Plan to the Congress but that the necessary authorizations "would not be in accord with the program of the President, at least at the present." He stressed the need for an estimate of benefits to compare with costs and he stated that no action would be taken by the president until he could review Reclamation's report, due on May 1, 1944.[101]

Congressional Conflict and the Creeping Commerce Clause

Stimson then submitted to the Speaker of the House on February 28, 1944 the Chief's report along with the responses including Smith's letter. In spite of Smith's statement that the Pick Plan would be only for the information of Congress, the House Flood Control Committee, under the Chairmanship of Mississippi representative William Madison Whittington, held hearings throughout February, when a House bill was framed. It clearly reflected the primacy of the Corps' wishes and those of Whittington, for whom lower-river-state flood control was of overpowering concern.[102] But the bill brought to the surface the long standing worry of western politicians that the federal government would take control of the river waters in their states. The Bureau of Reclamation had, since its founding in 1902, appropriated water for irrigation in accord with state water laws, whereas the Corps had operated, since the landmark court decision of 1824, under the commerce clause of the Constitution, which gave federal law precedence over state laws in regard to interstate commerce.[103]

Fort Peck Reservoir was, therefore, a symbol of this conflict between the upper Missouri Basin states which supported irrigation and the lower Missouri Basin states which supported navigation and flood control. How would that huge reservoir of water be used? The Corps wanted it for the nine-foot channel from Sioux City to the mouth, and Reclamation wanted it for irrigation in the arid regions. Whittington gave the upper basin people and Reclamation the opportunity to make their case before the committee, but such was his prestige that the bill reported out to the House on March 29, 1944, was still centered on Pick's Plan.[104] President Roosevelt, reiterating what the Budget Director had written, urged the committee to consider reclamation provisions because he thought small farms with irrigation the best way for returning veterans and war workers to reestablish themselves in that region.[105] Whittington listened but went ahead with the Corps' plan, which came to the House floor on May 8. After one day of debate, it passed and went to the Senate.[106]

However, the issues were too complex for quick passage of the bill by the Senate. In addition, the war still engaged the nation's attention, and 1944 was an election year. Moreover, on May 1, the Secretary of the Interior sent the long-awaited Reclamation Report to President Roosevelt, and the Sloan Plan became public.[107] No construction could have begun, anyway, in 1944 unless it could have been closely connected to the war effort. Despite those obstacles, the key senator, John Holmes Overton of Louisiana, would work to guide legislation and would ultimately succeed. Overton obviously had a major interest in flood control for the lower Mississippi. Moreover, he chaired Senate subcommittee hearings for both Commerce and Irrigation and Reclamation Committees.[108] Both he and Whittington sought to keep control of river planning with Congress rather than with the executives; thus, they were opposed to Reclamation in so far as it would take away control from the Corps. They both recognized, however, the need for Reclamation and the Corps to cooperate, and particularly Overton needed to work with senators from the arid regions to get his own flood control measures passed. And there was one other potential challenge to the Congress, the creation of a Missouri Valley Authority, which was President Roosevelt's idea for river basin development. But before that hurdle would be faced, Overton had to deal with the Sloan Report, which was of a quite different character from the Pick Plan.

The Sloan Report

There is no doubt that Reclamation rushed to finish its report once Pick's Plan appeared, and especially once Reybold sent it to Stimson. As 1944 began, Sloan put his report into final form. After passing through various governmental agencies, on May 5, Wyoming Senator Joseph C. O'Mahoney presented it to the full Senate.[109] The House passed the Pick Plan three days later and sent it to the Senate where, by mid-May, both proposals collided on the desk of Senator Overton. But was it really a collision?

Sloan had the advantages of having worked on his study for four years, seen Pick's plan, and heard objections to it. His report, therefore, focused not on transportation or flood control but rather on water as a resource. Entitled "Conservation, Control, and Use of Water Resources" his report emphasized agriculture and irrigation along with power.[110]

Sloan divided the basin into six sub-basins: the Yellowstone, the upper Missouri (above Fort Peck), minor western tributaries, the Niobrara, Platte, and Kansas, Fort Peck to Sioux City, and Sioux City to St. Louis. He described each one, analyzed its present development, and outlined future plans. He compared Corps and Reclamation plans, focusing on areas of agreement and coordination. Although Sloan addressed the many differences, the overall tone was conciliatory. Sloan emphasized that the two plans could be integrated. He ended the report with a detailed discussion of the power potential, of the geography, and of the economy in the basin. In the first appendix, he detailed the costs, totaling $1,258 million.

In the opening summary, Sloan had given that total cost, the annual running costs ($65 million), the annual benefits ($168 million), and the repayments and returns. These last figures add up to the total construction costs by allocating a large amount for flood control and navigation ($517 million) and adding them to the annual payments by users of irrigation ($298 million), users of power ($423 million) and municipalities ($20 million).

The most significant differences in the two plans were in the main river dams between Fort Peck and Sioux City. Sloan proposed only Oahe, Big Bend, and Fort Randall whereas Pick had Garrison, Oak Creek, Oahe, Fort Randall and Gavins Point. The Reclamation plan envisioned extra storage reservoirs on tributaries to mostly make up for the lesser storage on the mainstream. Since the Corps' major interest lay in those mainstream dams, Sloan's reductions there would be a major issue later on.

In fact, General Reybold made it such an issue in his letter of April 25 to Commissioner Bashore. These mainstream dams "are vitally needed for cyclic storage. . . . I consider that the maximum practicable amount of storage must be provided . . ." he wrote. Reybold also questioned sending water for irrigation out of the basin, as Sloan proposed, and he further objected to the high cost allocated to flood control and navigation. For, as he noted, the benefits estimated for irrigation were very large compared with those for combined flood control and navigation, whereas the costs allocated to irrigation are much less. In spite of these disagreements, Reybold offered a compromise by stressing that the Corps plan was open to augmentation by other bureaus.[111]

Pick's plan stimulated action, and Sloan's report provided support for action, but the Congress still had to debate two significantly different proposals. On May 12 the Sloan Report formally went to the Senate Committee on Irrigation and Reclamation, but before any action could occur, the Senate Commerce Committee sent the report to its two subcommittees, one on Rivers and Harbors and the other on Flood Control, both chaired by Senator Overton. He held hearings in May and June as O'Mahoney and others tried to amend the Pick Plan to give states jurisdiction over the waters. On June 13, President Roosevelt, in a letter to Senator Overton, seemed to support O'Mahoney, but the Senator's subcommittee went ahead on June 22 and affirmed the House bill, which was essentially the Pick Plan.[112]

But in June 1944, the war was the major item as the allies landed in Normandy and the Japanese offered stubborn resistance to island invasions. Legislation on the Missouri River Basin could not compete, so the complex problem of three different proposals (Pick, Sloan, and the Missouri Valley Authority) had to wait until after the summer recess. That did not prevent fervent discussions, proposals, and debates.

On August 7, Roosevelt wrote Senator Overton again, calling for a compromise between the supporters of irrigation and those of navigation. Senator James Edward Murray of Montana drafted a bill dated August 18 to create a Missouri Valley Authority, which is what the president really preferred.[113]

A crucial factor in the summer debates was the Missouri River States Committee, essentially a governor's committee, which met in early August, and began lobbying for a coordinated plan between the Corps and Reclamation.[114] The Committee feared that interagency quarrels or intra-basin disputes would prevent speedy congressional approval. When Congress reconvened in September, Roosevelt sent a message proposing an authority like TVA and attaching a resolution from the Committee that inferred its support. That inference was quite wrong. Rather, the Committee had urged Reclamation and the Corps to work out a unified plan, but O'Mahoney, as chair of the Senate Subcommittee on Irrigation and Reclamation, supported by Ickes, pushed for Reclamation's plan while Overton favored the Pick Plan.

Finally, encouraged by the Missouri River States Committee and by Sloan's response at a Senate hearing, the two groups met in Omaha on October 16 and 17 and quickly agreed on a combined plan.[115] This would be known as the Pick-Sloan Plan and would serve as framework for future developments. But most important to the Congress and to the States Committee, this plan would help determine river basin developments for the postwar period, and in that process block the president's plan.

Pick-Sloan and the Valley Authority

On October 25, 1944, Commissioner Bashore and General Reybold, together, submitted both to the Secretary of War and to the Secretary of the Interior, a five-page letter calling for the plan resulting from the October 16-17 meetings in Omaha. They also enclosed the plan that had been sent to them on October 17 and signed by General R. C. Crawford, Division Engineer (replacing Pick); Gail A. Hathaway, Head Engineer of the Corps; William A. Sloan; and John R. Riter, acting Director, Branch of Project Planning, Bureau of Reclamation. The substance of that plan, with all its immense consequences, was just one page long. It began by dividing the basin into Sloan's same six subdivisions and disposed of three of them as not being in dispute. In other words, everything proposed by either earlier plan would remain. The framers of this Pick-Sloan Plan agreed to use Sloan's project for the Yellowstone River Basin, to make minor revisions in the Niobrara, Platte, and Kansas River basins, and to accept Pick's five mainstream reservoirs but to eliminate the Oak Creek Dam (flooded out by the enlarged Oahe Dam) and to replace it with the Big Bend Dam proposed by Reclamation. Essentially, Pick got his five dams and Reclamation got its new irrigation acreage (4,760,400 acres). The much longer letter from the Commissioner and the Chief of Engineers elaborated on the justification for the combined plan and identified three "basic principles:" (a) the U.S. Army Corps of Engineers should have responsibility for the mainstream dams (all six, including Fort Peck), (b) Reclamation should have responsibility for all irrigation issues, and (c) both agencies would recognize the importance of hydroelectric power.[116]

Senator Overton decided to attach the Pick-Sloan Plan to the Flood Control Bill in the Senate, which began deliberation on November 21. This new plan entered the Congressional Record as a supplement both to the House Document on the Pick proposal and to the Senate Document on Sloan's report. The following week, President Roosevelt sent a letter of agreement to Congress, but with the cautionary note that a basin authority would be needed. The Senate passed the bill on December 1, reached agreement with the House on December 9, and sent it to Roosevelt four days later. The president signed it on December 22, 1944, with the statement that the bill was "not to be interpreted as jeopardizing in any way the creation of a Missouri Valley Authority."[117]

Congress and the president had settled the central issue of getting congressional approval for projects to be built after the war, but they had not established any new management plan. The Corps and Reclamation were still in charge of their mandated projects. Senator Murray, as Roosevelt's strongest Senatorial ally on the Missouri Valley Authority, persisted in introducing legislation but he was never successful. The Corps, Reclamation, the Congress, and even the states were unwilling to give up power over their local works. The strongest argument Murray had, of course, was the highly publicized 1933 Tennessee Valley Authority. It was Roosevelt's pride, and it appealed to people impressed by the need for neat, clear, bureaucratic solutions to messy management problems. In 1944, the TVA seemed to be a huge success under the dynamic leadership of David Lilienthal.

To a large extent, the TVA had been the brainchild of Senator George N. Norris (1861-1944), who had died on September 2, just as Pick-Sloan was making its way to the Senate. He was from the only state, Nebraska, that is fully in the Missouri Basin; his presence was felt in the Senate considerations of

Murray's efforts. But TVA was of a different time and in a very different region. When it was created in 1933 the country was open to innovation; it was during Roosevelt's first 100 days. Compared to the Missouri, the Tennessee Valley was much smaller and the river much shorter. The people in the Tennessee Valley were eager for help from the federal government. In addition, the TVA mechanism removed the ambiguity surrounding Wilson Dam–whether it should be public or private and its role in selling electricity.

The Missouri Basin was far more complex and had split into two political factions; and there were two powerful agencies there in competition.[118] Still, the idea for a regional authority had never died. In the 1930s, there had been a movement to divide major parts of the country up into valley authorities, removing the Corps and Reclamation from much of their territories and missions.[119] This would have changed the political nature of the United States and vested far more power in the federal government at the expense of the individual states.

Figure 6-25: Colonel Lewis A. Pick (second from right) and Bureau of Reclamation engineer Glenn Sloan (far right) examine the future site of Yellowtail Dam in Montana. Source Bureau of Reclamation.

With the Pick-Sloan Plan as law and the war coming to an end, the U.S. Army Corps of Engineers began the design work for the first major dams; Garrison, Fort Randall, and Oahe. All three were to be earthfill embankment dams designed with the experience of Fort Peck in mind.

GARRISON DAM

The Garrison Design Controversies

Roosevelt died on April 12, 1945, and the war ended in late summer. Corps engineers began to return from overseas. The Omaha District Office started to develop a Project Report on Garrison Dam, which it submitted to the Chief of Engineers on February 8, 1946. The Chief of Engineers held a conference in Washington on March 27-29, with representatives from the Omaha District and the Division offices. The design could now begin.[120] In June, the Corps began to build the new town of Riverdale, North Dakota, to house dam workers. Learning from Fort Peck Town, the Corps designed Riverdale for families and included schools, a shopping center, eight lane bowling alley, and even a beer parlor. At the height of construction, the Riverdale High School Knights went for three seasons undefeated in football.[121]

In late 1945, there had been much public discussion in North Dakota about the dam, about congressional funding, and about the height limitations that the House had put on the dam and reservoir.[122] Then, the originator of the dam returned after a 27-month absence to take his old job as Division Engineer and to push hard for full funding and immediate commencement of his project. Although he had been far away, building the 483-mile Ledo Road between China and India through Burma, Pick assured the Bismarck Chamber of Commerce, as 1946 began, "I've never lost interest in this valley; coming back here was like coming home."[123]

Dam construction needed power, and Garrison's predecessor, Fort Peck, would supply it via a new transmission line.[124] One dam would help build the next. The *Saturday Evening Post* ended a long article on the proposed Missouri Valley Authority by focusing on General Pick:

> Pick of the Ledo Road, Pick the dreamer, the engineer, the author of the first plan for the Missouri Valley. He too, is a peacemaker. Of course, he'll have help. There's always that Thing, outside in the dark: MVA. Just standing there, it scares up a lot of action.

It did not scare President Harry S Truman; he, following Roosevelt, still believed in the Missouri River Authority in spite of congressional rejection. Other supporters, however, saw MVA as dead and, therefore, moved over the divide to work for a Columbia Valley Authority, being pushed in the House by a young representative from Washington, Henry "Scoop" Jackson.[125]

Meanwhile, the Corps proceeded with the design and with contracts not only for the town but also for a road and Missouri River bridge to the site. However, a new obstacle became public in May, when General Pick appeared before leaders of the Indian tribes in the Fort Berthold Reservation, much of which was to be flooded by Garrison Reservoir. The affected people were to be moved to comparable land, but tribal leaders did not trust the Army, really the U.S. Government, to treat them fairly. Pick listened, saying little. The protests continued through 1946.[126] Then, in November, Pick was forced to plan for laying off a large part of his staff in the district offices because of Truman's order to put a ceiling on public works. Pick predicted a years' delay on the dam.[127] It did not happen.

A major issue for the Corps and the Indians lay in an amendment by Senator O'Mahoney to the bill allocating funds for the dam. This amendment stipulated that no money would go for the dam until the government and the Indian Tribes reached an agreement. Senator O'Mahoney intended the amendment to give the Indians bargaining power in their negotiations with the government.[128] In May 1946, the tribal leaders confronted General Pick in Elbowoods, North Dakota, and the debate, well publicized in local newspapers, built up throughout 1946.[129]

In early 1947, Pick appealed to the State Legislature in Bismarck, North Dakota, and warned that if the same amendment appeared in the next appropriations bill from Congress, "work on the dam must stop." He continued, "If this should occur it would be most difficult to revive the plan." He noted that "if the Garrison Dam goes forward it will be because people want it. If you don't want it, tell the Appropriations Committee and no money will be spent." Several days later, North Dakota Senator Milton R. Young announced that there would be no such amendment on future appropriations. That did not quell objections to the dam, certainly not those of the tribes nor those from the Farmers Union, the Stockmen's Convention, and the many supporters of the Missouri Valley Authority.[130] Then came the kind of event that moved politicians to action—a flood in the Missouri Basin.

Just as Pick's original plan came at the time of the 1943 flood, so the funds for Garrison Dam became quickly unlocked after the flood of 1947. Three million acres were flooded, twenty-six lives were lost, and damages of $111 million led the Missouri River States Committee on July 2 to call on the federal government to quickly appropriate adequate funds for Garrison Dam, with no restrictions imposed that would interfere with immediate work on the dam.[131] Local newspapers called for fast action, and the news made it to Washington. On July 16, President Truman sent a strong message to Congress requesting a ten-year flood control program with $250 million to be allocated immediately to the

Mississippi-Missouri watershed. The *St. Louis Post-Dispatch* carried the story and ran a huge headline, "The PLIGHT of the MISSOURI VALLEY." President Truman mentioned the Missouri Valley Authority only in passing, and, the following week, Senator Young could announce that the dam would get $20 million, of which $5.1 million was for the Indians, but there would be no holdup pending agreements. The Corps lost no time in letting its first big contract for the dam in early September.[132] Garrison Dam was now secure.

But the concerns of the Indian tribes would not go away quietly. The three affiliated tribes at Garrison (Mandan, Arikara, and Hidatsa) were the first to negotiate with the U.S. Army Corps of Engineers following the Pick-Sloan legislation. The tribes had to relocate 325 of their families, about 80 percent of tribal membership. The Indians objected to the compensation and requested just under $22 million; Congress, in 1949, did agree to $7.5 million more, for a total of $12.6 million. "Although the additional $7.5 million was too little to be fully effective, and the help provided by the Army was unsatisfactory, these gestures symbolized an attempt by Congress to acknowledge its moral and legal duty to the Indians."[133] Soon after Garrison Dam construction began, tribes on reservations downstream were confronted with problems similar to the three affiliated tribes.

Garrison Dam Design and Construction 1948-1950

The Corps designed the 12,000-foot-long embankment dam, influenced by Fort Peck, to have a slope of one to eight over much of both the upstream and downstream sections. The designers made the maximum design height 210 feet and the base an average of about 2,500 feet wide, plus 1,250 feet of impervious blanket extending from the upstream toe (elevation 1710) out into the reservoir. This blanket reflects a precaution stemming from the Fort Peck slide. But the most significant change from its predecessor was the decision to build the dam by dry-land methods using rolled earthfill. There was no mention of hydraulic fill. Nearly all of the earth would come from excavation for the intake, powerhouse, outlet, and spillway structures. Indeed, these major works would require 1.5 million cubic yards of concrete, almost half of what was used in the entire Hoover Dam.

The Corps planned the intake channel further upstream than in Fort Peck, to divert the river through eight intake tunnels, five to serve as penstocks leading water to the powerplant and the other three for regulating the reservoir height. The design called for the spillway to be adjacent to the dam on the left bank. The powerplant was to have a capacity of 400,000 kW.[134]

In early February 1948, excavation uncovered large seams of lignite, which the Corps began to stock pile for fuel to produce power and heat on the project. Huge earthmoving equipment roamed over the site in the spring, and in May, recognizing the inevitability of the dam, the Fort Berthold Indian Tribal Business Council signed a contract turning 155,000 acres of its reservation over to the federal government for a cash settlement of just over $5 million. There had, as yet, been no agreement on the relocation plan, but the government had set a deadline of June 30. The pressure of the Missouri River States Committee, stimulated by the 1947 flood, led to the defeat of Senator O'Mahoney's amendment favoring the Indians.[135]

Work accelerated as summer approached, and on July 24, President Truman announced his request for additional funds for Garrison Dam. The Corps let a $13.6 million contract in late September. While Garrison boomed along, the Sloan part of Pick-Sloan seemed stalled.[136]

In early December, Glenn Sloan told the Association of Commerce, at their annual dinner in Minot, North Dakota, that the Missouri-Souris reclamation project must overcome three obstacles, but that "We are going on despite all obstacles. Nothing is going to stop us." This immense project, outlined in Sloan's 1944 report, included pumping water from Fort Peck Lake over to the Souris River to irrigate 1,275,100 acres of land in northwestern North Dakota. Because Garrison Dam would supplement Fort Peck as a flood control dam, the older reservoir could be put to the service of irrigation. The projects' obstacles were the legal requirements for Reclamation to form irrigation districts before getting approval, their need to restrict farms to 160 acres, and the difficulty of selling Congress on pieces of the Pick-Sloan Plan rather than the whole plan. These obstacles, despite Sloan's optimism, would prove insurmountable, especially the last one.[137] And while Sloan urged the plan on, other forces raised more obstacles.

Senator Murray had announced in November that he would renew his call for a basin authority. His hope was that the newly elected Congress, being now "more liberal," would favor the TVA. He also believed that "there was a growing dissatisfaction throughout the Midwest over the way the Army Engineers are carrying out the Pick-Sloan Plan. . . ." But a potentially bigger threat came from a different quarter, the Hoover Commission on governmental reorganization. Hoover's view began to be public in late 1948, and, by the spring of 1949, strong criticisms surfaced essentially arguing that the Pick-Sloan Plan was poorly conceived and that the river work should be removed from the Corps and combined with Reclamation into one, presumably new, agency. In addition, the Commission stated clearly that there should be no more valley authorities; TVA was an experiment that was not to be repeated.[138]

On March 1, 1949, Lewis A. Pick became the Chief of Engineers. Shortly thereafter, the Missouri Basin Interagency Committee elected William Sloan to succeed Pick as its chairman. The two former partners soon clashed over the height of Garrison Dam. Appearing before the House Public Works Committee, Sloan argued that the plan was always for a pool elevation of 1830 feet above sea level. Pick, following Sloan to the witness stand, stated in his usual blunt manner, that plans for the dam "never have been for a pool level of 1830 feet." It was, he stressed, "unanimously agreed" at the Omaha meeting in 1945 that the dam should be built for a level of 1850 feet. This extra height was important to the Corps because it meant substantially more water available for power and navigation and more capacity available for flood control. Reclamation's interest was to use as much water as possible from the river for irrigation rather than have it sit in storage. But of greatest immediate concern to local politicians supporting Reclamation was the flooding of three irrigation districts and the town of Williston if the reservoir rose the additional 20 feet.[139]

Meanwhile the dam continued to rise rapidly and in late August the House, siding with the Corps, rejected the proposal to limit the pool elevation to 1830 feet. Pick's replacement as division engineer told a convention of the Association of Western State Engineers, that the higher level was largely for rural electrification. The convention earlier had passed a resolution opposing any valley authority. Again, the fear was that the federal government would take over state functions. In reality, most of the local leaders preferred to deal with the Congress and the Corps rather than with the executive and Reclamation. Not all Congressmen agreed with the pro-Corps position, but they were fighting a losing battle.[140]

Huge trucks carrying up to 54 cubic yards of earth moved across the Garrison site as the dam became visible even for the "casual observer" that fall. The heavy equipment allowed contractors to build the dam with only one-third the manpower estimated before construction.[141]

While the dam amazed visitors, it never ceased to agitate politicians. Governor John Bonner, of Montana, demanded assurance that no water would leave his state without its permission, and Governor Forrest Smith, of Missouri, stressed his state's need to be certain of having enough water for drinking and sanitation. President Truman maintained his preference for a valley authority to take on such priority issues. The same arguments persisted while the dam rose, oblivious to politics. It impressed international visitors from Iraq to Japan.[142]

Two opposing viewpoints continued to appear in the press reports and in the politicians' pronouncements. Some extolled the great feats that were unfolding in the Dakotas—the greatest such events in their recent history; while others questioned or condemned the wasteful project and the infighting between states and between bureaus. By year's end, one issue appeared to be settled, the immense Missouri-Souris diversion was dropped.[143]

Garrison Dam Completion 1951-1955

In April 1951, concrete casting began, and, soon thereafter, the earthwork recommenced after the winter shutdown. Construction of the intake structures and the powerhouse got underway in the spring, and, in mid-May electricity from Fort Peck began powering the construction of Garrison. In June 1951, testimony before the House Appropriations Committee revealed that the 1945 estimate of $130 million for the dam was low. The cost was projected to be $268 million when the dam was completed in 1955.[144]

Then came the flood of 1951. It brought in its wake a renewed call for a valley authority. Congressman John Elliot Rankin, from Mississippi, stated that "This disaster could have been prevented if Congress had passed my bill to create a Missouri Valley Authority." Representative Clair Magee from Missouri went even further during a committee hearing when he challenged a Colonel of the Corps by stating that "Power companies are generally opposed to the MVA plan, and I can state definitely the Army engineers are under the influence of the Power Companies." Later, Magee confronted General C. H. Chorpening, "Isn't it a fact that your superior, Lt. Gen. Lewis A. Pick, is opposed to public power, opposed to giving people cheap power?" "We have no position on public power," General Chorpening replied, "that is for Congress to decide."[145] Clearly, Pick was a focus of the hostility from proponents of a valley authority. But the work continued, and by year's end, the intake structure was more than half done and the powerhouse foundations prepared.[146]

In 1952, as the dam proceeded so did the debate over the ability of large dams to control floods. The Corps defended its huge mainstream dams, as Lt. Col. R. J. B. Page, Garrison District Engineer put it, "It is clear that the only measure which would have prevented the flooding along the Missouri River (in spring 1952) is the construction of large dams on the main stream. . . ."[147] Major General Donald G. Shingler, the Missouri River Division Engineer, pointed out how Fort Peck Dam and levee work had saved $50 million in damages around Omaha.[148] The floods, the costs, and the mammoth works all led *Time* to publish an article on the Peter Kiewit Sons Company that focused on the company's work under its Garrison Dam contract, the second largest construction contract ever awarded. Jointly with the Morrison-Knudsen firm, Kiewit held contracts of more than $60 million for the North Dakota project. *Time* detailed the background, the controversies, the present status of the Pick-Sloan program, and the competing views for a valley authority. Not only was there national interest but foreign engineers kept coming to Garrison to see the vast works.[149]

As the 1952 building season ended, the contractor prepared the river for diversion from its natural bed into the eight big tunnels on the right (west) bank. To avoid the dangers of an upstream slide, the engineers had designed a cofferdam to protect the upstream toe after diversion. That fall, the contractor built lumber mattresses to be placed over a rock dike across the old channel to help redirect the Missouri.[150]

In January 1953, the newly elected president, Dwight Eisenhower, received a recommendation from the Missouri River States Committee to form a Missouri River Basin Compact, but he declined to act on it. With a Republican Administration now in office, hopes for a valley authority also ended. About this time, before construction recommenced on Garrison, Reclamation announced that its large Missouri-Souris irrigation project would take water from the North Dakota Reservoir rather than from Fort Peck Lake.[151]

In April, with the cold weather lifted, the big event occurred: "Garrison Dam Closed," the commanding headline announced. From a temporary steel girder bridge over the Missouri parallel to the dam axis, a dozen huge trucks lined up and dumped rock across the open channel between the two arms of the embankment. This cofferdam dike, after two days of rock dumping, closed off the river, which obediently followed a newly dredged channel leading directly to the intake tunnels. The river was cooperating. After the floods of the two previous years, the river was much lower in 1953, so that the diversion went more

Figure 6-26: Spillway gate structure at Garrison Dam. Source: U.S. Army Corps of Engineers.

Figure 6-27: Water intake structure at Garrison Dam. Source: David P. Billington.

smoothly and quickly than expected. The cofferdam began to extend through the old channel and would soon serve as a wall to permit water between it and the dike to be pumped out. Once dry, the contractor would fill the closure with nearly 30 million cubic yards of earth to complete the dam. In May, *Life* ran a big story on Garrison, covering more on the construction and less on the boom town than their Fort Peck story had portrayed sixteen years earlier.[152]

Public interest shifted in 1953 from fascination with earthwork and the redirection of the river to the emergence of massive concrete structures: the spillway and powerhouse. But the controversies never completely disappeared. Plans went forward in June for President Eisenhower's visit to the Garrison Dam for the formal closure ceremonies; the affiliated Indian tribes participated, sitting on the platform with the president and giving him a ceremonial gift. Earlier in June, the tribes had called for retention of hunting, fishing, and grazing rights to the reservoir area in their reservation.[153]

In July of 1952, General Pick appointed General William E. Potter Division Engineer. These two were close friends, having served together after the war in the Missouri District and again in Washington after Pick became

Figure 6-28: General William E. Potter. Source: U.S. Army Corps of Engineers.

285

Chief of Engineers. After being Division Engineer for over a year, Potter decided to integrate the dams through computers, a novel idea at the time; and in December 1953, he announced a plan "where we can effectively utilize this new ultra-modern facility in our management program." He began with telephone connections to keep track of the water levels, flows, and power in the dams as they went on line. The system still operates effectively today with, as Potter predicted, computers.[154]

Figure 6-29: The riprap on the upstream face of Garrison Dam. Source David P. Billington.

Figure 6-30: The switchyard at Garrison Dam. Source: David P. Billington.

After the excitement of the closure, dam construction proceeded predictably to completion in 1955. Pick, meanwhile, had retired from the Army in early 1953. His favorite projects and the heart of his 1944 plan seemed safely in place, even if they would never be free from controversy. The reservoir began to be discussed seriously for recreation as placement of its rolled earthfill ended in late 1954. During 1955, the builders completed the spillway and the powerhouse, and in early 1956 the Corps officially dedicated the dam. Reclamation's gigantic Missouri-Souris Project never was funded, although a much smaller version, using Garrison's reservoir to irrigate 250,000 acres in North Dakota was approved.[155]

As the twentieth century ended, Garrison Dam was intact. It exhibited no slides, and its power production was well above the initial design. On a typical day in July 1996, it generated 495 megawatts of power from a flow of 36,600 cubic feet per second with a head of nearly 175 feet. Its efficiency was over 91 percent. The huge intake structure in the lake, reached by a bridge, its high surge tanks, and its curved concrete spillway all contrast vividly with the long embankment that serves to hold back the river and provide the head for its power.

Spillway Designs

For earthfill embankment dams, the concrete spillway is essential to prevent even the remotest possibility for overtopping during an extreme flood. For Garrison Dam, as for most such major structures, the engineers dimensioned the spillway to carry a huge flood, one never before recorded on the river at that location. The design required laying out a series of openings, each fitted with one tainter gate, such that the maximum design spillway discharge can safely pass. For Garrison, the Corps assumed such a discharge to be 827,000 cubic feet per second, even though the maximum discharge ever recorded at the site, up to 1947, equaled 260,000 cubic feet per second.[156]

Advances in hydrology have influenced the design of spillways, in general, resulting in structures with a greater capacity. The first formalized approach to spillway design came in 1913 with the publication of Fuller's formula that estimated a maximum discharge of Q_T in terms of the average yearly maximum $Q_a(1+0.8 \log_{10} T)$. Such an expression failed to predict very large floods, and, in 1932, the unit hydrograph was introduced to account for characteristics of a river basin and to be used in connection with a design storm. This storm, estimated for the region but not tied to the location of the dam, led to larger spillways than those based upon the return period formulas. In 1944, Gail Hathaway contributed to the development and implementation of spillway design based on the use of the probable maximum precipitation. The probable maximum precipitation then served to develop an upper bound on the flow possible where the spillway was to be constructed. For example, at the Garrison Dam location, the largest reservoir release would have been needed during the flood of 1881, and that would have required a spillway discharge of about 100,000 cubic feet per second. Using probable maximum precipitation contours, the designers determined an upper bound of over 800,000 cubic feet per second as the flow for which the spillway was designed.[157]

The plan to accommodate the design discharge included 28 openings, each with length, L = 40 feet and of height, H = 33.55 feet when the tainter gates are fully opened. The standard formula for discharge Q in cubic feet per second through such openings is given by $Q = C(L-nKH)H^{3/2}$ where n equals twice the number of openings and K is a coefficient that accounts for the contraction in the flow due to the piers that form the openings above the dam crest. These piers also support the gates. For Garrison, the engineers took C = 3.9, L = 40x28, n = 56, K = 0.015, and H = 33.55 feet, which gave the discharge Q = 827,000 cubic feet per second.

The Corps chose the shape of the spillway section, called a shallow ogee section, because of its hydraulic efficiency, which the coefficient of 3.9 reflects. On other types of sections, that coefficient is more commonly as low as 3.33. Repeated tests by the Corps and others established the coefficient K, and the height H comes from the maximum pool level of the reservoir.

The 24.5 million acre foot reservoir formed by Garrison Dam is the third largest in the United States, just behind Lake Mead at Hoover Dam and Lake Powell at Glen Canyon Dam. This great size means that the spillways would hardly ever be needed for floods, even though they have been used to regulate the reservoir level. Indeed, up to 1952, the greatest reservoir releases necessary, during the 1881 flood, were only about 100,000 cubic feet per second, and this discharge could have been easily taken by the three regulating tunnels at Garrison Dam. The conservatism in spillway design was typical of all those mainstream embankment dams across the Missouri River.

PICK'S PLAN AND THE MISSOURI BASIN

From Garrison to Big Bend

Meanwhile, downstream Pick's other four dams likewise block the river, produce power, and create lakes. Fort Randall, started a little before Garrison began operating in 1953; Oahe, begun in 1948, went into service in 1962; Gavins Point was in operation by 1955; and Big Bend, the last, started in 1959, and completed the entire project in 1964.

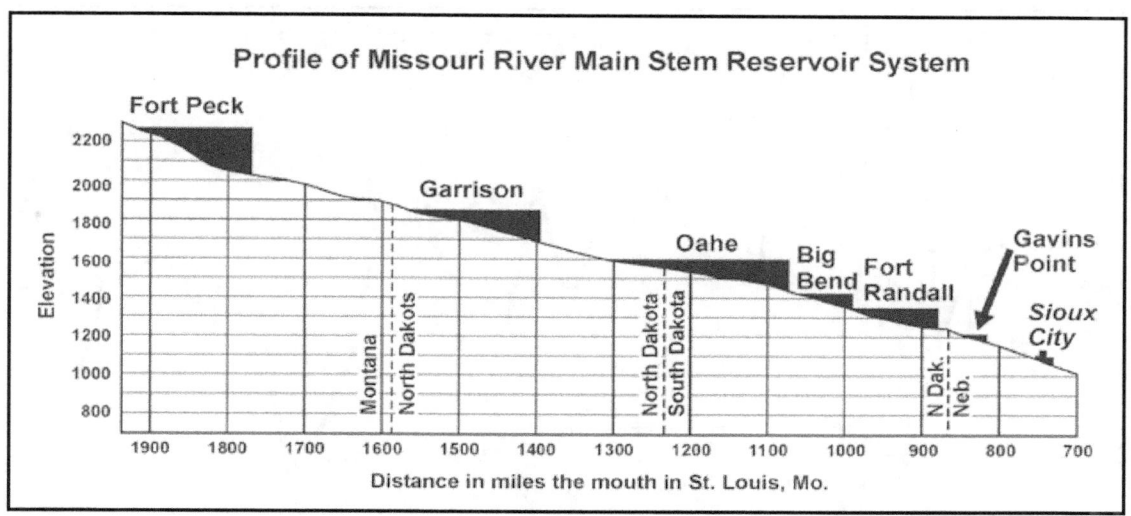

Figure 6-31: Profile of the Missouri River Mainstem Reservoirs. Source: *Big Dam Era*. Missouri River Division, U.S. Army Corps of Engineers.

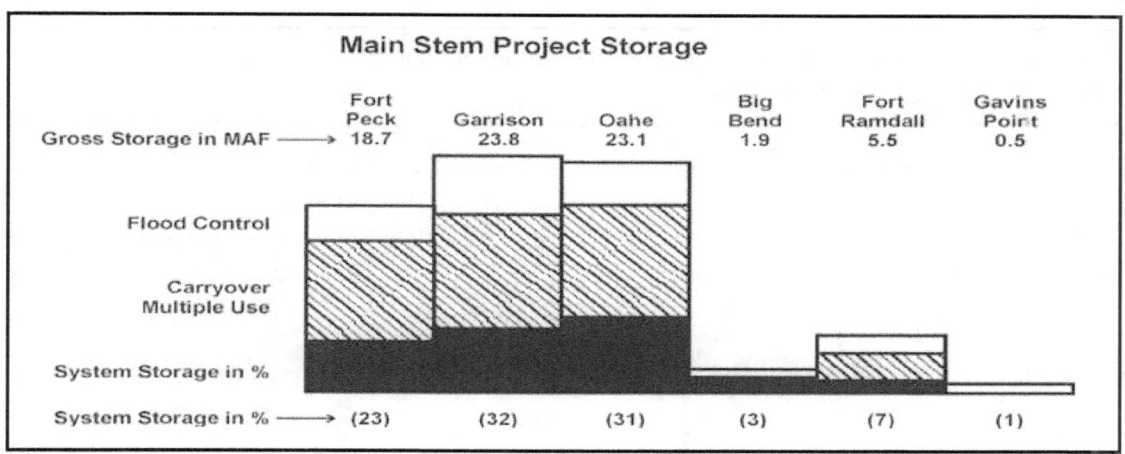

Figure 6-32: Missouri River Mainstem Storage. Source: *Big Dam Era*, Missouri River Division, U.S. Army Corps of Engineers.

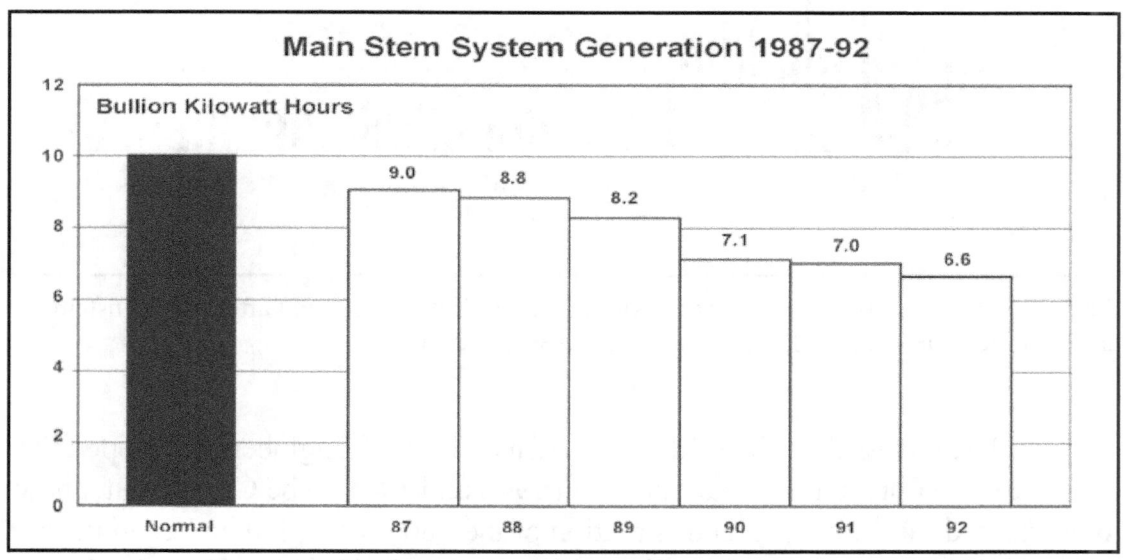

Figure 6-33: Missouri River Mainstem hydroelectric generation. Source: *Big Dam Era*. Missouri River Division, U.S. Army Corps of Engineers.

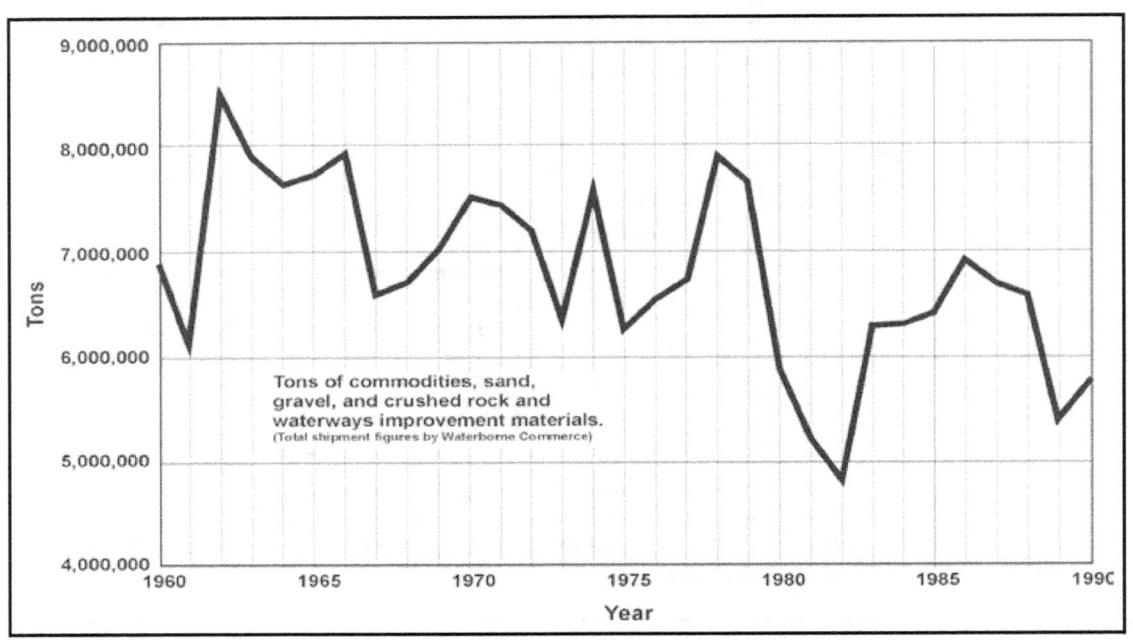

Figure 6-34: Commercial tonnage on the Missouri River. Source: *Big Dam Era*, Missouri River Division, U.S. Army Corps of Engineers.

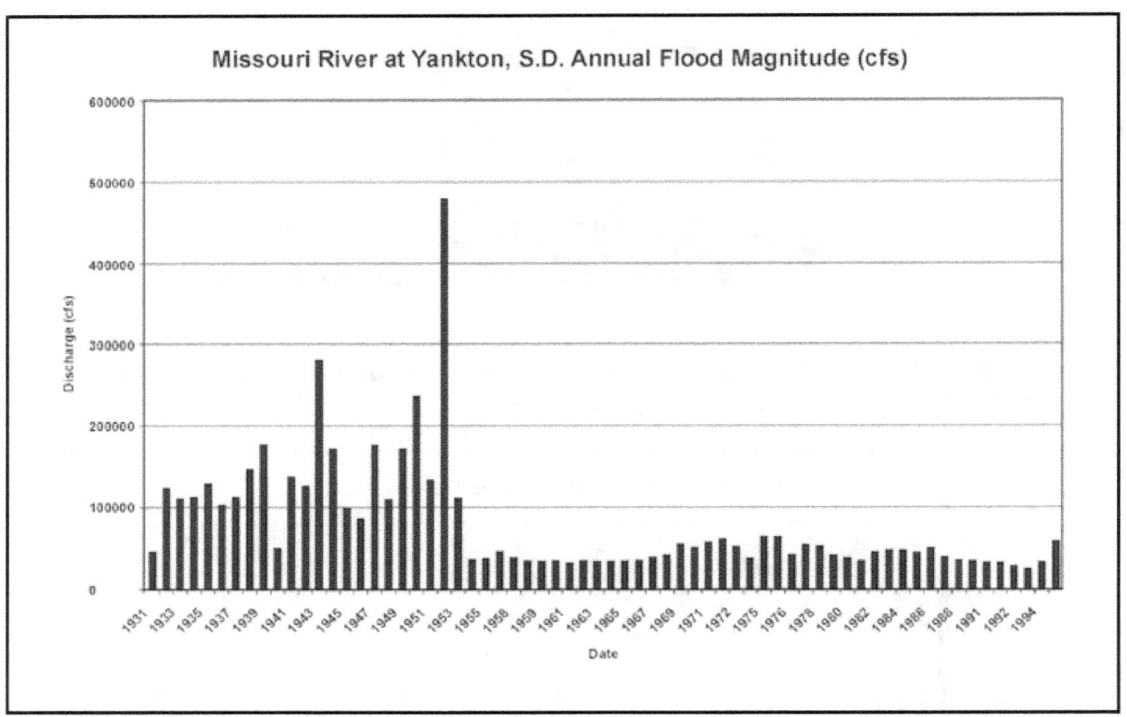

Figure 6-35: Missouri River flow at Yankton, South Dakota, before and after the mainstem dams. Source: Drawn from U. S. Army Corps of Engineer records.

The Omaha District of the U.S. Army Corps of Engineers developed the design for the Fort Randall Dam between 1946 and 1948. The Corps built a new town, named Pickstown, and construction of the 160 foot high dam began in early 1948. The dam required 28 million cubic yards of rolled earthfill and, when completed in mid-1956, had cost $183 million, almost 2.5 times the original cost estimate. President Dwight D. Eisenhower dedicated the first power unit in 1954, and by the early 1970s, the powerhouse was producing more than 2 billion kilowatt-hours yearly. After the gates had closed on July 21, 1952, the 107-mile-long reservoir, Lake Case, began to fill and eventually flooded 22,091 acres of Sioux land.[158]

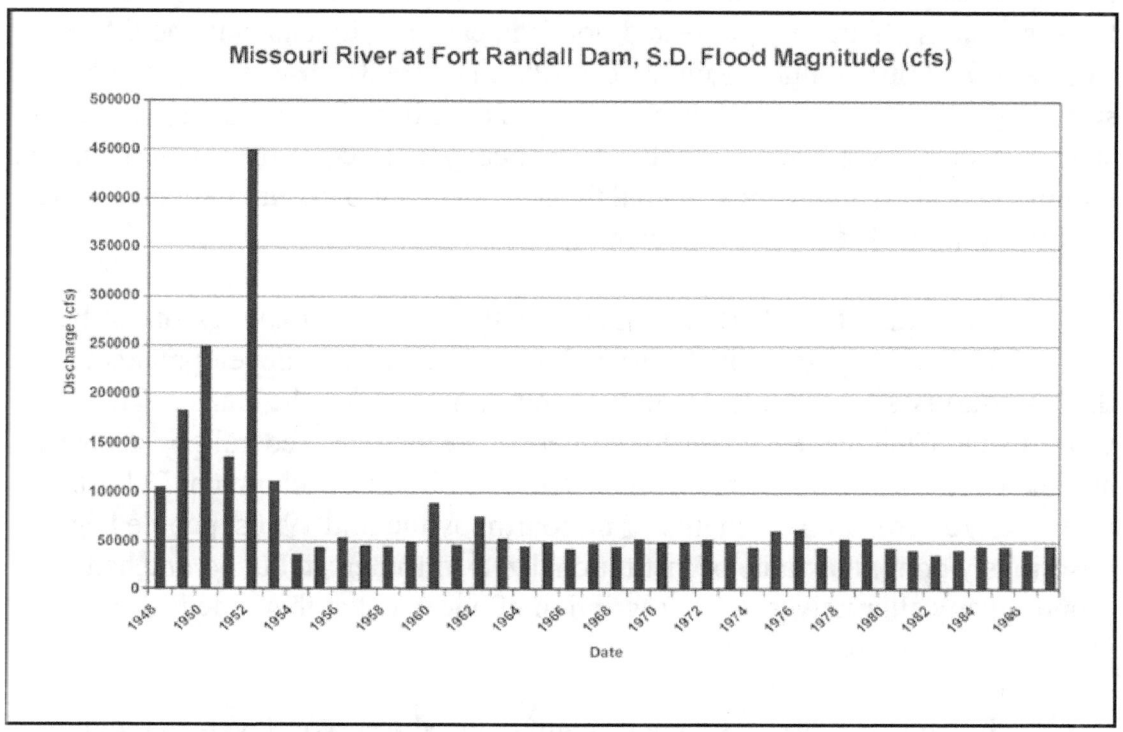

Figure 6-36: Missouri River Flow at Ft. Randall, South Dakota, before and after the mainstem dams. Source: Drawn from U.S. Army Corps of Engineers records.

Pick's plan was in place by 1964; how has it been perceived since then? A half a century after Pick's plan, the big dams represent the realities of the depression and the immediate postwar era in the United States. Political leaders in the basin states used the realities of navigation channels, flood control, agriculture protection, and large power to lobby Congress for action. Action meant primarily construction of large, multipurpose dams. Along with these realities there came the progressive ideals of regional development, the use of natural resources, and mastery over nature.[159]

Recent studies of the river have challenged these ideals and have focused on the ecosystem, the American Indians, and the need for new political instruments. The Tennessee Valley Authority, once heralded as the model for all basins, has not succeeded sufficiently to stimulate such authorities elsewhere.

A 1993 summary of the mainstem dams illustrates the planned uses for stored water in the six reservoirs. Clearly, the upper three dominate with eighty-nine percent of the storage and most of the flood control; i.e., that amount of storage space reserved for holding back water in times of flood. Between 1987 and 1992, a period of relative drought, the supply for power dropped. Also, from the completion of the six dams to 1990, the commercial navigation tonnage has shown an overall downward trend. On the other hand, the completion of Fort Randall Dam alone in 1953 stopped all floods since then at Yankton, and the system has effectively prevented floods in eastern Montana and the Dakotas. The dams had little influence on the great flood of 1993 that was largely the result of heavy rain in the lower basin.[160] In holding back heavy floods, the mainstem Missouri River dams have also dramatically affected sedimentation patterns within the basin, significantly altered flora and fauna growth, and affected water quality along the length of the stream.

According to Robert K. Schneiders, there are four main lessons to be learned from a study of the Pick-Sloan Plan and subsequent developments on the Missouri Basin. First, local citizen groups, in this case the Kansas City Commercial Club, have a powerful influence. Second, the federal government and the Corps never arbitrarily directed Missouri River development. Third, the costs are greater than the benefits. And fourth, by the mid 1990s, people began to seriously doubt the value of institutionalized self-interest, but no one had yet proposed any alternatives. His description of one event at this time illustrates the problem of basin management.

In the late 1980s, a severe drought reduced reservoir levels in the five Dakota dams endangering the fish and hence the $67 million tourist trade. Politicians from those states pressured the Corps not to release so much water from its dams and immediately the lower states cried foul. Water was needed below Sioux City for navigation and although only bringing in some $14 million, the uproar from farmers and barge interests was enough to stop the Corps' plan. Schneiders' conclusion from this story is that "the Missouri compelled people to look at themselves and to question their faith in technology, their commitment to progress, and their motives." He concludes that chapter:

> Yet few individuals entertained any thought of simply letting the river go, of taking down the dams and training structures and abandoning the traditional paradigm, even though that seemingly radical alternative possessed many obvious, tangible advantages.[161]

It would have been a useful exercise to explore just what those advantages would be. Perhaps the despair that some thoughtful people feel about the management of structured river basins springs from the inability of our political system to devise clear, rational solutions to problems of regional conflict. Since the Constitution supplies little direct help, each generation must think out its own approach and devise short-term means to put it into effect.

Schneiders ends his book with the conclusion that:

> No one will ever know enough about the Missouri River to regulate its flows effectively: it possesses too many characteristics; it is influenced by innumerable external factors; and its waters behave in an unpredictable fashion.[162]

This might be a good characterization of modern civilization itself, the buildng of which always requires going into unpredictable terrains and attempting to structure nature in a way conducive to an improved life in an industrialized society.

Endnotes

1. This is also known as the Pick-Sloan Missouri Basin Program.
2. U.S. Congress, House of Representatives, "Missouri River," *House of Representatives Document No. 238*, 73rd Congress, 2nd Session, February 5, 1934, 1, 70-9.
3. Lewis, Meriwether, and William Clark, editor Bernard DeVoto, *The Journals of Lewis and Clark* (Boston: Houghton Mifflin Company, 1953), xxiv.
4. A. W. Kendall, "Keeping Up with the 'Big Muddy'," *The Military Engineer* 30 (March-April 1938), 111-3.
5. J. M. Young, "River Planning in the Missouri Basin," *The Military Engineer* 22 (March-April 1930), 152-8.
6. *House of Representatives Document No. 308*, 69th Congress, 1st Session, April 13, 1926, 2.
7. "Work Begun on Fort Peck Dam to Regulate Missouri River," *Engineering News-Record* No. 9, (1933), 576.
8. "Missouri River," *House of Representatives Document No. 238*, 573-605.
9. "Missouri River," *House of Representatives Document No. 238*, 605.
10. "Missouri River," *House of Representatives Document No. 238*, 12. For an analysis of how an economically unjustified project became "a make-work project," see "Fort Peck Dam," *Engineering News-Record* (November 29, 1934), 697-8.
11. C. H. Chorpening, "Fort Peck Dam—Progress of Construction," *The Military Engineer* 27, No. 151 (January-February 1935, 36. Department the Army, U.S. Army Corps of Engineers, "Fort Peck Dam," *The Federal Engineer: Damsites to Missile Sites: A History of the Omaha District* (Omaha 1984).
12. Kollgaard and Chadwick, editors, *Development of Dam Engineering in the United States*, 3, 674.
13. Schnitter, *A History of Dams*, 18-21.
14. Schnitter, *A History of Dams*, Table 8, 60-1; Figure 89, 99; Tables 19, 20, and 21, 108, 110, 114.
15. Schnitter, *A History of Dams*, 156-7.
16. Schnitter, *A History of Dams*, 158-60.
17. Henry. Darcy, *Les Fontaines publiques de la ville de Dijon: Exposition et application des princpes à suivre et des formules à employer dans les questions de distribution d'eau.* (Paris: Victor Dalmont, 1856).
18. P. Forchheimer, "Über die Ergiebigkeit von Brunnenanlagen und Sickerschlitzen," *Zeitschrift, Architekten-und Ingenieur-Verein*, Hannover, 32, No. 7 (1886), 539-64.
19. Karl Terzaghi, *Erdbaumechanik* (Vienna: Franz Deuticke, 1925). An English version appeared as *Theoretical Soil Mechanics* (New York: John Wiley and Sons, 1943).
20. A. Casagrande, "Seepage through Dams," Contributors to Soil Mechanics 1925-1940, *Boston Society of Civil Engineers*, (1940).

21. T. A. Middlebrooks, "Fort Peck Slide," *Transactions of the American Society of Civil Engineers* 107 (1942), 723-42.

22. Schnitter, *A History of Dams*, 165.

23. Joel D. Justin, "The Design of Earth Dams," *Transactions of the American Society of Civil Engineers* 87 (1924), 1-61. Quotation on 3-4.

24. Joel D. Justin, "Discussion [of Middlebrooks]," *Transactions of the American Society of Civil Engineers* 107 (1942), 744-8.

25. James D. Schuyler, "Recent Practice in Hydraulic-Fill Dam Construction," *Transactions of the American Society of Civil Engineers* 63 (June 1907), 196-252 with discussion 253-77. Schuyler's closure is from 269-77. Quotation is on 198.

26. Schuyler, "Hydraulic-Fill Dam Construction," 207, 241. Schuyler's explanation reads as follows: "That the inner third of the dam should be composed of impervious material, or material which, by drainage and natural settlement, should consolidate into a mass which will become impervious to water, and remain in a moist, semi-plastic condition; that the outer half of each of the other thirds should be coarse, porous, open material, through which water, draining from the interior, would pass freely; while the inner halves of the outer zones should be a mixture of the coarse and fine, or a semi-porous material, in condition to act as a filter so as to prevent the escape of any of the fine particles from the inner third, but at the same time allow of the slow percolation of water through it."

27. Allen Hazen, "Hydraulic-Fill Dams," *Transactions of the American Society of Civil Engineers* 83 (1919–1920), 1713-45. Hazen drew on his earlier paper, "On Sedimentation," *Transactions of the American Society of Civil Engineers* 53 (December 1904), 45-71. Hazen had begun his career in sanitary engineering by "epoch-making" contributions to water purification. As a consulting engineer he designed the sewerage disposal facilities for the 1893 Chicago World's Fair and the water filter plant for Albany, New York. His paper on the latter won him the 1900 Rowland prize from the ASCE. From sand filters he became interested in sediment and hence hydraulic-fill dams. He was a director of the ASCE, 1907-09, and Vice President, 1926-27.

28. Charles H. Paul, "Core Studies in the Hydraulic-Fill Dam of the Miami Conservancy District," *Transactions of the American Society of Civil Engineers* 85 (1922), 1181-202. Quotation is on 1181.

29. J. Albert Holmes, "Some Investigations and Studies in Hydraulic-Fill Dam Construction," *Transactions of the American Society of Civil Engineers* 84 (1921), 331-58. Quotation is on 358.

30. Henry Petroski, *Engineers of Dreams: Great Bridge Builders and the Spanning of America* (New York: Vintage Books, 1995), 101-21.

31. Justin, "The Design of Earth Dams," 59-61. Four of these successful dams did have slides, Justin notes, and three of those were hydraulic fill. Edward J. Wegmann, *The Design and Construction of Dams*, 8th edition.

32. Harry H. Hatch, "Tests for Hydraulic-Fill Dams," *Transactions of the American Society of Civil Engineers* 99 (1934), 206-47. Quotation on 206.

33. Glennon Gilboy, "Mechanics of Hydraulic Fill Dams," *Journal of the Boston Society of Civil Engineers* 21, No. 3 (July 1934), 185-205, with a "Discussion" by Arthur Casagrande, 146-7. Quotation on 199.

34. "Men and Mechanization," *Engineering News-Record* (May 9, 1935), 684. See also the article "Three Big Dam Operations Begun in the Northwest," *Engineering News-Record* (April 5, 1934), 444-5, where Bonneville, Grand Coulee, and Fort Peck are described as showing "evidence of the strenuous efforts of those in immediate charge to employ men at the earliest possible moment . . . (as a consequence) field work has preceded the final development of design details."

35. *Engineering-News Record* (November 29, 1934), 698.

36. C. H. Chorpening, "Fort Peck Dam . . .," 36-41.

37. "Deep Sheetpile Cutoff Well for Fort Peck Dam," *Engineering News Record* (January 10, 1935), 35-9.

38. Chorpening, "Fort Peck Dam . . .," 40.

39. H. W. Richardson, "Dredge Plant Characteristics and Hydraulic Fill Operation," *Engineering News-Record* (August 6, 1936), 188-94.

40. "Diversion-Tunnel Driving," *Engineering News-Record* (August 29, 1935), 296-300.

41. H. W. Richardson, "Tunnel Enlargement and Lining," *Engineering News-Record* (August 13, 1936), 238-43. J. D. Jacobs, "Grouting the Tunnels," *Engineering News-Record* (August 20, 1936), 274-7. "Fort Peck Diversion Tunnels Redesigned for Faster Driving," *Engineering News-Record* (March 26, 1936), 450-1.

42. A. W. Pence, "The Fort Peck Dam Tunnels," *The Military Engineer* 29, No. 163, 18-23.

43. James J. Halloran, "Earth Movement at Fort Peck Dam," *The Military Engineer* 31, No. 175, 5.

44. Joel Justin, *Earth Dam Projects* (New York: John Wiley, 1932), 55-9. Justin gave the formula Q = Q(ave) (1 + 0.8LogT) where Q is the greatest average rate of flow for 24 consecutive hours during a period of years, Q (ave) is the average yearly 24 hour flood and T is the number of years in the period considered. Thus for Q (ave) = 65,500 cubic feet per second and T = 100 then Q = 65,500 (1 + 0.8 log 100) = 65,500 (2.6) = 173,000 cubic feet per second. For Wyman's values, see "Missouri River," *House of Representatives Document No. 238*, 860.

45. Henry C. Wolfe, "The Fort Peck Dam - The Project," *The Military Engineer* 27, No. 151, (January-February 1935), 31-5.

46. Justin, *Earth Dam Projects*, 52-3. Justin notes the disadvantages of an earth dam where the spillway requirements are excessive.

47. Wolfe, "Fort Peck Dam," 33. See also "Concrete Structures for Spillway," *Engineering News-Record* (August 29, 1935), 303-4. See also John R. Hardin, "Fort Peck Dam Spillway," *The Military Engineer* 29, No. 163 (January-February 1937), 24-8.

48. John Gage, *Turner: Rain, Steam and Speed* (New York: Allen Lane, 1972), 99 pages.

49. Theodore E. Stebbins Jr., and Norman Keyes, Jr., *Charles Sheeler: The Photographs* (Boston: New York Graphic Society, 1987), 162 pages.

50. M. R. Montgomery, *Saying Goodbye* (New York: Alfred A. Knopf, 1989), 54-63.

51. "Introduction to the First Issue of Life," *Life* 1, No. 1 (November 23, 1936), 3.

52. "10,000 Montana Relief Workers Make Whoopee on Saturday Night," *Life*, 1, No. 1, (November 23, 1936), 9-17.

53. Ewart G. Plank, "The Town of Fort Peck," *The Military Engineer* 28, No. 161, (September-October 1936), 326.

54. Jerold B. Van Faasen, *Making It Happen: A Sixty-Year Engineering Odyssey in the Northwest* (Seattle: 1998), 58.

55. T. B. Jefferson, "Longer Life for Dredge Pumps," *Engineering News-Record* (December 2, 1937), 896.

56. Van Faasen, *Making It Happen*, 60. Halloran, "Earth Movement at Fort Peck Dam, 5-6.

57. Van Faasen, *Making It Happen*, 61.

58. Ray Kendall, deposition, quoted in Montgomery, *Saying Goodbye*, 87-8.

59. Ralph E. Anglon, deposition, quoted in Montgomery, *Saying Goodbye*, 89.

60. Van Faasen, *Making It Happen*, 61.

61. Halloran, "Earth Movement at Fort Peck Dam," 7-8.

62. War Department, U.S. Army Corps of Engineers, *Report on the Slide of a Portion of the Upstream Face of the Fort Peck Dam* (Washington, D.C.: U.S. Government Printing Office, July 1939), 14 pages with 29 drawings and charts plus three appendices. The Report by the Board is only two pages long. The remainder is a description of the dam, details of the slide, the immediate reconstruction, and the investigation, including tests.

63. C. L. Sturdevant to the District Engineer at Fort Peck, October 1, 1938; T. A. Middlebrooks to Major Clark Kittrell, October 10, 1938; and C. L. Sturdevant to the Chief of Engineers, October 29, 1938. National Archives, College Park, RG77, Entry 111, Folder 7245 Part 3. The Board met from November 15 to 17, to review test results and to request further investigations. In a further meeting from January 31 to February 4, 1939, the Board discussed all the studies and entertained various viewpoints from which a draft report emerged.

64. Thaddeus Merriman to Major Clark Kittrell, October 19, 1938, National Archives, College Park, RG77, Entry 111, Folder 7245 Part 4.

65. *Report on the Slide . . .*, July 1939, 8-9.

66. *Report on the Slide . . .*, July 1939, 9.

67. The Classical Coulomb equation for stability expresses the shear resistance, s, as $s = c + p \tan \varnothing$ where c is the cohesion, p is the pressure perpendicular to the sliding surface, and tan f is the angle of internal friction. The factor of safety is thus $F.S. = \frac{c + \tan \varnothing}{s}$ which can be rewritten as $F.S. = \frac{cl + W \cos \alpha \tan \varnothing}{W \sin \alpha}$ where l is the length of the potentially sliding surface and dW is the weight of material above that surface. Where c=o the safety factor $F.S. = \frac{W \cos \alpha \tan \varnothing}{W \sin \alpha} = \frac{\tan \varnothing}{\tan \alpha}$. See Karl Terzaghi and Ralph B. Peck, *Soil Mechanics in Engineering Practice* (New York: John Wiley and Sons, 1948)., 181-4. For a clearly worked example, see Spencer J. Buchanan, "Levees in the Lower Mississippi Valley," *Proceedings of the ASCE* 63, No. 7, (September 1937), 1304-21. (Buchanan studied under Terzaghi.)

68. Department of the Army, U.S. Army Corps of Engineers, *Garrison Dam and Reservoir: Analysis of Design, Excavation and Main Embankment*, Garrison District (May 1948), plate 28. See also Department of the Army, U.S. Army Corps of Engineers, *Oahe Reservoir: Definitive Project Report*, Office of the District Engineer, Omaha (February 1948), Drawing 68-412.
69. Memorandum from F. D. Roosevelt to the Chief of Engineers, November 2, 1938, National Archives, Archives II, College Park, Maryland, RG77, Entry 111, Box 490, Folder 7245 Part 4.
70. Memorandum from M. C. Tyler, Acting Chief of Engineers to the Secretary of War, November 3, 1938, National Archives, RG77, Entry 111, Box 490, Folder 7245 Part 4.
71. Thaddeus Merriman to Major Clark Kittrell, October 19, 1938, National Archives, College Park, RG77, Entry 111, Folder 7245 Part 4. Here, Merriman asks for more information about events "before the slide on the upstream face."
72. Thaddeus Merriman to Major Clark Kittrell with copies to each member of the board, November 22, 1938, National Archives, College Park, RG77, Entry 111, Box 490, Folder 7245 Part 4.
73. "Thaddeus Merriman (Memoir)," *Transactions of the American Society of Civil Engineers* 107 (1942), 1799-1804. In 1934, S. O. Harper, then Acting Chief Engineer of the Bureau of Reclamation in Denver, had recommended Merriman as a consulting engineer for the Grand Coulee Dam, but Elwood Mead, the Commissioner, disapproved with the comment, "I have no objection to him either on personal or professional grounds, but his headquarters are now in Kansas City and he is the chief engineering advisor on the Fort Peck project, which competes with what we are doing in Montana." Elwood Mead to S. O. Harper, March 3, 1934, National Archives, Denver, RG115, Entry 7, Grand Coulee Dam Project files.
74. Memorandum from H. M. Westergaard to Major C. Kittrell, "Ratio of the overall force of friction to the total weight of material moved during the partial failure of Fort Peck Dam, with consideration of the dynamics of the motion," January 28, 1939, 12 pages; enclosed with W. F. Cummins Jr., to Major C. Kittrell, February 6, 1939, National Archives, College Park, RG77, Entry 111, Folder 7245, Part 5.
75. Thaddeus Merriman to Major C. Kittrell, February 7, 1939, and Warren J. Mead to Major C. Kittrell, February 28, 1939, enclosed with Board of Consultants on the Fort Peck Dam to General J. L. Schley, March 3, 1939, National Archives, College Park, RG77, Entry 111, Box 491, Folder 7245, Part 5.
76. Board of Consultants to General J. L. Schley, March 3, 1939, National Archives, College Park, RG77, Entry 111, Box 491, Folder 7245, Part 5.
77. Memorandum from General Tyler to General Schley, March 7, 1939, with an approval at the end by Schley, National Archives, College Park, RG77, Entry 111, Folder 7245, Part 5.
78. *Engineering News-Record* (November 4, 1940), 3.
79. Joel Justin to Carlton Proctor, July 7, 1939. Proctor had written on July 5 to confirm the plan to have a session devoted to Fort Peck. Justin notes that he, Casagrande, Gilboy, Merriman, and Gerig would be discussors. Only the discussions of Gilboy, Gerig, and Justin were published in the *Transactions* (1942).
80. Karl Terzaghi, "Soil Mechanics–A New Chapter in Engineering Science," *Journal, Institution of Civil Engineers* London (June 1939), 118-9.
81. T. A. Middlebrooks to Karl Terzaghi, August 30, 1939, Thomas A. Middlebrooks Papers, Office of History, Headquarters, U.S. Army Corps of Engineers. Middlebrooks enclosed the report of the Board.
82. Karl Terzaghi to Thomas A. Middlebrooks, September 8, 1939, Thomas A. Middlebrooks Papers, Office of History, Headquarters, U.S. Army Corps of Engineers. A biography, Richard E. Goodman *Karl Terzaghi: The Engineer as Artist*, ASCE Press, 1998, 340 pages, makes no mention of Fort Peck, of Middlebrooks, or this correspondence on the slide. "More Light Needed," *Engineering News-Record* 122, No. 19 (May 11, 1939).
83. T. A. Middlebrooks, "Fort Peck Slide," *Transactions of the American Society of Civil Engineers* 107 (1942), 723-42. Quotation on 735.
84. Glennon Gilboy, "Discussion on Fort Peck Slide," of Middlebrooks' "Fort Peck Slide," *Transactions of the American Society of Civil Engineers* 752-5. Quotation on 753.
85. William Gerig to Thomas A. Middlebrooks, January 27, 1941, Thomas A. Middlebrooks Papers, Office of History, Headquarters, U.S. Army Corps of Engineers.
86. Thomas A. Middlebrooks to William Gerig, February 26, 1941, Thomas A. Middlebrooks Papers, Office of History, Headquarters, U.S. Army Corps of Engineers.
87. James L. Sherard, Richard J. Woodward, Stanley F. Gizienski, and William A. Clevenger, *Earth and Earth-Rock Dams* (New York: John Wiley and Sons, Inc., 1963), 316-22. The sheet pile cutoff wall corroded quickly, and it did not prevent excessive seepage, anyway, so it was replaced by a drainage system with wood pipe, 293, 318.
88. Thomas A. Middlebrooks, "Progress in earth-dam design and construction in the United States," *Civil Engineering* 22 (September 1952), 118-26.

89. Sherard et al., *Earth and Earth-Rock Dams*, 144. Sherard was a student of Terzaghi and Casagrande at Harvard.
90. David P. Billington, "History and Esthetics in Suspension Bridges," *Proceedings of the American Society of Civil Engineers. Journal of the Structural Division*, ASCE 103, ST8 (August 1977), with Discussions and Closure 105, ST3 (March 1979).
91. *Official Register*, ASCE (New York, 1982), 193-4.
92. Joel D. Justin to General M. C. Tyler, March 4, 1939, with a note on the letter written by Tyler telling of his phone instructions to Justin.
93. John R. Ferrell, *Big Dam Era: A Legislative and Institutional History of the Pick-Sloan Missouri Basin Program* (Omaha: Missouri River Division, U.S. Army Corps of Engineers, 1993), 40.
94. Ferrell, *Big Dam Era*, 2, 9.
95. *House Document No. 475*, 78th Congress, 2nd Session, "Report" from Colonel Lewis A. Pick, 19.
96. Ferrell, *Big Dam Era*, 8-9.
97. *House Document No. 475*, 78th Congress, 2nd Session, 24-6.
98. *House Document No. 475*, 78th Congress, 2nd Session, 27-8.
99. *House Document No. 475*, 78th Congress, 2nd Session, 28-9.
100. *House Document No. 475*, 78th Congress, 2nd Session, 7.
101. *House Document No. 475*, 78th Congress, 2nd Session v-ix, 1-5.
102. Ferrell, *Big Dam Era*, 27.
103. Ferrell, *Big Dam Era*, 22.
104. Ferrell, *Big Dam Era*,, 27-31.
105. Ferrell, *Big Dam Era*, 24.
106. Ferrell, *Big Dam Era*, 37.
107. *Senate Document No. 191*, 78th Congress, 2nd Session, "Missouri River Basin: Conservation, Control, and Use of Water Resources of the Missouri River Basin in Montana, Wyoming, Colorado, North Dakota, South Dakota, Nebraska, Kansas, Iowa, and Missouri," April 1944, 1-2.
108. Ferrell, *Big Dam Era*, 19.
109. *Senate Document No. 191*, 1-16.
110. *Senate Document No. 191*, 16-211.
111. *Senate Document No. 191*, 6-8.
112. Ferrell, *Big Dam Era*, 47, 54.
113. Ferrell, *Big Dam Era*, 54-5.
114. Ferrell, *Big Dam Era*, 11, 55.
115. Ferrell, *Big Dam Era*, 60, and *House Document No. 680*, 78th Congress, 2nd Session, "Message from the President of the United States," September 21, 1944, with a "Resolution of the Missouri River States Committee to Secure a Basin-Wide Development Plan" (sent to the President on August 21, 1944).
116. *Senate Document No. 247*, 78th Congress, 2nd Session, November 21, 1944, 6 pages.
117. Ferrell, *Big Dam Era*, 66.
118. Ferrell, *Big Dam Era*, 75.
119. John E. Thorson, *River of Promise, River of Peril: The Politics of Managing the Missouri River* (Lawrence: University Press of Kansas, 1994), 61.
120. *Garrison Dam and Reservoir: Analysis of Design, Excavation and Main Embankment*, Department of the Army, U.S. Army Corps of Engineers, Garrison District, May 1948, 1-2.
121. "The Garrison Story," in Sheila C. Robinson (compiler), *The Story of Garrison Dam: Taming the Big Muddy* (Garrison, North Dakota: BHG, Inc., 1997), 55-66, copy in the archives of the U.S. Army Corps of Engineers, Garrison Dam, Riverdale, North Dakota.
122. "Amended Height of Garrison Dam Attacked on Three Fronts," *Bismarck Tribune* (December 6, 1945).
123. "Pick Calls for Unity in Valley," *World Herald* (January 3, 1946).
124. "Power Line to Garrison Dam Advanced," *Fargo Forum* (January 8, 1946).
125. Wesley Price, "What You Can Believe About MVA," *The Saturday Evening Post,* (January 19, 1946),
132. Truman's view appeared in "MVA In 1946 Truman Hope," *World Herald*, (January 20, 1946). For Jackson's activity "Valley Plan Fight Shifts to Northwest," *World Herald* (January 27, 1946).
126. "Pick Hears N.D. Tribes Plead for Lands Dam Would Flood," *Bismarck Tribune* (May 28, 1946). A series of articles appeared in late 1946, for example "Strenuous Opposition to Relocation Program For Indians Develops," Mandan (North Dakota) *Daily Pioneer* (Thursday, December 5, 1946).
127. Max Coffey, "Engineers' Lay-Off Curbs River Job," *Rocky Mountain Empire Magazine* (November 10, 1946).

128. Lorne Kennedy, "Fund Restored for Projects on Missouri River," *World Herald* (November 30, 1945). "Army Plans to Begin Building Town in Spring," *Bismarck Tribune* (December 21, 1945). "Senate Gives OK to O'Mahoney's Dam Amendment," *Bismarck Tribune* (March 15, 1946). This last reports on a second amendment attached to a bill that includes money to begin building the dam. A similar amendment reported in December was bypassed because the first appropriation was to build the town.

129. "Pick Hears N.D. Tribes Plead for Lands Dam Would Flood" and "Berthold Tribe Did Not Always Object to New Location," *The Fargo Forum* (December 15, 1946).

130. "Relocation Issue Vital to Project," *Grand Forks Herald* (January 26, 1947), and Richard P. Powers, "Future Appropriations Won't Have Restrictions Such As Indian Land Clause Which Hits Garrison Dam," *Minot Daily News* (January 29, 1947). For objections see *The Leader* (January 30, 1947); *The Williston Daily Herald* (April 17 and May 29, 1947); and *The Bismarck Tribune* (June 11, 1947).

131. "Ten River States Want Garrison Dam Built Fast," *Minot Daily News* (July 7, 1947). "Erosion Losses," *St. Louis Globe Democrat* (July 17, 1947).

132. "Ten Year Plan for Flood Control," *St. Louis Globe Democrat* (July 17, 1947). Richard G. Baumhoff, "Record Flood Damage This Year Emphasizes Need for Action Now," *St. Louis Post Dispatch* (July 20, 1947), and "$20,000,000 Approved for Garrison Project Includes Indian Funds," *Minot Daily News* (July 22, 1947). Contracts were announced in the *Bismarck Tribune* (September 8, 1947).

133. Michael L. Lawson, *Dammed Indians: The Pick-Sloan Plan and the Missouri River Sioux, 1944–1980* (Norman: University of Oklahoma Press, 1994), paperback, 59-62.

134. J. S. Seybold, "Constructors Roll Nearly One Million Yards a Week Into Garrison Dam," *Civil Engineering* (October 1949), 28-32, 85.

135. *Valley City Times Record* (February 7, 1948). "Big Equipment Being Used at Garrison Dam Site," *Minot Daily News* (April 29, 1948). "Indians' Fight Over Big Dam at End as Contract is Signed," *Fargo Forum* (May 21, 1948).

136. "Giant Embankment Now Taking Shape," *Grand Forks Herald*, July 4, 1948. "Truman Asks More Money for Garrison," *Minot News* (July 24, 1948). "Stage Two Garrison Bids Let," *Mandan Pioneer* (September 22, 1948).

137. "Sloan Asks Solution to Souris Problems," *The Minot Daily News* (December 3, 1948). *Senate Document No. 191*, 111-114.

138. "Sen. Murray to Try MVA Bill Again," *The Minot Daily News* (November 17, 1948). "Senator Murray Will Introduce TVA Bill," *North Dakota Union Farmer* (December 6, 1948). Peter Edsom, "Army-Interior Duplication Doomed," *St. Paul Pioneer Press* (December 9, 1948).

139. For Pick's appointment see: Department of the Army, U.S. Army Corps of Engineers, *The History of the U.S. Army Corps of Engineers* (Alexandria, Virginia, 2nd edition 1998), 149. For the Sloan-Pick debate see "Pick-Sloan Disagree," *Bismarck Tribune* (June 1, 1949). For the concerns of Williston see "The Truth About the Army Engineers and Garrison Dam," *The Williston Daily Herald* (August 22, 1949). The article castigates the Corps and especially General Pick.

140. "Garrison Dam Pool Limit Rejected," *St. Paul Pioneer* (August 23, 1949). "Army General Blasts At Lower Pool Level," *The Bismarck Tribune* (August 25, 1949). Usher L. Burdick, "The Truth About the Army Engineers and Garrison Dam," *The Williston Daily Herald* (August 22, 1949).

141. "'Pup' Used at Garrison Dam Only One of its Kind," *The Minot Daily News* (September 27, 1949). "Garrison Dam Takes Shape For the Casual Observer," *The Fargo Forum* (October 9, 1949). "Garrison Tunnels Under Construction," *McLean County Independent* (October 6, 1949).

142. Gordon Mikkelson, "Debate Rages Over Ninth of River Plans," *Minneapolis Morning Tribune* (January 21, 1950).

143. "Garrison Dam Featured by Contractors Magazine," *The Bismarck Tribune* (August 1, 1950). "Missouri Basin is Major Problem in Nation's Economy," *The Leader*, Bismarck (September 7, 1950). Richard Baumhoff, "Huge Missouri-Souris Diversion Project Apparently to be Dropped," *St. Louis Post-Dispatch* (October 12, 1950).

144. "Concrete Flows as Big Year Begins at Garrison," *The Minot Daily News* (April 9, 1951). "Garrison Dam Activity Centered in Powerhouse Site," *The Minot Daily News* (April 28, 1951). "Ft. Peck Power Flows to Dam," *The Williston Daily Herald* (May 17, 1951). "Present Garrison Dam Cost Up to $268 Million," *The Bismarck Tribune* (June 7, 1951).

145. "Mississippi Congressman Calls for Missouri Valley Authority," *The Leader* (July 26, 1951). "Charges Army Influenced by Private Power," *The Fargo Forum* (August 5, 1951).

146. John D. Paulson, "Great Amounts of Concrete, Steel Go into Dam," *The Fargo Forum* (December 2, 1951).

147. "Dakota Floods Share Nation's Top Picture News Spotlight," *The Fargo Forum* (April 13, 1952). The 1951 and 1952 floods are described in Robert L. Branyan, *Taming the Mighty Missouri: A History of the Kansas City District, Corps of Engineers: 1907-1971* (Kansas City, Missouri: U.S. Army Corps of Engineers, 1974), 49-61. Page was correct about the dams' ability to prevent flooding just below their locations, but farther downstream, where large tributaries enter the Missouri and where heavy rainfall can occur, large floods such as the one in 1993 can still result.

148. "General Says Pick-Sloan Plan Would have Averted '52 Flood," *The Fargo Forum* (May 7, 1952).

149. "Big Mattresses Built For River Diversion," *Minot Daily News* (September 15, 1952). John D. Paulson, "Garrison Dam to Force Missouri to 'Detour' in '53," *The Fargo Forum* (November 1952).

150. Thorson, *River of Promise*, 137-9.

151. "Garrison Dam Closed," *The Bismarck Tribune* (April 16, 1953). *Life* (May 18, 1953).

152. "Members of Tribes Losing Land to Participate in Dam Closure Ceremony on June 12," *Williston Daily Herald* (June 3, 1953). "Indians Want Retention of Rights to Garrison Land," *New Town News* (June 4, 1953).

153. "Electric Brains to Guide Basin Reservoir Operation," *Minot Daily News* (December 3, 1953). William E. Potter, *Engineering Memoirs* (Alexandria, Virginia: U.S. Army Corps of Engineers, July 1983). Oral history interview by Martin Reuss, 111-2.

154. Harold Ickes, "The Wide Missouri Gets Wider," *Field and Stream,* (March 1954). Cal Olson, "Garrison Dam Fill Job Completed," *The Fargo Forum* (March 5, 1954). Department of the Army, *The Federal Engineer*, 130; "The Missouri Moves Over," *Life*, (May 18, 1953), 45.

155. Department of the Army, U.S. Army Corps of Engineers, "Analysis of Design: Spillway Structures and Gates," *Missouri River Garrison Reservoir*, North Dakota, Office of the District Engineers, Garrison District, Bismarck, North Dakota, April 1952, Chapter II. For Laboratory Research on Spillways see "Laboratory Research Applied to the Hydraulic Design of Large Dams," Bulletin No. 32, Waterways Experiment Station, Vicksburg, Mississippi, June 1948.

156. Department of the Army, U.S. Army Corps of Engineers, "Analysis of Design: Spillway Structures and Gates," *Missouri River Garrison Reservoir*, Chapter II. For Laboratory Research on Spillways see "Laboratory Research Applied to the Hydraulic Design of Large Dams," Bulletin No. 32, Waterways Experiment Station, Vicksburg, Mississippi, June 1948.

157. Lawson, *Dammed Indians*, 47-9 and Department of the Army, *The Federal Engineer,* 105-17.

158. Thorsen, *River of Promise*, 186.

159. Ferrell, *Big Dam Era,* 123-45.

160. Robert Kelley Schneiders, *Unruly River: Two Centuries of Change Along the Missouri* (Lawrence: University Press of Kansas, 1999), 250-1.

161. Schneiders, *Unruly River*, 258.

CHAPTER 7

THE CENTRAL VALLEY PROJECT: SHASTA AND FRIANT DAMS

WATER ISSUES IN CALIFORNIA

Agriculture and Topography of the Central Valley

Since the days of the Gold Rush, California has stood as the most economically important state in the West. And for almost as long, it has been the most important agricultural producer in the West, first because of huge wheat harvests shipped by sea to international markets and later because irrigated fruit, nut, and citrus crops could reach national markets via transcontinental rail connections.[1] The importance of water to agriculture remains strong, even at the beginning of the twenty-first century, when tens of millions of Californians live in urban/metropolitan centers. More than 80 percent of the state's annual water supply still goes into crop production.[2]

In the mid-nineteenth century, California farmers initially relied on winter and early spring rainfall to nourish large wheat crops.[3] When it became apparent that natural precipitation varied too much from year-to-year to ensure bountiful harvests, farmers sought ways of insuring the presence of sufficient moisture during critical stages of their wheat crop's growth cycle. Irrigation based around the diversion of streams and rivers began to flourish in the late 1860s and 1870s, but it was recognized that much flood water passed to the sea (or into low-lying, seasonal, freshwater lakes) without being diverted into canals or ditches. In essence, the only way to capture excess flow resulting from spring snowmelt and from storm surges was to build storage dams to "capture the floods." In California, dams in the gold-producing region north of Sacramento pioneered the practice of large-scale water storage for hydraulic mining in the state.[4] By the 1880s, dams specifically focused on (and financed by) irrigation interests were in operation and offering important precedents for future initiatives.[5]

The topography of California presents a few distinct regions that played a role in the growth of irrigation and water development projects. In the extreme southeastern corner of the state, the Imperial Valley draws water from the Colorado River via the All-American Canal to support an agricultural empire adjoining the Mexican border; greater Los Angeles and the southland below the Tehachapi Mountains (including communities in the Santa Ana River Valley, such as San Bernardino, Riverside, and greater Orange County) also developed tremendously productive irrigated farms in the late nineteenth and early twentieth century.[6] But the huge, 450-mile-long, 50-mile-wide, Central Valley in the heart of the state represents California's most important agricultural region.[7]

Figure 7-1: The Bureau of Reclamation's Central Valley Project showing the primary features. Source: Bureau of Reclamation.

Encompassing several million acres, the Central Valley is bounded on the east by the Sierra Nevada, on the west by the less lofty Coastal Ranges, and to the north and south by imposing escarpments connecting the Sierra Nevada to the Coastal Ranges. The northern half of the Central Valley, known as the Sacramento Valley, is drained by the Sacramento River and numerous tributaries that flow out of the surrounding mountains. These tributaries include major streams such as the American, Yuba, Feather, and Pit Rivers, flowing out of the Sierra Nevada and smaller streams such as Stony Creek that flows eastward out of the Coastal Ranges north of San Francisco. After joining with the San Joaquin River in the low-lying "Delta" region that forms about 50 miles downstream from Sacramento, the Sacramento River flows into San Francisco Bay via Suisun Bay and the Carquinez Strait. The combined Sacramento/San Joaquin River is the largest and most important source of freshwater flowing into San Francisco Bay.

The southern half of the Central Valley, known as the San Joaquin Valley, is drained largely by the San Joaquin River and the numerous tributaries, including the Merced and Tuolumne Rivers, that flow out of the southern Sierra Nevada; only a few small—and often dry—creeks flow eastward out of the southern Coastal Ranges. Upstream from its confluence with the Sacramento River in the "Delta," the San Joaquin River flows for several score miles through a flat, low-lying expanse of riparian land that, prior to the construction of upstream storage and diversion dams, was frequently inundated by annual spring/early summer floods. Beyond this riparian land (that during the latter nineteenth century came to comprise the heart of the Miller & Lux agriculture and ranching empire), the valley includes vast tracts of flat land that—while not subject to natural flooding—are gently sloped and wonderfully suited to agriculture if properly irrigated.

North (or downstream) from Fresno, regional streams feed into the San Joaquin River, and accumulating precipitation flows north toward the Delta and San Francisco Bay. But south of Fresno, the flow of major rivers does not usually reach the San Joaquin River. This is because the topography of the extreme southern section of the Central Valley traps the flow of rivers within seasonal lakes and marshy areas and does not allow it to reach the Pacific Ocean via San Francisco Bay. The Kings River flows through the region immediately south of Fresno, and, in some wet years, heavy floods can reach the San Joaquin River proper; but in the wake of extensive water storage and diversion systems, the Kings River usually does not contribute to the flow of the San Joaquin and instead collects in a large shallow pond known as Lake Buena Vista. South of the Kings River, the Kaweah, Tule (or Tulare), and Kern Rivers always remain landlocked and—at a time before their flow was largely diverted for irrigation—they were responsible for creation of the shallow and seasonal Tulare Lake, located near the city of Bakersfield. Usually the term "San Joaquin Valley" is used to refer to the entire expanse of the southern Central Valley; however, at times, the area that drains into Lake Buena Vista and Tulare Lake has been called the Tulare Valley.

Figure 7-2: Early irrigation diversion along the Kings River in the San Joaquin Valley, circa 1900. Source: Donald C. Jackson postcard collection.

The overall topography of the Central Valley is critically important in terms of understanding how the two major dams built across the Sacramento and San Joaquin Rivers relate to one another and to the land irrigated by water impounded in their reservoirs. Also important to understanding the origin and purpose of these two dams—Shasta Dam across the Sacramento and Friant across the San Joaquin—is an appreciation of how precipitation is distributed across the California landscape. In the northern half of the Central Valley, there is a relatively large quantity of water made available by nature that fills the Sacramento River and makes possible large-scale river boat navigation up to the city of Sacramento. The southern half of the valley is much drier, and the average flow of the San Joaquin River is much lower than its brethren to the north. As settlers in the late nineteenth century and early twentieth centuries came to realize, much of the land in the San Joaquin Valley is ideally suited to irrigation, and there is more land susceptible to intensive irrigation than can be cultivated using the flow of the southern rivers.

"Move the Rain": The Central Valley Project

This circumstance of excess water supplies to the north and thirsty, potentially productive lands to the south comprises the foundation for what became known as the Central Valley Project, built by the Bureau of Reclamation starting in the depression-wracked 1930s. Unabashedly characterized by its supporters as a project that would "Move the Rain" by diverting huge quantities of water from north to south, the Central Valley Project was not the first federally financed reclamation project within California.[8] But its scale dwarfed those that preceded it, and, as built, the scope and scale of the multipurpose Central Valley Project is comparable to the gigantic Hoover Dam in the Southwest and the Grand Coulee and Bonneville Dams built contemporaneously. Recognizing the importance of the project in the development of California, this chapter lays out the origins of the Central Valley Project, the role of Shasta (originally called the Kennett) and Friant Dams within the project, and the process involved in transforming these two mnumental structures from drawing board to reality.

In recounting the story of the Central Valley Project, it is important to stress that the project was not born out of a vision of federal engineers who perceived a singular and essential place for the federal government in the development of California agriculture. Neither did the project spring forth fully-conceived by state engineers focused on a Progressive Era mission to maximize economic use of the state's water. Nor did the scheme to transfer huge quantities of water hundreds of miles within the confines of the Central Valley derive from the efforts of private land-holding interests seeking to increase the productivity (and economic value) of their property. In truth, the Central Valley Project resulted from a long and complex interaction among federal, state, and private entities who shared a general interest—not to mention a competitive rivalry—in developing California's water resources. At times, the specific desires of particular actors within the state's economic and political milieu clashed dramatically; at other times, remarkable congruence transpired and, ultimately, led to powerful alliances sharing goals that, only a short time earlier, might have appeared impossible to achieve.

The story of the Central Valley Project extends back more than sixty years prior to the time when Reclamation, within the U.S. Department of the Interior, formally initiated work on what came to be known as Shasta Dam. During those intervening sixty-plus years, the "thread" tying together events often became quite thin and tenuous, although in retrospect, it might be tempting for a historian to overlay a patterning that makes everything appear oh-so-logical and inevitable. A classic example of this phenomena concerns the famous "Marshall Plan" proposed by Robert Marshall of the U.S. Geological Survey in 1919.[9] Conceived as a project to develop irrigated land in the Sacramento and lower San Joaquin River Valleys, the Marshall Plan did propose a major storage dam on the upper Sacramento River that bears striking similarity to the Shasta Dam in both location and function. But many other aspects of the Marshall Plan bear scant similarity to the Central Valley Project, as built. And there are key components of the Central Valley Project that were completely absent from the Marshall Plan. Thus, while it is reasonable to look to the Marshall Plan as a key part of the foundation for the Central Valley Project, its importance in terms of what was actually built can be easily overemphasized.

Figure 7-3: An irrigation canal in the San Joaquin Valley about 1900. Source: Donald C. Jackson postcard collection.

Similarly, a survey of Central Valley water resources administered by the U.S. Army Corps of Engineers in the mid-1870s certainly bears relevance to events in the 1920s and 1930s, but the linkages are less direct than one might surmise. In essence, the story of the Central Valley Project is one filled with serendipity and contingency that belies the notion that western water resources development springs forth from a single vision of federal, state, or private preeminence. Rather, it represents a complex conjoining of interests that eventually found expression in a public works initiative fostered by President Franklin Roosevelt's New Deal of the 1930s.

Miller & Lux and Early Agriculture

Early commercial Anglo-American agriculture in the Central Valley did not depend upon irrigation. In the 1850s, large wheat farms sprang up that depended upon the nutrients and moisture that had accumulated in the soil during the previous decades and centuries. Cattle ranches in the low-lying grasslands of the San Joaquin Valley also proliferated as entrepreneurs sought ways to supply meat to (and make money from) men working in the goldfields. By the 1860s, non-irrigated wheat farming began to decline as the moisture and nutrients were depleted from the soil. Contemporaneously, cattle ranching became more integrated into the meat marketing industry and ranch lands became concentrated under the control of a relatively few businessmen. The most important of these cattle ranchers were Henry Miller and Charles Lux, who (operating as the partnership Miller & Lux) purchased large tracts of riparian land in the lower San Joaquin Valley and in the Kern River basin near Bakersfield.[10]

Low-lying riparian land considered so desirable by Miller & Lux was, at least prior to the construction of large-scale dams and diversion works, naturally irrigated by spring floods that overflow the banks of the river for a few weeks, soak into the soil, and then gradually recede, allowing grass to flourish as the land slowly dries. By the late 1860s, the possibilities of artificially irrigating much larger tracts of land than could be watered naturally were apparent to farmers and potential entrepreneurs. The first person to actively pursue the construction of a large irrigation system focused around large-scale diversion of water from the lower San Joaquin River was John Bensley who formed the San Joaquin & Kings River Canal Company in the late 1860s. Bensley's scheme proposed construction of a 40-mile long canal that would carry water along the western rim of the valley below the settlement of Firebaugh's Ferry. Bensley did not own the land that would lie below his company's canal—much of it was controlled by Miller & Lux—but he believed landowners would be willing to buy water from him in order to irrigate their lands and thus increase economic productivity.

Of course, 40 mile-long canals are not cheap, and Bensley experienced difficulty in raising sufficient investment capital to construct the canal. Large landowners in the lower San Joaquin Valley, including Miller & Lux as well as other California businessmen, appreciated the possibilities that Bensley's canal presented, but they were not eager to become dependent upon a canal company that they could not control. The result was a new enterprise called the San Joaquin & Kings River Canal and Irrigation Company (SJ&KRC&IC) that absorbed Bensley's earlier outfit and included several San Francisco businessmen as owner/investors. These investors included Miller & Lux and William Ralston, the president of the California National Bank, who took over from Bensley as president and principal advocate of the new company.

The new San Joaquin & Kings River enterprise sought to carry out Bensley's original project, but it also possessed a wider vision that encompassed development of the entire San Joaquin basin as well as the Kings, Tulare, and Kern River basins. To help plan for the future, Ralston hired Richard Brereton to serve as the company's chief engineer. Brereton was a British engineer who had worked for the previous 15 years in India helping to build large-scale irrigation canals and water distribution systems. In coming to California in 1871, he brought with him engineering skills largely absent in western America. He also brought a sense that irrigation development should be accomplished on a grand scale and with the tacit—if not overt—support of government. Brereton criticized Bensley's canal construction efforts as inferior to the work he was familiar with, and he set out to plan a system of wider scope than the company was presently pursuing. As part of this, he conceived of a major dam along the shore of Lake Buena Vista that could store floods from the Kings and Tulare Rivers and—via a lengthy canal extending north along the west side of the San Joaquin Valley—foster water distribution over vast areas of dry but irrigable land.

The implementation of Brereton's proposed system would require millions of dollars, much more than Ralston, Miller & Lux, and their partners could (or would) raise on their own. After failing to find support among British financier/capitalists, Ralston and Brereton traveled to Washington, D.C., in early 1873 seeking federal support for the company's private irrigation project. During President Ulysses S. Grant's administration, the notion of utilizing the resources of the federal government for private gain was hardly noteworthy. But, even within the context of a Republican Congress and Republican President who were disposed to help American business enterprise, Ralston and Brereton had difficulty attracting much enthusiasm for any scheme that would draw upon government financial resources or land grants to directly support a private irrigation project. Specifically, any work of this kind was opposed by small-farming proponents, including newly organized chapters of the Grange, which were gaining support in California and other agricultural states.

In place of a land grant that could have been used as security to attract private investment, Ralston and Brereton were, instead, offered a modest government-sponsored investigation of the irrigation possibilities present within the Central Valley. Presumably, such an investigation could offer a means for rationalizing subsequent financial support of irrigation work in the region; it could also offer a potential rationale for implementing irrigation projects that were not necessarily oriented toward serving the interests of large landowners. As such, it represented a much less controversial (and far less expensive) alternative to direct federal aid or land grants for the SJ&KRC&IC.

The Board of Commissioners' Survey

In March 1873, Congress and President Grant authorized an expenditure of $6,000 (of this, only $5,000 was actually spent) from the War Department budget to carry out a year-long survey of the Central Valley and report on the feasibility of irrigation within its confines. The survey was to be conducted by a board of commissioners comprised of two officers from the U.S. Army Corps of Engineers, a representative of the U.S. Coast Survey, a representative from the California state government, and a private sector citizen and engineer knowledgeable about conditions in the valley. The board was headed by Lt. Colonel B. S. Alexander and staffed by Major George Mendell and Professor George Davidson of the U.S. Coast Survey. Attempts were made to recruit Josiah D. Whitney, the California State Geologist, and Robert Brereton, chief engineer of the SJ&KRC&IC, but both declined the honor and the board functioned as a three-person entity for the duration of its work. Their effort commenced in

April 1873 and extended until the next spring. At that time, the board submitted a report to President Grant, subsequently published as *Report of the Board of Commissioners on the Irrigation of the San Joaquin, Tulare and Sacramento Valleys of the State of California*.

Limited in both time and funding, the board carried out a significant reconnaissance of the greater Central Valley, but it was incapable of undertaking any in-depth studies of particular watersheds or of making any detailed recommendations concerning the design of specific dams or water control systems. In large part, the board relied on survey data and hydraulic measurements supplied by other groups and individuals, although it did attempt to examine in person as much of the Valley as possible so that it could speak with first-hand experience about the character of the land and water resources covered in the report. Eschewing any attempt to make detailed suggestions as to the exact form and location of future canals, the board's report stressed more fundamental issues, not least of which was simply that much land in the Central Valley (especially on the west side of the San Joaquin and Tulare Valleys) was well suited to agriculture but could become productive only if an outside source of irrigation water could be developed. In essence, the board did not assume that a national (as opposed to western) audience would automatically recognize the need for irrigation systems to bring western land into cultivation.

Although a substantial amount of data specific to California was included in the board's report, there was also an enormous amount of descriptive material related to foreign canals in France, Italy and—most significantly—British India. In retrospect, one of the most noteworthy aspects of the report was the extent to which it stressed the value of British built and designed canal systems in India as models for large-scale irrigation development in the Central Valley. Unimpressed by the success of privately-financed irrigation projects in India, the board also stressed the importance and value of large-scale work that lay beyond the capabilities of individual farmers. Specifically, the Board proclaimed that:

> The experience of other countries appears to prove that no extensive system of irrigation can ever be devised or executed by the farmers themselves.... That while small enterprises may be undertaken by the farmers in particular cases, it would not be in accordance with the experience of the world to expect of them the means or inclination to that cooperation which would be necessary to construct irrigating-works involving large expenditures. That enterprises of this character, if built at all, must be built by the state or by private capital.[11]

Overall, the board recognized the paucity of precipitation in the southern half of the Central Valley and the relative abundance of water in the northern half. No grand scheme was offered that could rectify these inequities, but the report did note that a large dam at the northern end of the Sacramento Valley (in essence, the equivalent of what became Shasta Dam) could feed water into a lengthy canal extending down the western side of the Sacramento River Valley that could irrigate thousands of acres of land. But in keeping with the modest budget allocated to the Board, little in the way of firm, practical recommendations were made, and the Sacramento Valley canal was discussed in only the most schematic manner.

Miller & Lux and Riparian Rights

In immediate terms, the Board's report spurred no federal legislative reaction, either pro or con. In fact, the report's emphasis on government-sponsored activity—at least if it was interpreted as involving the federal government—seemed to render it irrelevant in terms of what actually occurred in the Central Valley during the latter nineteenth century. Contemporaneously with the report's preparation and subsequent submittal to President Grant, the nation as a whole fell into one of the worst economic depressions of the nineteenth century. Ralston and most of the other businessmen associated with the SJ&KRC&IC saw their financial empires collapse (Ralston's Bank of California failed in 1875 and he died within a year). Only Miller & Lux survived the lengthy recession intact and, as one result of this, assumed control of the canal and related holdings that been been developed to irrigate land in the lower San Joaquin Valley (land that belonged primarily to Miller & Lux). The stage was now set for a protracted battle between Miller & Lux (who controlled tens of thousands of acres of irrigated land bordering both the lower San Joaquin River and the Kern River near Bakersfield) and smaller-scale farmers seeking to develop irrigated settlements in the areas of the valley that were at higher elevations than the vast tracts controlled by Miller & Lux.[12]

From the 1870s through the 1930s, the dichotomy between large "agribusiness" and smaller-scale "family farms" formed the political backdrop for intense legislative and judicial maneuvering within the California political landscape. Ripe with nuance that makes broad-brush generalizations a dangerous business for historians, nonetheless, the essence of the "big business versus family farm" conflict revolved around a few issues and circumstances.[13] For example, significant irrigation in the region surrounding Fresno occurred in the 1870s and 1880s through the development of privately-financed "colonies" that were promoted by large-scale business interests but which led to a proliferation of relatively small-scale farms possessing secure water rights attached to the flow of the Kings River.

First of all, Miller & Lux were fanatically persistent in their defense of riparian water rights attached to their low-lying lands adjoining the San Joaquin, Kern, and other rivers in the state.[14] Although western America is popularly characterized as a region in which the "Doctrine of Appropriation" is the legal foundation of water rights, in California the "Riparian Doctrine" (by which landowners could secure water rights simply as a result of their owning land along a stream bank) was also recognized by the state constitution and upheld by state courts. In asserting their "riparian rights" to the San Joaquin and other streams, Miller & Lux necessarily challenged the legality of any upstream water diversions that would reduce the flow of water past (and, in times of flood, over) their low-lying lands. Anyone seeking to appropriate water for the purpose of irrigating land that lay above the bottom lands that Miller & Lux controlled would unavoidably encroach upon the partnership's riparian rights.

Even brief reflection upon the conflict attending any effort to administer appropriated and riparian rights simultaneously will lead an observer to the conclusion that they are inherently incompatible. And the legal wranglings that define California water law in the late nineteenth and early twentieth centuries would bear out the correctness of such a conclusion. Armed with the weapon that riparian water rights represented a "property right" that could not be taken away except through "due process" of law, as administered by the courts (not the state legislature or the federal government), Miller & Lux vigorously defended themselves in regard to any possible diminution of water flow for their lands. Their success in this endeavor was reflected in the 1886 decision of the California Supreme Court in the case of *Lux v. Haggin*, which upheld the riparian doctrine and precluded diversion from the Kern River by a consortium led by Ali Haggin who wished to irrigate land upstream from Miller & Lux's property.[15]

This ruling did not destroy all possibility of appropriating water out of California's rivers (e.g., after winning in court, Miller & Lux struck a deal with Haggin allowing a certain amount of water to be diverted upstream from their holdings, but the terms of the agreement were set by Miller & Lux in light of the power that accrued to them as a result of the court ruling), and it prompted the state legislature to pass the Wright Irrigation District Act (named after Modesto assemblyman C. C. Wright) authorizing the formation of irrigation districts that allowed for an alternative to irrigation development focused around large-land owning interests. In the years after 1887, the success of irrigation districts in California waxed and waned, but by the second decade of the twentieth century, irrigation districts (which relied upon enabling legislation that had evolved significantly since the enactment of the original Wright Act) became a powerful force in the state's agricultural economy.[16]

In the decades after completion of the board of commissioner's report in 1874, California agriculture continued to grow, but the nature and character of the development was not guided by a strong governmental authority (whether federal or state). Instead, large land-owning interests, acting to protect riparian rights largely developed in the years following the Gold Rush, exerted enormous influence over how and where—and to whose benefit—irrigation would develop. Spurred by prodigious amounts of legal wrangling, the California state courts played a critical role in determining regional water allocations. In essence, the ideal of strong government planning and guidance expressed in the board of commissioner's report held only minimal relevance to actual development.[17]

The Federal Hiatus: 1873–1934

After the conclusion of the board of commissioners investigation in 1874, the federal government assumed a secondary status in regard to water resources development in the Central Valley (and in all of California for that matter). Water rights were an issue of state law, and the federal government was largely excluded from the legal wrangling attending the conflict between the riparian and appropriation doctrines. Similarly, property rights were also largely an issue of state concern once land had been transferred out of the public domain and into private control.

Of course, land within the public domain remained under the control of the federal government, and this could have provided an entree for bringing the federal government into the field of irrigation development. In fact, the notion of utilizing the public domain constituted a key rationale underpinning the creation of the U.S. Reclamation Service in 1902; proceeds derived from the sale of public lands were originally conceived as the source of funding for Reclamation Service projects, and it was anticipated (or hoped) that small-scale farmers obtaining land directly from the federal government would become primary beneficiaries of federal irrigation initiatives. But, in California, little good agricultural land remained within the public domain at the start of the twentieth century, and the Reclamation Service played only a minor role in California agriculture prior to authorization of the Central Valley Project in the 1930s.

The most visible role played by the federal government in California's economic development derived from an activity that is not often considered a primary concern of western water development: navigation. In the arid West, water is often considered too scarce and valuable to utilize simply as a means of floating ships and barges. In the humid East, of course, the use of rivers as transportation corridors is long-standing, dating to the early years of the Republic; the role of the federal government in protecting river navigation as an essential component of interstate commerce was sanctioned by the U.S. Supreme Court in the 1820s, and during the nineteenth century, the U.S. Army Corps of Engineers assumed the lead in maintaining the navigability of major streams such as the Mississippi River and its tributary, the Ohio River.

The Sacramento River is completely confined within the boundaries of California. However, during the early years of the gold rush, steamships began plying the waters of the lower Sacramento River and providing a direct connection with transportation facilities in San Francisco Bay. Thus, the Sacramento River became recognized as a navigable stream tied into a larger transportation system connected to other states (along the Pacific Coast as well as those with Atlantic ports) and other countries.

Initially, navigation within the Sacramento River Valley did not become an object of concern because it required huge quantities of water to support large-scale river traffic. Rather, controversy arose because of hydraulic mining techniques practiced by mining companies in Northern California. To increase production from the mid-1850s onward, gold mining companies focused on large-scale techniques that entailed dumping huge amounts of waste rock and soil–often termed "debris"–into river beds below their processing facilities. Over time, this debris accumulated in the Sacramento River and many of its major tributaries to such an extent that it both fostered heavy, uncontrolled flooding of surrounding agricultural land and clogged the Sacramento River so that ships risked damage as a result of running aground.[18]

The conflict that erupted in the Sacramento Valley in the 1880s among mining, farming, and navigation interests attracted the attention of the federal government because it involved commercial river traffic, and this was an activity that, in constitutional terms, was recognized as falling under federal jurisdiction. As a result, in 1893 a federally supported California Debris Commission (with ties to the U.S. Army Corps of Engineers) was authorized by Congress to regulate California's hydraulic mining industry and ensure that debris would not be allowed to build up within the Sacramento River and impede navigation. This action formally acknowledged that use of the Sacramento River–at least in part–involved commercial navigation. This acknowledgment would eventually play a role in the events leading up to the authorization of the Central Valley Project.[19]

In the years following the formation of the California Debris Commission in 1893, the issue of federal involvement in maintaining the navigability of the lower Sacramento River remained politically active. And it encompassed more than simply holding back mining debris in the mountains and foothills. By 1904, a report by a board of engineers headed by T. G. Dabney recommended a plan for flood control based largely on the construction of permanent levees designed to confine flood flows in the Sacramento River's main channel.

In 1910, a new, more comprehensive report was prepared under the direction of District Engineer Thomas H. Jackson of the U.S. Army Corps of Engineers (hence its name: the "Jackson Report") that expanded upon the levee plan and included extensive dredging and the formation of weirs and "bypass" channels to disperse high flood waters. This report also acknowledged the possible utility of upstream reservoirs for controlling floods, but averred that:

> While favoring the use of reservoirs as far as possible, and considering that one of the advantages of the project herein proposed is that it lends itself to future storage possibilities, the [California Debris] Commission believes that it is not economical to construct reservoirs for flood control, but that such construction should be deferred until these reservoirs prove desirable for power and irrigation purposes.[20]

The "Jackson Report" provided a basis for subsequent Corps involvement in the lower Sacramento and helped spur inclusion of the drainage basin in the 1917 Flood Control Act. Although focused on flood control and navigation concerns, this act further extended the federal government's involvement—in concert with state and local interests—in the development of water resources in central California and helped define what eventually became the Central Valley Project in the 1930s.[21]

The Early Reclamation Service in California

A casual observer might logically presume that, with the authorization of the Reclamation Service in 1902, California would be a key focus of the new federally sponsored irrigation projects. After all, the state already possessed an economically productive agricultural industry focused primarily on irrigation development. However, the introduction of the federal government into the business of irrigation would come with a price, and existing farming interests in California feared that they would surrender some of their autonomy if they relied upon a federally sponsored (and federally administered) system for their water supply. This fear was exacerbated by concern that the law authorizing the Reclamation Service also stipulated that farms served by federal reclamation projects were to be limited in size to only 160 acres.

Commonly characterized as the "160 acre limit" (although it was quickly amended to allow a husband and wife to operate, in tandem, a 320 acre farm), this stipulation reflected congressional concern that federal irrigation somehow be focused on "small farmers" and not become a mechanism for providing subsidies to large landowners (or, even worse, large-scale, land-grabbing "speculators"). Clearly, as a result of the acreage limitation requirements, organizations such as Miller & Lux possessed little incentive to become associated with Reclamation Service projects. And many other farmers, such as those residing in the irrigation colonies surrounding Fresno, already held relatively reliable water rights that did not need to be bolstered by the storage of flood waters behind federal dams.

As a result, the early efforts of the Reclamation Service in California became concentrated on projects that lay at the periphery of the state in either a physical or economic sense. For example, two of the most prominent early federal initiatives involved interstate streams that were far removed from the Central Valley: these were the Yuma Project, along the lower Colorado River in the far southeast corner of the state, and the remote Klamath Project, along the Oregon/California border.[22] The political nature of Reclamation Service projects was also reflected in the decision to abandon planning for federally sponsored irrigation development in the Owens Valley once it became clear that the City of Los Angeles wished to tap the Owens River as a source of municipal water supply.[23] In this context, it is not so surprising that the one early Reclamation Service project in the Central Valley represented a relatively minor endeavor, at least in comparison to the totality of irrigation in the valley and in comparison to the Central Valley Project of the 1930s.

The Orland Project

In 1906, the Reclamation Service received authorization for construction of a project to divert the flow of Stony Creek and allow for irrigation development of about 14,000 acres in the region around the town of Orland, located about 100 miles north-northwest of Sacramento. Stony Creek is a relatively minor tributary of the Sacramento River that flows eastward out of the Coastal Ranges (rather than westward out of the much more imposing Sierra Nevada Mountains) and its impoundment and diversion for irrigation was not considered significant enough to alter the overall flow of the lower Sacramento River. The primary component of the Orland Project was the East Park Dam, a curved gravity concrete dam, completed in 1911 across Little Stony Creek (a tributary of Stony Creek). In a technical context, the East Park Dam and associated facilities worked well, but the productivity of the land around Orland proved to be relatively weak and the financial condition of the farmers responsible for paying for the project languished. By the early 1920s, farmers dependent upon the Orland Project had largely resorted to growing low-value forage crops (primarily alfalfa) as feed for dairy cattle.[24]

In the mid 1920s, concern over the plight of the Orland Project farmers led Congress to authorize construction of an additional storage dam across Stony Creek for the purpose of increasing water supplies and thus improving the economic viability of the overall project. Or, as a Bureau of Reclamation publication later characterized it:

> The heavy shortages and the financial losses suffered by the water users on the project during the years 1918, 1920, and 1924 resulted in an appropriation by the Congress in the spring of 1926 for starting construction of the Stony Gorge Dam.[25]

The hope was that increased water supplies would foster the opening up of additional land to irrigation and this would increase agricultural revenues sufficiently to allow repayment of the project's cost. Completed in 1927, Stony Gorge Dam is the first and only large-scale, flat-slab, buttress dam built by Reclamation. With a maximum height of 142 feet and a length of 868 feet, the dam impounds a reservoir with a storage capacity of about 50,000 acre-feet. Although this is certainly not of inconsequential size, compared to Shasta Dam, which Reclamation commenced building a decade later, it is extremely small (specifically, Stony Gorge Dam provides only about 1.1 percent of the storage capacity provided by Shasta Dam).[26]

The Iron Canyon Project and the Sacramento Valley

While the Orland Project represented but a small portion of the potential productive capacity of the Sacramento Valley, in the early twentieth century, efforts were made by boosters in the valley to get the federal government to support construction of a dam across the Sacramento River near the town of Redding. Usually referred to as the Iron Canyon Dam, this structure was promoted by the Sacramento Valley Development Association—and its offshoot, the Iron Canyon Project Association—as a means of irrigating more than 100,000 acres of land in the upper Sacramento Valley along the western side of the valley in the Orland region. Initial studies were carried out by Reclamation Service engineers in 1904-05, and, in 1913, a large-scale, yet hardly comprehensive, investigation was completed by the Reclamation Service with direct financial support provided by the Iron Canyon Project Association.[27]

This report is instructive in that it highlights the possibilities that could accrue in terms of regional growth if a large-scale storage dam were built at the upper end of the Sacramento Valley. Yet, it also draws attention to the difficulties that would attend such a project in terms of accommodating the needs and rights of navigation interests seeking to maintain river borne commerce up to (and beyond) Sacramento. In addressing navigation issues, the report highlighted conflicts inherent in a legal structure that grants states the right to administer and control water rights and water usage but simultaneously empowers the federal government to oversee all issues related to commerce on navigable rivers.

The State of California might well exercise control over the water use in the Sacramento River Basin, but this control could prove less than authoritative in the face of federal navigation requirements.[28] The 1914 report sponsored by the Reclamation Service and the Iron Canyon Project Association could not resolve the conundrum posed by the status of state and federal authority over use of water within the navigable and flood prone Sacramento River. But, in concert with the California Debris Commission's "Jackson Plan," discussed earlier in this chapter and first publicly presented in 1910, it brought the issue into the public limelight. In the words of the report:

> Reference is made herein to possibilities of cooperating with flood control and navigation interests in the lower Sacramento Valley. No conclusion is possible on these important matters save after careful and thorough investigation. . . . The relative importance of the waters of the Sacramento to transportation or to agricultural development in the Sacramento Valley is a question that should be decided by the state and nation at the earliest date possible, in order that the various improvements proposed can be brought into harmony . . .[29]

In terms of dam technology, the 1914 Iron Canyon Project report devoted considerable attention to the geological condition of the Iron Canyon site and how this might affect design issues. Specifically, the Iron Canyon site was found to be relatively weak, porous, and susceptible to the flow of subsurface water through the foundation.[30] Conditions were not considered bad enough to warrant rejection of the site as completely unsuitable, but the porous character of the foundation clearly concerned the Reclamation Service engineers. In offering a possible design, they made clear that it represented only a preliminary proposal and that further geological testing of the site would be necessary before developing a final design and construction program. Nonetheless, the 1914 report laid out two basic design options: "The earth and rockfill type and the solid gravity masonry type. . . ."[31]

In terms of cost, the two design options explored by the Reclamation Service "show[ed] little difference in cost" (both would have required an overall project expenditure of around $15 million), and the rockfill design was recommended because of concerns related to the suitability of the Iron Canyon foundation. However, the Reclamation Service made a point of developing a design and cost estimate for the masonry gravity structure, and the 1914 report specifically opined that:

> An arched type of masonry dam across the Sacramento has certain attractive features, due to its compactness, solidity, beauty

> of design and the firm manner in which it can be keyed into the rocky abutments . . .[32]

A rockfill embankment design may have been favored in 1914 because of the porous conditions at the site and fears that percolation could undermine a masonry gravity design through sliding and uplift. In fact, the report specifically acknowledged that:

> In a gravity masonry dam designed to act as a more or less absolute closure the introduction of percolating water under the dam foundation may under certain conditions result in upward pressure threatening the stability of the structure itself . . . .[33]

In light of this concern, a rockfill design was preferred over a masonry gravity alternative. But interest in design of the latter type did not disappear.

After the end of World War I, another joint study of the Iron Canyon Project was initiated in May 1919. Sponsorship of, and funding for, the study came from the Reclamation Service, the Iron Canyon Project Association, and the State of California (acting through the State Department of Engineering). In May 1920, a *Report on Iron Canyon Project California* was published by the U.S. Department of the Interior laying out the results of this new study.[34] An extensive set of additional geological test borings of the Iron Canyon site were made at three potential damsites within the canyon. Plans were developed for three dam design alternatives with a maximum height of 170 feet and an overall length of about 5,000 feet. These alternatives consisted of: (1) a concrete gravity design extending the entire length of the structure, (2) a concrete buttress design featuring a gravity section at the spillway and the powerhouse, and, (3) an earth embankment with concrete corewall design featuring a concrete gravity section at the spillway and powerhouse.[35]

In a "Report of board of engineers upon Iron Canyon Dam Sites and Type of Dam" that was published as an appendix to the larger report, the Reclamation Service made a point of recommending that Location III (the furthest downstream of the three potential sites) be considered the most advantageous (Location I was the focus of the 1914 report).[36] But in terms of making a recommendation regarding which of the three design alternatives was best, the Reclamation Service declined to offer an opinion. All three were considered to cost about the same (around $17 million for the dam itself). But accuracy of such cost estimates were downplayed, as the Reclamation Service made clear that actual costs could turn out to be significantly higher:

> We conclude, therefore, that while conditions for a dam at the
> best site available are far from ideal a safe dam can be con-
> structed at this point, Location III, but it must be admitted that
> the item of contingencies to guard against all dangers which
> may become apparent upon opening the foundations may be
> greater than usual and that for this dam, including also overhead
> expenses, estimated at 25 percent, may be exceeded.[37]

Over the course of the next few years, the Iron Canyon Project remained a possibility. And in 1925, yet another report on its feasibility was prepared by the Reclamation Service under its new name, the Bureau of Reclamation.[38] However, by the time this 1925 report appeared, the Iron Canyon Project had already been superseded by a proposed 500-foot-high dam at the Kennett Dam site (about 50 miles upstream from Iron Canyon) that had become the focus of considerable study by the California Department of Engineering. The Iron Canyon Project never bore fruit in terms of its specific plan for developing a reservoir in the upper Sacramento Valley. But it, nonetheless, played a significant role in the planning that preceded authorization of the Central Valley Project in the 1930s.[39]

Rice and Drought

While the Iron Canyon Project languished for lack of a political consensus that could align support behind the project and its less than ideal dam site, the notion that huge quantities of water flowing in the Sacramento River could be captured in a major reservoir near the confluence of the Sacramento and Pit Rivers remained strong in the mind of northern California boosters. As it turned out, a key impetus for large-scale storage along the Sacramento River came from an agricultural industry that had not existed in California prior to the early twentieth century. The product of this new industry was rice, a commodity that required huge amounts of fresh water and was well suited to the natural conditions of riparian land along the length of the Sacramento River. In 1910, only about 100 acres of rice were planted in California, but, by 1914, 15,000 acres were in production, and, by 1920, the industry had exploded to encompass more than 160,000 acres (or about 25 percent of all irrigated land in the Sacramento Valley).[40]

Rice culture consumes huge quantities of water, and, by the early 1920s, it was perceived as a key reason that the flow of fresh water into San Francisco Bay was rapidly decreasing and "salt water intrusion" was gradually working its way up into the delta land surrounding the confluence of the Sacramento and San Joaquin Rivers.[41] In addition to the dramatic growth in rice production, the flow of the Sacramento River also suffered because of droughts that afflicted much of California in the period 1917 to 1920 and in 1924.[42] The combination of increased irrigation diversions, low amounts of natural precipitation, and continuing concern that navigation not be impeded provided a powerful impetus for finding a way to take full advantage of the Sacramento River in a manner that would not require any existing uses of the river to be abolished or significantly curtailed. This appeared possible only if a large dam on the upper reaches of the river could store seasonal flood waters and make them available throughout the year.

STATE PLANS FOR CALIFORNIA

The "Marshall Plan" and the Evolution of a State Water Plan

Political interest in developing California's water resources under the aegis of a coordinated state program was evident in the late nineteenth century, but reluctance on the part of large, land-owning interests to embrace such initiatives (e.g., funding a strong and active State Engineer's office) prevented them from attaining much significance.[43] With the ascendance of progressive reform movements in the early twentieth century, the notion of state guided and administered water development took on greater political acceptability. Although the water and property rights of existing land-owners and irrigation interests could not be obviated by legislative action, the creation of a State Water Commission in 1914 to oversee the legal process by which water rights were granted, administered, and adjudicated marked an important step in increasing the role of state government in water development.[44]

As the State Water Commission gradually gained more acceptance of its authority over myriad issues involving water rights, it became less controversial for groups to propose a strong state role in implementing large-scale water development projects. There was no single moment when, suddenly, everyone in California enthusiastically embraced the idea of government-sponsored projects; rather, government sponsorship involved a process of gradual acceptance based upon incremental change and adaptation to evolving political interests. A key step in this process came in 1919 when a former federal official in the U.S. Geological Survey began promoting a bold scheme to make fuller use of the Sacramento River. Robert Marshall had had a long and distinguished career in western resource planning before he proposed what became known as the "Marshall Plan." Although his scheme did not immediately win popular acceptance, it started the ball rolling in the direction of a government-sponsored plan for developing the Central Valley.[45]

In essence, Marshall proposed construction of a large storage dam across the upper Sacramento River that would serve three primary functions: (1) it would allow the diversion of water into large irrigation canals running along the west and east sides of the Sacramento Valley and southward into the San Joaquin Valley; (2) it would ensure a minimum flow in the Sacramento River, thus allowing for continued navigation and combating the influx of salt water into the delta region; and (3) it would facilitate the generation—and sale—of large quantities of hydroelectric power that would largely pay for the overall development of the system. The plan also proposed a canal diversion to carry water south from the Stanislaus River (located about fifty miles south of the city of Sacramento) and into San Joaquin Valley. This water supply was intended to replace water from the Kern River that was to be diverted over the Tehachapi Mountains and into Los Angeles and other areas of Southern California. In addition, a branch canal extending from the San Joaquin Valley was to carry water to users in the greater San Francisco metropolitan region.[46]

The Marshall Plan was certainly bold and, with a price tag estimated to exceed $700 million, expensive.[47] Designed to attract support from constituents throughout the state (this is why it included the Kern River diversion to Southern California and the connecting canal to the San Francisco Bay area), Marshall's scheme garnered political support from a consortium of "progressive" leaders opposed to the hegemony of private power companies. As the California State Irrigation Association, the progressive leaders lobbied for the plan. At the same time, the plan was opposed by private power companies (including Pacific Gas and Electric Company, the firm that controlled the electricity market for most of Northern California) and by engineers who questioned the financial feasibility of building the various technical components of the system.[48] Undeterred, Marshall and his supporters pushed to get the California state legislature to authorize and fund the plan, and, as part of this, they aggressively promoted the project and made enormous strides in raising public awareness of the possibilities that existed in terms of damming and developing the Sacramento River as a regional water supply and as a source of electric power.

Over the course of the 1920s, the Marshall Plan was discussed and considered by California's legislators and its electorate, and it endured as a possible scheme. After failure to win support from the state legislature in 1921, approval for the Marshall Plan was sought in the form of an amendment to the state constitution, submitted to California voters in 1922, 1924, and 1926, but each time it failed to garner sufficient votes for passage. Gradually supplanted by other proposals that drew from its basic idea of building a large dam across the upper Sacramento River, by the end of the 1920s, the Marshall Plan had been superseded by other studies authorized by the legislature and developed under the direction of the state engineer. Interestingly, Marshall himself—who was so instrumental in the initial promotion of his plan—proved a quixotic figure, who at times in the mid-1920s, publicly advocated rejection of the proposed amendment designed to implement the Marshall Plan.[49] Nonetheless, the political stir resulting from the publicity surrounding Marshall's original proposal helped prompt the state legislature in 1921 to fund engineering investigations into possible water development projects within the state.[50] And these initiatives eventually led to a politically viable plan for distributing water in the Central Valley.

The Early Studies of the State Engineer

In the wake of the state legislature's funding for a study of California's water resources, the state engineer set out to examine conditions in a less politically charged context than that associated with the Marshall Plan or the "Water and Power Act," which voters rejected consistently. As one historian has phrased it, the state engineer (acting through consultants and the staff of the Department of Engineering) conducted a survey directed toward:

> Gauging stream flows; searching for reservoir sites; classifying reservoirs according to cost and benefits; mapping the land irrigated in 1920; determining the total amount of irrigable land in California and classifying it according to quality and yield; determining the water requirements of that land; investigating the feasibility of diversions of water from water-rich to water-deficient areas; estimating the future water needs of California cities and possible sources of supply; determining the effect of reservoir construction on flood control; estimating the potential power development on California's streams; recommending ways to prevent saltwater encroachment; and assessing the effects of deforestation on stream flow.[51]

Taken in aggregate, this task was daunting indeed. And probably the only reason that the investigation generated any report at all within a two-year time period was because no consideration was given to existing water development systems or to the legal implications of possible future projects. In other words, the study was to consider California as some kind of environmental "blank slate" awaiting water resources development. This limit on the study's conceptual parameters meant that no immediate practical proposal could result from the investigation, but it also offered significant political "protection" because it eliminated the need to try to resolve or reconcile conflicts over water and property rights. Thus, the report could avoid stirring up immediate opposition.

The 1923 report issued by the state engineer (in response to the legislature's 1921 directive) offered an extensive hydrographic study of precipitation and streamflow patterns and presented basic data regarding more than a 1,000 possible reservoir sites in the state. However, specific recommendations concerning these myriad storage sites were scant, with the most concrete proposal in the report focusing on a low barrier or dam at the head of San Francisco Bay that would be designed to prevent saltwater intrusion into the Sacramento-San Joaquin Delta. No large dam was proposed across the upper Sacramento River. But the study did acknowledge the disparity in precipitation between the Sacramento and San Joaquin Valleys and proposed a canal and pump system running up the San Joaquin River that could carry water 200 miles from the delta to Tulare Lake, in the extreme southern reaches of the San Joaquin Valley. From Tulare Lake, it presumably could be distributed to a wide swath of land extending from Fresno to Bakersfield.[52]

The "Water Surge" and Riparian Rights

The concept of a "water swap," whereby water from the San Joaquin River would be transferred southward toward Bakersfield (in the Kern River watershed), did not appear in the 1923 state engineer's report, but found expression in a special study authorized a year later in response to pleas from farmers in the Tulare River watershed (just north of Bakersfield), who were concerned about dwindling water supplies. In this new report (released in 1925), a scheme was proposed that would involve construction of major storage dams across both the San Joaquin and Kings Rivers. Water from the San Joaquin River could then be transported southward by a gravity canal to be used by agricultural interests that previously had relied on water from the Kings River. In turn, water stored behind the Kings River dam could be transported by gravity canal to the Tulare and Kern River basins farther to the south. There—due in no small part to the extensive water rights claimed by Miller & Lux—irrigable land was kept out of production because of shortages in the local water supply. Drawing from the scheme outlined in the 1923 report, fresh water from the delta was to be pumped up the lower San Joaquin River and made available to agricultural interests that otherwise would be relying on the undammed flow of the San Joaquin River; however, it would not be pumped as far as Tulare Lake. Also, like the 1923 report, no mention was made of a large storage dam across the upper Sacramento River.[53]

In 1926, the California State Supreme Court once again (as it had in the case of *Lux v. Haggin* in 1886) affirmed the validity of riparian water right claims and even went so far as to endorse the notion that riparian use of water did not need to adhere to a standard of "reasonable" or "beneficial" use.[54] The public reaction to the seeming inequity in this ruling was strong, and—while legislative action could not directly obviate the implications of the ruling insofar as they involved existing property rights—the ruling furthered interest in development of a plan that would guide and administer water use in the state.[55] Thus, the next report on state water planning issued under the aegis of the state engineer in 1927 took on additional significance because of the concern generated by the recent court ruling on riparian rights and the desire of many California citizens to seek ways of insuring regional water supplies in the wake of the ruling.

The 1927 State Engineer's Reports and Kennett Dam

In 1927, the California State Engineer's Office proposed plans that dropped a salt water barrier across the lower Sacramento River in favor of a large storage dam across the upper Sacramento at the Kennett Dam site. Other storage dams across tributaries of the Sacramento were also proposed. The hydroelectric power generated at Kennett Dam would be utilized to pump water up the San Joaquin River from the delta to facilitate the "water swap" southward toward the Kern and Tulare basins. The importance of this "water swap" was highlighted by noting that it would allow a reduction in total pumping height of 360 feet compared to what would be needed if the water were pumped directly from the delta to the Bakersfield area. This savings in water pumping would allow more of the electricity generated at Kennett Dam to be sold into the regional electric power market and used to pay for the overall project (estimated at more than $350 million). The importance of electric power and its sale was fully acknowledged as an essential component in the dam's financing.[56]

Beyond the question of revenue derived from hydroelectric power generation, the 1927 state engineer's report is noteworthy on three accounts: (1) it stresses the importance of a large, multipurpose storage dam across the upper Sacramento that could impound flood waters, generate hydroelectricity, provide for navigation, protect the delta from salt water intrusion, and supply water to facilitate a "water swap" from the San Joaquin River southwards; (2) it describes the Kennett Dam site in detail and establishes the basic form and layout of the concrete curved gravity dam that was subsequently built as Shasta Dam; and (3) it recognizes the importance of, if not the outright need for, federal assistance in financing the project. As perceived by state officials in 1927, federal support would most likely derive from a direct financial allocation relating both to flood control benefits derived from the construction of storage dams and to benefits accruing to navigation along the Sacramento River.[57]

What occurred over the next four years was refinement of the basic plan of the project and building of political coalitions strong enough to carry the project through the legislative approval process. In 1929, a key committee in the California State Legislature voted to accept the 1927 state engineer's report as the basis for a state water plan. As part of this, the salt water barrier to protect the delta was excluded from the proposed work, and, in a politically sensitive decision, the committee recommended that power generated at Kennett Dam (and other dams) be sold at the powerhouse to commercial companies (most notably, Pacific Gas and Electric) which could then market it to businesses and consumers through transmission and distribution systems that would be financed by private investment (and *not* by taxpayers). This action largely neutralized opposition from private power interests (that had blocked passage of the Marshall Plan/Water and Power Act), because the water plan now presented little overt threat to investor-owned utilities.[58]

While the 1927 state engineer's report set the stage for the eventual political resolution of the state and federal interaction necessary to bring about implementation of a large scale reservoir on the upper Sacramento River, it also established the technical template for what became Shasta Dam. In rejecting the Iron Canyon Dam as an inadequate solution to the water problems besetting the Central Valley, the State Engineer's Office proposed construction of a substantially higher dam at the Kennett site than had been proposed at Iron Canyon (upwards of 600 feet high compared to a height of less than 200 feet at Iron Canyon). The report carefully considered the costs and advantages of a dam at the Kennett site for five distinct heights (i.e., 220 feet, 320 feet, 420 feet, 520 feet, and 620 feet), while acknowledging that "the most desirable capacity of the Kennett Reservoir . . . is not subject to an exact analysis" in the sense that selection of an appropriate design depended on a determination of what the long-term market for electricity might be over the next several decades.[59] The higher the dam, the more it would cost to build and the more it would cost to clear the reservoir area and relocate highways and the Southern Pacific Railroad right-of-way; but a higher dam allowed for more assured electric power production at "primary" rather than "secondary" rates, and this would translate into higher revenues from the sale of hydroelectric power. The report concluded that a 420-foot-high dam with a generating capacity of 400,000 kW and a storage capacity of almost 3 million acre-feet represented the design offering the "greatest capacity commensurate with reasonable production costs."[60]

The report revealed that:

> foundation conditions at the Kennett dam site, as disclosed by the diamond drill borings . . . are suitable for a high dam of any type. Topographic features and the absence of earth in large quantities, however, limit considerations to a concrete-gravity or a rock-fill dam.[61]

Although the report acknowledged that "construction of a rock-fill dam could proceed under usually favorable conditions" because an ample supply of rock was available at a high elevation in relation to the dam site, other factors caused the state engineers to reject a rock-fill design even though "preliminary estimates indicate that a rock-fill dam may be constructed for somewhat less cost than a gravity-concrete dam." The mitigating factor involved the greater thickness of the rock-fill design and "the added cost of power and flood control outlets" through the thicker rock-fill structure "makes the total cost about the same for either type."[62]

After eliminating the rock-fill design as a possible design choice, the report focused exclusively on the details of a concrete gravity design that, in its curved alignment across the site, was extremely similar to the design used for Shasta Dam a few years later. In contrast to the Shasta Dam design, the 1927 Kennett Dam design featured a powerhouse about 1,800 feet downstream from the dam proper (the Shasta powerhouse is only a few hundred feet below the dam) and two spillways are featured at each end of the Kennett design (Shasta Dam has one large spillway in the center of the structure). But these differences represent design variations that do not affect the basic form of the concrete curved gravity structure that was established in the 1927 State Engineer's report as the most appropriate design for the site.[63] And it was the 1927 report that would be sent to the state legislature and become the focus of, and platform for, all subsequent political debate.

The Federal/State Interaction

The state legislature reacted favorably to the committee report, but, before proceeding to enact authorizing legislation, it sought a more detailed proposal and it sought interaction with the federal government as a possible source of financial support. President Hoover (in concert with California Governor C. C. Young) subsequently appointed a commission to assess the proposed plan offered by the State Engineer's Office. This commission endorsed the plan and, in noting that the project's financial feasibility was tenuous if the state could not borrow money at an interest rate of 3.5 percent (which seemed unlikely), recommended that "the project be constructed by the Federal governmen t. . . and that the works, when completed, be operated by the State as far as practicable. . . ."[64]

Nothing definitive came of this federal/state commission's report upon its release in 1930, and the idea of federal financing for the project did not immediately win widespread support; however, it was an important step in the development of the project. An even more important step came in March, 1931 when the State Engineer's Office submitted a detailed report to the legislature that expanded upon the report submitted in 1929; and called for construction of Kennett Dam, Friant Dam, a 120-mile long pump/canal system to deliver water from the delta into the San Joaquin Valley, and a canal from Friant Dam southward toward Bakersfield that would facilitate the north-to-south "water swap."[65] In August 1933, this report comprised the basis of a state law authorizing issuance of $170 million in state bonds to finance the initial stages of the project (including the two major dams at Kennett and Friant). It also authorized construction of

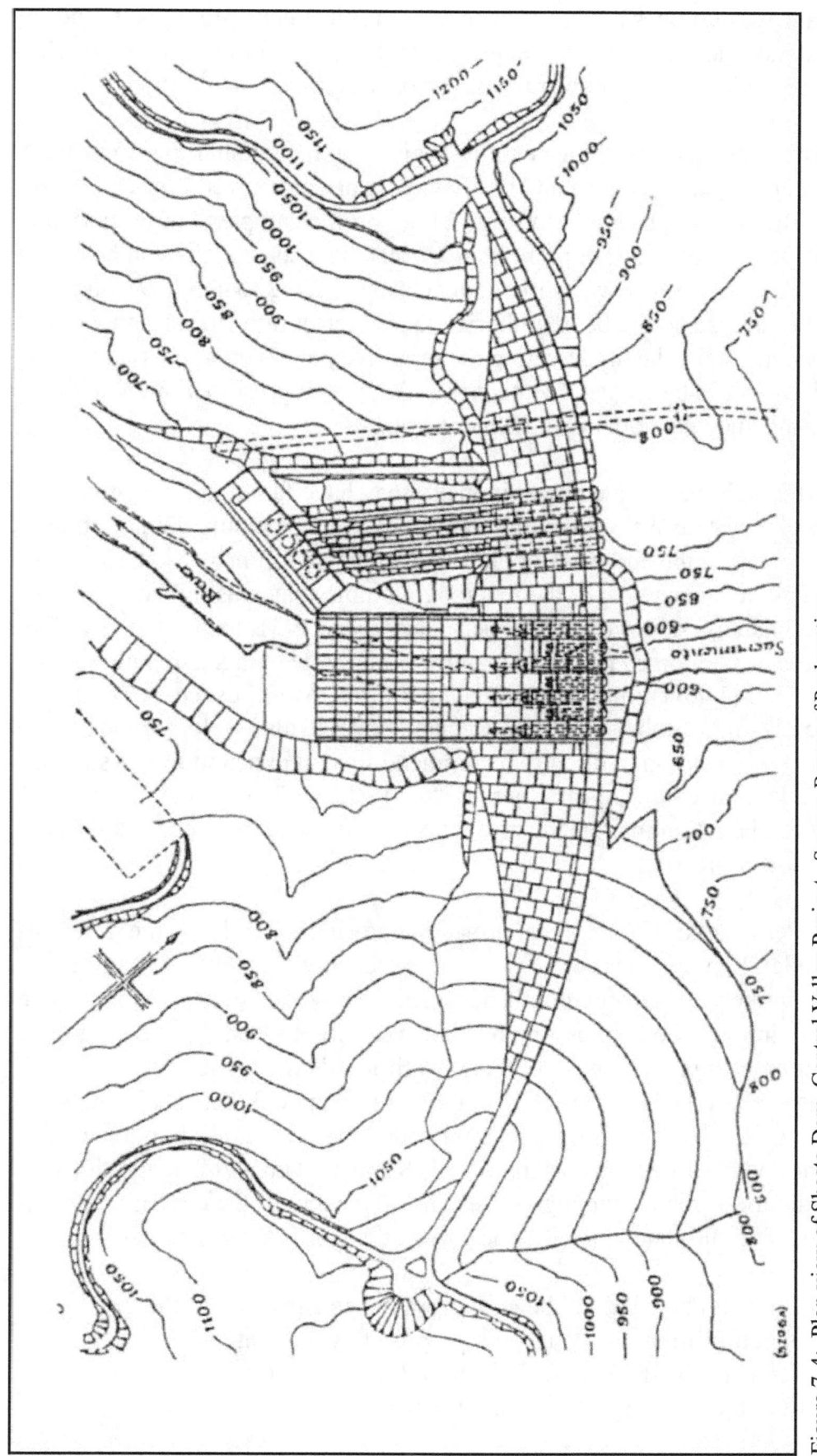

Figure 7-4: Plan view of Shasta Dam, Central Valley Project. Source: Bureau of Reclamation.

electric transmission lines connecting the major project generating plants and pumping stations associated with the project and allowed for distribution of electricity via a publicly owned power transmission system.

This latter feature of the enabling legislation ran counter to provisions of the 1929 state engineer's report and prompted the private power industry (in particular, Pacific Gas and Electric) to wage a high profile campaign to rescind the act via an initiative placed before the state's voters in a special election held in December 1933. The vote was remarkably close (459,712 to 426,109), but voters chose not to rescind the bond authorization. Interestingly, the argument that apparently carried the day in defeating the initiative played upon the notion that valuable jobs would be created by the project, and this was something that the depression-plagued economy could benefit from.[66]

Although the private power industry may have failed to block implementation of the State Water Plan, the state's political leadership did not immediately move to sell the authorized bonds and commence construction. Instead, the involvement and support of the federal government increasingly came to be seen as the financial savior of the project. As early as 1931, the U.S. Army Corps of Engineers had endorsed the basic features of the state plan and recommended that the federal government contribute $6 million toward the cost of building Kennett Dam.[67] In the same year, Reclamation had also endorsed the plan and, while in no position to directly recommend federal expenditures to support the project, had raised the possibility that, "the Federal Government through the Bureau of Reclamation may be the most appropriate agency to undertake the execution of this project. . . ."[68]

After Franklin Roosevelt assumed the presidency in 1933, the possibility of direct federal involvement increased simply because Roosevelt was more accepting of public works expenditures designed to provide "relief" during the economic hard times of the Depression than was Herbert Hoover. However, questions about the nature of such support and California's willingness to accept federal "strings" as a condition for receiving it were not quickly resolved. In 1933, the U.S. Army Corps of Engineers recommended that $7.7 million in federal funds be provided to support construction of Kennett Dam and that additional consideration be given to funding as much as 30 percent of the cost of other features included in the state plan. The next year, the Corps suggested that:

> the general and Federal benefits from the construction of the Kennett Dam on the plans now proposed by the State warrants a special direct participation of the Federal Government of $12,000,000 in the cost of this structure.[69]

The Corps identified potential navigation, flood control, power, irrigation, and salinity control benefits in its plan. The state of California quickly followed the Corps' report with a request, in September 1933, for an emergency loan and grant of $166.9 million from the Federal Emergency Administration of Public Works[70] to finance what was now officially called the "Central Valley Project." No immediate action was taken to approve this request, but, in 1935, the "Central Valley Project Works" were authorized for construction by the U.S. Army Corps of Engineers as part of the Rivers and Harbors Act of 1935.[71] However, in September 1935, President Roosevelt signed an Executive Order that offered conflicting guidance as to which federal bureau would be responsible for implementing the Central Valley Project. In this order, he specifically directed that $20 million dollars from the Emergency Relief Appropriation Act of 1935 "be transferred. . .to the Department of the Interior, Reclamation Service. . .for the construction of works of irrigation and reclamation . . . in accordance with reclamation laws."[72] Interestingly, this $20 million appropriation was to be spent only on the construction of Friant Dam and on canals designed to transport water from the reservoir (Millerton Lake) impounded behind Friant. And the U.S. Army Corps of Engineers had specifically noted in earlier reports that Friant Dam played no role in the navigation improvements along the Sacramento River, placing it in a different category than Kennett Dam.

A "Federal Reclamation Project"

In December 1935, following a recommendation by Secretary of the Interior Harold Ickes, President Roosevelt approved "the Central Valley development" as a "Federal Reclamation Project" and signaled that Reclamation would be the sole federal bureau responsible for implementing the project. In August 1937, the primary authority of Reclamation was unequivocally instituted by the 1937 Rivers and Harbors Act, which stipulated that:

> the $12,000,000 recommended for expenditure [in the 1935 River and Harbor Act] for a part of the Central Valley project, California . . . [shall be expended] in accordance with the said plans by the Secretary of the Interior instead of the Secretary of War. . .

Just as importantly, the 1937 act affirmed that, in placing responsibility for the project under the control of the Bureau of Reclamation, the "provisions of the reclamation law, as amended, shall govern the repayment of expenditures . . ." relating to works deemed essential to the entire project.[73]

The significance of this latter stipulation related to the acreage limitation requirements that had been a part of projects since 1902. Although the enforcement of acreage limitation regulations (allowing farmers only 160 acres of irrigable land, or 320 acres for married couples) had proved problematic throughout the history of Reclamation, the specific reference to "reclamation law" in the 1937 act was the one big "string" that came along with federal financing and construction of the Central Valley Project. And this "string" proved contentious enough that development of the Kings River (and the possible construction of Pine Flat Dam) were completely excluded from the project and left to be funded by some other method or means. Nonetheless, issues related to acreage limitations were perceived by many Californians as being political concerns that could be resolved or litigated at some future date when dams and canals were ready to be used. Before such concerns could assume practical importance, it was first necessary to actually build the two big dams that lay at the heart of the project.

SHASTA AND FRIANT DAMS

Design of Shasta and Friant Dams

Shortly after authorization of the Central Valley Project in August 1937, the name of Kennett Dam (which had been taken from the name of a small mining town near the site) was formally changed to Shasta Dam as a way to draw attention to the new character of the structure as a federal and, more specifically, a Reclamation project. And now that the dam was officially to be built by the Bureau of Reclamation, the design also became Reclamation's responsibility. In its particular details and ultimate size, the design built to impound the upper Sacramento River was a product of Reclamation's design staff; this was made clear in a November 1937 press release from the Department of the Interior noting that Reclamation's Chief Engineer, R. F. Walter, had approved a 560-foot-high concrete gravity design (with a reservoir capacity of 4.5 million acre-feet) that surpassed the 500-foot high design proposed as part of the original state plan.[74] Nonetheless, because of the lengthy process that preceded passage of the Central Valley Project as a federal project and the desire to initiate construction as soon as possible, it appears that many considerations involving basic design issues (not least of all the site of both Shasta and Friant Dams) were determined by work undertaken by the State Engineer's Office during the preceding decade.

Both Shasta (with a crest length of approximately 3,500 feet) and Friant (with a crest length of approximately 3,400 feet) are concrete gravity dams incorporating ample cross-sections capable of accommodating even the most deleterious effects of "uplift" along the foundation. Shasta features a maximum base thickness of 580 feet, which actually exceeds the maximum height of the structure (560 feet); Friant features a maximum thickness of 268 feet, which is over 80 percent of the maximum height. Based upon information provided by

Kenneth Keener of the Bureau's engineering staff, the journal *Engineering* reported that:

> The decision to build a concrete gravity-type dam [at Shasta] was not reached without consideration of the possibility of employing an earth and rock-fill structure . . . it was necessary that the dam should have a height of 500 ft. above the [original] river bed. No earth and rock-fill dam of such a height has ever been constructed. . . . This did not, in itself, rule out the possibility of building a 500-ft. structure of this type, but other considerations favored a concrete [gravity] dam . . . [particularly the fact that] a concrete dam would [provide the most convenient and cheapest arrangement for outlet pipes and] the spillway and outlet conduits could be incorporated in a single structure . . . suitable materials for a rock-fill dam were also available, but investigation showed that the cost of providing for a spillway, as well as making arrangements for the outlet conduits with a dam of this type would make it more expensive than the concrete structure. The nature of the [wide canyon] ruled out consideration of a single-arch concrete dam. A multiple-arch dam might have been possible, but there was no experience available with a dam of that type of the dimensions necessary at Shasta, and in view of the paramount necessity for absolute security . . . a gravity structure was determined on.[75]

In rejecting any possibility of utilizing a multiple arch design at Shasta, Reclamation's rationale derived from the tremendous height of the spillway section of the proposed dam. At Friant (with a maximum height not much more than 300 feet), using a multiple arch design was more plausible, but there does not appear to be any evidence that this option was ever seriously considered. In fact, as early as January 1937, many months before Reclamation was given authority over the entire Central Valley Project, Sinclair O. Harper of Reclamation's Denver staff informed Reclamation's Acting Commissioner, John Page, that Friant was to be built as a "straight concrete gravity type."[76] In this context, Reclamation opted for a design that was in full accord with the recommendation of a special Multiple Arch Dam Advisory Committee formed by the California State Engineer in May 1931 (just after submittal of the final state report detailing what would become the Central Valley Project). The final report of this committee, issued in 1932, denigrated multiple arch technology, and, since that time, no new multiple arch dam has been erected in the state.[77]

The design selection process for both Friant and Shasta Dams reflected a perspective that favored concrete gravity dams with a standing that accorded

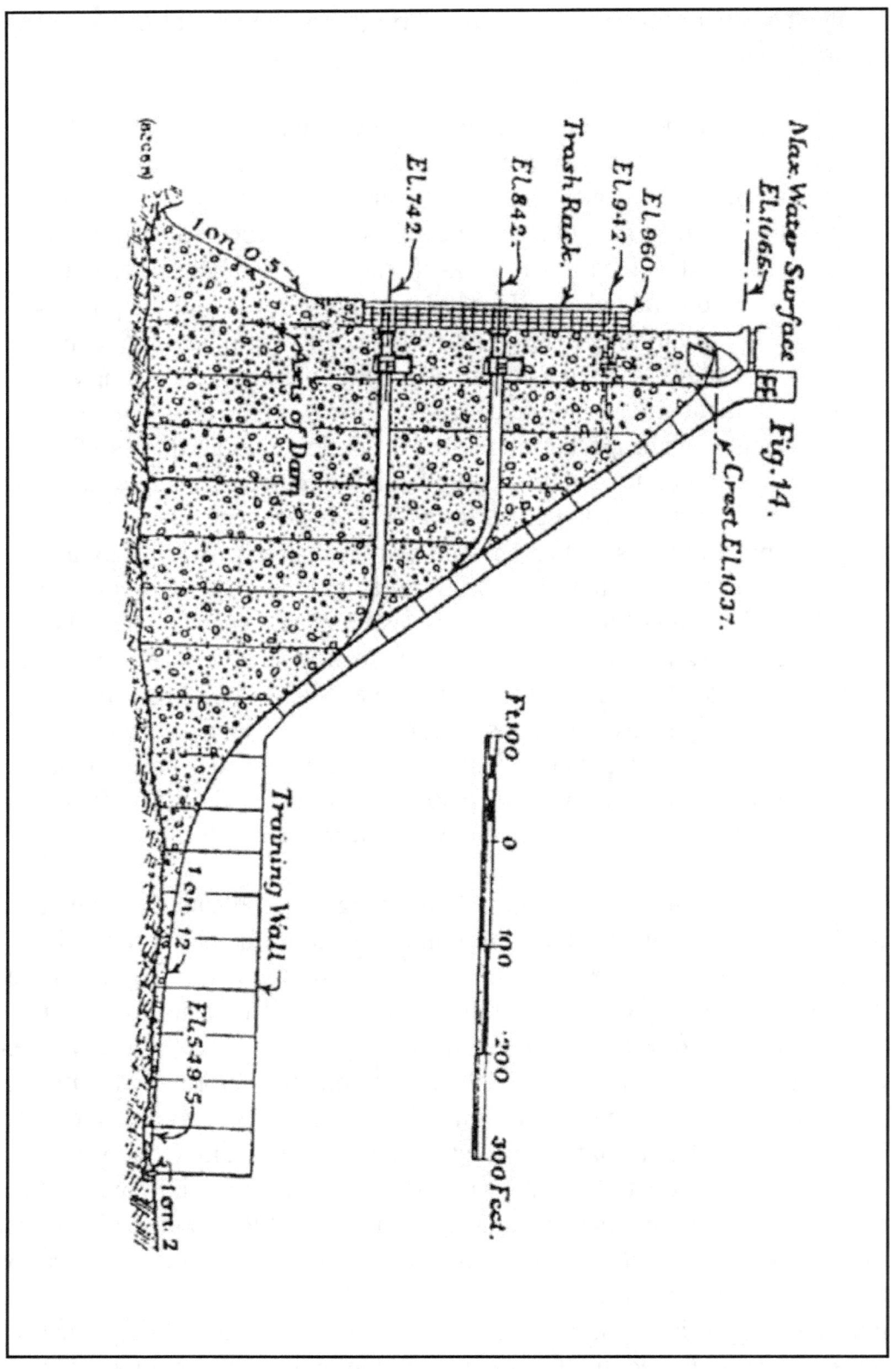

Figure 7-5: Cross-section of Shasta Dam showing the gravity profile. Source: Bureau of Reclamation.

them great strength and stability. As such, there seemed little reason to explore options that appeared more risky, regardless of whether or not they might prove economically advantageous. Although Shasta Dam is built with a distinct upstream curve, this curvature was ignored in determining the design's gravity cross-section. As reported in *Engineering*: "All resistance to loading is dependent upon gravity action and that in working out the design no account was taken of the extra strength provided by the slightly arched form."[78]

Shasta Dam: Planning and Bidding

Once it became clear in the late summer of 1937 that the Bureau of Reclamation would be responsible for building Shasta Dam, events moved quickly in terms of preparing the site and advertising the primary construction contract. By January 1938, money had been allocated to commence work in relocating and reconstructing the Southern Pacific Railroad line that ran directly through the reservoir "take" area.[79] Contemporaneously, work was also started on the establishment of camp communities that could house the thousands of workers (many with families) who would soon be drawn to the site.[80]

By spring of 1938, the Bureau of Reclamation was ready to let the primary construction contract. As detailed in the May 1938 issue of *Western Construction News*, the federal government was to be responsible for providing "processed aggregate to the contractor, in carload lots, as required. This is a different procedure from the corresponding work at Hoover Dam and Grand Coulee Dam, where the securing and processing of aggregate was the work of the contractor." To fulfill this responsibility, Reclamation subsequently let a separate contract for construction of a 9-mile-long conveyor belt system to deliver sand and gravel excavated at a quarry downstream from the dam site.[81]

The primary construction contract called for "a standard form of contract with the U.S. Government, which permits wide latitude in the procuring of labor. Practically the only restriction applies to the maintaining of a 40-hour week, and the wage scale." This wage scale listed more than 100 different job categories (ranging from blacksmith, to blacksmith's helper, to hoist operator, to powderman, to dump truck driver, and a multitude in between) with wages ranging from $.75 per hour (for a general laborer) to $1.50 (for a power shovel operator). The bidding schedule also broke down the various components of the building process into 175 separate tasks/items that were to comprise the means for developing a final, total bid for the job as a whole.[82]

Bids were formally advertised on April 1, 1938, and opened on June 1 in the Bureau of Reclamation's offices in Sacramento. The huge job was beyond the capacity of any single contractor, and, similar to what had transpired at

Hoover and Grand Coulee Dams, it prompted the formation of groups of contractors who perceived an advantage in the pooling of their skills and financial resources in attempting to take on the task. For the Shasta Dam contract, only two bids were filed, one from the Shasta Construction Company (which comprised many of the same firms that had previously formed Six Companies Inc. and successfully built Hoover Dam) and one from Pacific Constructors Inc. Led by L. E. Dixon Company, a prominent Los Angeles contracting outfit, this latter consortium had been organized in 1937 to bid on the Grand Coulee Dam contract. Failing in that effort, the group subsequently expanded in size (taking in the large and successful Arundel Corporation of Baltimore) and focused on developing as competitive a bid as possible to win the Shasta Dam contract. In the end, Pacific Constructors Inc.'s (PCI) efforts to streamline its bid proved prescient because, when the bids were opened in Sacramento in June, PCI's bid of $36,939,450 undercut the Shasta Construction Company by less than $275,000.[83]

After winning the contract, PCI immediately set out to organize itself in order to handle the tremendous job it was now committed to completing. While searching for a general superintendent to oversee construction, it received notice from Frank Crowe indicating his interest in filling this position. Crowe had started work for the Reclamation Service as a junior level engineer in 1905 and, as a government employee had been involved in the construction of several prominent dams including Arrowrock and Tieton. In the 1920s, he left government service to join the Idaho-based contracting firm of Morrison-Knudsen; in this capacity, he came to take charge of building Hoover Dam in 1931, after Six Companies Inc. (which included Morrison-Knudsen as a partner) was low bidder on the construction contract. After completion of Hoover, he took charge of erecting Parker Dam for Six Companies Inc.,

Figure 7-6: Commissioner of Reclamation John C. Page speaking on October 22, 1938, at the ceremony marking the beginning of heavy construction on Shasta Dam. Source: Bureau of Reclamation.

and he would have become the general superintendent for Shasta Construction Company if that firm had won the Shasta Dam contract. But once PCI won the contract, Crowe opted to leave behind his association with Six Companies and Morrison-Knudsen and take on the challenge of building Shasta Dam for PCI.

Although Crowe had not been associated with PCI during the project planning and conceptualizing that went into the development of the final bid, once he became "general superintendent," he played a key and prominent role in formulating how the construction would proceed. In fact, because of his background and the knowledge accrued during the construction of Hoover, Parker, and a multitude of other dams, he was given wide latitude by PCI to administer the construction process. As the company acknowledged upon completion of the dam:

> Pacific Constructors was really a Frank Crowe organization—
> veteran builders who knew their job.... We owe Frank Crowe
> and the entire construction organization a vote of thanks for
> their untiring work, devotion and loyalty, and those of us who
> were in constant touch with the operation give it herewith in full
> measure.[84]

The Construction Process

The building of Shasta Dam involved an enormous amount of work, but it was organized into a relatively few basic types of tasks, including:

1. River diversion
2. Foundation excavation
3. Concrete production
4. Formwork construction
5. Concrete placement

Implementing these tasks could vary significantly, depending upon their particular place in the chronology of construction, but the tasks nonetheless represent a basic way of organizing the construction process.[85]

River diversion was important because, although the dam site is more than two hundred miles upstream from the delta, the Sacramento River carries a significant amount of flow through the site no matter what the season. Of course, the heaviest flow occurs during spring floods, but control of water was always an important component of the project's success. Control and diversion of the river was handled by two complementary means. During the initial stages of

Figure 7-7: This 1942 construction photo shows the right abutment on the downstream face of Shasta Dam, during the placement of penstocks. Source: Bureau of Reclamation.

Figure 7-8: Night view of operations at Shasta Dam in 1941, seen from near crest elevation on the left abutment. Source: Bureau of Reclamation.

construction, the riverflow was concentrated within the deepest part of the river gorge.

Excavation and construction activities focused on the two sides of the structure flanking the gorge. Once the two sides of the dam had assumed sufficient size to withstand possible overtopping without damage, it became possible to switch to the second phase of the diversion plan.

Phase two involved use of the 1,800-foot-long tunnel excavated through the site's west abutment during the initial construction stage. At first, this tunnel was used by the Southern Pacific Railroad line which had been forced to abandon its original right-of-way in the gorge to allow for the commencement of construction. Once a completely new right-of-way around the reservoir had been completed, the tunnel became available for water diversion. This came during the latter stages of construction and when an earth and rockfill cofferdam diverted water out of the streambed and into the tunnel. As a result, the deepest section of the gorge could be "unwatered" and the foundation prepared for the placement of concrete.

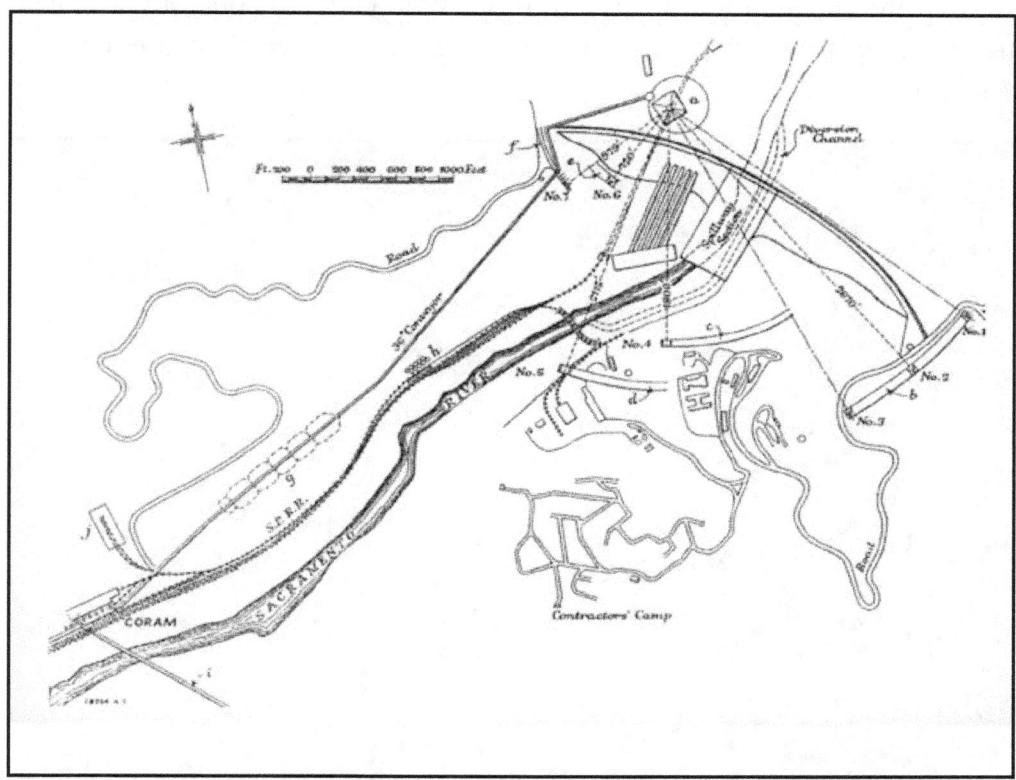

Figure 7-9: Plan view of Shasta Dam showing the construction plant and aerial cableway system. Source: Bureau of Reclamation.

Excavation was important because of the desire to establish the concrete structure on a foundation that was as solid and impervious as possible. To help clear the site of earth and less-than-solid-rock "overburden," some blasting was done. Removal of loosened earth and rock debris was handled by steam shovels, bulldozers, large gasoline-powered dump trucks, and other forms of mechanized equipment. Much of the excavated material was used to build-up the height of the western abutment downstream from the damsite to help support the penstocks that would later be built between the dam and the powerhouse. Once the excavation had proceeded to the point of reaching a solid base of bedrock, extensive efforts were made to drill holes for injection of pressurized grout (i.e., water and cement) into the foundation. Grouting represented an attempt to fill in, as much as possible, any deep foundation cracks or seams that might allow water to seep under the structure and push up on the base of the gravity dam, thus acting to reduce its stability.

Once the foundation had been prepared, workers could begin to erect the wooden formwork that would hold the wet concrete in the proper shape while it hardened. But before this could occur, cement had to be delivered to the site via the Southern Pacific rail line running through the site and aggregate had to be transported via a conveyor belt running from a quarry located nine miles downstream. The cement, aggregate, and Sacramento River water were brought together in a mixing plant upstream from the dam site proper. Great care was taken to insure the quality of the 6.5 million cubic yards of concrete produced during construction of the dam, and it was constantly tested to insure that it met specifications.

Figure 7-10: A March 6, 1942, view of Shasta Dam under construction. Source: Bureau of Reclamation.

Delivery of the mixed concrete—which had to be placed quickly once water had been introduced to the cement—was handled by a huge cableway system built to convey material to various parts of the dam site. This cableway system utilized a large steel headtower[86] just upstream from the dam site, adjacent to the western abutment. This primary tower supported seven separate cableways that spread out and provided coverage of the entire dam site. The concrete mixing plant was next to the headtower and connected to it by a circular rail and tramway system. Cars capable of holding two eight-cubic-yard "loads" of concrete were operated as part of this electric-powered tramway system; the cars would be filled with concrete at the mixing plant and then moved along the circular track to a location where they could be dumped into one of the eight-yard buckets attached to each of the seven cableways. The cableways transported the wet concrete to the placement location of that moment, and the empty cars would return to the mixing plant (via the circular tramway) where they would be filled with another load of concrete, and the process would be repeated.

Figure 7-11: A vertical section construction join with keyways and galleries in Shasta Dam. March 7, 1942. Source: Bureau of Reclamation..

Each bucket had a capacity of eight-cubic yards which meant that, for a structure that required over 6 million cubic yards of concrete to complete, the filling/lifting/dumping process had to be repeated over a half million times during construction. The plant constructed by PCI was designed to mix, transport, and place up to 10,000 cubic yards per day (the equivalent of more than 1,200 batches a day). Placement of the concrete occurred in a very controlled manner. The dam was divided into a series of interlocking blocks measuring 50 feet by 50 feet and shaped by wooden forms. Concrete was poured into each of the blocks in a series of five-foot lifts, and great care was taken to "vibrate" the wet concrete to ensure that it completely filled the forms without leaving any voids or airspaces

Figure 7-12: Shasta Dam with penstocks and powerhouse on May 5, 1976. Source: Bureau of Reclamation.

Figure 7-13. The Shasta Dam spillway. Source: Bureau of Reclamation.

that would tend to weaken the structure. Once the concrete in a lift had hardened sufficiently, usually within 48 hours, the form work was loosened and configured to handle the next five-foot lift of concrete. Because a multitude of blocks were under construction at any one time, work could proceed continuously.

Construction Schedule

Pacific Constructors Inc. was formally awarded the primary construction contract for Shasta Dam in July 1938, and, by the end of that summer, work at the site was well underway. The first tasks involved construction of housing facilities and the blasting and construction of the railroad bypass tunnel through the west abutment. During 1939, the bulk of work focused on the blasting and excavation of the foundation and preparation of the myriad facilities necessary to carry out construction. In October 1939, PCI and Superintendent Crowe began overseeing erection of the 460-foot-high steel headtower that would support the seven cableways to be used to transport concrete. The headtower was not complete until July 1940; and placing of concrete commenced that month. By August 1942, four million cubic yards had been placed, and, by February 1943, five million cubic yards had been placed. By the summer of 1943, the structure was approaching its final shape, and no water flowed through the dam site except through the tunnel/bypass diversion tunnel. In February 1944, the contractor blocked flow through the diversion tunnel, and water storage began in Lake Shasta. The last bucket of concrete was poured on January 2, 1945, and, during the next few months, PCI handled clean up and salvage of equipment.

The contractor completed Shasta Dam at the beginning of 1945. The structure began operating as soon as the diversion tunnel was closed and natural flow of the river came under the control of Reclamation's operators, who determined when (and to what extent) water would be released from the reservoir. In terms of hydroelectric power production, completion occurred in 1950 when the 375,000 kW generating plant came on line, and the dam was officially declared finished. But, by

Figure 7-14: An aerial view of Shasta Dam. Source: Bureau of Reclamation.

341

this time, the dam had been functioning as a civil engineering structure for several years.

Friant Dam Construction

Although Friant Dam was a structure of major proportions, compared to Shasta Dam it was relatively small (approximately 2.15 million cubic yards of concrete vs. 6.5 million cubic yards), and it attracted much less attention than the structure to the north. In addition, the start of construction at Friant was delayed until October 1939 (more than a year after Shasta) and occurred after the Nazi invasion of Poland, which marked the beginning of World War II in Europe. Taken in aggregate, all these factors contributed to its lower public profile. Nonetheless, Friant was still a significant engineering project that used a sophisticated construction plan.[87]

Friant Dam is across the San Joaquin River approximately twenty miles northeast of downtown Fresno. Authorization for the Bureau of Reclamation to build Friant Dam occurred in 1935, two years before Reclamation received authorization to build Shasta, but planning lagged because, as phrased by Bureau of Reclamation Commissioner John Page in the summer of 1938:

> After long and arduous labor, we have brought the Miller & Lux matter to a point where a conclusion of these negotiations [over water rights issues] seems to be imminent. . . . In all the history of the Bureau of Reclamation, we have had no project involving so many and so complex legal problems as those encountered on the Central Valley Project. . . .[87]

Finally, more than a year later, the contracting partnership formed between the Griffith Company and Bent Brothers of Los Angeles, was authorized to begin work based on its low bid on the primary construction contract. The first steps taken were to excavate and clear the dam site of earth and rock that lay atop bedrock. Once a significant portion of the foundation along the southern side of the site had been cleared, "a temporary diversion flume was built across the foundation excavation and the river diverted into the flume in the late Spring of 1940." This allowed the river section of the dam site to be cleared, and, once the foundation had been properly prepared, this section was where the first concrete was placed. Three 14-foot-diameter temporary openings were left in the lower part of the structure that arose within the river section. Subsequently, the contractor diverted the river through these temporary openings beginning in March 1941; six months later the overall dam had been completed to the point that the river was diverted into the permanent outlets built into the base of the dam and the temporary openings were filled with concrete. As *Civil Engineering* soon reported, during the construction process "no special difficulties were encountered in handling the river."[88]

After "light blasting," 14-cubic-yard self-loading scrapers cleared topsoil from the dam site. Once harder rock was encountered, jackhammers drilled into the foundation and facilitated the blasting of weakened or "rotten" rock. This material was removed by diesel-powered shovels (varying in capacity from 1.25 to 5 cubic yards) that filled 8-cubic-yard capacity gasoline dump trucks. These trucks carried most of the debris to deep portions of the reservoir site that lay below the elevation of the lowest outlet pipes, and some was stored downstream for use in backfilling and grading once the concrete structure was complete. Before concrete was placed upon the foundation, an extensive program of grouting was undertaken that forced a rich, fluid cement (under a maximum pressure of 150 pounds per square inch) deep into bedrock. With the grouting of 725 holes (requiring more than 21,500 sacks of cement) an attempt was made to minimize water seepage through the foundation and eliminate any chance that "uplift" might destabilize the massive concrete structure.

Figure 7-15: Friant Dam under construction showing the trestle and crane construction system used to deliver concrete and form work used in building the dam. Source: Donald C. Jackson collection, courtesy of John Snyder.

Cement for the concrete was delivered to the site via a spurline of the Southern Pacific Railroad. Aggregate (both sand and carefully sized gravel) was taken from a quarry along the banks of the San Joaquin River about three miles downstream from the dam site and delivered via a special railroad built and operated by the contractor. Water was drawn from the river and precooled in special refrigerating equipment that helped to reduce the amount of heat generated while the wet concrete "set." Concrete was mixed in a plant on the south side of the San Joaquin River about 200 feet downstream from the dam site proper. Delivery of freshly mixed, wet concrete was handled by four-cubic-yard buckets that fit into special diesel-electric railcars. These cars moved along the massive steel trestle that extended across the entire length of the dam. Sitting atop the trestle were two 294-foot-long "hammerhead" cranes and two revolver cranes with 137-foot-long booms. The cars moved under the cranes and the buckets were lifted and transported to wherever the concrete placing operations were underway at that moment. Dual trackage allowed movement of the empty car back to the mixing plant without interfering with the movement of filled buckets to other cranes.

In a system similar to that employed at Shasta, concrete was poured in discrete fifty-foot-wide blocks formed in a sequence of five-foot lifts. After the wet concrete was allowed to harden for a minimum of twenty-four hours during the summer and thirty-six hours during the rest of the year, wooden panels used to form these blocks were reused. The first bucket of concrete was placed on July 20, 1940, and work proceeded at a steady rate for the next two years. The daily record for concrete placement totaled 9,059 cubic yards, and in August 1941, the monthly peak reached a little over 228,000 cubic yards. By January 1942, the structure was 90 percent finished, and the contractor completed it that summer, less than three years after concrete placement began.

Contemporaneously with the construction of Friant Dam, canals were built to carry water to agricultural users to the north via the Madera Canal and to the south via the Friant-Kern Canal. Because these canals were designed to transport the bulk of water stored behind the dam, no hydroelectric powerplant was erected in conjunction with the dam. It took several years before the Delta-Mendota Canal delivered water to the upper San Joaquin watershed and began to replenish the natural flow that was now being diverted southward toward Bakersfield. By the time this component of the long anticipated "water swap" occurred, Friant Dam had long since proved itself a formidable and imposing component of California's hydraulic infrastructure.

Aftermath

Upon completion of Shasta Dam in early 1945, the first phase of the construction involved in implementing the Central Valley Project came to completion. In the ensuing decades, various components of the project were brought on-line (including the Delta-Mendota Canal, which delivered water up the San

Joaquin River and effectuated the "water swap" that sent water stored behind Friant Dam southward toward Bakersfield, the Contra Costa Canal that delivered Sacramento River water to residential and agricultural communities east and north of San Francisco Bay, and the Trinity River Diversion that delivered water into the upper Sacramento River that otherwise would have flowed westward into the Pacific Ocean). Technically, the Central Valley Project functioned with remarkable efficiency and generated no controversy relating to the quality of its design and construction. But in a political context, the project attracted continued attention and controversy.

Figure 7-16: Friant Dam and Millerton lake on July 10, 1980. Source: Bureau of Reclamation.

The focus of this attention concerned the "acreage-limitation" restrictions that accompanied federal authorization of the project in 1937. Because this authorization included the specific stipulation that the operation adhere to "reclamation law," it appeared, to some people, that any farmer or agricultural producer who benefitted from water supplied by the project would be constrained to use this water on only a very limited amount of land (i.e., a maximum of 320 acres for a wife and husband partnership). But many large land-owners were involved in agricultural production in the region served by the project, and they viewed the reclamation law stipulation in a different light.[89]

In many ways, the struggle that ensued over the applicability of "reclamation law" simply reflected conflicts dating back to the 1870s and 1880s when Miller & Lux fought to protect their riparian rights in the face of upstream appropriators seeking to develop agricultural land focused around smaller-scale farms. The struggle also reflected the tension that existed within the corpus of water law, a tension intensified by the fact that water rights were essentially a matter of state law and, regardless of how the federally administered Central Valley Project might have introduced new and valuable quantities of water into the region's agricultural economy, significant water rights pre-dated implementation of the Central Valley Project. This complicated any easy resolution to the problem of trying to enforce an "acreage limitation" requirement.

Resistance to the Bureau of Reclamation's efforts to invoke federal control over the size of farms eligible for CVP water came even before the formal authorization of the project in 1937. Specifically, it was evident in the exclusion of the Kings River and associated agricultural areas from inclusion in the CVP. During the 1920s, the Kings River area had been considered within the State Water Plan investigated by the State Engineer's Office, and water from the Kings River was included as part of the north-south "water swap." But by the late 1930s, agricultural interests in the Kings River Valley perceived association with the larger CVP as problematic, if not outright undesirable, and they opted to seek other ways of building the Pine Flat Dam that would store the flood waters of the Kings River.

By 1940, it became evident that landowners in the Kings River watershed perceived the U.S. Army Corps of Engineers as a more financially attractive federal program in terms of dam building and reservoir operation; a key attraction of the Corps' flood control projects was that they were 100 percent funded by the federal government. Over the course of the next decade, the Bureau of Reclamation and the Corps struggled over the issue of who would build the Pine Flat Dam and whether or not use of water resources developed as a result of this large-scale concrete gravity dam would be governed by "reclamation law" and the "acreage-limitation" restriction. In the end, the Corps won the battle to build Pine Flat Dam, although language in the authorizing legislation called for adherence to the principles of "reclamation law" and signaled that California's large landowning interests were not content to accept federal restrictions without a fight.[90]

Resistance to Reclamation's insistence on enforcement of the "acreage limitation" also became manifest in the late 1940s with the publication of Sheridan Downey's *They Would Rule the Valley*, an artfully written diatribe that castigated the federal government as some kind of socialist invader of the American heartland which disrupted sacrosanct patterns of private property rights in the name of bureaucratic empire building. As one of California's two U.S. Senators, Downey brought significant attention to the problems inherent in trying to administer a federal reclamation project in the midst of large agricultural interests who had access to water resources bearing no association to the federal program. He especially highlighted how groundwater pumping complicated any simple approach to allocating water within the San Joaquin Valley. His book served as a harbinger of efforts in the 1950s to have California "buy-back" the Central Valley Project from the federal government, an effort that eventually proved ineffectual.

The uneasy legacy of the federal-state partnership that spawned construction of the Central Valley Project eventually found expression in important legislation governing western water development. First, in planning for construction of what came to be known as the California State Water Project in the 1950s, California's legislators opted to avoid any involvement with the federal government in the huge undertaking. Rivaling (if not surpassing) the Central Valley Project in scope and significance, the California State Water Project is responsible for transferring huge quantities of water from behind Oroville Dam, on the Feather River north of Sacramento, hundreds of miles southward to greater Los Angeles and San Diego. As part of this project—which was largely completed in the 1970s—water is pumped more than 3,000 feet in height over the Tehachapi Range at the southern end of the Central Valley to reach metropolitan users in Southern California. Water from the project is also used to support agriculture in the southern reaches of the San Joaquin Valley, and it was largely for this reason that state politicians wished to avoid any association with the federal government—and "reclamation law"—in financing the project.

Concern over "acreage restrictions" led to important legislative action in 1982 when, during Ronald Reagan's first term as President (and subsequent to his tenure as a two-term governor of California), legislation was enacted that increased the "acreage limitation" for federal reclamation projects from 160 acres per individual to 960 acres per individual. Although this dramatic adjustment did not completely stem the controversy over the participation of agribusiness in federally financed and subsidized water development projects, it made it much easier for the *status quo* system of large-scale farming to be accommodated in federal reclamation projects and to participate in the benefits of the Central Valley Project without concern over possible conflict with "reclamation law."[91] With this legislation, the system of regional water development—with origins that could be traced back to the 1870s—finally attained a legal standing that acknowledged the economic power and political standing of western agricultural interests. These agricultural interests operated in a much different political and economic arena than they had in 1902, and they represented a distinctly separate cultural strain far different from the "yeoman farmers" who figured so prominently in rhetoric championing irrigation, reclamation, and the social value of "making the desert bloom." The new acreage limitation represented a major evolution in American society and in the reclamation program.

Figure 7-17: Drying almonds at the Banes Almond Orchard on the Orland Project, California in 1916. Source: Bureau of Reclamation.

Endnotes

1. The agricultural history of California—with special emphasis on the role of irrigation in this history—is well described in Donald J. Pisani, *From the Family Farm to Agribusiness*.
2. Fred Powledge, *Water: The Nature, Uses, and Future of Our Most Precious and Abused Resource* (New York: Farrar Straus Giroux, 1982), 120-1. In the early 1980s, California's average annual water consumption from all sources was approximately 36 million acre-feet with 32 million acre-feet, (85 percent) devoted exclusively to agriculture. This percentage has been dropping gradually but steadily as urban growth continues to increase.
3. Pisani, *From the Family Farm to Agribusiness*, 5-11, describes the state's early wheat industry.
4. A good reference on early California mining dams is A. J. Bowie, *A Practical Treatise on Mining in California* (New York: D. Van Nostrand, 1905), 94-113.
5. For example, the masonry arch Bear Valley Dam was completed in 1884 to store water for irrigation in the San Bernardino/Redlands region of Southern California. See Brown, "The Bear Valley Dam," 513-4.
6. Irrigated agriculture in Southern California is well described in Carey McWilliams, *Southern California Country: An Island on the Land* (New York: Duell, Sloan, & Pearce, 1946).
7. The essential topographic character of the Central Valley is well described in U.S. Congress, House of Representatives, *U.S. Board of Commissioners on the Irrigation of the San Joaquin, Tulare and Sacramento Valleys of the State of California,* 4, 3rd Congress, 1st Session, 1874, H. Ex. Doc. 1, pt. 2, Serial 1745. This report was reprinted by the U.S. Army Corps of Engineers in 1990 as Jackson, et. al, *Engineers and Irrigation*. Subsequent page references to this report will correlate with the 1990 reprint edition.

8. The notion that the Central Valley helped to "Move The Rain" is specifically noted in the title of the initial chapter of the 1945 book *Shasta Dam and its Builders* (San Francisco: Schwabacher-Frey Company, 1945), written and distributed under the aegis of Pacific Constructors, Inc. The chapter is called "The Central Valley Project 'Moves the Rain.'"

9. In 1920, Colonel Robert Bradford Marshall, Chief Hydrographer of the United States Geological Survey in California, began promoting a plan to more fully utilize the waters of the Central Valley. Acting as a private citizen rather than as a representative of the Geological Survey, Marshall's proposal was published in the pamphlet "California's Greatest Opportunity," distributed by the privately-financed California State Irrigation Association. The text of this pamphlet is reprinted in U.S. Congress, House of Representatives, Clair Engle, Chairman, Committee on Interior and Insular Affairs, *Central Valley Project Documents, Part One: Authorizing Documents* (Washington, D.C.: Government Printing Office, 1956), 139-150. [84th Congress, 2nd Session, H. Doc. 416]. Subsequent reference to this publication will be: *Central Valley Project Documents*.

10. The history of the Miller & Lux ranching and agricultural empire is sympathetically portrayed in Charles Treadwell, *The Cattle King* (Santa Cruz, California: Western Tanager Press, 1981).

11. Jackson, et. al, *Engineers and Irrigation*, 144.

12. The demise of Ralston's financial empire is recounted in Pisani, *From the Family Farm to Agribusiness*, 118-9, and in Jackson, et. al, *Engineers and Irrigation*, 3.

13. The proliferation of irrigation colonies around Fresno is described in Pisani, *From the Family Farm to Agribusiness*, 122-5. Also see U.S. Department of Agriculture, *Report of Irrigation Investigations in California* (Washington D.C.: Government Printing Office, 1901), 259-326.

14. The history of Miller & Lux's efforts to defend their claim to riparian water rights is the focus of M. Catherine Miller, *Flooding the Courtrooms: Law and Water in the Far West* (Lincoln: University of Nebraska Press, 1993).

15. For more on the significance of the ruling in *Lux v. Haggin*, see Pisani, *From the Family Farm to Agribusiness*, 191-249. Robert Dunbar, *Forging New Rights in Western Waters*.

16. The history of the Wright Act and subsequent development of the irrigation district as a viable component of California's political economy is documented in Pisani, *From the Family Farm to Agribusiness*, 250-82, 352-8.

17. The complexity of water rights litigation in the San Joaquin Valley at the start of the twentieth century—and the prominent role played by Miller & Lux in this litigation—is made clear in Jackson, et.al, *Report of Irrigation Investigations in California*, 237-46.

18. The political firestorm that erupted from the accumulation of debris in northern California's rivers is recounted in Robert Kelley, *Gold vs. Grain: The Mining Debris Controversy* (Glendale, California: Arthur Clarke, 1959).

19. For the history of the California Debris Commission, see Robert Kelley, *Battling the Inland Sea: American Political Culture, Public Policy, and the Sacramento Valley, 1850-1886* (Berkeley: University of California Press, 1989), and Joseph J. Hagwood Jr., *The California Debris Commission: A History* (Sacramento: U.S. Army Corps of Engineers, 1981).

20. From the "Jackson Report," as quoted in Hagwood, *California Debris Commission: A History*, 52.

21. A good synopsis of the Corps' and the Debris Commission's involvement in flood control and navigability efforts along the Sacramento River from the 1890s through the 1920s is provided in Hagwood, *California Debris Commission: A History*, 49-56.

22. The troubled histories of the Yuma and Klamath Projects are described in Pisani, *From the Family Farm to Agribusiness*, 313-24.

23. The story of the federal government's initial interest in, and subsequent abandonment of, a reclamation project in the Owens Valley in 1904-1906 is best told in William Kahrl, *Water and Power*.

24. The early history of the Orland Project is succinctly summarized in Pisani, *From the Family Farm to Agribusiness*, 329-31.

25. U.S. Department of the Interior, Bureau of Reclamation, *Dams and Control Works* (Washington, D.C.: Government Printing Office, 1938), 76.

26. Bureau of Reclamation, *Dams and Control Works*, 76-84.

27. Pisani, *From the Family Farm to Agribusiness*, 331-4, describes the issues surrounding the conception and ultimate demise of the Iron Canyon Dam project.

28. Pisani, *From the Family Farm to Agribusiness*, 35-54.

29. U.S. Department of the Interior, U.S. Reclamation Service, in cooperation with Iron Canyon Project Association, *Report on Iron Canyon Projects by the Office of the U.S. Reclamation Service at Portland, Oregon, October, 1914*, and Report by Board of Review Appointed by the Secretary of the Interior, November, 1914, 14 and 53-4.

30. *Report on Iron Canyon*, 62-4. The report summarized conditions as follows: "The situation may be briefly summed up thus: A porous underbody of unknown depth capped by fairly tight volcanic tuffs and agglomerates from 40 to 60 feet thick in the main gorge. . . ." (64).
31. *Report on Iron Canyon*, 71.
32. *Report on Iron Canyon*, 71.
33. *Report on Iron Canyon*, 65.
34. Homer J. Gault and W. F. McClure, *Report on Iron Canyon*. Gault was an engineer with the Reclamation Service while McClure was the California State Engineer. The report is dated May 1920, but it was not published until 1921.
35. Gault and McClure, *Iron Canyon Project*, 23.
36. Gault and McClure, *Iron Canyon Project*, 62-9.
37. Gault and McClure, *Iron Canyon Project*, 68.
38. Walker Evan's October 1925 Bureau of Reclamation report entitled "Report on Iron Canyon Project — California" was printed as chapter five in Paul Bailey's, *The Development of the Upper Sacramento River* (Sacramento: California State Printing Office, 1928).
39. Pisani, *From the Family Farm to Agribusiness*, 331-4, discusses the Iron Canyon Project and its eventual demise.
40. Pisani, *From the Family Farm to Agribusiness*, 383.
41. Pisani, *From the Family Farm to Agribusiness*, 384-6.
42. Pisani, *From the Family Farm to Agribusiness*, 382.
43. Pisani, *From the Family Farm to Agribusiness*, 154-90, describes the work—and ultimate demise—of William Hammond Hall as California's State Engineer between 1878 and 1888.
44. Pisani, *From the Family Farm to Agribusiness*, 364-80, describes the origins and activities on the State Water Commission (renamed the Division of Water Rights in 1921).
45. Pisani, *From the Family Farm to Agribusiness*, 394-6, describes Marshall's early career and his reputed inspiration for conceiving his "plan."
46. See *Central Valley Project Documents*, 139-50, for a complete version of the Marshall Plan text as printed in 1920 by the California State Irrigation Association.
47. Pisani, *From the Family Farm to Agribusiness*, 398.
48. Early opposition to the Marshall Plan is described in Pisani, *From the Family Farm to Agribusiness*, 400-1.
49. Pisani, *From the Family Farm to Agribusiness*, 438-9, describes the latter part of his life and manner in which he was separated from anything having to do with the Central Valley Project prior to his death in 1949.
50. The 1921 legislation authorizing a comprehensive study of California's water resources is reprinted in *Central Valley Project Documents*, 150-3.
51. Pisani, *From the Family Farm to Agribusiness*, 403.
52. Pisani, *From the Family Farm to Agribusiness*, 408-10. Central Valley Project Documents, "Investigations of the State Engineer" (Bulletin No. 4) 150-9.
53. Pisani, *From the Family Farm to Agribusiness*, 411-2. This report is synopsized in *Central Valley Project Documents*, "Investigations of the State Engineer (Bulletin No. 9)," 159.
54. The significance of *Herminghaus v. Southern California Edison Company* is explicated in Pisani, *From the Family Farm to Agribusiness*, 412-5.
55. The public reaction to the Herminghaus decision was reflected in the 1928 enactment of an amendment to the California State Constitution that required riparian water rights to adhere to the same types of "reasonable" use that governed appropriated water rights claims. See Pisani, *From the Family Farm to Agribusiness*, 412-5. Miller, *Flooding the Courtrooms*.
56. Pisani, *From the Family Farm to Agribusiness*, 416-7. A detailed synopsis of the 1927 report is provided in *Central Valley Project Documents*, "Investigations of the State Engineer" (Bulletin No. 12), 168-83.
57. Pisani, *From the Family Farm to Agribusiness*, 417, specifically notes that "most members of the legislature assumed that the federal government should pay at least part of the project's cost, if only for flood control and navigation benefits." In the federal legislation authorizing construction of the Boulder Canyon Project in late 1928, $25 million of the construction cost was covered by direct federal aid for flood control.
58. Pisani, *From the Family Farm to Agribusiness*, 419-22.
59. Paul Bailey, *Development of the Upper Sacramento River*, investigations of the State Engineer 50-1.
60. Bailey, *Development of the Upper Sacramento River,* 14, 53.
61. Bailey, *Development of the Upper Sacramento River,* 47.
62. Bailey, *Development of the Upper Sacramento River,* 47.

63. Bailey, *Development of the Upper Sacramento River*, 47-9. The report recognized the potential danger posed by uplift and specifically stipulated that, "drain pipes along the full length of the dam downstream from the grout holes would connect through vertical pipes to drainage tunnels in the dam . . . under drains would relieve upward water pressure from percolating water below the concrete lining should there be any," see page 48.

64. *Central Valley Project Documents*, "Federal-State Commission on Water Resources, 1930," 249-52. This commission was known as the "Hoover-Young Commission" because appointments were made by President Herbert Hoover and California Governor C. C. Young.

65. *Central Valley Project Documents*, "State Water Plan (Bulletin No. 25), 1931," 256-349. Pisani, *From the Family Farm to Agribusiness*, 434.

66. Pisani, *From the Family Farm to Agribusiness*, 434-6.

67. *Central Valley Project Documents*, "Report of the Corps of Engineers (H. Doc. No. 791, 71st Congress), 1931," 363-97.

68. *Central Valley Project Documents*, "Bureau of Reclamation, Bissell Report, 1931," 397- 403. Quote from page 403.

69. See *Central Valley Project Documents*, "Report of the Corps of Engineers (H. Doc. No. 191, 73rd Congress), 1933," 505-35, 549. "Report of the Corps of Engineers (H. Doc. No. 35, 73rd Congress), 1934," 544-55.

70. This was later known as the Public Works Administration.

71. *Central Valley Project Documents*, "Rivers and Harbors Act (Initial Authorization of Central Valley Project Works), 1935," 558-9.

72. *Central Valley Project Documents*, "President Roosevelt to the Secretary of the Treasury, September 10, 1935," 559.

73. *Central Valley Project Documents*, "Rivers and Harbors Act (Reclamation Project Authorization)," 568-9.

74. "Memorandum to the Press, November 13, 1937" Box 165, Project Correspondence File, 1930-1945, Folder 301.1, Dams and Reservoirs, Shasta Dam - 1938, Bureau of Reclamation, RG115, National Archives, Denver.

75. "The Shasta Dam on the Sacramento River, California," *Engineering* 157 (January 7, 1944), 2-3.

76. S. O. Harper to John Page, January 15, 1937, National Archives, Denver, RG115, Entry 7. Bureau of Reclamation Project Correspondence File, 1930-1945, Box 163, Folder: 301.1, Dams and Reservoirs, Friant Dam through 1939, Bureau of Reclamation, Record Group 115, National Archives, Denver.

77. See Jackson, *Building the Ultimate Dam*, 241-2, for more on the origins and influence of this committee.

78. "The Shasta Dam on the Sacramento River, California," 42.

79. "Program for Beginning Construction on the Central Valley Project," *Western Construction News* (January 1938).

80. "Review of Shasta Dam Plans, "*Western Construction News* (May 1938); reprinted in Allen, *Redding and the Three Shastas*, 77.

81. "Review of Shasta Dam Plans," *Western Construction News* (May 1938), 79.

82. Pacific Constructors, Inc., *Shasta Dam and its Builders*, 43, 46.

83. The description of the construction process draws from a variety of sources including: *Shasta Dam and its Builders*. Al M. Rocca, *America's Shasta Dam: A History of Construction, 1936-1945* (Redding, California: Redding Museum of History, 1995). Ralph Lowrey, "Construction Features of Shasta Project Part 1: General Layout and Aggregate Production," *Civil Engineering* 11 (June 1941). Ralph Lowrey, "Construction Features of Shasta Project Part 2: Conveying Aggregate, Mixing and Placing Concrete," *Civil Engineering* 11 (July 1941). C. S. Rippon, "Methods of Handling and Placing Concrete at Shasta Dam," *Journal of the American Concrete Institute* 14 (September 1942). "The Shasta Dam on the Sacramento River, California," *Engineering* 157 (January 7, 1944). "The Shasta Dam on the Sacramento River, California," *Engineering*, 157 (January 21, 1944). "The Shasta Dam on the Sacramento River, California." *Engineering* 157 (February 4, 1944). "The Shasta Dam on the Sacramento River, California," *Engineering* 157 (February 25, 1944).

84. The description of the construction process employed at Friant is taken from the following sources: R. B. Williams, "General Features of Friant Dam," *Civil Engineering*2). C. T Douglas, "Concrete Mixing Plant for Friant Dam," *Civil Engineering* 12 (April 1942). D. S. Walter, "Construction Plant and Methods—Friant Dam," *Civil Engineering* 12 (May 1942).

85. John C. Page to B. W. Gearhart (House of Representatives), July 27 1938, National Archives, Denver, RG 115, Entry 7, Bureau of Reclamation Project Correspondence File, 1930-1945, Central Valley 301. Box 152, Folder 301, Central Valley Engineering Reports, Jan. 1938-Dec. 1938, Bureau of Reclamation, RG115, National Archives, Denver. The Miller and Lux interests held large tracts of land in the lower San Joaquin River Valley that retained riparian water rights. See more on their early history in earlier parts of this chapter.

86. Hoover Dam and most other large concrete gravity arch dams to this time used cableway systems between the canyon abutments of the dam for purposes of construction. Because of configuration of the canyon at Shasta that was not possible. Frank Crowe, and his nephew, John Crowe, both trained civil engineers, created the idea of using the headtower system to support seven cableways.

87. Williams, "General Features of Friant Dam," 83.

88. The conflict over the acreage limitation is discussed in Erwin Cooper, *Aqueduct Empire: A Guide to Water in California, its Turbulent History and its Management Today* (Glendale California: Arthur H. Clark Company, 1968), 154-61. Also, see Hundley, *The Great Thirst*, 257-68.

89. The battle between the Corps and Reclamation over construction of Pine Flat Dam is discussed in considerable detail in Arthur Maass, *Muddy Waters: The Army Corps of Engineers and the Nation's Rivers* (Cambridge, Massachusetts: Harvard University Press, 1951). The controversy is also summarized in Marc Reisner, *Cadillac Desert*, 181-7.

90. Hundley, *The Great Thirst*, 380-5.

CHAPTER 8

DAMS FOR NAVIGATION AND FLOOD: TYGART AND MAINSTEM DAMS ON THE OHIO, UPPER MISSISSIPPI, AND TENNESSEE RIVERS

The Ohio River Floods and Tygart Dam

On the Ohio River and the upper Mississippi from Cairo to Minneapolis, the dams of the twentieth century have all been oriented primarily to navigation. Keokuk, on the Mississippi, is a multipurpose dam (navigation and hydroelectric power) built privately, but the many locks and dams on these mainstreams were conceived as single purpose structures. Tygart Dam, begun in 1934 and completed in 1939, was a pioneering project for the U.S. Army Corps of Engineers because its goal was to build a reservoir that would prevent late winter flooding and provide for late summer navigation. It was, for the Corps, a new type of dam, truly multipurpose (yet without power generating facilities) and built higher than any of its previous dams. Many issues of dam design and reservoir operation had to be thought out afresh at Tygart.

The Ohio River Valley, running from the headwaters of the Allegheny and Monongahela Rivers to the junction at Cairo, Illinois with the upper Mississippi is a major water transportation system carrying (in 1992) over forty percent of the nation's inland waterborne commerce. Its 2,584 miles of navigable rivers comprise only eleven percent of the national system, but its position within the vast coal fields of Appalachia and a heavy industrial region give it economic prominence out of all proportion to size. The Ohio River, itself, is 981 miles long and drains an area of 209,900 square miles. Over twenty-six million people in parts of fouteen states live in its drainage basin.[1]

The maximum flow at Pittsburgh was estimated in an early 1927 article to be 440,000 cubic feet per second, with a minimum of about 1,600 cubic feet per second, whereas at Cairo the maximum estimate was 1,500,000 cubic feet per second and the minimum 27,500 cubic feet per second. In the major flood of 1937, the flow into the Mississippi at Cairo reached almost two million cubic feet per second, making the Ohio the major arm of the Mississippi.[2]

The flooding of the Ohio River had caused more costly damage than occurred in any other river basin, culminating with the floods of 1935 and 1936. Pittsburgh was particularly hard hit. Of the 107 people killed nationwide in the March 1936 flood, forty-six were in Pittsburgh, where nearly 3,000 buildings were damaged or destroyed. The Golden Triangle (the point of land where the

Monongahela and Allegheny Rivers meet to form the Ohio) was under sixteen feet of water. The river crested at forty-six feet above normal flow at Pittsburgh, surpassing all previous measured heights.[3] These disasters helped lead Congress, after long debate, to pass the 1936 Flood Control Act.[4]

One immediate result of the 1936 Act for the Corps was authorization for construction of nine, high, flood control dams to protect Pittsburgh. In 1937, work began on two dams, and in 1972, during Hurricane Agnes, those projects, plus Tygart Dam, prevented estimated damages that were higher than the total cost of construction.[5] This flood control mission was still new to the Corps in the 1930s especially as it conflicted with its traditional mission of river navigation. To see that change most clearly requires an appreciation of the beginnings of the Corps' river activity, which initially centered on the Ohio Basin and led to new forms in mainstem dams.

Figure 8-1: Tygart Dam construction near Grafton, West Virginia, on July 6, 1936. Source: Records of the U.S. Army Corps of Engineers, Record Group 77, National Archives and Records Administration, College Park, Maryland.

Figure 8-2: This trestle, shown on April 26, 1936, provided quick access for concrete placement during construction. Source: Records of the U.S. Army Corps of Engineers, Record Group 77, National Archives and Records Administration, College Park, Maryland.

Navigation from New Orleans to Monongahela

The Supreme Court decision in *Gibbons v. Ogden*, the growing clout of newly created states in the Trans-Appalachian West, and President Monroe's continuous vacillation on the subject encouraged Congress to assume a more activist role in issues dealing with interstate commerce. As we have seen in chapter one, within three weeks' time in 1824, Congress passed both the General Survey Act and an act authorizing the removal of navigation obstacles on the Mississippi and Ohio Rivers. A few months after passage of the second act, the U.S. Army Corps of Engineers was building wing dams to improve navigation on the Ohio River by allowing a two and one-half foot channel for shallow-draft steamboats. This was soon followed by the construction of snagboats to remove floating timber and limbs sticking out of the water. Then in 1879 the Corps embarked on a program to build locks on the Ohio to establish a six-foot channel. By 1929, the Ohio had been converted to a nine-foot navigation channel 'staircase" of locks, dams, and navigation pools that resulted in a nine-foot channel. So, in 1824, the Corps began to clear the Ohio and Mississippi of debris and to open a two-and one-half foot channel for shallow-draft steamboats. In 1879, the Corps had embarked on a program to build locks and dams so that the Ohio, by 1929, had become more a series of narrow lakes than a free running river. With industries such as steel and coal growing rapidly, river transportation was crucial to the region.[6]

The 308 reports gave the Corps the opportunity to review navigable rivers in terms of flood control and power as well as navigation, and the Ohio River 308 report, published in 1935, laid out these issues, eliminated irrigation as unnecessary for the rainy region, and presented a comprehensive plan for developing the entire basin. The report was essentially completed before the new administration took power in early 1933.[7] The Corps had expected the 308 reports to lead to its control over developments in all major river basins where navigation played a part, except the Colorado, which was left for the Bureau of Reclamation.[8] TVA presented a deep threat not only to the Corps' work on the Ohio Basin, but also, as described in chapters 5 and 6, in other major basins where large, multipurpose dams seemed feasible.

To understand the situation in 1933 when the Corps lost the opportunity to design and build control over multipurpose dams in the Tennessee Valley and when it began to design Tygart Dam, we need to go back into the history of flooding in the Ohio River Basin.

Floods, Droughts, and Power

Floods had caused major property damage in the vast Mississippi River Basin throughout the nineteenth century. The solution established by the U.S. Army Corps of Engineers, following the huge Humphreys and Abbot report of 1861, had been to build higher and higher levees. Charles Ellet Jr., a brilliant civilian engineer, produced a report in 1852 calling for a combination of levees and reservoirs on tributaries to control floods. It took many years for the federal government to begin taking the idea of reservoirs seriously.[9]

In 1889, the South Fork Dam at Johnstown on the Conemaugh River (a tributary to the Allegheny River) had collapsed because of overtopping, killing over 2,000 people. A "levees only" policy still seemed best to the Corps. But the disastrous flood of 1907, which nearly destroyed Pittsburgh's center, led to the creation of the Pittsburgh Flood Commission, and, in 1912, its report returned to Ellet's idea of a diversified flood control program that included dams and reservoirs. The next year, the Ohio flooded wildly, causing damages of $142 million, but damages on the lower Mississippi were only $61 million. The populous regions in the north began to clamor for flood control measures.[10]

The most telling early example of Ellet's idea for dams and reservoirs on tributaries came in the Great Miami River Basin, a part of the Ohio River Valley in southwest Ohio just above Cincinnati. There, the 1913 flood killed over 360 people, and the total dead in the surrounding region was 467. Local people took the initiative, formed an association, raised over $2 million for engineering and legal studies, got state legislation passed, and formed the Miami Conservancy District in 1915. The district assessed the direct beneficiaries of the flood control structures so that it had enough money to begin construction in 1918. The district plan, designed and directed by Arthur E. Morgan, included five large retaining reservoirs around Dayton, the valley's principal city. Earthen dams built with hydraulic fill created the reservoirs, which had a combined storage capacity of 841,000 acre-feet (365.6 billion cubic feet).[11]

Figure 8-3: This October 22, 1935, view shows the baffles and stilling dam at Tygart Dam under construction with the cofferdam on the right. Source: Records of the U.S. Army Corps of Engineers, Record Group 77, National Archives and Records Administration, College Park, Maryland.

The water level was kept low behind the dams during the winter to prepare for the large runoff from spring rains. In April of 1922, a severe flood tested the newly built system and it passed with no difficulty. Since then, it has proven able to resist all floods including the winter and spring of 1959 when rainfall was nearly the same as in the killer flood of 1913.[12]

The federal government moved more slowly than the Miami River organization, but it did move, thanks to President Theodore Roosevelt, who after the 1907 Pittsburgh flood, appointed a commission to study water resources. The commission's leading personality, Nevada Republican Senator Francis G. Newlands, (primary sponsor of the 1902 Reclamation Act), tried without success to get Congress to transfer rivers and harbors work away from the Army. In 1916, the House formed a committee on flood control, and, the next year, Congress passed the Flood Control Act of 1917, covering the lower Mississippi and Sacramento Rivers. This act left the Corps in control and implied a shift from single purpose projects (navigation) to multipurpose projects that now included flood control. The 1920 Water Power Act added hydroelectric power but Congress did not provide any way to integrate those purposes until the 1927 Rivers and Harbors Act (January 1927) authorized the Corps to perform the surveys identified in 1926 in House Document 308.[13]

Because of the earlier work of Major Fiske (see chapter 3), Major Lewis H. Watkins' 75-page "Report on Survey of the Tennessee River and Its Tributaries" became one of the first 308 documents. Attached to it were 630 pages of appendices plus 199 plates published in a separate volume. This pioneering report gave, for the first time, a clear picture of a major river valley from the perspective of multipurpose development. Ironically, this report by the U.S. Army Corps of Engineers eventually provided Nebraska Republican Senator George W. Norris with the data to make a successful case for taking the development of the Tennessee River away from the Corps.[14]

Figure 8-4: The downstream face of Tygart Dam during construction. Source: Records of the U.S. Army Corps of Engineers, Record Group 77, National Archives and Records Administration, College Park, Maryland.

Figure 8-5: Completed Tygart Dam. Source: Records of the U.S. Army Corps of Engineers, Record Group 77, National Archives and Records Administration, College Park, Maryland.

In his letter approving Watkins' report, the Chief of Engineers, Major General Lytle Brown (1872-1951), himself a native of Tennessee, concluded that the river be improved to give a nine-foot depth for navigation either by many low dams (as on the Ohio) or by fewer high, hydropower dams, provided that they be "built by private interests, States, or municipalities." The federal government should only contribute to the cost in an amount equal to that required to build locks and dams solely for navigation. This proviso was similarly expressed in Butler's 308 report for the upper Columbia River. Thus the Corps, following the Republican administrations of Coolidge and Hoover, favored private or at least nonfederal power developments.[15]

But in this 308 report for the Tennessee River Valley, the Corps had projected multipurpose dams of the same type as Wilson and Tygart Dams, making preliminary designs of 23 dams and more detailed plans for a major high dam on the Clinch River at Cove Creek (later to be built by the TVA and named Norris Dam). This dam project was singled out in the 308 report for special funding in 1927 so that the Corps could make core borings and a detailed investigation of the dam site as well as a full stability analysis of the dam structure. It was clear in the report that the Corps was preparing to build Cove Creek Dam, whereas all other aspects of the survey were "not sufficiently definite to permit the exact determination and detail layout of the various projects." The entire survey of all tributaries and the Miami River, plus preliminary designs, cost just over $910,000, exclusive of Cove Creek, which cost just above $180,000.[16]

359

The two central features of the 308 Tennessee River Report were Wilson Dam, which began operating in 1925, and Cove Creek, the precursor to Tygart Dam. C. M. Hackett, the principal Corps engineer in Nashville, completed the preliminary design of Cove Creek in December 1928. It followed the precedent of Wilson Dam in its stability analysis, but otherwise it represented a departure for the Corps. It was not a mainstem river dam like Wilson, but, rather, it was a high storage dam of straight gravity section. The highest section of the dam was to be 247 feet high, and it would be 185 feet wide at its base. Overall, it would have a crest length of 2,009 feet and a reservoir storage capacity of 3,224,400 acre-feet. This huge capacity would have been the largest east of the Mississippi. The Corps had designed its power capacity to be 165,000 kilowatts, slightly less than the 185,000 kW then installed at Wilson Dam. The estimated cost was to be $37.5 million.[17]

Because of the federal government's policy favoring private power in 1930, General Lytle Brown requested proposals from nonfederal groups for constructing and operating high power dams. He got no proposals largely because of the depression and the concern over markets for all the newly created electric power. He, therefore, recommended the plan for low dams with locks primarily for navigation. The Corps did, however, begin construction of the first of these low dams 15.5 miles above Muscle Shoals at the head of the lake formed by Wilson Dam.[18] Little other work went forward, and the Corps did no further substantial work on Cove Creek.

Figure 8-6: The reservoir filling behind Tygart Dam. December 21, 1937. Source: Records of the U.S. Army Corps of Engineers, Record Group 77, National Archives and Records Administration, College Park, Maryland.

Meanwhile, 1930 brought not only the 308 Report on the Tennessee, but also a severe drought which almost reduced the Monongahela River to a dry creek. Navigation was impossible, and the Pittsburgh District called on the West Penn Power Company to release water from their Lake Lynn on the Cheat River. The power company agreed, and navigation on the heavily traveled river resumed. The Corps now saw the need for a big tributary reservoir to keep boats moving in summer and early fall.[19] Flood and then drought combined to make a high dam necessary, but the funding was still an insuperable problem in 1930.

Cove Creek and Tygart

The Corps had made its complete investigation of Cove Creek in 1927, and the consulting geologist had recommended a site in September 1927. The Corps made a design and cost estimate that year, which subsequently appeared in the 308 report.[20] That report stimulated congressional interest to which General Lytle Brown responded in 1930. He gave detailed dimensions of the dam, which are remarkably similar to those eventually built by the TVA.[21] No further work proceeded on Cove Creek until after the inauguration of Franklin D. Roosevelt in early 1933.

Figure 8-7: Built under contract through the Public Works Administration (PWA), Tygart Dam was completed in 1937. Source: Records of the U.S. Army Corps of Engineers, Record Group 77, National Archives and Records Administration, College Park, Maryland.

361

On May 18, 1933, the Congress approved the Tennessee Valley Authority Act, and on June 8, President Franklin D. Roosevelt, by executive order, placed the proposed Cove Creek Dam under the authority of Arthur E. Morgan, Chairman of the three-man Board of Directors of the TVA. In July, the board approved construction of the dam, renamed it Norris Dam, and asked the Bureau of Reclamation to make the final plans. In its summary report of 1934–35, the board refers to these events, but does not mention the name Cove Creek nor the U.S. Army Corps of Engineers, which had already made a detailed design. This obvious rebuff, probably led by Morgan who had developed a deep dislike of the Corps, led to some design changes but not to a change in its basic dam type; a massive straight-crested concrete gravity structure.[22] It was this type that the Corps used for its design of the West Virginia dam on the Tygart River.

As Tygart Dam got underway in early 1934, Pittsburgh District Engineer, Major Wilhelm D. Styer (1893-1975), felt the need for as much help as possible in designing this pioneering project, and, he, therefore, turned to the Corps' own precedent, Cove Creek. On January 10, 1934, he wrote to the Division Engineer in St. Louis requesting drawings of Cove Creek. The St. Louis Division Engineer responded on January 25, sending "one complete set of photostats of the Cove Creek Dam drawings as they were when the data were turned over to the Tennessee Valley Authority." He also sent Styer a bibliography of high dams and told him of the existence of complete drawings for Hoover Dam in the office of the Chief of Engineers.[23]

Early in February, Styer traveled to the site of the Cove Creek Dam, now renamed Norris Dam, to talk to the field engineer there and gain more information on that project, which had begun construction in October of 1933. Styer knew the similarity between the Cove Creek design and that being prepared in Pittsburgh for Tygart.[24]

The original Cove Creek design had a spillway separated from the river section of the dam, whereas the redesigned and renamed Norris Dam had the spillway within the river section, a solution which Styer adopted for Tygart. In addition, Norris had three drum-type gates, each fourteen feet high and one hundred feet long, placed above the spillway crest to allow for a greater reservoir storage.[25] Crest gates would also be taken over by Styer for Tygart, although in later discussions with the designers and the consultants they would be eliminated. Clearly, the Cove Creek Project was in the minds of the Pittsburgh District engineers as they prepared to design a high dam for the Monongahela Valley.

TYGART DAM

The Planning and Design of Tygart: 1933-1934

Tygart Dam was a new venture for the Corps at the same time that it was embarking on those other pioneering dams at Bonneville and Fort Peck. But unlike those western dams, Tygart would be conceived of as a navigation and flood control project without power. Moreover, for the Corps it was to be the largest dam east of the Mississippi. Unlike Wilson Dam, on the Tennessee River, Tygart would be designed entirely by permanent Corps personnel (except for minor architectural treatment by Paul Cret), and it would be set in motion quickly once the new administration took office in early 1933.

In earlier studies, the Tygart reservoir was to impound only 127,600 acre-feet rather than the 327,500 acre-feet in the final design. This more modest earlier design would have required a dam of little over 120 feet in height. However, in 1933, activity began to proceed with the larger project having a spillway height of 209 feet.[26] The Pittsburgh District Office of the U.S. Army Corps of Engineers, under Major Styer, began design and, by October, was able to send estimates of costs to the newly appointed Chief of Engineers, Major General Edward M. Markham (1877-1950). Styer noted that some of his estimates were rough because of "the absence of sufficient time to make more accurate determination." His October estimate was $11,853,238, of which about $2.4 million was to relocate a rail line.[27] Three weeks later, he sent a more detailed summary to the Chief of Engineers that included an estimate of men employed, totaling 2,200 during the twenty-nine months that the dam construction would take place. It is clear from the correspondence that the Chief of Engineers was under pressure from the new administration to begin construction quickly.[28]

Styer's figures satisfied Congress, which made $3 million available to begin work. Then, on January 11, 1934, the Public Works Administration authorized the project with the entire cost to be paid for by the federal government.[29]

There followed a brief dispute in which some local people argued for dams across tributaries other than the Tygart River, all based on mine-runoff pollution and on superior aesthetics. The Corps rejected the arguments for changing the dam site, but it did take seriously the concern over pollution and the aesthetics of the dam design.[30] Technical issues of dam design, however, were more pressing if workers were to be put into the field soon, so the Corps quickly turned to the foundations and the spillway. The Pittsburgh District hired Warren Mead, a consulting geologist. Mead was already working with the Corps at Fort Peck. The district also negotiated a contract with the Carnegie Institute of Technology

363

to model the spillway and develop experiments to find a good way of dissipating the energy of water discharged from the dam.[31]

Styer was anxious to get all the information he could for this new venture, so, in early April, he wrote to John Savage, Chief Design Engineer for the Bureau of Reclamation, requesting data on one of the most recent Reclamation studies for the Madden Dam spillway. This was a strange incident because the Madden Dam was then under construction for the Panama Canal, and with the U.S. Army Corps of Engineers in charge. The tests carried out for Reclamation by the Colorado Agricultural College (now Colorado State University) at Fort Collins, Colorado, were completed in late 1931 and reported on in the *Engineering News-Record* in mid 1932. Yet, it was another consultant hired by the Pittsburgh District, L. F. Harza, who called Styer's attention to the Madden Dam tests. Clearly, all the new dam studies pressing forward at this time made difficult a detailed investigation of the literature. Savage responded quickly, sent Styer the test report along with a 1933 technical memorandum on spillway design. Charles Wellons, the chief designer of Tygart Dam, thanked Savage warmly for the reports, which he found "very valuable to us in preparing for our tests. . .."[32]

Two additional planning issues were power and aesthetics. In early July, Morris L. Cooke, a noted reformer and engineer, wrote to Harold Ickes urging him to give instructions for the Army engineers to include provisions for future power development in Tygart Dam. Roosevelt had appointed Cooke as chairman of a Mississippi Valley committee of the Public Works Administration to advise on federal aid to public works, and it was in this position that he wrote the Secretary of the Interior.[33] Nothing significant came of that request, although the issue would be raised again. But the aesthetic issue did receive considerable attention.

The district had hired Paul Philippe Cret (1876-1945) to study the visual aspects of the dam complex. Cret, trained in his native France, had immigrated in 1903 and was well known as a classical architect and as a teacher at the University of Pennsylvania. He had a reputation for working on engineering projects, including such major works as the 1926 Delaware River Bridge (now called the Ben Franklin Bridge), and in mid-1934, he began a correspondence with Major Styer on the Tygart design. As Cret expressed it, "Very bold and simple forms are the most effective for concrete construction." The final design clearly reflects Cret's work only in the arcaded roadway on either side of the spillway and in the end piers that flank the spillway crest. However, as detailed in the correspondence of August and September of 1934, many small items received serious attention from Major Styer, Cret and his partners, and Styer's engineers.[34] Styer urged Cret to move quickly because the district was in a great hurry to ready the plans for bid advertisement.

To review the final plans, the Chief of Engineers directed the district to appoint a board of consultants, as had been done at Fort Peck and Bonneville. Styer named Louis C. Hill (formerly a supervising engineer for the Reclamation Service on Roosevelt Dam), L. F. Harza, and James B. Growden, along with William McAlpine, William Gerig, and himself from the U.S. Army Corps of Engineers. Meetings in August settled many minor issues plus the major one of removing the crest gates planned above the spillway. Omitting these gates greatly simplified spillway construction but did remove the possibility of storing more water in the reservoir. Finally, on September 4, the board met, recommended a series of minor changes, and cleared the way for issuing the bid plans two weeks later.[35]

The contractors had a month to make up their bids, and, on October 23, the district received five bids. The lowest bid was submitted by the Frederick Snare Corporation of New York at just over $6.3 million, or $832,000 below the Corps' estimate. After several meetings to discuss changes issued in early October, the government awarded the contract to Snare on October 28.[36] Although a contractor was now engaged, the design was far from final. Because of the political pressure to put workers in the field and because money was now available, the careful studies of major design issues could not be completed before construction. For example, the spillway tests at the Carnegie Institute of Technology had yet to begin, the foundation analysis was still being discussed in November, and a flood frequency analysis appeared only on August 31, when it was too late to make major design changes in the bid documents.[37]

In November, Styer sent one of his engineers to Cleveland to observe some hydraulic tests, and, on November 15, his consultant, Louis Hill, wrote him about the spillway test results for the Bonneville Dam, on which he was also a consultant. Styer knew nothing about this and immediately wrote to the Portland District engineer for information, which arrived on December 11. Apparently, the Portland District Engineer offered to do tests for Tygart, but Styer wrote that the Carnegie facilities were sufficient. The Corps was still decentralized with no formal interconnections for those pioneering design studies. By December 14, Styer could report to the division engineer that some hydraulic results were already available. At the very end of the year, Snare began to work at the site, but the discussions about design would continue for many more months.[38]

Design During Construction at Tygart: 1935–1936

Two issues dominated the design discussions during 1935: foundation safety and dam shape. The former is often somewhat unsettled until construction begins, but the latter, depending less upon local conditions, can often be well determined before. However, these huge dams, arising all at once in the 1930s, confronted the designers with difficult questions.

At a major meeting of consultants and Corps personnel on March 28 and 29, 1935, apparently there was discussion of the dam form, including earth-fill embankment, flat slab with buttresses (Ambursen), and massive head buttress (Noetzli). Presumably, this part of the discussion was rather perfunctory because no mention of it appears in the official meeting report. The group agreed on the massive gravity concrete form. The main agenda item, however, was the foundation.[39]

The foundation conditions at the dam site consisted of layers of sandstone and shale, but it was impossible to chart those conditions with any precision before excavation revealed the depths and locations of the layers. Several meetings of consultants in March of 1935 at the site resulted in disagreements. Finally, a meeting on March 28 and 29 considered conditions "satisfactory for the construction of a dam of the size and general type contemplated."[40]

This conclusion, however, did not go unchallenged. Two of the seven board members present at that meeting (McAlpine and Harza) disagreed with the report. William McAlpine, head engineer in the office of the Chief of Engineers, wrote a strong letter in early April stating "I do not think that the foundation condition so far disclosed by excavation should be classified as 'satisfactory'. . .[because the further] excavation may disclose the presence of other plastic [hence dangerous] strata." McAlpine was emphasizing the uncertain nature of local materials that can be determined only by intensive explorations that clear away rock and soil and drill holes (borings) to withdraw samples from deeper layers.

Several days later, L. F. Harza wrote a milder but still critical response to the report. He echoed the same theme: not enough was yet known to ensure a satisfactory foundation. He objected to the implication "that information is complete and confirmative."[41] These objections did not stop work but did cause the engineers to make more borings and to continue discussions of proper foundation design for the massive concrete structure.

Work went forward with more borings and continuing excavation all in preparation for casting concrete, which began in late May. By then, the Corps had a complete dam design, even though many details were not fully developed. This design consisted of a final form as well as a method for building it.

The form consisted of a roughly triangular solid concrete section across the valley with a crest length of 1,850 feet and a maximum height on either side of the spillway of 232 feet above the foundation rock. The base width there is 195 feet, whereas it is 207 feet for the 490 foot long spillway section whose height is 209 feet or twenty-three feet lower than the concrete dam on either side. The Corps designed this spillway to discharge 200,000 cubic feet per second with the reservoir filled to the height of the side walls and, hence, twenty-three feet above the spillway crest. At this water height the eight steel-lined outlet culverts (located 167 feet below the spillway crest) could discharge another 45,000 cubic feet per second.[42]

Because the maximum recorded stream flow of 65,100 cubic feet per second had occurred in 1912, the Corps spillway design was clearly conservative. For flood control, the reservoir would be kept as low as possible, consistent with enough flow for navigation between mid-December and the end of March. Thereafter, the reservoir could store more water to keep navigation viable from April 1 to the end of November. Relatively large releases during the first half of December then prepare the reservoir for the next yearly cycle.[43]

All these hydraulic aspects influenced the spillway and outlet works and were vigorously studied by the Corps during 1935. Although concreting had begun in late May, the precise concrete form of the dam still occupied the consultants and the Corps. In early June, Styer issued a change order to the Snare Company ordering them to build a greatly revised structural system at the downstream base of the spillway because "further model tests have shown that [the originally designed system] will not adequately dissipate the energy of the discharge flows. . .." The new system consisted primarily of baffles to deflect the flow; a low auxiliary dam thirty-three feet high, 489 feet long, and 223 feet downstream from the spillway toe; and two sixty-seven-foot-high walls parallel to the flow extending from the spillway to either end of the low dam. This structure created a basin of water thirty-three feet deep to act as a cushion absorbing the energy of the turbulent water from the outlet works and the spillway overflow. This energy had caused cavitation of earlier concrete dams by pulling out material due to the suction (or reduced pressure) arising from local increases in velocity of the water.[44]

This major change, coming so late, caused an increase in cost and illustrated the effect of proceeding with construction before the design could be completed. Meanwhile, the foundation design for the dam itself was still being argued when a slide occurred as the contractor was excavating on the left bank. Furthermore, excavations on the right bank led the engineers to change the design of the main dam sections there and to do further explorations to ensure against the danger of sliding and of seepage beneath the dam.[45]

As concrete casting continued in the late summer of 1935, word came from the Panama Canal of troubles at the Madden Dam where the outlet conduits had developed severe cavitation. Since the conduits at Tygart had followed the design at Madden, the district engineer arranged to have Carnegie Tech build and test a model of the entrance conduits designed for Tygart. The Corps decided to use model tests of Madden to be able to compare small-scale experiments with observed full-scale cavitation and also to use small-scale tests for Tygart to decide on any design changes. The results for Madden generally confirmed the validity of model studies, and the Tygart tests indicated a lesser problem of cavitation. However, unlike at Madden, the Corps had designed the Tygart outlets to have steel liners. Thus, the Carnegie tests did not signal the need for modifications, although these lined conduits would now be subjected to close and frequent inspection once in service. All these hydraulic studies were pioneering, and their results were in demand by other dam designers during the late 1930s.[46]

Meanwhile, the foundation design remained an issue as concreting went forward in the summer of 1935. In early August, one consultant warned of dam failure resulting from foundation stresses caused by the concrete pushing down on the sandstone and shale. He argued for a change in shape of the dam to reduce this effect. This opened other issues, such as sliding and uplift caused by seepage under the concrete structure. For sliding, the design was to embed the concrete into the natural foundation so that "it would be necessary to shear so large an area that it would be impossible to apply [water induced] force enough. . ." as the consultant Louis Hill put it. But to reduce seepage, the consultants had insisted that a series of holes, drilled into the foundation and filled with grout under pressure, be placed near the heel of the dam to provide a curtain by filling any spaces within the stone and shale layers.[47]

A flurry of further board meetings occurred in the early fall in which the consultants worried openly about the foundation, commented on new boring results, and ordered minor changes in the dam shape. Of great significance to the board was the method of casting the lowest layers of concrete into spaces excavated from rock so that a strong surface would exist to prevent both seepage and sliding. The board was thus continually giving new instructions to the contractor in the middle of his project. Of course, the contractor "was greatly concerned . . . [and felt that] the job was being seriously delayed by the nature and progress of the foundation preparation."[48]

The mixing of design and construction was getting out of hand because of the board continually raising new design issues, the contractor complaining about construction delays, and the Corps worrying about cost overruns. Particularly disturbing was the continued dispute over the dam shape and hence how to form monoliths that were already being concreted. So, on October 28, 1935, Division Engineer Colonel R. G. Powell stepped in, called a meeting of Corps people, but did not include the consultants or the contractor. The group reviewed four different designs for the concrete dam recommended at different times by the board and agreed on the original plan with only minor modifications. In Powell's letter to the Chief of Engineers, he argued that the other plans, while improving safety, would add substantially to the cost and that the original plan was already safe enough. The central technical issue was whether compression stresses at the heel of the dam, when the reservoir is empty, might break the grout wall which prevents leakage. (See Sketch dated October 28, 1935). Powell rejected that argument and ordered the contractor to proceed as planned, thereby going against the board. Actually, only one member, James Growden, seriously objected to Powell's decision, although the others shared some of Growden's reservations. Both Hill and Harza agreed with Powell.[49]

In mid-December, the district sent the consultants a detailed memorandum outlining all their previous major design changes and the subsequent instructions given to Snare. The Corps wanted the consultants to recognize the confusion raised by mixing design and construction. At its next meeting, three days later, the board seemed much more willing to accept the Corps' direction, although the lone dissenter, Growden, still objected frequently. The cracking noted in one monolith during the high pressure grouting of the foundations raised a new concern. Undisturbed, the board observed that this cracking was "found in every dam similar to Tygart. . .and that there was apparently nothing that could be done about them."[50]

Still, worry over cracking persisted through the next year and led to reducing both the pressure used for grouting and the cement content of the concrete. Reduction lessened the heat built up in massive concrete shapes and, hence, reduced the danger of thermal cracking. But the locus of design authority had shifted from the consultants to Corps personnel who, of course, had made the design in the first place. In May, the division engineer ordered the grouting pressure reduced and the board agreed to that order in a mid-July meeting in spite of the dissent of Growden, who a month later wrote that the new grouting procedure is "inadequate and grouting done in accordance therewith will not be satisfactory." The Corps went ahead with its procedure, nevertheless, reaffirming it in meetings that fall.[51]

Two last design issues raised by Louis C. Hill concerned uplift and sliding. He worried about the sliding in regions where the rock solidity was uncertain and had raised this issue as early as September of 1935 and again in his February 3, 1936, letter. A Corps officer, Edward Daly, responded with a clear calculation explaining how the backing of rock at the toe of the dam would be made at least 15 percent of the dam height ($d = 0.15H$) so that the resulting stress against that backing would be low even if sliding friction between dam base and rock were neglected. Daly's response satisfied Hill, but he also worried about the influence of seepage beneath the dam, which could exert an upward pressure and, thus, reduce the safety.[52]

This uplift force, by reducing the effect of dam weight, thereby would reduce the friction force between dam and foundation and increase the danger of sliding. The St. Francis Dam failure of 1928, attributed in part to uplift, focused engineers' attention on the problem, which led to two major lengthy papers in the ASCE *Transactions* just at the time Tygart was beginning. The first one (1934), of 20 pages, stimulated a discussion of 62 pages, while the second one, by L. F. Harza, was 33 pages with 21 pages of discussion. It was in this context that Hill raised the issue of seepage and uplift and urged the Corps to install gauges to measure uplift under the dam.[53]

By August 31, 1936, the contractor had cast over 700,000 cubic yards (out of a final total of 1,251,550 cubic yards) of concrete so that there could be little changing of the design that summer. This was an immense work for West Virginia, and by that date, 326,208 people had visited the site since concreting had begun in late May 1935. Then, on March 12, 1936, a massive flood occurred on the Ohio that Tygart, only 42 percent completed, could not mitigate. The great property damage caused by the previously noted flood shook the Congress, which quickly authorized the nine-reservoir system in the Allegheny-Monongahela basin to protect Pittsburgh and below.[54]

Tygart Completion and Operation: 1937–39

The Corps' design changes ceased in 1936 and construction continued through 1937, so that in February of 1938, the dam was officially completed. In July 1937, the Secretary of War sent Senator Rush Holt of West Virginia an estimated budget $18.3 million for the entire project, almost identical to the final figure of $18.4 million sent to the House Rivers and Harbors Committee on May 23, 1940. The 1933 estimate had been $11,853,238. The increase resulted from the design changes that raised dam costs from $8,208,000 to $12,432,000 and a much more expensive railroad relocation that drove costs from $2,860,000 to $4,617,000. The other big change was in the cost of land, estimated in 1933 to be $113,000 but finally amounting to $1,049,000.[55]

During the first two years of operation, there were two major floods; one in February 1939 and one in April 1939, and the flood damage reduction attributed to Tygart Dam totaled $2,818,000. In addition, the Tygart reservoir supplied extra flow into the Monongahela during the summer drought of 1939, thereby preventing a reduction and possible cessation of commercial navigation on the river. A complete halt would have cost the larger companies about $1,230,000 per month for the two month drought.[56]

The Corps quickly identified another benefit because of the reservoir's ability at low water to release stored water and flush out pollution from sanitary and industrial wastes. The Corps evaluated this benefit for 1938-1939 to be $281,000, so that the total two-year savings to the public was $3,379,000. Corps employees Robert M. Morris and Thomas L. Reilly worked out this benefit analysis and published it in the ASCE *Transactions* of 1942.[57] The prominence of this detailed article shows the pioneering work of the Corps in developing a program for multiuse dams without power. Part of the motive of this article was to argue for the newly approved (1936) nine-dam system of reservoirs in the Allegheny-Monongahela basin to function with Tygart in protecting both rivers and Pittsburgh.

The only system of dams in operation in 1936 were those in the Miami Conservancy District, but they were single purpose reservoirs devoted to flood prevention. Similarly, in 1933, local people in eastern Ohio formed the Muskingum Watershed Conservancy District, modeled after Miami. But being unable to raise the needed funds in the midst of the Great Depression, they entered into a partnership with the U.S. Army Corps of Engineers. Again, this system was single purpose, aimed at flood control. All 14 dams were completed successfully by 1938, and, in 1939, the Corps took over operation of the entire system.[58]

Flood control and Pittsburgh

Following World War II when the Corps turned back to civilian projects more intensely, the flood control issues once again brought politics and engineering together, especially after the October 1954 Hurricane Hazel disaster. The nine new reservoirs plus Tygart held the Pittsburgh flood down by about nine feet, but the water still caused great damage. Local leaders began to campaign hard for construction of another major flood control dam, this one on the Allegheny nine miles upstream from Warren, Pennsylvania. This new dam was about 140 miles north of Pittsburgh, where the Kinzua Creek joins the Allegheny.[59] Despite significant opposition, partly because the project involved the taking of land from the Seneca Indians, the Kinzua Dam was completed in 1966.

In 1972, Hurricane Agnes wreaked havoc in Western Pennsylvania, but the Kinzua Dam, which cost $108 million, saved the people downstream an estimated $247 million in damages.[60] Although such damage costs are difficult to evaluate with any accuracy, it seems clear that the Kinzua Dam has served its flood control purpose well in spite of the controversy. But on the Ohio, navigation has always been a major factor for the Corps.

Figure 8-8: Seepage into the earthfill section of Kinzua Dam was prevented by constructing a trench three feet wide, 1100 feet long, and up to 160 feet deep on the upstream side. This was then filled with bentonite slurry to stabilize the side walls and then filled with concrete placed through pipes from the bottom up displacing the bentonite slurry. Source: Pittsburg District, U.S. Army Corps of Engineers.

SLACKWATER NAVIGATION SYSTEMS

Modernization of the Ohio

By 1929, the Corps had completed a nine-foot channel on the Ohio River from Cairo to Pittsburgh by means of 53 locks and dams the earliest of which had opened in 1904. These dams were navigable in the sense that during high water the dam wickets could be lowered to pass the flood and to allow boats to go over them. These movable dams, many of which were Chanoine wicket structures, were costly and unreliable to operate.[61]

During the depression years the Corps, stimulated greatly by floods in 1935 and 1936, centered its attention on flood control as first detailed in the 308 reports. Not until after the war did the issue of navigation resurface politically with the greatly increased river traffic. From 1930 to 1955 the traffic increased tenfold (from 1,474 million ton-miles to 14,901 million ton-miles) and the wicket gates were obsolete.[62] The Corps needed a new plan, and it began in the 1930s to replace the low movable dams with higher dams with crest gates.

By 1979, this modernization plan had been nearly completed so that twenty fixed dams now replaced the 53 older and lower ones. The last such dam, near Cairo, is due to open in 2005. Between 1955 and 1977 the traffic had more than doubled (up to 37,476 million ton-miles).[63] Those newer dams are primarily for navigation, but on the upper Mississippi, where navigation intensity is roughly half that on the Ohio, factors such as flooding and wildlife protection have gained prominence.

The Upper Mississippi Basin

In 1929, the Congress authorized the nine-foot channel for the upper Mississippi where previously the Corps had operated under the assumption of open water navigation. Plans went ahead for dams in spite of the deepening depression, but the entire project nearly lost out in the newly elected Roosevelt administration's first 100 days.[64]

Figure 8-9: The Kinzua Dam on the Allegheny River during construction of the powerhouse in 1968. This is a private power project for pumped storage. Source: Ohio River Division, U.S. Army Corps of Engineers.

Saved by intensive lobbying headed by the governor of Minnesota, the project went ahead, not so much for river improvement, as for large-scale employment. By 1940, twenty-seven dams crossed the river from Minneapolis to Granite City, Illinois, just north of St. Louis.[65] Taken as a whole, these dams, like the ones across the Ohio, represented a major project for the U.S. Army Corps of Engineers.

The Design of Dam Gates

Having so many dams roughly the same size, the Corps took the opportunity to evolve some new designs especially for the dam gates. Beginning with roller gates, moving to a combination of roller and tainter gate dams, and finally developing an elliptical tainter gate system.

Roller gates, invented in Germany early in the twentieth century and widely used in Europe, had been only sparingly used in the United States prior to 1930 (Reclamation's Grand Valley Diversion Dam near Palisade, Colorado, completed in 1916, used roller gates) when the Corps decided to adopt them for Lock and Dam No. 15 near Rock Island in its first project for the nine-foot channel.[66] When completed in 1934, Dam No. 15 was the largest roller gate project in the world. There are eleven roller gates each 99.3 feet long between concrete piers. Nine of the gates are 19 feet 4 inches in diameter and provide wide openings when the river is in flood but otherwise in low water periods they are closed to allow for a navigable pool upstream.[67]

At the same time that the Corps was building Dam No. 15, it was also constructing roller gate dams in the Ohio Valley, especially on the Kanawha River which joins the Ohio north of Huntington, West Virginia, and the Gallipolis Locks and Dam (now known as The Robert C. Byrd Locks and Dam) 13.5 miles downstream from the mouth of the Kanawha.[68] The Corps chose roller gates over wicket ones because of reliability and durability, even though they were more expensive to build. Furthermore, in 1935 the Corps considered roller gates superior to tainter gates because they could be built to span over twice the distance of the latter.[69]

The roller gates set between two concrete piers are long cylinders which are hoisted by chains in one pier that pull the rotating gate up over racks like a giant bicycle chain. In the raised position the high water flows beneath it, but when lowered onto the concrete sill it retains the river. In November of 1933, the Corps began construction of Dam No. 4 where the roller gates were submersible thus allowing ice to pass easily over the top. Many other modifications ensued so that by the late 1930s innovations by the Corps in partnership with contractors had led to "a decidedly American style submersible roller gate."[70]

Figure 8-10: The locks of Mississippi River Dam #15 in November 1938. Source: U.S. Army Corps of Engineers.

While roller gates evolved, the Corps also developed the structure of tainter gates to such an extent that by the end of the nine-foot channel project in 1940 the roller gates had become obsolete. The Corps used both types of gates on many dams in the upper Mississippi in the 1930s. However, in Dam No. 24, completed in 1939, the Corps designed tainter gates eighty feet long and of sufficient strength and economy to render roller gates unnecessary. These tainter gates were the largest built up until that time.[71]

Tainter gates originated in the United States. Many types of such gates had been used throughout the nineteenth century, and in 1886 Jeremiah Burnham Tainter patented a design he had purchased from Theodore Parker. The Corps first used these tainter gates in 1889; they were easy to build and to operate, but they seemed to be too small and too unreliable for the large openings needed for major rivers such as the Mississippi. However, during the 9-foot channel project on the upper Mississippi, the Corps developed new shapes and new supporting structures that made much larger tainter gates possible.[72]

Tainter gates are made of curved steel plates convex on the upstream side where their arch-like form effectively resists the upper pool water pressure. The plates are stiffened by curved beams and supported by trussed radial arms that converge to a pivot on the downstream side. The gates rotate about this

375

pivot to open or close. Tainter gates are lighter, less costly, and easier to maintain than roller gates and, once they proved reliable in the upper Mississippi, they became a standard for the Corps. Again during the 1930s, the Corps looked at the large number of dams to be designed and to the opportunity to keep changing the gates through innovations.[73] The basic dam structure did not much change, but many significant details did. The upper Mississippi system was a gigantic laboratory for hydraulics, structures, mechanisms, and environmental impact.

Shipping, Fish and Wildlife

Navigation was clearly the central purpose for the dams; from 1939 to 1976 the tonnage on the river increased by over 28 times.[74] Nevertheless, the creation of slack water rather than a free flowing river had significant effects on fish, wildlife, municipal sewage and drainage facilities. Regulating the flow actually improved some conditions for fish and wildlife but made the disposal of sewage more difficult. These issues have led local people to secure federal funding for coordinated planning by the Upper Mississippi Basin Commission.[75] The nine-foot channel came into being quickly due not only to the federal policy of using public works to put people to work, but also to the commercial importance of river navigation.

Figure 8-11: Mississippi River Dam #15 showing the roller gates in the raised position on September 23, 1938. Source: Records of the U.S. Army Corps of Engineers, Record Group 77, National Archives and Records Administration, College Park, Maryland.

The dams that created the nine-foot channel, although conceived as a navigation project, nevertheless have developed into multipurpose projects that include environmental management. As recently as 1993, "conservation and

Figure 8-12: Mississippi River Dam #17 has Tainter Gates at the center and roller gates on each side. November 22, 1938. Source: Records of the U.S. Army Corps of Engineers, Record Group 77, National Archives and Records Administration, College Park, Maryland.

Figure 8-13: Mississippi River Dam #24. Source: Records of the U.S. Army Corps of Engineers, Record Group 77, National Archives and Records Administration, College Park, Maryland..

navigation interests are still defining their visions for the upper Mississippi River, and they are still compromising. . . . The origins of this compromise lay in the forces that created the Mississippi's landscape at Guttenberg, Iowa." There the Corps worked closely with the Iowa Department of Natural Resources and the U.S. Fish and Wildlife Service to protect the fish and the migratory water fowl.[76] This cooperation and compromise represents the major challenge to federal dams for the new century.

Figure 8-14: The gates were lowered for the first time to start creating a navigation pool at Mississippi River Dam #24, March 13, 1940. Source: Records of the U.S. Army Corps of Engineers, Record Group 77, National Archives and Records Administration, College Park, Maryland.

Endnotes

1.   Leland R. Johnson, *The Ohio River Division: U.S. Army Corps of Engineers* (Cincinnati: U.S. Army Corps of Engineers, 1992), 94, 1-3. See also W. A. Johnson, "Report of the District Engineer," October 31, 1933, in *Ohio River*, House Document No. 306, 74th Congress, 1st Session, August 23, 1935, 43.
2.   Edmund L. Daly, "The Mastery of the Ohio River," *The Military Engineer* 19 (May-June 1927), 188. For the 1937 flood, see Paschal N. Strong, "The Great Flood of the Ohio," *The Military Engineer* (May-June 1937), 161.
3.   Arnold, *Evolution of the 1936 Flood Control Act,* 63. Leland R. Johnson, *The Headwaters District* (Pittsburgh: U.S. Army Corps of Engineers, C. 1978), 203-7.
4.   Arnold, *Evolution of the 1936 Flood Control Act,* 88-90. The Act was passed by the Senate on May 21, 1936, by the House on June 3, and signed by President Roosevelt on June 22.
5.   Johnson, Leland R., and Jacques S. Minnotte, *The Headwaters District Roundtables: An Eyewitness History of the Pittsburgh District, United States Army Corps of Engineers: 1936-1988* (Pittsburgh: U.S. Army Corps of Engineers, 1989 [Revised edition of *The Headwaters District*, c. 1978 which is cited in second previous footnote]), 2.
6.   L. R. Johnson, *The Ohio River Division*, 1-13.
7.   W. A. Johnson, "Report of the District Engineer," 41.
8.   W. A. Johnson, "Report of the District Engineer," 2. It is clear from Major Fiske's reports and those of Nashville Division engineers in the 1920s that the Corps expected to develop the Tennessee River (see Chapter 3).
9.   Arnold, *Evolution of the 1936 Flood Control Act*, 6-7. For the Humphreys-Ellet controversy, see John M. Barry, *Rising Tide, the Great Mississippi Flood of 1927 and How it Changed America* (New York: Simon and Schuster, 1995), 13-91. For a detailed discussion of the Humphreys report, see Reuss, "Andrew A. Humphreys and the Development of Hydraulic Engineering: Politics and Technology in the Army Corps of Engineers, 1850-1950," 1-33.
10. Arnold, *Evolution of the 1936 Flood Control Act,* 11.

11. Arthur E. Morgan, *The Miami Conservancy District* (New York: McGraw Hill, 1951). Miami Conservancy District, *Official Plan Miami Conservancy District* (Dayton, Ohio: Miami Conservancy District, May 1916).

12. A clear and concise description of the Miami Conservancy District appears in Armstrong, Robinson, and Hoy, editors, History of Public Works in the United States 1776-1976, 275-7.

13. Arnold, *Evolution of the 1936 Flood Control Act,* 11-7.

14. U.S. Congress, House of Representatives, *Tennessee River and Tributaries: North Carolina, Tennessee, Alabama, and Kentucky,* House Document No. 328, 71st Congress, 2nd Session, Washington, March 24, 1930. For the activities of Senator Norris, see Hubbard, *Origins of the TVA*, 275-6.

15. *Tennessee River,* House Document No. 328, 7.

16. *Tennessee River,* House Document No. 328, 29, See also Part 2, plates 74 and 81.

17. C. M. Hackett, "Report on Cove Creek Dam Site and Recommendations for Design of Dam, Power House, Barge Lift, and Spillway," *Tennessee River,* House Document No. 328, 60, and Appendix F, 469-78.

18. Leland R. Johnson, *Engineers on the Twin Rivers,* 184-6. See also Tennessee Valley Authority, *Annual Report of the Tennessee Valley Authority for the fiscal year ended June 30, 1935*, Washington, 1936, 10. See also *Tennessee River,* House Document No. 328, part 2, plate 101.

19. Johnson and Minnotte, *Headwaters District Roundtables,* 8. See also the earlier volume by L. R. Johnson, *Headwaters District*, 201.

20. Major Lewis H. Watkins to the Chief of Engineers, June 2, 1927, memorandum with appendices A (geology) and B (alternate sites). See also Watkins to the Chief of Engineers, September 27, 1927, memorandum with a geologist recommending one site location for the Cove Creek Dam, National Archives, College Park, RG77, Entry 111, Folder 7493.

21. In later reports to political figures, Major General Lytle Brown gave the cost, which was developed in 1927, and reported in the 308 report. See Major General Lytle Brown to Representative Lewis W. Douglas, February 10, 1930, and to Senator William E. Brock, May 15, 1930, National Archives, College Park, RG77, Entry 111, Folder 5221. See also Brown to George Akerson, Secretary to the President, December 6, 1930, National Archives, College Park, RG77, Entry 111, Folder 7493.

22. Arthur E. Morgan, Harcourt A. Morgan, and Tennessee Valley Authority, David E. Lilienthal, *Annual Report of the Tennessee Valley Authority for the Fiscal Year Ended June 30, 1935*, Washington, 1936, 4.

23. Major W. D. Styer to the Division Engineer in St. Louis, January 10, 1934; and Lieutenant Colonel E. L. Daley to Major Styer, January 25, 1934, with enclosures, National Archives and Records Administration, Philadelphia, Records of the U.S. Army Corps of Engineers, RG77, Entry 1290, Box 12, Files No. 4, Doc. No. 5.

24. John F. Barksdale to Major Styer, February 6, 1934, with enclosures of a speech by Ross White, Construction Superintendent of TVA, on construction of Norris Dam (delivered on November 1, 1933), National Archives, Philadelphia, Record Group 77, Entry 1290, File No. 4, Doc. No. 9.

25. Tennessee Valley Authority, *The Unified Development of the Tennessee River System,* Knoxville, 1936, 95.

26. W. A. Johnson, *Ohio River,* House Document No. 306, 182. This is the 308 Report which appeared while Tygart Dam was under construction, but Johnson completed his report before the Corps had made the final design. For the relationship between reservoir elevation (and hence dam height) and storage capacity, see Major W. D. Styer to the Chief of Engineers, December 1, 1933, National Archives, Philadelphia, RG77, Entry 1290, Box 12, File No. 4, Doc. No. 3.

27. Major W. D. Styer to the Chief of Engineers, October 17, 1933, National Archives, Philadelphia, RG77, Entry 1290, File No. 4, Doc. No. 1.

28. Captain Lucius Clay to the District Engineer (Styer) in Pittsburgh, November 6, 1933; and Major W. D. Styer to Clay on November 8, 1933, National Archives, Philadelphia, RG77, Entry 1290, File No. 4, Doc. No. 2.

29. W. E. Potter, "Tygart River Reservoir Dam," *The Military Engineer* 28, No. 161, (September-October 1936), 331. "Tygart River Reservoir Dam," U.S. Army Corps of Engineers, Pittsburgh District, Revised November 1938, 2.

30. W. F. Tompkins to the Pittsburgh District, January 29, 1934, enclosing a memorandum by Clem Shaver, a local Democratic Party leader, urging either of two different sites; and W. R. Winslow and H. E. Anderson to the District Engineer, February 8, 1934, recommending against changing the location, National Archives, Philadelphia, RG77, Entry 1290, File No. 4, Doc. No. 7.

31. Major W. D. Styer to Major T. B. Larkin, February 7, 1934, and Larkin to Styer, February 17, 1934, National Archives, Philadelphia, RG77, Entry 1290, File No. 4, Doc. No. 10. See also Styer to the Division Engineer, March 13, 1934, memorandum requesting model tests, National Archives, Philadelphia, RG77, Entry 1290, File No. 4, Doc. No. 13. These tests were approved by G. B. Pillsbury, Acting Chief of Engineers, March 30, 1934, on the foregoing document.
32. Major W. D. Styer to John Savage, April 7, 1934; Savage to Styer in Pittsburgh, April 17, 1934, with two enclosures; and Charles Wellons to Savage, May 7, 1934, returning one report. National Archives, Philadelphia, RG 77, Entry 1290, Box 12, File No. 4, Doc. No. 16. For a detailed presentation of the Madden Dam tests, see Richard R. Randolph, "Hydraulic Tests on the Spillway of the Madden Dam," *Transactions of the American Society of Civil Engineers* 103 (1938), 1080-112.
33. Morris L. Cooke to Harold L. Ickes, July 2, 1934, memorandum attaching a further memorandum from J. C. Dickerman of the Federal Trade Commission to M. L. Cooke, June 29, 1934. For Cooke see Jean Christie, "Morris L. Cooke and Energy for America," in Carroll W. Purcell Jr., editor, *Technology in America: A History of Individuals and Ideas*, 2nd edition (Cambridge, Massachusetts: MIT Press, 1991), 244.
34. See the letters between Major W. D. Styer and J. H. Dodds for the Corps and Paul P. Cret and his partners John Harbeson and William Livingston in July, August, and September 1934, National Archives, Philadelphia, RG77, Entry 1290, File No. 4, Doc. No. 32.
35. Major W. D. Styer to the Chief of Engineers, August 15, 1934, enclosing memorandum dated August 13 of meeting, August 10-11, 1934, on plans and specifications for the Tygart Dam, National Archives, Philadelphia, RG77, Entry 1290, File No. 10, Doc. No. 2. Styer to the Chief of Engineers, August 27, 1934, reported approval from the Acting Chief to issue plans for bidding once the Board of Consultants had reviewed them. The board recommendations appeared in a report signed by Hill, Harza, Growden, Gerig, McAlpine, Powell, and Styer and sent to the District Engineer on September 5, 1934, National Archives, Philadelphia, RG77, Entry 1290, File No. 10, Docs. Nos. 3-4.
36. Major W. D. Styer to the Chief of Engineers September 15, 1934, also National Archives, College Park, RG77, Entry 111, Folder 3524, Part 1. Styer estimated the cost of the dam to be $10,200,000 with $2,812,000 for materials furnished by the Government. He thus gave the bid estimate to be $7,388,000 whereas the published figure at the opening was $7,137,737.13. For additional changes, see H. A. Montgomery to the Chief of Engineers, October 1, 1934, with enclosure dated September 29, 1934.
37. H. E. Anderson to the District Engineer, August 31, 1934; and "Conference on Design and Specifications for the Tygart River Dam," by the Board of Consultants, November 13, 1934, National Archives, Philadelphia, RG77, Entry 1290, File No. 10, Doc. No. 5. The latter discusses Warren Mead's analysis of the dam foundations.
38. Louis C. Hill to Major W. D. Styer, November 15, 1934, Styer to Hill, November 20, 1934; and Styer to the Portland District Engineer, November 20 and December 1l, National Archives, Philadelphia, RG77, Entry 1290, File No. 4, Doc. No. 24. For the hydraulic results, see Styer to the Division Engineer, December 14, 1934, National Archives, Philadelphia, RG77, Entry 1290, File No. 4, Doc. No. 13, For the starting date of construction, see Potter, "Tygart River Reservoir Dam," 332.
39. James P. Growden to Major W. D. Styer, August 6, 1935, described the issues on form that he said were discussed at a meeting on March 28-29, 1935, National Archives, Philadelphia, RG77, Entry 1290, File No. 10, Doc. No. 10. The issue of form does not, however, appear in the official report of that meeting (see note 42).
40. "Report on Conference of Consultants for the Tygart River Dam, March 28 and 29, 1935," Pittsburgh, March 30, 1935. National Archives, Philadelphia, RG 77, Entry 1290, Box 32, File No. 10, Doc. No. 8. An earlier meeting on March 5 had begun those discussions but not reached such a general conclusion, see "Conference held March 5, 1935 at Tygart River Dam," W. E. Potter, 1st Lieut. C. E., March 7, 1935, National Archives, Philadelphia, RG77, Entry 1290, Box 32, File No. 10, Doc. No. 7.
41. W. H. McAlpine to Major W. D. Styer, April 9, 1935; and L. F. Harza to Styer, April 12, 1935, National Archives, Philadelphia, RG77, Entry 1290, Box 32, File No. 10, Doc. No. 8.
42. For dimensions and discharges, see "Tygart River Reservoir Dam," (1938) cited in note 31 above. The Spillway Capacity $Q = CLH^{3/2}$ where $L = 490$ feet, $H = 23$ feet and $C = 3.7$ from which $Q = 200,000$ cubic feet per second. For C, see Leonard Church Urquhart, editor, *Civil Engineering Handbook* 3rd edition (New York: McGraw Hill, 1950), 348-9. Data are all prior to 1933. For the outlet culverts, see Urquhart, 329, where $Q = CAV$ and $V = 2HG = 111$ feet per second and $A = 5.67 \times 10 \times 8$ (for eight rectangular openings) $= 455$ square feet and $C = 0.9$ so that $Q = 0.9 \times 455 \times 111 = 45,000$ cubic feet per second.

43. Potter, "Tygart River Reservoir Dam," 9-11. The benefits of the dam were already apparent during late 1937, after closure, when potential floods would have wreaked substantial damage in the Monongahela Valley. In addition, see 11-2 for a discussion of water supplied by the reservoir during the dry period from mid-August to mid-September 1938, which permitted continuous navigation and also improved the quality of the water for industrial and municipal consumption.

44. Major W. D. Styer to the Frederick Snare Corporation, June 5, 1935, enclosing detailed costs; and Styer to the Chief of Engineers, June 8, 1935, National Archives, College Park, RG77, Entry 111, Box 1679, 3524, Part 1. The Board of Consultants had already approved the design change to the stilling basin in their meeting of March 28-29, 1935. C. W. Wellons sent results of the tests to set the shape of deflectors below the outlets. See C. W. Wellons to W. H. McAlpine, August 22, 1935, National Archives, Philadelphia, RG77, Entry 1290, File No. 4, Doc. No. 30.

45. "Report on Conferences of Consultants for the Tygart River Dam, July 17 and 18, 1935," signed by W. D. Styer, William Gerig, and W. H. McAlpine, July 18, 1935; "Memorandum on Conference at Tygart River Dam, Grafton, West Virginia, on July 25, 1935," signed by L. C. Hill, W. J. Mead, W. D. Styer, and C. M. Wellons, July 26, 1935; and letters to Major W. D. Styer from W. H. McAlpine, July 30, 1935, and from Louis C. Hill, August 2, 1935, National Archives, Philadelphia, RG77, Entry 1290, File No. 10, Doc. No. 10.

46. Harold A. Thomas and Emil P. Schuleen, "Cavitation in Outlet Conduits of High Dams," *Transactions of the American Society of Civil Engineers* 102 (1942), 421-2 and 455-6. In a memorandum to the Chief of Engineers, September 6, 1935, Major W. D. Styer requested funds for the Tygart tests, National Archives, Philadelphia, RG77, Entry 1290, File No. 4, Doc. No. 13. In a memorandum to the Division Engineer, Missouri River Division, January 31, 1936, Styer referred to other cavitation studies and to the Tygart spillway and outlet works and how their design has remained "substantially as shown in the folio [of May 1, 1935]." National Archives, Philadelphia, RG77, Entry 1290, Box 3, File No. 1, Doc. No. 58.

47. James P. Growden to Major W. D. Styer, August 6, 1935, with responses from Warren J. Mead on August 9, W. H. McAlpine on August 16, W. Gerig on August 26, and Louis C. Hill on September 3 and 4, 1935, National Archives, Philadelphia, RG77, Entry 1290, Box 32, File No. 10, Doc. No. 10. Quote is in Louis C. Hill letter of September 3, 1935.

48. L. C. Hill, W. J. Mead, W. Gerig, and W. H. McAlpine to the District Engineer, Pittsburgh, September 18, 1935, memorandum of a meeting September 17-18, 1935, with a detailed drawing showing the locations of the 30 inch exploration holes. See also Mead, Gerig, Growden, and McAlpine to the District Engineer, Pittsburgh, October 26, 1935, with *Sketch 1* "Excavation and placing of concrete section founded on shale strata underlying thick sandstone and thin clay seam," and *Sketch 2* "Method of placing base lifts: monolith in which foundation strata rises," U.S. Engineer Office, October 21, 1935, National Archives, Philadelphia, RG77, Entry 1290, File No. 10, Doc. No. 13. For the contractor's reactions, see Lieutenant W. E. Potter to the District Engineer, Pittsburgh, October 15, 1935, National Archives, Philadelphia, RG77, Entry 1290, Box 32, File No. 10, Doc. No. 12.

49. Captain H. A. Montgomery to the Division Engineer, Cincinnati, October 31, 1935, with two sketches; and Colonel R. G. Powell to the Chief of Engineers, November 23, 1935, National Archives, Philadelphia, RG77, Entry 1290, File No. 10, Doc. No. 14.

50. W. J. Mead, J. P. Growden, and W. H. McAlpine to the District Engineer, Pittsburgh, December 21, 1935, National Archives, Philadelphia, RG77, Entry 1290, Box 33, File No. 10, Doc. No. 16.

51. Louis C. Hill to the District Engineer, Pittsburgh, February 3, 1936, National Archives, Philadelphia, RG77, Entry 1290, File No. 10, Doc. No. 17; and L. F. Harza to Major W. D. Styer, March 17, 1936, File No. 4, Doc. No. 36. See also Captain B. C. Fowlkes to the District Engineer, Pittsburgh, May 20, 1936; and Louis C. Hill to Captain R. M. McCutchen, June 10, 1936, National Archives, Philadelphia, RG77, Entry 1290, File No. 10, Doc. No. 18. Powell's order appears in a memorandum to the District Engineer, Pittsburgh, May 15, 1936, attached to the report of a Board of Consultants meeting held on July 21. See Captain B. C. Fowlkes to the District Engineer, Pittsburgh, August 19, 1936; and Growden's response in a letter to Captain Albert H. Burton in the district office, August 28, 1936, National Archives, Philadelphia, RG77, Entry 1290, Box 33, File No. 10, Doc. No. 19. Further meetings occurred on October 23 and on November 19 during both of which the cracks were discussed. See Captain B. C. Fowlkes to the District Engineer, Pittsburgh, October 24, 1936 and November 20, 1936, the latter signed by Growden, Lieutenant Colonel W. E. R. Covell, and C. M. Wellons, National Archives, Philadelphia, RG77, Entry 1290, File No. 10, Doc. No. 21.

52. Louis C. Hill to the District Engineer, Pittsburgh, February 3, 1936; Lieutenant Edward G. Daly to Hill February 13, 1936; and Hill to Daly on February 13, 1936, concurring, National Archives, Philadelphia, RG77, Entry 1290, File No. 4, Doc. No. 36.

53. Louis C. Hill to the Division Engineer, January 31, 1936 (misdated February 31), enclosing a diagram of uplift measurements on the Pine Canyon Dam near Pasadena, California; Hill to Major W. D. Styer, March 2, 1936; and Hill to Captain R. M. McCutchen June 10, 1936, National Archives, Philadelphia, RG77, Entry 1290, File No. 4, Doc. No. 36. The published papers are: D. C. Henny, "Stability of Straight Concrete Gravity Dams," *Transactions of the American Society of Civil Engineers* 99 (1934) 1041-123 (includes discussion). L. F. Harza, "Uplift and Seepage Under Dams on Sand," *Transactions of the American Society of Civil Engineers* 100 (1935), 1352-406 (includes discussion).

54. W. E. R. Covell, "Flood Control in the Pittsburgh District," *The Military Engineer* (May-June 1939), 177-8.

55. Harry H. Woodring, Secretary of War, to Senator Rush Holt, July 19, 1937, National Archives, College Park, RG77, Entry 111, Folder 7235 Part 2. Brigadier General Thomas M. Robins to Representative Vincent F. Harrington, May 23, 1940, National Archives, College Park, RG77, Entry 111, Folder 5221.

56. Robert M. Morris and Thomas L. Reilly, *Tygart Dam: Review of Operations 1938-1939* (Pittsburgh: U.S. Army Corps of Engineers, April 1940), 51-4. This report gave the cost savings, due to flood prevention, as $3,001,000 (see 46) but a later article (see note 59) revised this downward.

57. Robert M. Morris and Thomas L. Reilly, "Operation Experiences, Tygart Reservoir," *Transactions of the American Society of Civil Engineers* 107 (1942), 1349-78 (with discussions). This paper is a revised version of the report referred to in note 58.

58. For the Miami project, see Charles H. Paul, "Flood Control in the Miami Valley, Ohio," *Transactions of the American Society of Civil Engineers,* 85 (1922), 1503-22. For the Muskingum, see Francis X. Purcell Jr., "Muskingum River Flood Control," *The Military Engineer* 28, No. 159 (May-June 1936), 184-8. The Corps completed all 14 dams in the Muskingum project from 1934 to 1938, L. R. Johnson, *Headwaters District,* 200. See also L. R. Johnson, *Headwaters District,* 1992, 141-4.

59. L. R. Johnson, *Headwaters District,* c. 1978, 276.

60. L. R. Johnson, *Headwaters District*, 1992, 300.

61. Department of the Army, U.S. Army Corps of Engineers *Ohio River Navigation* (Cincinnati, 1979), 2. O'Brien, et. al, *Gateways to Commerce*, 41.

62. Department of the Army, *Ohio River Navigation*, 22. See also Francis H. Oxx, "The Ohio River Movable Dams," 49-60.

63. Department of the Army, *Ohio River Navigation,* 22, and L. R. Johnson, *Headwaters District*, 1992, 13.

64. O'Brien, *Gateways*, 47-55.

65. O'Brien, *Gateways*, 13.

66. The first roller gates in the United States, completed in 1915, were for the Grand Valley Diversion Dam near Grand Junction, Colorado. There were six roller gates, 70 feet long for the Reclamation Service project. See U.S. Department of the Interior, Water and Power Resources Service, *Project Data* (Washington, D.C.: Government Printing Office, 1981), 513-7.

67. O'Brien, *Gateways*, 67, and Tweet, *History of the Rock Island District*, 260-71.

68. "Roller-Gate Dams for the Kanawa River," *Engineering News-Record* 111, No. 12, (September 21, 1933), 337-42, and W. H. McAlpine, "Roller Gates in Navigation Dams," *The Military Engineer* 26, No. 150, 419-23. For the Gallipolis plan, see W. A. Johnson, "Report of the District Engineer," 148-9.

69. J. C. Harrold, "Outline of the General Principles Used in the Design of Dams on the Upper Mississippi River," *Engineering As Applied to the Canalization of a River* (St. Louis: September 1935), 34. See also a comparison of the Ohio and Upper Mississippi in the same volume by L. Ylvisaka, 19.

70. O'Brien, *Gateways,* 64-71. The quote refers to the article "Roller-Gate Dams for the Kanawa River," *Engineering News Record*, 111, No. 12 (September 21, 1933), 337-42, which *Gateways* states is written by S. G. Roberts (see O'Brien, *Gateways,* 44).

71. O'Brien, *Gateways*, 73-9.

72. O'Brien, *Gateways*, 76-7.

73. O'Brien, *Gateways*, Chapter VII.

74. O'Brien, *Gateways*, 14.

75. Merritt, *The Corps, the Environment, and the Upper Mississippi River Basin,* 53-63 and 91-7.

76. John O. Anfinson, "Commerce and Conservation on the Upper Mississippi River," 388-417. Quote on page 417.

CHAPTER 9

THE ENVIRONMENTAL IMPACT OF THE BIG DAM ERA

Conservation and Controversies

In a 1941 book, *American Bridges and Dams*, Paul Zucker wrote: "Of all works of engineering, the perfect bridge most nearly approaches the realm of art—the dam, the realm of nature." Dams compared with "God's immovable mountains." Zucker went on to suggest that, "No other achievement of peaceful civilization during the last two decades on this war-torn earth has contributed more to the welfare of future generations than the building of dams in this country." The dam's legacy consisted of storing water, averting and controlling floods, irrigation, conservation of soil, improving navigation, and generating electrical energy. "It is natural," he concluded, "that a vast country of mighty rivers, like the United States, should excel in the construction of dams."[1]

About forty years after Zucker's book appeared, an international movement against current dam-building practices emerged, comprised of numerous environmental, human rights, and social activist groups from a variety of local, regional, national, and international anti-dam campaigns. Criticisms abounded: dams flooded valleys, displaced farmers, and blocked fish migrations. Dams reduced water quality. Dams changed forever natural riverine flows. Dams were short-sighted structures that drew funds away from other potentially sounder technologies. For some, the conclusion was obvious—destroy or decommission the dams and free the rivers.[2]

The story of dam building and its environmental impacts is not as neatly explained as the opposing views imply. However, it is safe to say that the perception of large dams, their economic and societal value, and their environmental implications have undergone considerable change in recent years from the heyday of the big dam era in the middle of the twentieth century. These changes in perceptions are at once obvious, but there are complex undertones.

Harnessing rivers into service to humankind was the major objective of the big dam era, which lasted from about 1935 to 1965. Construction in this period ran the gamut from the 26 dams of the Tennessee Valley Authority in the East to the pre-World War II giants Hoover and Grand Coulee Dams, and Glen Canyon Dam (opened in 1964) in the West.[3] In this sense, the big dam era elaborated on the legacy of Progressive Era water policy, bent on maximizing the 'efficient use' of a vital natural resource. Sometimes the aim was to reduce flooding and to improve navigation. In other cases, flowing water was manipulated to provide abundant sources for irrigation and urban use or to produce electricity. Multipurpose projects attempted to accomplish all of these goals. To the builders of Shasta Dam, for example, controlling "the temperamental waters of the Sacramento area" meant an opportunity to build an agricultural and industrial empire in the Central Valley of California.[4] As historian Mark Harvey observed, "Dams, the traditional dictum went, harnessed 'wild' and 'untamed' rivers and transformed them into calm, docile waterways."[5]

Dams also were great symbols of American achievement. In the 1950s, the American Society of Civil Engineers listed Hoover Dam and Grand Coulee Dam as two of the seven civil engineering wonders of the United States; in 1964, it crowned Glen Canyon Dam as the outstanding engineering triumph of the year.[6] The mighty structures were viewed at once as symbols of the coming of age of the modern West, as the product of American engineering know-how, and as sublime renderings that improved rivers and complemented nature's aesthetic. Power production, in particular, gave dams a great reach well beyond the site of their construction, transforming hinterland into cities. Several conservationists also envisioned dams as "the cornerstone of social policy extending well beyond cheap hydroelectricity." Regional development programs, they believed, would have economic and social benefits, especially for the underprivileged.[7]

Dam builders and a variety of other interests also had pragmatic reasons for extolling the virtues of big dams. Especially in the years of the Great Depression, massive federal construction projects, like dams, meant jobs and long-lasting financial security for farmers tending irrigated lands and for cities in search of stable and plentiful water supplies. The likely maximum number of workers employed at any one time at Hoover Dam was 5,200 workers; more than 7,000 for Grand Coulee; and 10,500 for Fort Peck Dam. Several commentators argued that the electricity generated by big dams showed its worth particularly in World War II and in the Cold War.[8] One set of estimates stated that water development projects in the United States resulted in 26,000 miles of channeled waterways, 58,000,000 acres of irrigated land, 30,000,000 kilowatts of hydroelectricity, and flood control through 400 large dams.[9]

Despite the enthusiasm over dam-building from the 1920s to the early 1960s, some people—namely the rural poor, small farmers, Native Americans, Hispanic communities, and even some farm families of means—failed to enjoy fully the benefits of the new construction. Displacement of people prior to inundation or construction often affected those with little political or economic leverage. Since World War II in particular, several Indian reservations were impacted by the construction of federal dams on major rivers. Nine thousand acres of Seneca land were taken along the Allegheny River for the building of the Kinzua Dam. Fort Mohave, Chemehuevi Valley, Colorado River, Yuma, and Gila Bend reservations in Arizona and California lost land to dam projects in the Colorado River Basin. In 1960, 190 Indian families on the Standing Rock Sioux Reservation lost land to proposed flooding for the Oahe Dam in the Missouri River Basin. Bonneville Dam displaced the Indian dip-net fishery at the Cascades; Dalles Dam, also on the Columbia, eliminated the prized fishing site of Celilo Falls. And the list goes on.[10] While not directly affected in the way some Native Americans were, some easterners, nevertheless, resented federal dollars flowing to the West to underwrite the big dam projects, at least until dam-building in the East began to accelerate after 1945. In the Midwest, where few deep canyons made giant dams possible, or where dams threatened marshlands necessary to sustain bird hunting, the "big dam era" was not always welcomed. Even in the West, a variety of terrains and economic interests influenced support for these massive public works projects. Some small farmers in the West, for example, who may have looked to the dams and reservoirs as a source of plentiful water, ultimately lost out to corporate farms and big irrigators, and others saw the water directed toward urban development instead of rural preservation.[11]

Early indicators also appeared that raised questions about the physical implications of the big dams before the emergence of the modern environmental movement in the 1960s. As one historian noted, dam building "was conceived within American conservation tradition—the tradition of utilitarianism which stressed efficient control of nature in the public interest. . . ."[12] However, some conservation-minded people held reservations about big dam projects. Karl E. Mundt, congressman from South Dakota, noted in 1943 that, ". . . dams which are properly designed and wisely built can expand recreational and conservation assets just as certainly as improper and unwise dams can destroy them. . . ."[13] Other conservationists questioned dam building for power production and irrigation as a "national mania" or even a "national menace." Field naturalist William L. Finley argued, "Some of these projects destroy existent wealth or endanger some other resource."[14]

The tone and the focus of environmental concern over dam building evolved dramatically in the 1960s. What had been viewed as projects of great economic hope and possibility were now being critiqued more closely in terms of erosion of downstream channels, changes in fish population and riparian vegetation, water evaporation loss, displacement of native peoples, dwindling scenic wonders, and urban sprawl. Even the utilitarian value of dams came into question as dam infrastructure began aging and began raising growing concern about its performance and safety. Within environmental circles, especially, emphasis shifted from the value of dams to the value of scenic and wild rivers. The realization that the United States was the second most dammed country in the world and that most major rivers in the lower 48 states were regulated by some combination of dams, locks, or diversions, made the preservation of undammed rivers more important.[15]

The story of the environmental implications of big-dam building in the United States during the twentieth century is at once a story of the contested use of natural resources and changing perceptions, values, and symbols of human-made structures. In some sense, the story is an apt summation of the dramatic impact of large federal dam construction presented in this book.

The Economic and Social Impacts of Dams

In the 1920s and 1930s, the economic potential of big dams appeared so great that it was difficult to look beyond the expectation of material progress. Hoover Dam was an archetype for such speculation. Constructed between 1931 and 1935 on the lower Colorado River, the massive arch gravity dam was not only meant to protect downstream communities from flooding, but to stimulate economic growth through the production of hydropower for southern Arizona, Nevada, and southern California, and to provide bountiful supplies of irrigation water for the Imperial Valley. The cost of construction would be recouped by the generation of electrical power to be produced by 16 main turbines with a rated capacity of 1,735,000 horsepower. Los Angeles and surrounding communities also would reap the benefits of the dam through the construction of the Colorado River Aqueduct. Water, indeed, was a crucial factor in the rapid growth of the city, which was transformed from a community of 500,000 in 1920 to a major urban center of 2 million in 1952. During the war, electricity generated at Hoover Dam (at the time, still called Boulder Dam) provided power for steel and aluminum mills and for Douglas, Lockheed, and North American aircraft plants—which accounted for one-fifth of the nation's aircraft production. In the postwar years, Hoover Dam power helped stimulate the growth of tourism, especially centered in the burgeoning gambling mecca of Las Vegas.[16] As one writer noted in 1936, water was "the miraculous developer . . . of the Southwest."[17]

Water and hydroelectricity were intimately woven with urban growth in general in the twentieth century. In the big dam era, rural life became less attractive as agribusiness came to marginalize small farmers, more power encouraged industrial growth, and metropolitan areas looked ever outward for new sources of water and energy. Between 1920 and 1940, the urban population of the United States increased from 54.2 million to 74.4 million.[18] While metropolitan growth was less uniform in the West than in the old manufacturing belt, for instance, metropolitan population growth as a proportion of the region's population exceeded other U.S. regions by 1960 (64 percent compared to 62.5 percent).[19] Between 1930 and 1970, striking expansion took place in several western and southern cities, in particular. San Diego increased from 94 to 307 square miles;

San Jose from 8 to 117; Phoenix from 10 to 247; Houston from 72 to 453; and Jacksonville from 26 to 827.[20]

In volume, consumption of water by irrigation and industrial purposes greatly exceeded municipal uses. In the West, nondomestic uses of water–particularly irrigation–outstripped all others.[21] But it was the consumption of electrical power where cities–along with industry–dominated the market from the power production of big dams. As historian Carl Abbott stated, "Easily transmitted over long distances, hydroelectric power facilitated the industrial growth of cities from Spokane to Los Angeles to Austin."[22]

In the Pacific Northwest, it was believed that Grand Coulee Dam could transform the regional economy. This New Deal-era project was meant to spearhead a "planned promised land" along the Columbia River. Completed in 1941, Grand Coulee was the centerpiece of an enterprise that resulted in the largest single reclamation project undertaken in the United States. The project was projected with an intended irrigated area of more than 2.5 million acres and was, for some time, the world's largest hydroelectric power generator. The dam would generate more than 6.18 million kilowatts of electricity, dwarfing even the awe-inspiring production of Hoover Dam, and eventually irrigating more than 556,000 acres of land (far less than originally contemplated). It returned more than $4 billion directly through the generation of power and became well known for providing electricity for aircraft manufacturing, the production of magnesium and aluminum, and powering the Hanford atomic installation near Richland, Washington, central to the building of the first atomic bomb toward the end of World War II.[23] Some contemporaries and a later generation of historians and others questioned this unbounded enthusiasm for the economic benefits of dam building and were convinced that too little thought went into long-range regional planning. But euphoria over economic gain wavered very little among those prepared to reap the benefits of increased power production, better navigation, municipal water supply, and irrigated farmlands.[24]

Environmental Threats: Flooding and Silting

Environmental implications of dam building were always part of the equation, but the vantage point in the big dam era strayed little from the desire to control nature and manage its resources. Indeed, dam building focused on environmental issues linked directly to resource development or a selective form of preservation. Flood control, for example, rested upon the assumption that rivers needed to be predictable and not threaten the built environment.[25] In constructing Hoover Dam, calculations revealed that previous river discharge had varied from a few thousand to more than 300,000 cubic feet per second, depending on the amount of runoff or the occurrence of floods. In the lower Colorado River

Delta area, protective levees had been constructed along a 150-mile stretch, but half of the headworks and levees had been destroyed by the river. In 1905, the river had breached the levees inundating the Imperial Valley, and discharging millions of gallons of water and creating the Salton Sea over a period of 18 months. Reclamation construction engineer, Walker R. Young, stated in 1937 that, "Without regulation, the river had little value to the lower basin area."[26] In the official history of Hoover Dam, it was noted that Lake Mead "swallows the floods."[27] Few had to be convinced of the benefit.

In the mid-1930s, the orientation of many engineers, and the policy of government agencies responsible for flood-control measures, remained focused on structural solutions to water problems, meaning the building of dams, levees, and floodwalls to curb flooding. Indeed, states, levee boards, cities and counties, railroads, and other groups also built levees and other water control works making uniform flood-control policy—beyond the building of structures at least—especially difficult. The 1928 Boulder Canyon Project Act was among the first large multipurpose water projects to include substantial flood control measures. The 1936 Flood Control Act established flood control for navigable waters and their tributaries as an essentially federal activity.[28] Despite the increased federal authority over flood control and the urgings of some experts, acceptance of flood plain management beyond a simple structural approach was slow in developing.[29] Debates over cost-benefit guidelines only complicated the picture, and, into the 1960s, reliance on structures to prevent flooding dominated policy decisions. Yet, some critics argued, "despite the billions invested in flood control dams, levees, and other works, flood losses had been steadily mounting."[30]

Like flood damage, silting was recognized as an environmental danger to be reckoned with. On the Hoover Dam project, it was estimated that the Colorado River carried annually from 88,000 to 137,000 acre-feet of silt to the delta. Concern centered on obstructing the diversion works and the effects that silt-laden water had upon irrigation, especially the building up of the ground elevation with material of "questionable value." Trapping the silt in the reservoir appeared to be a workable solution, since it was believed that at the rate of buildup it would take many years. E. W. Lane, a consulting engineer with Reclamation and former head of its Hydraulic Lab, and J. R. Ritter, chief of the Hydrology Division, Branch of Project Planning in Denver, noted that, "A number of different people have studied the rate at which Lake Mead will be filled with sediment. They are unanimous on the point that it will have a long, long life."[31]

Unfortunately, an understanding of the broader implications of silting was not well appreciated in the big dam era. Building dams, indeed, trapped silt in reservoirs, reducing the tons of deposits. However, silt contained nutrients essential to the survival of fisheries downstream and in coastal waters. Reducing silt, therefore, had lasting ecological consequences. In a similar sense, focus on the "silt menace" was not matched with an understanding of the potential problems caused by salinity. Reservoirs and other water development structures led to an increase in the salinity of many rivers, especially by the 1950s, due, in part, to highly saline runoff from irrigated land and also to reduced river flows. Increased upstream water diversion also led to seawater intrusion into estuaries and river deltas.[32] Silt "control," like flood "control," was, therefore, a kind of early environmental or non-ecological approach to the physical impact of dam building–a commitment to harnessing natural events through human technologies.

Dam Failures and Dam Safety

In the big dam era, dam safety was essentially regarded as an engineering issue linked to proper design and construction. Faith in the safety of well-designed dams was not misplaced, especially for major dams. Nonetheless, the potential for disaster or even a near failure rested on several possible factors: the unpredictability of flooding, uncertainties in the geologic setting, seepage through foundations and embankments, defects in design and construction, and liquefaction under earthquake conditions.[33]

Embankment dams, especially in comparison with masonry and concrete structures, were a particular concern in the early twentieth century. Into the 1920s, some engineers believed that restrictions on height should be placed on embankment dams in case of overtopping. Deficiencies of hydraulic fills also became an area of concern. Sheffield Dam, near Santa Barbara, a twenty-five-foot high earthfill structure, failed as a result of an earthquake in June 1925. The dam was seven miles from the epicenter of a 6.3 Richter magnitude quake. The embankment and the foundation were comprised primarily of loose silty sand, and seepage had saturated the foundation and the lower part of the fill, resulting in liquefaction of the foundation during the earthquake. Such events led to increasing interest in using instruments to study the performance of dams. The Engineering Foundation, for example, sponsored experiments on Stevenson Creek Dam (a thin arch 60-foot-high structure) in California. It had been constructed in 1926 especially for testing.[34]

The failure of St. Francis Dam, completed by the Los Angeles Department of Water and Power in 1926 to the northwest of the city, at about midnight on March 12, 1928, set off intense efforts to improve dam safety in California, especially with respect to new dam construction. Regarded as one of the worst civil engineering disasters in the twentieth century, the failure of St. Francis Dam resulted in over 400 people dead or missing and $15 million in property damage. The collapse occurred just before midnight.

Billions of gallons of water rampaged down the San Francisquito Valley, carrying concrete blocks as large as 10,000 tons downstream for a mile or more, destroyed a concrete powerhouse, roared through Saugus, and crashed through a work camp at Kemp and several other communities, before spilling into the Pacific Ocean fifty miles from the damsite.[35] Although not on a par with the destruction caused by the great Johnstown flood in Pennsylvania, which resulted in 2,200 deaths, the failure of St. Francis Dam was significant because it was a concrete structure, 205 feet in height, rather than an earthen structure. There had been some workers and ranchers who had said the dam was unsafe, but the Chief Engineer of the Los Angeles water works, William Mulholland of Owens Valley fame, had inspected the dam on the very day of the disaster and had not been concerned about some cracks in the concrete and some slight leakage. Mulholland's assessment and his close association with the design and building of the dam shattered his reputation and ended his career. The more than a dozen investigations after the disaster, however, were inconclusive and often conflicting. Most of them agreed, nonetheless, that the design had not been sufficiently reviewed by independent experts and that the foundation of the dam was weak.[36] Beyond the public outcry about the disaster, the dam-building industry was jarred by the failure of a prominent gravity arch dam. In 1929, California passed a dam safety act which placed all dams within the state, except those owned by the federal government, under supervision of the state engineer. The supervision includes design, construction, operation, alteration, and repair.[37] Although other states followed California's lead, their efforts were relatively modest by comparison.

Engineers responded to the safety issue in new dam construction by attempting to improve engineering practices. Beginning in 1929 and continuing into the 1940s, attention was given to developing a theoretical method of slope analysis for embankment dams and to further studying of slope stability and compaction of embankments. Additional attention was paid to compaction, moisture control, and liquefaction. Statistical methods were employed to estimate the flood potential of watersheds. H. M. Westergaard published an important paper in 1933 on earthquake effects on dams.[38] Nevertheless, safety problems periodically surfaced in the 1930s and early 1940s, although not on a par with the St. Francis Dam disaster. Seismic activity was discovered at Lake Mead in the mid- to late-1930s (the first shocks were felt in September 1936), and after a period of debate, many agreed that there was a connection between the reservoir and the seismic activity.[39]

Despite the increased study of dam safety measures, each case of failure posed some unique problems and often led to a variety of assessments. The safety issue in the big dam era was not used as a rationale for slowing the construction of dams or questioning their value to society. Building new, large dams—while most often incorporating, or at least taking account of, the latest safety research—remained a greater priority than making safety modifications on existing dams.[40]

Despite legal and professional efforts taken to ensure the dams would be built to be safe and secure, dam building in the mid- and late-twentieth century remained an art dependent upon the skills and judgments of human beings. This was brought home in a forceful–and tragic–manner on June 5, 1976, when the Bureau of Reclamation's newly completed Teton Dam, in southeastern Idaho, failed during the first filling of the reservoir. Built to provide hydroelectric power production and increased irrigation supply for the upper Snake River Valley around Idaho Falls and Rexburg, Teton Dam was a 305-foot-high, 3000-foot-long earthfill dam built across the Teton River. The central core-zoned earthfill dam failed and released eighty billion gallons of water which roared downstream in a 20-foot wave taking eleven lives, thousands of head of cattle, thousands of buildings, and tons of topsoil. At least 25,000 people were forced from their homes.[41] The failure occurred late on a Saturday morning, and, fortunately, many downstream residents were successfully warned of the impending disaster. The death toll could clearly have been much higher had the collapse taken place in the late evening/early morning hours.[42] Don Pisani has argued the failure severely embarrassed Reclamation and signaled that "the boom years were over" for western water development.[43]

In the aftermath of the Teton Dam failure, intensive studies were undertaken to determine the cause. Those studies were hampered by the fact that the rush of water down the canyon washed away geological evidence that would speak to the condition of the foundations and the actual effect of measures taken (e.g., the injection of a cement "grout curtain" wall into the cracked and relatively porous rock foundation) to prevent water from seeping through the foundation and to prevent erosion at the interface between the foundation and the earthen structure. After a six-month-long investigation, an independent panel of engineers appointed under the auspices of both the U.S. Department of the Interior and the State of Idaho reported that "because the failed section was carried away by the flood waters, it will probably never be possible to resolve whether the primary cause of leakage in the vicinity of Sta. 14+00 was due to imperfect grouting of the rock . . . or cracking in the key trench-fill, or possibly both." Specifically, the independent panel reported that:

> The fundamental cause of failure may be regarded as a combination of geological factors and design decisions that, taken together, permitted the failure to develop. The principal geologic factors were (1) the numerous open joints [i.e., cracks] in the abutment joints, and (2) the scarcity of more suitable

materials for the impervious zone of the dam than the highly erodible and brittle windblown soils. The design decisions included among others (3) complete dependence for seepage control on a combination of deep key trenches filled with windblown soils and a grout curtain . . . and (4) inadequate provisions for collection and safe discharge of seepage or leakage which inevitably would occur through the foundation rock and cutoff systems. . . . In final summary, under difficult conditions that called for the best judgment and experience of the engineering profession, an unfortunate choice of design measures together with less than conventional precautions was taken to ensure the adequate functioning of the Teton Dam, and these circumstances ultimately led to its failure.[44]

If nothing else, the Teton Dam failure served—and continues to serve—as a compelling reminder that every dam, and every dam site, presents a unique set of technological problems that must be addressed and solved on an individual basis in order to provide for a safe and enduring water storage structure. The failure also resulted in several acts and a number of federal programs, including at Reclamation, designed to better insure future dam safety.

A Fish Story: Conflicts over Resource Development[‡]

Along the Columbia River, in particular, dam building came into direct conflict with fish habitats. Today, one-third of the original Columbia Basin salmon habitat is blocked by dams.[45] Anadromous fish, including Pacific salmon and steelhead trout, must return to their spawning grounds to reproduce and depend on the free flow of rivers to accomplish that goal. The building of Bonneville and Grand Coulee Dams, in particular, intensified the growing debate over the effects of dams on migratory fish and their spawning sites. Contemporary conservationists and several historians have claimed that few people, particularly the dam builders, gave much thought to the plight of the salmon and steelhead, and that high dams virtually eliminated them from much of the Columbia and other rivers. While there is little doubt that the new high dams caused great loss to the fish population, the story is somewhat more complex.

The growing debate over dams impeding the migration of fish was not so much couched in environmental terms in the big dam era, but in terms of

---

[‡] The issues of anadromous and catadromous fisheries and the effects of dams upon them are exceptionally controversial. The literature, both technical and popular, is extensive and available for those who further wish to explore impacts on fisheries. Suffice it to say here that more detailed history could be done on the issues involved. That is another study for another location. This study has space only to touch on some of the issues, from fish ladders to hatcheries to trucking and barging to nitrogen "poisoning." (Editor).

competing resources. As an economic resource, rather than as an essential component of the riverine ecology of the Columbia and other rivers, anadromous fish gained significance in the eyes of the federal dam builders of the era.[46] In addition, dams were only one cause, albeit a major cause, of declining fish populations along the Columbia and other rivers. On this latter point, several forms of resource exploitation—mining, cattle grazing, lumbering—created water pollution problems on the Columbia and Snake Rivers because of runoff which adversely affected fish stocks. Irrigated agriculture increased demand for water and killed salmon with its diversion dams and unscreened ditches. Small dams, going back to the late nineteenth century, also were destructive to fish populations. By the 1930s, vast areas of spawning grounds had already disappeared, and catches had declined dramatically through intensive commercial harvesting and canning operations. Efforts at conservation had been meek at best.[47]

The multipurpose dams, without a doubt, raised the stakes for anadromous fish in the Pacific Northwest and elsewhere. Fear of the eradication of salmon and other migratory fish became a galvanizing issue in the early fight for fish and wildlife conservation. Nascent conservation groups, such as the Izaak Walton League (founded in 1922) and later the National Wildlife Federation (formed in 1936), battled for the protection of aquatic habitats throughout the country. In 1934, the Fish and Wildlife Coordination Act required the builders of federal or federally licensed dams to consult with the Fish and Wildlife Service to prevent loss or damage to wildlife resources. This was the first law intended to protect wildlife in water projects.[48]

Along the Columbia River, the most concern focused on the fate of the salmon; the issue was most especially the survival of a vital industry and a fragile resource. Those directly benefitting from the survival of the salmon—sportsmen, canners, Native Americans—loudly objected to the potential effects of the high dams on fish migration. And conservationists chimed in to support those interests. An article in *Nature Magazine* claimed, not entirely accurately, in 1938, on the heels of the construction of Bonneville Dam:

> The sad part of projects of this sort [dams and inland waterway and irrigation projects on the Columbia, Willamette, and Snake Rivers] is that the surveys upon which they are based are concerned only with the feasibility of the construction. Little or no attention is paid to what resources may be destroyed in the process.... On the destruction side, the Columbia and Willamette dams mean the eventual destruction of the salmon industry.... [The industry] supports thousands of families and keeps them off the relief rolls.[49]

The development of fishways at Bonneville, designed to allow the fish to migrate around the dam, was viewed by many as a workable solution to the problem, but, nonetheless, led to some controversy. While the Corps recognized the need for passages to protect anadromous fish well before Bonneville's construction, the rush to begin the work to provide unemployment relief resulted in a lag in refining details of a fish-passage plan.[50] In the end, Bonneville included three concrete "fish ladders" and two sets of fish lifts.

Some were impressed with the success of these measures. A 1951 study proclaimed: "Perhaps the most amazing feature to the tourist is not Bonneville's powerhouse or even its giant lock, but the amazing fish ladders. . . ."[51] Others were cautiously optimistic that the ladders and locks would have some benefit to the now struggling fisheries industry.[52] Still others were dubious, arguing either that the cost per fish was too high to merit success, or the innovative technology failed to deliver on its promise. Assistant Chief of Engineers Thomas Robins (North Pacific Division Engineer in the 1930s, when Bonneville was being constructed), among others, defended the Corps against charges that juvenile fish mortality in dam turbines was high and thus threatened the salmon population. More problems arose, he argued, when juvenile fish were flushed down spillways, where they encountered water saturated with nitrogen gas which could cause gas bubble disease.[53] While more intensive study continued on fish ladders and locks, juvenile fish and turbines, and other threats to anadromous fish, uncertainty remained as to the effectiveness of the plan at Bonneville to protect the salmon and the fisheries industry and still to produce hydroelectric power.[54]

Grand Coulee Dam posed a related, but different, problem for the salmon and the fisheries industry. The Bureau of Reclamation, which had taken charge of the dam project in 1933, faced the same sort of scrutiny over its fish policy as had the Corps at Bonneville. Critics claimed that Reclamation cared little about salmon if there was a choice to be made between fish and other economic contributions of the project. However, Reclamation initially recommended the development of a flume and a mechanical elevator to carry fish around the dam. The plan proved unsatisfactory because Grand Coulee would be too high. Alternatives were considered, but without resolution. Temporary fish ladders were authorized during the early construction of the dam, but finally, in 1939, an emergency plan for capturing chinook and blueback salmon was announced. In May, fish traps at Rock Island began operating and the first trucks carrying salmon motored around the dam site. Focus turned to hatcheries, and in August, 1941, 50,000 chinook salmon raised at a hatchery in Leavenworth were released into the Entiat River. Although Reclamation announced that the salmon transplant experiment was a success, many remained skeptical.[55]

The dams on the Columbia River had caused significant ecological changes, such as increasing the temperature of the water, and had, at the very least, accelerated the decline of the salmon runs. At Grand Coulee, which now sealed off the upper Columbia to salmon, a make-shift solution to preserving the fish and the fisheries had replaced the natural migration with questionable results. The diversion of salmon to other locations may have preserved some salmon, but above the dam the species could not survive as before.[56]

The Echo Park Controversy

Until some years after World War II, concern about the environmental implications of dams had focused largely on site-specific issues: flooding and silting; dam failures as opposed to general attention to dam safety; and salmon on the Columbia and Snake Rivers. Questioning the inherent value of dams and their impact on the environment rarely entered the discourse. The Echo Park controversy changed this. It helped move along a process that inevitably resulted in fracturing a rather uneasy consensus on the conservation value of dams, polarizing the sides, and increasingly casting the Corps and Reclamation as anti-environmental. It also saw the coming of age of several environmental organizations, and suggested that the public could become involved in environmental issues as never before.

At the heart of the Echo Park controversy was the effort to save national parks and national monuments from inundation and resource exploitation. Echo Park became the biggest battle over wilderness preservation since Hetch Hetchy. By the 1950s, the old conservation movement was maturing, broadening its perspective to embrace a range of issues including wilderness preservation, wildlife habitat protection, outdoor recreation, a burgeoning interest in pollution, and the perpetuation of national parks and monuments. The Izaak Walton League was the largest and most politically active group. The National Wildlife Federation was linked closely with hunters and fishermen. The Audubon Society broadened its scope to embrace wildlife conservation in general, and the Wilderness Society, which had begun in the East, became known for its advocacy of wilderness protection in the West. The Sierra Club, along with the Wilderness Society, was beginning to develop a reputation as an outspoken critic of government dam builders, among others, and to expand its interests well beyond its California roots.[57]

Big dams became the harbinger of a threatened public landscape. At first, a dam at Echo Park—located at the point where the Yampa River joins the Green River along the Utah-Colorado border—was just another proposed project for the Bureau of Reclamation. It was part of the Colorado River Storage Project initiated in the 1940s to make additional water available for rural and urban residents in four states—Wyoming, Colorado, New Mexico, and Utah—in the Upper Colorado River Basin. It also was part of a major "participating project," the Central Utah Project. The construction of the dam, however, would inundate Echo Park, a scenic valley upstream from the proposed dam, and miles of nearby Lodore and Yampa Canyons, all of which lie within Dinosaur National Monument. In 1915, President Woodrow Wilson had designated 80 acres of the area, where dinosaur bones had been discovered, as a national monument. In 1938, President Franklin Roosevelt expanded the site to 210,000 acres.[58]

The Green River, a major tributary of the Colorado, narrowed as it passed through the cliffs at the core of Dinosaur National Monument, immediately below Echo Park, and offered an attractive dam site for generating hydroelectric power for the surrounding states. Since the national monument was located in a remote spot on the Colorado Plateau and had seen few visitors, the dam builders saw no reason why the National Park Service, which had jurisdiction over Echo Park, would object to the proposed construction. During World War II, before Reclamation made a formal claim to the site, the National Park Service displayed no strong reaction to the idea. However, the issue became a jurisdictional and a political matter after the war. Reclamation had enthusiastic support for the dam from the Upper Basin states of New Mexico, Colorado, Utah, and Wyoming, but the National Park Service refused to support it. Secretary of the Interior Oscar Chapman now had to mediate the dispute, and if he agreed with Reclamation, legislation could be written incorporating the new dam and then sent off to Congress for its approval. After public hearings in April 1950, Chapman was persuaded that the proposal had merit and approved the building of Echo Park Dam.[59]

The decision to build Echo Park Dam produced energetic opponents, none more vocal and committed than preservationist groups. It is not a cliche to suggest that the controversy exposed a clash of environmental values, resulting in the galvanizing of a preservationist coalition that would continue to question the river management policies of Reclamation and the Corps. Echo Park Dam would not violate Reclamation's commitment to resource conservationism. To the contrary, the new dam would help to tame an unruly river and to maximize the use of riverine resources in the region. Preservationists were particularly concerned that building the dam would not only destroy a unique wilderness area, but would set a terrible precedent for exploiting resources in America's national parks and monuments. They had reason to be alarmed because between 1945 and 1950, Olympic National Park, Grand Canyon National Park, Glacier National Park, Jackson Hole National Monument, Superior National Forest, and the Adirondack Forest Preserve had been threatened by various economic interests. But at Echo Park, a federal bureau was making plans to "invade" a national monument.[60] U. S. Grant III, a former general officer with the Corps and president of the American Planning and Civic Association, chided Reclamation for the decision to build the Echo Park Dam and Split Mountain Reservoir: "The Trojan Horse in our national park system, model 1950, is now driven by electricity supplied from water power impounded behind great dams."[61] Bernard DeVoto, then a writer for *Harper's Magazine*, wrote an exposé in the July 22, 1950, issue of the *Saturday Evening Post,* of the decision to build the Echo Park Dam: "Shall We Let them Ruin Our National Parks?" DeVoto served on the Secretary of the Interior's Advisory Board and was privy to a debate over the building of the dam between Bureau Commissioner Michael Straus and Park Service Director Newton Drury. Since Drury could not take his feelings public, DeVoto decided to air the issue and also attacked Reclamation and the Corps for their proposals to put dams in national parks.[62] The article and the decision to go ahead with the dam mobilized wilderness advocates. Eventually, the leaders of 32 opposition organizations created a lobbying group, the Citizens Committee on Natural Resources, to take the battle to Washington.[63]

Political allies "proved vitally necessary" to the campaign to save Echo Park because neither the modern regulatory apparatus for environmental protection nor access to a wide array of media outlets were available to wilderness advocates at mid-twentieth century. Allies would have to be courted in eastern, midwestern, and southern states, where skepticism was high about the viability of hydropower to pay for multipurpose projects. In addition, concerns were deep about adding more acreage to the agricultural economy. On another front, California might be counted on for opposition to the dam because of its claims on Colorado River water. Other potential allies, but equally strange bedfellows, were fiscal conservatives opposed to federal public works and private utility companies fearing public power. In Congress, the strongest opposition to Echo Park came from the House of Representatives.[64]

For much of the first half of the 1950s, the future of the Echo Park Project was uncertain as government officials vacillated amidst a blizzard of reports. Opponents stepped up their protests, and lawmakers squabbled. Bending to pressure, Secretary of the Interior Chapman reversed his previous pronouncement and established a task force in November 1951 to explore alternative sites to Echo Park. When he left office in January 1953, newly appointed Secretary of the Interior Douglas McKay would have to decide where the Eisenhower Administration would stand on the issue. McKay did not favor withholding natural resources from economic development, even if they were in national parks and monuments. However, before taking a position on the specific proposal of Echo Park, he dispatched Undersecretary of the Interior Ralph Tudor (a former U.S. Army Corps of Engineers officer) to the Dinosaur National Monument site to reevaluate Reclamation's calculations about the rates of evaporation at the proposed reservoir.[65] Such figures were vital to determine the ability of a damsite to store a maximum amount of water at a reasonable cost. Tudor reported that the Echo Park site was vastly better than alternatives. Based on the findings, McKay approved Echo Park Dam as part of the Colorado River Storage Project in December 1953.[66]

Congress took up the authorization of the project, with the inclusion of Echo Park Dam, in 1954, and with bipartisan support, the chances for passage seemed good. However, protests of the proposed legislation elevated the issue to national status, and the intensity of the debate started several years before had not been quelled. Fred M. Packard, executive secretary of the National Parks Association, a leading conservationist group, strongly asserted, "The issue is clear-cut, in spite of the fog of technical data and irrelevant side issues that have confused its comprehension by Congress and the public."[67] Probably the most devastating critique came from Sierra Club Executive Director David Brower, who challenged Reclamation's evaporation figures. Using rather elementary math and a few charts, he proved to a congressional subcommittee that Echo Park was not likely to save as much water as supporters' argued. Also other evidence was surfacing that the Colorado River Storage Project was more of a power-generation project than a water storage project as argued by Reclamation. In this context, power anticipated from Echo Park would be expensive, making the site even less economically desirable.

All this, plus aesthetic arguments about the inundation of a national monument, worked against support for Echo Park in Congress. The final straw, however, may have been the threat by the wilderness groups that they would fight the project as a whole if Echo Park was not removed from the proposed legislation. An open letter to the "Strategy Committee" of the upper Colorado River Storage Project, written by the Council of Conservationists, stated:

> We want you to know that we will fight with every honest device at our command if the Upper Colorado bill continues to include, or require--now or at some later date--a dam at Echo Park, or elsewhere in a national park or monument.

What had been a loose-knit group of local and regional conservation groups had acquired a loud, national voice. On April 11, 1956, President Dwight Eisenhower signed Public Law 485 authorizing the Colorado River Storage Project–without Echo Park Dam.[68]

The conventional wisdom has it that the price the preservationists paid for saving Echo Park was a concession on the Glen Canyon Dam, which would substitute for it in the project. From the start, however, Echo Park Dam and Glen Canyon Dam had been interconnected in Reclamation's plans to develop the Colorado. In addition, the aesthetics of Glen Canyon were not well known in the 1950s, and after all, opponents of the Echo Park Dam clearly focused on saving Dinosaur from intrusion. In the debate over evaporation, Brower had made the strong case against a high dam at Echo Park, believing a high dam at Glen Canyon—a site barely within the Upper Colorado River Basin—was one more argument against building a dam at Echo Park. At the time, Reclamation preferred the high dam at Echo Park and a low dam at Glen Canyon because the Upper Basin states wanted a site "safely within their territory" to protect their water rights. A dam at Glen Canyon would likely supply power downstream to Arizona and California rather than in the Upper Basin, but with the addition of the Echo Park Dam, California would not be viewed as singularly benefitting from the project. In Congress, Reclamation could make that case and likely garner broader support for the overall project.[69]

Reclamation did not get all it wanted in Congress, but neither did the wilderness advocates. Although the Echo Park case was a major victory for national park and monument preservation and a galvanizing event in the burgeoning modern environmental movement, regrets remained over Glen Canyon. Storage produced by the dam sometimes intruded into the Rainbow Bridge National Monument. A new round of litigation and debate followed, caustic enough to create major tensions within the environmental community over what compromise may have been struck in 1956 and the potential loss of scenic lands.[70] David Brower certainly had regrets, believing that he was partly responsible for the "death" of Glen Canyon.[71] There were to be no clear victories for either side.

Dams and the Modern Environmental Movement

As the big dam era was coming to an end in the 1960s, controversy surrounding dam building took a decidedly different tone. The substantive issues around which debate had occurred prior to that time largely remained the same—flooding, silting, and salinity; dam safety; dam siting and displacement of people; and threats to natural environments. But the context was different. Wallace Stegner noted in 1965:

> Water, once paramount, has become secondary. The questionable dams are never simple water holes. What dictates the damsite is as often as not the power head: efficient generation of power calls for a higher dam, and hence a bigger lake, than a simple waterhole does.[72]

Redefining "risk" in an increasingly sophisticated and complex environmental era meant greater challenges to building and utilizing dams, not simply criticisms of specific dams. The act of dam building was increasingly coming into question, with trade-offs that were much more general in perception than the displacement of a special group of people or the inundation of a particular valley. Dam building was being questioned for threatening wild rivers and endangered species, for overbuilding structures at the expense of natural sites, and for placing too much emphasis on unrelenting economic growth. In addition, as historians have noted, "Beginning in the 1960s, an increasingly urbanized, educated society focused more on recreation, environmental preservation, and water quality than on irrigation, navigation, or flood control."[73]

In this setting, Reclamation and the Corps were increasingly scrutinized. Beginning in the late 1960s, it became more common to question their organization, their mission, and their cost/benefit ratios. Elizabeth B. Drew, in a 1970 issue of the *Atlantic*, expressed the opinion that the Corps was unwilling to change with the times, and she quoted its chief, Lieutenant General F. J. Clarke, as saying "With our country growing the way it is, we cannot simply sit back and let nature take its course." She added that despite the rigidity of Corps policy, few politicians were willing to criticize the Corps publicly because "almost all of them want something from it at some point—a dam, a harbor, a flood-control project." A *Nation* article in 1966 branded the Corps as "the pork-barrel soldiers" with anti-environmental aims.[74] How would dam builders respond to new challenges in an era when dams and reservoirs were no longer praised unreservedly for harnessing untamed rivers?

What made the modern environmental movement so remarkable was the speed with which it gained national attention in the late 1960s. Nothing epitomized that appeal better than Earth Day. The idea began as a "teach-in" on the model of an anti-Vietnam War tactic. In *Earth Day—The Beginning*, the staff of Environmental Action (formerly Environmental Teach-In, Inc.) declared:

> On April 22, [1970,] a generation dedicated itself to reclaiming
> the planet. A new kind of movement was born—a bizarre alli-
> ance that spans the ideological spectrum from campus militants
> to middle Americans. Its aim: to reverse our rush toward
> extinction.

Across the country, on 2,000 college campuses, in 10,000 high schools, and in parks and various open areas, as many as 20 million people celebrated what was purportedly "the largest, cleanest, most peaceful demonstration in America's history." In form, Earth Day was so much like a 1960s-style peace demonstration that the Daughters of the American Revolution insisted that it must be subversive. In fact, however, it was pitched at moderate activists, somewhere between the New Left and the older conservationist groups, such as the Sierra Club and the Audubon Society. As a symbol of the new enthusiasm for environmental matters, and as a public recognition of a trend already well underway, Earth Day served its purpose.[75]

The Richard Nixon Administration gave its blessings to Earth Day. In his first State of the Union message, the President declared, "Clean air, clean water, open spaces—these should be the birthright of every American." On January 1, 1970, four months earlier, Nixon signed the National Environmental Policy Act (NEPA) of 1969. While opposing the bill until it cleared the congressional conferees, the administration ultimately embraced it as its own. By identifying his administration with environmentalism, Nixon wanted to address the issue on his own terms. Many people trumpeted their approval of the President's gesture; others reserved judgment or remained cynical.[76]

NEPA was far from "the Magna Carta of environmental protection" that some people proclaimed, but it nonetheless called for federal bureaus and agencies to consider environmental effects, and ways of reducing those effects, before funding, approving, or carrying out projects. NEPA did not mandate particular results, but it did arguably promote efforts to preserve and enhance the environment. It particularly emphasized the application of science, disclosure, and public participation in the decision making process and in the search for solutions. With respect to integrated river basin management, the National Environmental Policy Act encouraged Reclamation and the Corps to give more attention to environmental considerations and also gave environmental agencies more say in the process.[77]

NEPA, in addition, required federal agencies to prepare environmental impact statements (EIS) for those projects that would significantly impact the environment. These provisions, for instance, stipulated that early public notification should be given that an EIS was being prepared and that citizen comments should be part of the final statement. An EIS could lead to decisions to place limits on construction and could lead to resiting of structures. The new law gave opponents of dams a tool to slow down or impede construction.[78] According to one source, by the mid-1970s seven Corps projects were halted as the result of environmentalist litigation. Of the remaining 61 environmental suits being litigated, 27 involved alleged violations of NEPA.[79]

NEPA provided substantial opportunity for citizen participation, especially through the review of EISes and other environmental documents. It established the Council on Environmental Quality (CEQ) to review government activities pertaining to the environment, to develop impact statement guidelines, and to advise the president on environmental matters.[80] The CEQ was essentially a presidential instrument, and governmental environmental programs remained widely dispersed. In early 1970, the Council recommended the establishment of a Department of Natural Resources and the Environment to centralize several departments and bureaus into one agency, but the departments of Interior, Agriculture, and Commerce resisted such a consolidation of programs.

In June it was announced that pollution control programs and the evaluation of impact statements would be the responsibility of a new body—the Environmental Protection Agency (EPA), which began operations in December 1970 under the direction of William Ruckelshaus. Initially, it included divisions of water pollution, air pollution, pesticides, solid waste, and radiation. Other natural resource and environmental programs remained in other agencies, especially the Departments of Commerce and Interior. More significantly, EPA did not have single overall statutory authority for environmental protection; it simply administered a series of specific statutes directed at particular environmental problems.[81]

Despite some of the tentative steps of NEPA and limits in EPA's authority, national environmental policy was undergoing a substantial change in the 1970s. As early as the late 1940s and 1950s, social scientists and others already had begun to attack government agencies with responsibility for conservation programs. They found them self-serving and denied that their experts had special knowledge of what constituted "the public interest." The practice of delegating discretionary authority to administrative elites in the Progressive Era was questioned severely. In its place, calls came for greater political accountability among government bureaucrats, more congressional control, public access through the courts, and opening of the decision making process to any affected interest.[82]

By the mid-1970s, environmentalism was a solidly fixed national movement. Mainstream environmental groups responded by taking greater initiative in helping to draft new legislation, pressing for the implementation of existing legislation, focusing on the environmental impact review process, and monitoring government agencies. Demand rose to strengthen conservation laws for managing resources and to step up efforts in nature protection. Criticism rose against the traditional government role of promoting economic growth at the expense of resource depletion. In addition, the courts became an important battleground as more litigation tested key regulatory provisions. Inevitably, such extensive changes in environmental policy-making brought about significant changes in the relationship between the environmental community and agencies like the Corps and Reclamation.[83]

Not surprisingly, Reclamation and the Corps balked at the rising environmental criticism of their efforts in the 1960s and 1970s, often resisting the conclusion that the context in which water and dam projects found themselves had changed. Justifying new dams on the grounds of economic growth, traditional flood control concepts, recreational opportunities, and so forth had much less resonance in an era when charges of pork-barrel projects and environmental degradation of natural landscapes were increasingly heard not only by environmental interest groups, but by a public increasingly suspicious of federal programs of any kind. Supporters of Reclamation and the Corps, however, claimed that the agencies made real efforts to adjust and to change in the face of intensifying criticism. For example, researchers Daniel Mazmanian and Mordecai Lee asked the question of the Corps: "Can Bureaucracies Change?" Their conclusion was that they can. Their assessment was that the Corps was aware of its "lack of capacity" to confront broad ecological questions as early as the mid-1960s. In response, it established an Environmental Resources Branch within the Planning Division in 1966.

In April 1970, the Chief of Engineers established the Environmental Advisory Board (EAB) to examine policies and programs, to identify problems, and to recommend changes. The board was unique in the sense that it was not established as an in-house body, but composed of members of environmental groups, albeit relatively moderate groups, outside the Corps. Among the board's activities was recommending the establishment of the Environmental Reconnaissance Inventory, a comprehensive resource inventory initially implemented in four locations in the mid-1970s.[84] The board concept certainly has had limits, however. Not surprisingly, consensus on issues was not easy to reach as the board periodically found itself divided on policy. Also, attention to pressing immediate issues, such as lawsuits or congressional deadlines, impeded effort at long-range planning.[85] The relationship between the Corps and the environmental community remained cautious and often adversarial in the 1970s despite the internal changes in the bureau. Environmentalists often discovered that projects stopped by court injunctions were ultimately under construction after revised EIS reports were prepared. As historian Jeffrey Stine has observed, in the 1970s "The Corps . . . regarded the environmental legislation as a mandate not to stop building, but to build in the best possible way. . . ."[86] Like the Corps, Reclamation had projects held up by environmental protests and lost some political support in Congress, but it managed to continue developing some important projects in the 1970s.[87] Neither bureau, however, rose to the heights of the big dam era.

Flood Control and Non-Structural Alternatives

As much as the environmental context had changed beginning in the mid-1960s, and as much as Reclamation and the Corps seemed willing to, or were forced to, bend to changing national environmental policy, divisive issues over the impact of dams remained. Familiar issues, however, were colored and flavored by growing resistance to the value of big dams promoted by an array of critics. The Corps continued to argue in favor of flood control projects as beneficial to local communities. But increasing pressure was being exerted on the Corps to consider non-structural alternatives to dams and levees as flood-control devices. Suggestions were made for the development of detention ponds for flood waters or for expanding green belts along flood-plain lands near streams. Section 209 of the 1970 Flood Control Act promoted "multiobjective planning," a form of planning that recognizes noneconomic values such as environmental quality along with economic interests.[88]

Water Quality and Other Environmental Impacts

A substantial context for change in dealing with repercussions of dam building focused on water quality and related issues in the new environmental era. Sentiment grew that dams were "the least reversible form of river alteration" and resulted in deleterious physical changes in the nation's river basins.[89] Whereas issues of silting and salinity received at best modest attention before the mid-1960s, a variety of more direct questions arose after that time about the quality of water resulting from dam and reservoir construction, the buildup of silt behind reservoirs, and the residual impacts of intense irrigation. As Donald Pisani noted, "Environmentalists saw clear limits to economic growth and worried about such problems as siltation, alkali buildup, and the poisoning of groundwater with herbicides and pesticides."[90]

By their very nature, dams and reservoirs changed the riverine ecology.[91] Certainly, reservoirs can improve water quality for many users, and dams of different design and operation can produce different effects downstream. "The ability of large dams to compensate for the unpredictability" of nature, one study noted, "is what makes them so attractive . . ."[92] However, changes occur from a free-flowing environment to a standing or lake environment, drowning a variety of native flora and fauna, ruining forests, altering or destroying riparian vegetation and habitat, encouraging evaporation which concentrates salts, and sometimes creating mud flats. Water released from dams is likely to be low in oxygen, thus threatening river life. In deep reservoirs, the water column can stratify by temperature. Little oxygen or light can reach the lower strata, and the upper stratum becomes warmer. This change can create a forbidding environment for cold-water fish and can allow for the habitat to be taken over by other species. Dams can alter water temperature in other ways as well. At Glen Canyon Dam, for example, water released into the Colorado River is approximately 20 degrees colder than would be natural, which destroys many native organisms. Much of the river cannot produce algae, which in turn disrupts the food chain.[93] On the other hand, such situations sometimes result in flourishing trophy cold water fisheries.

While having constant, predictable in-stream flows has been valuable for irrigators and other water users, artificially regulated flows produce a number of problems. Native river animals and plants have a difficult time adjusting to constant flows and constant temperatures when the natural rhythms of rivers are altered. Without high flows, silt does not get flushed from the streambed gravel, harming many species of fish and insects that depend on clean, oxygenated gravel for their eggs and larvae. Artificially and naturally low flows can cause back channels and sloughs to dry up, thus destroying primary spawning areas for trout. In 1987, a low flow from Palisades Dam on the Snake River killed approximately 600,000 cutthroat and brown trout, mostly juveniles, along with much of the aquatic food chain.[94] In spite of this problem the Snake River does support a blue ribbon trout fishery in the area.

Intensive irrigation has led to serious salinity problems in many agricultural regions. As Donald Worster stated, "What nature has taken geological eons to achieve, the leaching of salts from the root zone of plants, the irrigator undertakes to do in a matter of decades." Intensive irrigation can lead to a rising water table, bringing dissolved salts to the root zone or to the surface. Growers in the Imperial Valley in Southern California faced this daunting prospect, and by the early 1970s spent more than $66 million on tile drains and canal linings to capture saline runoff and to discharge it elsewhere. They also faced shifting to salt-tolerant crops, even though they yielded less income. An alternative was to consume more water, if possible, to flush the salt deposits.[95]

The salinity issue took on international proportions in the Colorado River Basin. A 1944 treaty had guaranteed Mexico 1.5 million acre-feet, but the agreement did not address water quality. Over time, Mexico was receiving heavily saline drainage from irrigated fields in the United States. In 1961, the Wellton-Mohawk Irrigation District, along the lower Gila River in Arizona, discharged drainage water rich in salt into the Colorado River, immediately above Mexico's diversion canal, and essentially doubled the average annual salinity of the flow across the border. The U.S. denied that its treaty included any obligation on water quality issues, but fresher water was released from American dams, and a channel was built to divert the drainage around the Mexican intake in 1965. This proved to be a temporary solution, and finally, in 1973, both countries signed an agreement to settle the dispute. Realizing that similar disagreements could break out again, Congress passed the Colorado River Basin Salinity Control Act in 1974. All along the Colorado River, use and reuse of the water had diminished the flow and contributed to degradation of the quality of water not only as it crossed the border into Mexico, but also in the Imperial Valley and for the Metropolitan Water District of Southern California.[96]

The environmental repercussions of dam building are complex and not easily resolved— some have argued they are irreversible. One solution was to cease building dams on the remaining free-flowing rivers in the country. River preservation was given a boost by the Wild and Scenic Rivers Act of 1968. While not giving natural features legal standing *per se*, it provided an alternative to resource development by protecting the shorelines of designated rivers from federally permitted development. The act was an important sign that the perception of rivers as a commodity in the traditional sense was changing. Yet by the 1990s, the mileage preserved in the system was less than one percent of the nation's natural river courses.[97]

Decaying Dams: The Impending Crisis in Dam Safety

In the years since the mid-1960s, concern about the safety of specific dams turned into uneasy anxiety about the safety of all existing dams. Those built since mid-century, especially federal dams, had a good safety record overall, but there was an increasing likelihood of potential disasters. Some had uncorrected safety problems that had been detected but not addressed. In addition, of the 49,422 large dams (twenty-five feet or more in height and impounding more than 16.3 million gallons) 39,000 dams had never been inspected by state or federal engineers. The largest percentage of these dams were non-federal structures, where the regulatory gap was the greatest. Before the mid-1970s indicators of bigger problems could be found in deterioration and corrective actions required at dams. For instance, in 1965 Lahontan Dam, a Reclamation dam in western Nevada, was found to have crumbling concrete in its spillways. While this was a serious problem which did not alone threaten the safety of Lahontan Dam, it took twelve years for the Bureau to work with the local irrigation district which managed the dam to produce a formal proposal to rectify it. Navajo Dam, a Reclamation structure in northwestern New Mexico, was completed in 1963 and was found to leak as much as 1.8 million gallons per day by 1977. Even though all dams leak, this was considered to be excessive and dangerous and required corrective action. Canyon Lake Dam in Rapid City, South Dakota, failed in June, 1972; Walter Bouldin Dam, an Alabama earthfill structure, failed by erosion in February, 1975; Bear Wallow Lake Dam, an embankment dam in North Carolina, failed in February, 1976. A small and antiquated earthen dam at Toccoa, Georgia, regarded as a serious hazard, failed in 1977 overwhelming Toccoa Falls Bible Institute and taking thirty-nine lives. And there were others.[98]

These dam failures, accentuated by the failure of Teton Dam, even more so than the St. Francis Dam disaster of 1928, were a wake-up call about the deteriorating condition of dam inventories as well as the safety of dams due to design and construction flaws. Dam safety engineering had led to improved approaches to embankment dam analysis in the 1960s, studies of embankment liquefaction (after the San Fernando earthquake in 1971), and to the National Dam Inspection Act in 1972.[99] After the Teton Dam disaster, federal agencies reviewed safety practices and established an interagency committee to coordinate dam safety programs, which evolved into the Interagency Committee on Dam Safety. It issued management guidelines for planning, design, construction, operation, and regulation of dams in the United States. The Corps began seismic investigations and established the Dam Safety Assurance Program in 1977. In 1978, Reclamation began its Safety Evaluation of Existing Dams Program, independent of other offices in the Bureau of Reclamation, to determine if it needed to move into a modification program to make a dam safe. The National Dam Inspection Program developed an inventory of about 76,000 dams, classified according to the potential for loss of life and property.[100] Between December 1977 and October 1981, approximately 8,800 "high hazard" dams were inspected, and specific actions were recommended, ranging from additional inspections to emergency repairs. (Subsequent inspections, investigations, and remedial work became the responsibility of the owners of the dam.)[101] In 1979, the government published *Federal Guidelines for Dam Safety* to encourage high standards among federal agencies. A Presidential Executive Order in July 1979, placed responsibility for coordinating dam safety in the Federal Emergency Management Agency.[102]

With the end of the "big dam era," dam safety was not so much a question of carefully monitoring new construction as it was being vigilant about the deterioration of a large and an increasingly aging inventory of dams and reservoirs. While the Corps and Reclamation moved to shore up safety programs for federal dams, they did not have jurisdiction over the thousands of non-federal dams throughout the country. A related question, which extended beyond specific considerations of safety, was: What is the useful life of these dams? This was both an engineering and an environmental issue.

Two Fish Stories: Pacific Northwest Salmon and the Tellico Snail Darter

In the wake of the big dam development along the Columbia and Snake Rivers, the Corps invested more than $60 million in fisheries research in an effort to have salmon and steelhead populations coexist with the multipurpose projects. In 1955, the large Fisheries-Engineering Research Laboratory was constructed at Bonneville, further committing the Corps to addressing the issue of adult fish passage beyond the dams. Research efforts also went into studying degradation of habitats and fish diseases, such as gas bubble disease, which had become a serious problem after the Corps completed the dams on the lower Snake River in the 1960s and 1970s. One estimate suggested that gas supersaturation killed

70 percent of the fish migrating downstream in the lower Snake. By 1971, the Corps organized the Nitrogen Task Force to confront the problem.[103]

Criticism of both the Corps' dams and its research continued. While the Corps' research in the 1950s focused on adult fish populations, issues involving juvenile fish were often neglected. For example, dams continued to be built without much information about their impact on young salmon. By the late 1960s, the Corps was involved in cooperative studies with the National Marine Fisheries Service and state fisheries to improve bypasses for juvenile fish, and, in the 1970s, it developed a new transportation program for juvenile passage. Critics believed these efforts at barging and trucking fish were ineffective and too manipulative of natural migrations. These complaints and accusations were superimposed over rivalries between Native American and commercial fishermen concerning the taking of fish, disagreements with fisheries agencies over the best methods of protecting salmon and steelhead, and resistance from those who opposed the building of additional dams.

Dealing with a new generation of environmentalists proved particularly vexing for the Corps. Its response to criticisms of conservationists in the wake of Bonneville had been to help preserve a precious natural resource—anadromous fish—while developing other benefits, such as water for irrigation, hydropower and flood control. The subsequent research programs that the Corps fostered or participated in were largely to meet that end. Now criticisms were being raised about keeping anadromous fish in their natural habitat–a concern shared by Native American groups–so as not to create "aquarium fish" or "token zoo runs." There were objections to building any additional dams because they could threaten fish habitats at a time when runs were disappearing from many rivers. In addition, a change in tone in the modern environmental era respecting "the water rights of fish" seemed to have changed the context of the debate.[104]

The substantial change in the environmental regulatory apparatus in the 1970s transformed fish from the victims of dam building into a weapon to fight dam construction. The Tellico Dam controversy is the most notable example. The Tellico Project, on the Little Tennessee River south of Knoxville, was originally suggested in 1936 as part of the TVA system. Initial appropriations from Congress were not approved until 1966; construction began the next year. At the time, opponents questioned the project at congressional hearings, pointing out that the river had unique natural characteristics and had cultural and historic value because of archeological artifacts left by the Cherokee and predecessor groups and had been the site for the early European occupation of Tennessee. Congress, however, had turned a deaf ear to pleas to remove the dam's authorization.

Armed with more substantial weaponry in 1971, opponents filed a suit in federal court contending that the TVA had not prepared an adequate environmental impact statement on the project. Two years later, TVA completed an EIS that never provided non-reservoir alternatives. Still, the project forged on, but this time opponents attempted a novel approach to stop the dam. In August, 1973, biologist David Etnier of the University of Tennessee discovered *Percina tanasi*—the snail darter—a species of fish found in the Little Tennessee River. Under provisions of the Endangered Species Act—the nation's first comprehensive law to protect species from extinction—the U.S. Fish and Wildlife Service listed the snail darter as an endangered species in 1975, and in 1976 listed the Little Tennessee River as a critical habitat for the fish. Proponents of the dam were outraged and the issue made its way to court.

In 1978, the U.S. Supreme Court heard the case, at which time Attorney General Griffin Bell belittled the effort to protect such an insignificant fish in the face of a major TVA project. The Court, however, upheld the Endangered Species Act, remarking that if Congress was unhappy with the decision it could change the law. TVA began studying alternatives to Tellico, but Congress did take up debate over the Endangered Species Act and ultimately passed amendments to exempt Tellico from it. The dam was completed in 1979.[105]

The snail darter was not the central issue in this story. And despite the fact that the dam was built, it was becoming clearer that opponents of big dams now had potent means, in the form of new environmental laws, of challenging dam builders. That these challenges were made indicated little willingness by some or no tolerance by others for multipurpose projects. The economic and conservation justifications of the past carried little weight in the 1960s and 1970s.

Environmentalism Comes of Age: Rampart Dam and the Grand Canyon Dams

That a variety of other battles over dams continued to rage in the 1960s and 1970s is further testament to a complete change in context about dam and reservoir construction since the waning of the big dam era. Marc Reisner argued "The battle over the Grand Canyon dams was the conservation movement's coming of age."[106] This most lively controversy was certainly a crucial bridge between the dispute over Echo Park and the wily use of new environmental legislation as manifest at Tellico. During the 1960s, substantial energy in the conservation community went into campaigns involving the Central Arizona Project. Among other things, the plan called for the building of two storage dams in Bridge Canyon and Marble Canyon, along the Colorado River. Both were meant to produce hydroelectric power. The major concern was that Bridge Canyon

Dam would back up water into Grand Canyon National Park's Inner Gorge, in what appeared to be clear violation of the 1919 law establishing Grand Canyon National Park, which provided for future federal dams if development did not compromise other purposes of the park. Opponents of the new hydroelectric dams recommended that coal-fired steam plants and nuclear plants be constructed instead. (This was trading one environmental threat for another, of course, and some conservationists ultimately protested this alternative.) The Grand Canyon dams were proposed as part of Reclamation's Central Arizona Project, which was considered essential to providing water and power for Arizonans. Arizona had been locked in a protracted struggle with California and the Upper Basin states over water rights to the Colorado River. Commissioner Floyd E. Dominy had branded opponents as "status quo conservationists" whose arguments were "frantic flak." David Brower, Executive Director of the Sierra Club, in turn called Dominy and Reclamation "the dam-it-all reclamationists." Ultimately, the project was authorized (1968), because of its local and regional significance, but without Bridge Canyon and Marble Canyon Dams.[107]

Although the fight against Rampart Dam did not have the vivid symbolism or emotional power of the fight over the Grand Canyon dams, it represented an extension of the battle over big dams for the first time beyond the lower forty-eight states. In the early 1960s, the Corps' Rampart Dam was proposed to be built on the Yukon River, ninety miles northwest of Fairbanks, Alaska. It was to be a 525-foot structure that would impound a body of water larger than Lake Erie. In the context of the Cold War, one congressional supporter urged construction on the grounds that it would be bigger than anything the Russians had built. Conservation organizations opposed the dam, protesting the inundation of about 11,000 square miles in the interior of Alaska, of which 8,000 square miles were waterfowl-producing habitat, and the blocking of salmon migration into a third or more of the upper Yukon watershed.

Rampart could provide electricity for six million people—but Alaska had only 253,000 people at the time, and the dam site was 2,000 miles away from the lower forty-eight states. In pre-energy-crisis America, such electrical power production seemed excessive. An article in *Natural History* stated in 1963 that:

> the whole concept of "the big dam," with its concomitant hydropower, water supply, flood control, and so on, is beginning to appear archaic, as well as too destructive of natural resources, many conservationists believe.

The "big dam philosophy," it went on "is in need of re-examination, preferably by the scientist and the conservationist. . . ." In 1967, Johnson Administration officials decided to curtail the project. It was a folly that, given its dubious economic benefits and waning support, would likely never have been completed.[108]

Not all battles over new dams were fought in the West. Efforts by the Corps to build a dam in Kentucky's Red River Gorge faced stiff opposition. Authorized by Congress in 1962, the plans called for a 141-foot-high dam creating a lake covering 1,500 acres 70 miles southeast of Lexington. A coalition of individual landowners, the Sierra Club, the Audubon Society, and "Save our Red River"—the Red River Gorge Legal Defense Fund—sued the Corps to stop the construction and carried the battle into the 1970s.[109] Farmers and the Sierra Club also fought to curtail the building of the Meramec Park Dam on the Meramec River, near St. Louis. The lawsuit, filed by the Sierra Club against the regional officers of the Corps, contended that the Meramec Basin Project did not comply with the National Environmental Policy Act.[110] Similar stories abound along other rivers throughout the country, including the Missouri.[111]

A well-publicized battle took place over a large dam project on the Delaware River running through Pennsylvania, New York, New Jersey, and Delaware. Although the Delaware's average discharge is less than 3 percent of the Mississippi's, it provides nearly 10 percent of the U.S. population with water. The Delaware, however, was one of the last major rivers in the nation without a dam on its main stem. Seven years after a severe flood in 1955, Congress authorized a multireservoir plan for the Delaware, with a federal-state water compact adopted to implement it. The centerpiece of the reservoir plan was a dam to be constructed at Tocks Island–the largest dam project to be carried out by the Corps east of the Mississippi. The debate over a dam across the Delaware began long before the 1955 flood. Four states and two large cities had deep interest in the water in the Delaware Basin, including New York City. Although outside the basin, it decided to tap the Delaware as early as 1928, prompting New Jersey and Pennsylvania to promote the need for a dam. Congress did not authorize the Tocks Island Dam until 1962, and several years of wrangling followed the authorization. Between 1975 and 1977, attempts by environmental groups and others to get the dam deauthorized failed, but general support for the project waned because of the Vietnam War, concerns over potentially large federal land acquisitions, extensive dislocations, and environmental concerns. Ultimately, the Tocks Island region of the Delaware River was added to the National Wild and Scenic Rivers System and an important water-management agreement for the Delaware River Basin, involving no dams on the Delaware, was adopted.[112]

Conclusion

Had the construction of new big dams become outdated by the 1960s and 1970s? Certainly, numerous environmentalists had come to believe so, but so had a variety of political supporters and even some officials within the federal bureaucracy. New large dams were just not going to be built any longer, at least not at the fervent pace of the big dam era. Was this a clear victory of an increasingly powerful and well-organized environmental lobby? Was it a result of the decreasing influence of once powerful federal agencies? Or did the "big dam" simply play itself out as a prized technology offering economic progress, wise use of resources, and protection from the vagaries of nature? An argument can be made for a little of all these elements. In a larger sense, however, a paucity of good dam sites and changes in the political, economic, social, and environmental context of the United States from the early twentieth century to the late 1970s ultimately doomed the big dam era. The economic opportunities offered by new big dams had diminished or were measured against undesirable trade-offs. Not everyone stood either with the dam builders or with the conservation organizations in every fight. Society was never that clearly polarized. It simply became difficult for dam builders to offer the benefits of new big dams with the same certainty or assurances in the 1960s and 1970s that they had offered in the 1930s and 1940s. Times had changed: the old constituencies championing dams had broken up, there were increased demands on the federal budget, and cynicism about the role of government accelerated. Also, as time went on, there were more skeptics who increasingly failed to share a common view about the desirability of "harnessing" rivers for human benefit. The changes in the political, economic, environmental, and social setting were matched by a change of heart or, at least, a change of priorities. The big dam era left a powerful structural and economic legacy, but it was a legacy that presents challenges to society in terms of making the best use of America's water resources.

Endnotes

1. Paul Zucker, *American Bridges and Dams* (New York: Greystone Press, 1941), 14.
2. Patrick McCully, *Silenced Rivers: The Ecology and Politics of Large Dams* (London: Zed Books, 1996), 281-311. Michael Collier, Robert H. Webb, and John C. Schmidt, *Dams and Rivers: A Primer on the Downstream Effects of Dams* (Tucson: U.S. Geological Survey, Circular 1126, June 1996), 6-8. Hundley, *The Great Thirst*, 359.
3. Collier, Webb, and Schmidt, *Dams and Rivers*, 2.
4. Pacific Constructors, Inc., *Shasta Dam and its Builders*, 9.
5. Mark Harvey, "Symbols from the Big Dam Era in the American West," (unpublished paper delivered at the ASEH conference, San Antonio, Texas, 1998), 3. Theodore Steinberg, "'That World's Fair Feeling': Control of Water in 20th-Century America," *Technology and Culture* 34 (April 1993), 401-2, took a more strident position, suggesting that "The guiding philosophy behind this profusion of dams was the simple will to control and dominate the natural world." In addition, he argued, "Perhaps never before has the will to conquer nature been so consciously and purposefully expressed, so matter-of-fact. What was being expressed here was arrogance by design."
6. Pitzer, *Grand Coulee*, 2. Steinberg, "'That World's Fair Feeling': Control of Water in 20th-Century America," 402.

7. See quote in Clayton R. Koppes, "Efficiency/Equity/Esthetics: Towards a Reinterpretation of American Conservation, *Environmental Review* 11 (Summer 1987), 135-6. The debate over public power was fought over several fronts, however. See Hundley, *The Great Thirst*, 224-5. Linda J. Lear, "Boulder Dam: A Crossroads in Natural Resource Policy," *Journal of the West* 24 (October 1985), 86, 88, 90-1. Wesley Arden Dick, "When Dams Weren't Damned: The Public Power Crusade and Visions of the Good Life in the Pacific Northwest in the 1930s," *Environmental Review* 13 (Fall/Winter 1989), 119, 122, 133-6. Mary Austin, "The Colorado River Controversy," *Nation* 125 (November 9, 1927), 511. James C. Williams, *Energy and the Making of Modern California*, 261-3. Lee, *Reclaiming the American West*, 40-1. "Who Benefits by Boulder Dam?" *New Republic* 63 (July 30, 1930), 310-2. Frank Bohn, editor, *Boulder Dam: From the Origin of the Idea to the Swing-Johnson Bill* (1927), 102. William E. Warne, *The Bureau of Reclamation* (Boulder: Westview Press, 1985; first published, 1973), 90-103. Kleinsorge, *Boulder Canyon Project*, 281-300.

8. Dick, "When Dams Weren't Damned," 123-4, 127, 129-30, 132-3, 138-9. Harvey, "Symbols from the Big Dam Era in the American West," 2-3, 6-8, 12-5.

9. Tim Palmer, *Endangered Rivers and the Conservation Movement* (Berkeley: University of California Press, 1986), 1.

10. Lawson, *Dammed Indians*, xx-xxi. Mike Lawson, "The Oahe Dam and the Standing Rock Sioux, *South Dakota History* 6 (Spring 1976), 203. Joy A. Bilharz, *The Allegheny Senecas and Kinzua Dam: Forced Relocation Through Two Generations* (Lincoln: University of Nebraska Press, 1998), xv. McCully, *Silenced Rivers*, 70-6. Fradkin, *A River No More*, 172. Peter Iverson, *"We're Still Here:" American Indians in the Twentieth Century* (Wheeling, Illinois: Harlan Davidson, Inc., 1998), 131. Sarah F. Bates, Lawrence J. Getches, Lawrence J. MacDonnel, and Charles F. Wilkinson, *Searching Out the Headwaters: Change and Rediscovery in Western Water Policy* (Covelo, California, Washington D.C.: Island Press, 1933), 125. Roy W. Meyer, "Fort Berthold and the Garrison Dam," *North Dakota History* 35 (Summer-Fall 1968), 220-1. Elizabeth S. Helfman, *Rivers and Watersheds in America's Future* (New York: David McKay Company, Inc., 1965), 169-79. Richard L. Berkman and W. Kip Viscusi, *Damming the West* (New York: Grossman Publishers, 1973), 151-96.

11. See Dick, "When Dams Weren't Damned," 125-6. Kollgaard and Chadwick, editors, *Development of Dam Engineering in the United States*, 1038. F. Lee Brown and Helen M. Ingram, *Water and Poverty in the Southwest* (Tucson: University of Arizona Press, 1987), 3-4. Donald Worster, *Rivers of Empire*, 210-1. See also Marc Reisner, *Cadillac Desert*. Critics of federal water policy in this period often point to special interests who benefitted most, and sometimes unfairly, from the big dams—particularly irrigators. One criticism is that irrigators do not pay a fair share of the price of impounding water. See Tim Palmer, *The Snake River: Window to the West* (Washington, D.C.: Island Press, 1991), 59, 63. Tim Palmer, *The Columbia: Sustaining a Modern Resource* (Seattle: The Mountaineers, 1997), 60-1. Edward Goldsmith and Nicholas Hildyard, *The Social and Environmental Effects of Large Dams* (San Francisco: Sierra Club Books, 1986), 51. Robert S. Devine, "The Trouble with Dams," *Atlantic Monthly* 276 (August 1995) 68.

12. Dick, "When Dams Weren't Damned," 118.

13. Karl E. Mundt, "Not All Dams are Damnable!" *Outdoor America* 8 (December 1943), 4. See also Joseph C. Goodman, "Build a Dam, Save a Stream," *American Game* 20 (March-April 1931), 23, 28.

14. William L. Finley, "Salmon or Kilowatts: Columbia River Dams Threaten Great Natural Resource," *Nature Magazine* 26 (August 1935), 107.

15. Palmer, *Endangered Rivers and the Conservation Movement*, 1-2. McCully, *Silenced Rivers*, 1-6. Collier, Webb, and Schmidt, editors, *Dams and Rivers*, 1-2. Guy LeMoigne, Shawki Barghoutti, and Herve Plusquellac, editors, *Dam Safety and the Environment* (Washington, D.C.: The World Bank, 1990), 3. Worster, *Rivers of Empire*, 310-1. Dick, "When Dams Weren't Damned," 118, 147-8. Lear, "Boulder Dam: A Crossroads in Natural Resource Policy," 91.

16. Lowitt, *The New Deal and the West*, 82-3. Ralph B. Simmons, *Boulder Dam and the Great Southwest* (Los Angeles: The Pacific Publishers, 1936), 141. U.S. Department of the Interior, Bureau of Reclamation, *The Story of Hoover Dam* (Washington, D.C., 1953), 40-3, 51-53, 65-6. Kleinsorge, *Boulder Canyon Project*, 246-311. Stevens, *Hoover Dam*, 259-61.

17. Simmons, *Boulder Dam and the Great Southwest*, 84. See also David O. Woodbury, *The Colorado Conquest* (New York: Dodd, Mead & Company, 1941), 358.

18. Carl Abbott, *Urban America in the Modern Age: 1920 to the Present* (Arlington Heights, Illinois: Harlan Davidson, 1987), 2, 5. Howard P. Chudacoff and Judith E. Smith, *The Evolution of American Urban Society*, 4th edition (Englewood Cliffs, New Jersey: Prentice-Hall, 1994), 4.

19. Carl Abbott, *The Metropolitan Frontier: Cities in the Modern American West* (Tucson: University of Arizona Press, 1998), xix.

20. See Miller and Melvin, *The Urbanization of Modern America*, 184-5. John C. Bollens and Henry J. Schmandt, *The Metropolis: Its People, Politics, and Economic Life,* 2nd edition (New York: Harper and Row, Publisher, 1970;), 17, 19. Alfred H. Katz and Jean Spencer Felton, editors, *Health and the Community* (New York: Free Press, 1965), 25. U.S. Department of Commerce, Bureau of the Census, *Historical Statistics of the United States, Colonial Times to 1970,* Part 1 (Washington, D.C.: Department of Commerce, 1975), 8, 11. Kenneth T. Jackson, *Crabgrass Frontier: The Suburbanization of the United States* (New York: Oxford University Press, 1985), 139-40.

21. See Bureau of the Census, *Historical Statistics of the United States* Part 2, 619, 621; Water Resources Council, *The Nation's Water Resources* (Washington, D.C.: Water Resources Council, 1968), 4-1-1 and 4-1-2. Murray Stein, "Problems and Programs in Water Pollution," *Natural Resources Journal* 2 (December 1962), 395. Jack Hirshleifer, James C. DeHaven, and Jerome W. Milliman, *Water Supply: Economics, Technology, and Policy* (Chicago: University of Chicago Press, 1960), 2, 26.

22. Abbott, *The Metropolitan Frontier*, 6. See also Williams, *Energy and the Making of Modern California*, 173-8, 206-7, 278-83.

23. Pitzer, *Grand Coulee*, xi-xii, xiv, 363. Robinson, *Water for the West*, 63-4. S. E. Hutton, "The Grand Coulee Dam and the Columbia Basin Reclamation Project," *Mechanical Engineering* 62 (September 1940), 651-2. See also Reuss and Walker, *Financing Water Resources Development*, 39-42. For World War II and hydropower, see Donald J. Pisani, "Federal Water Policy and the Rural West," in R. Douglas Hurt, editor, *The Rural West Since World War II* (Lawrence: University Press of Kansas, 1998), 119-38.

24. See Lowitt, *The New Deal and the West*, 116. White, *"It's Your Misfortune and None of My Own,"* 485-7. Pitzer, *Grand Coulee*, 364, 368. For figures on total installed hydropower by the early 1970s, see Peter H. Freeman, *Large Dams and the Environment: Recommendations for Development Planning* (A Report prepared for the United Nations Water Conference, Mar Del Plata, Argentina, March 1977), 3. See also Jan A. Veltrop, "Importance of Dams for Water Supply and Hydropower," in Asit K. Biswas, Mohammed Jellali, and Glenn E. Stou, editors, *Water for Sustainable Development in the Twenty-first Century* (Delhi: Oxford University Press, 1993), 102-15.

25. Steinberg, "That World's Fair Feeling," 403. See also Theodore Steinberg, *Slide Mountain: or the Folly of Owning Nature* (Berkeley: University of California Press, 1995).

26. Walker R. Young, "Boulder Dam Plays Its Part in Reclamation," *Reclamation Era* 27 (February 1937), 26-7.

27. U.S. Department of the Interior, *The Story of Hoover Dam*, 36.

28. The 1928 Flood Control Act, which authorized the Corps to use federal funds to unify the flood control system for the whole alluvial valley of the Mississippi River, led directly to the 1936 legislation. Once flood control on the Mississippi became a federal responsibility, it was just a matter of time before all flood control became a federal activity on all navigable rivers. Armstrong, Robinson, and Hoy, editors, *History of Public Works in the United States, 1776-1976*, 250-2, 257-8. Keith Petersen, "The Army Corps of Engineers and the Environment in the Pacific Northwest" (May 1982), Records of the Office of History, Headquarters, U.S. Army Corps of Engineers, Alexandria, Virginia. For information on flood control along the Missouri River, see Henry C. Hart, *The Dark Missouri* (Madison: University of Wisconsin Press, 1957), 89-90, 94-5, 120, 127, 133,150, 210. Marian E. Ridgeway, *The Missouri Basin's Pick-Sloan Plan: A Case Study in Congressional Policy Determination* (Urban: University of Illinois Press, 1955), 78-88, 93-4, 109-15, 139-53, 180-92, 210-47.

29. Martin Reuss, "Coping with Uncertainty: Social Scientists, Engineers, and Federal Water Resources Planning," *Natural Resources Journal* 32 (Winter, 1992), 117-27.

30. "Dams and Wild Rivers: Looking Beyond the Pork Barrel," *Science* 158 (October 13, 1967), 235.

31. E. W. Lane and J. R. Ritter, "The Life of Hoover Dam," *Reclamation Era* 34 (April 1948), 61. See also Young, "Boulder Dam Plays Its Part in Reclamation," 27-8. Simmons, *Boulder Dam and the Great Southwest*, 85. Donald Worster, *Under Western Skies*, 68. U.S. Department of the Interior, Bureau of Reclamation, *The Story of Hoover Dam*, (Washington D.C.: Government Printing Office, 1953) 46. Elmer T. Peterson, *Big Dam Foolishness: The Problem of Modern Flood Control and Water Storage* (New York: Devin-Adair Company, 1954), 69-85.

32. Goldsmith and Hildyard, *The Social and Environmental Effects of Large Dams*, 94. See also Worster, *Under Western Skies*, 75-6. Lear, "Boulder Dam: A Crossroads in Natural Resource Policy," 91-2. "Flood Control? Not by a Dam Site!" *Outdoor America* 16 (March-April 1951), 8.

33. LeMoigne, Barghoutti, and Plusquellac, editors, *Dam Safety and the Environment*, 1, 9-10. See also Robert B. Jansen, *Dams and Public Safety* (Washington, D.C.: U.S. Department of Interior, Water and Power Resources Service, 1980), 94. "High Dams in the United States," *Reclamation Era* 26 (January 1936), 22.

34. Kollgaard and Chadwick, editors, *Development of Dam Engineering in the United States*, 1035-6.

35. Kermit Pattison, "Why Did the Dam Burst?" *American Heritage of Invention & Technology* 14 (Summer 1998), 23-4. Peter Briggs, *Rampage: The Story of Disastrous Floods, Broken Dams, and Human Fallibility* (New York: David McKay Company, Inc., 1973), 21-2.

36. Pattison, "Why Did the Dam Burst?" 24-31. Briggs, *Rampage*, 22. See also Charles F. Outland, *Man-Made Disaster: The Story of St. Francis Dam* (Glendale, California: Arthur H. Clark Company, 1963). Reisner, *Cadillac Desert*, 100-4.

37. Gaylord Shaw, "The Search for Dangerous Dams - A Program to Head off Disaster," *Smithsonian* 9 (April 1978), 42-3. Pattison, "Why Did the Dam Burst?" 31. Kollgaard and Chadwick, editors, *Development of Dam Engineering in the United States*, 1036-7. Jansen, *Dams and Public Safety*, 96.

38. Kollgaard and Chadwick, editors, *Development of Dam Engineering in the United States*, 1038.

39. Goldsmith and Hildyard, *The Social and Environmental Effects of Large Dams*, 104-6. "WPA Dam Fails at Kansas City," *Engineering News-Record* 119 (September 23, 1937), 495. "Large Slide in Fort Peck Dam Caused by Foundation Failure," *Engineering News-Record* 122 (May 11, 1939), 55-8.

40. For a good case on the politics of dam safety, see David M. Introcaso, "The Politics of Technology: The 'Unpleasant Truth About Pleasant Dam,'" *Western Historical Quarterly* 26 (Autumn, 1995), 333-52.

41. Palmer, *The Snake River*, 29. USCOLD, *Lessons from Dam Incidents. USA-II* (New York: ASCE, 1988), 191-5. Kollgaard and Chadwick, editors, *Development of Dam Engineering in the United States*, 1041. Worster, *Rivers of Empire*, 308.

42. A description of the collapse and the human tragedy associated with Teton Dam is provided in Reisner, *Cadillac Desert*, 398-425.

43. Pisani, "Federal Water Policy and the Rural West," 139.

44. Independent Panel to Review Cause of Teton Failure, *Summary of Conclusions from Report to the U.S. Department of the Interior and State of Idaho on Failure of Teton Dam* (Washington, D.C.: U.S. Government Printing Office, December 1976), ix-x.

45. The issues of anadromous and catadromous fisheries and the effects of dams upon them are exceptionally controversial. The literature, both technical and popular, is extensive and available for those who further wish to explore impacts on fisheries. Suffice it to say here that more detailed history could be done on the issues involved. That is another study for another location. This study has space only to touch on some of the issues, from fish ladders to hatcheries to trucking and barging to nitrogen "poisoning." (Editor).

46. Palmer, *The Columbia*, 47.

47. Allan H. Cullen, *Rivers in Harness*, 116.

48. Richard White, *The Organic Machine*, 89-90. Mighetto and Ebel, *Saving the Salmon*, 7-8, 17-9, 30-2. White, *"It's Your Misfortune and None of My Own,"* 487. Harvey, "Symbols from the Big Dam Era in the American West," 31. See also Lisa Mighetto, "Salmon, Science, and Politics: Writing History for the U.S. Army Corps of Engineers," *Public Historian* 17 (Fall 1995), 21-22. Dick, "When Dams Weren't Damned," 144-6. Joseph E. Taylor III, *Making Salmon: An Environmental History of the Northwest Fisheries Crisis* (Seattle: University of Washington Press, 1999).

49. Palmer, *Endangered Rivers and the Conservation Movement*, 59-60.

50. "Dams and Destruction," *Nature Magazine* 31 (October 1938), 505. See also Finley, "Salmon or Kilowatts: Columbia River Dams Threaten Great Natural Resource," 107-8.

51. Mighetto and Ebel, *Saving the Salmon*, 53-5. See also Frank N. Schubert, "From the Potomac to the Columbia: The Corps of Engineers and Anadromous Fisheries," (Unpublished manuscript, December 1978), Office of History, Headquarters, U.S. Army Corps of Engineers, Ft. Belvoir, Virginia. Lowitt, *The New Deal and the West*, 158.

52. Albert N. Williams, *The Water and the Power: Development of the Five Great Rivers of the West* (New York: Duell, Sloan and Pearce, 1951), 281.

53. Pitzer, *Grand Coulee*, 223-30. See also Lowitt, *The New Deal and the West*, 159. "Dams and Destruction," 491. "Fish Stairways," *Literary Digest* 121 (May 30, 1936), 19.

54. Due to dam spillway design on the Columbia, some pools below dams became supersaturated with nitrogen, an issue which has been addressed by redesign of spillways to avoid entrainment of nitrogen.

55. Jim Marshall, "Dam of Doubt," *Collier's* 99 (June 19, 1937), 19-22. Mighetto and Ebel, *Saving the Salmon*, 71, 81, 84-8, 103. "Dams Threaten West Coast Fisheries Industry," *Oregon Business Review* 5 (June 1947), 3-4. Harvey, "Symbols from the Big Dam Era in the American West," 31.

56. Pitzer, *Grand Coulee*, 223-30. See also Lowitt, *The New Deal in the West*, 159. " Dams and Destruction," 491. "Fish Stairways," 19. R.G. Skerrett, "Fish Over a Dam," *Scientific American* 159 (October, 1938), 182-5.

57. Joseph T. Barnaby, North Pacific Fishery Investigations to Director, Fish and Wildlife Service, May 25, 1945, Thomas M. Robins File, Office of History, Headquarters, U.S. Army Corps of Engineers. See also White, *The Organic Machine*, 89.
58. Robinson, "The United States Army Corps of Engineers and the Conservation Community: A History to 1969."
59. Harvey, *A Symbol of Wilderness*, xi-xiii. Bates, *Searching Out the Headwaters*, 45.
60. Mark W. T. Harvey, "Echo Park Dam: An Old Problem of Federalism," *Annals of Wyoming* 55 (Fall 1983),10. Robinson, *Water for the West*, 92.
61. Mark W. T. Harvey, "Echo Park, Glen Canyon, and the Postwar Wilderness Movement," *Pacific Historical Review* 60 (February 1991), 48. Jon M. Cosco, *Echo Park: Struggle for Preservation* (Boulder: Johnson Books, 1995), xii-xv. Wallace Stegner, "Battle for the Wilderness," *New Republic* 130 (February 15, 1954), 13. Michael P. Cohen, *The History of the Sierra Club, 1892-1970* (San Francisco, Sierra Club Books, 1988), 143-4.
62. U. S. Grant III, "The Dinosaur Dam Sites Are Not Needed," *Living Wilderness* 15 (Autumn, 1950): 17. See also "Dams or Wilderness Areas?" *Audubon Magazine* 52 (September-October 1950): 287.
63. Harvey, "Echo Park. . .," 48-50.
64. Robinson, *Water for the West*, 93. Worster, *Rivers of Empire*, 274. See also Richardson, *Dams, Parks & Politics*, 57, 60-61.
65. Harvey, *A Symbol of Wilderness*, xiii, xviii. See also Robinson, *Water for the West*, 93. Harvey, "Echo Park . . .," 13, 15. Rich Johnson, *The Central Arizona Project, 1918-1968* (Tucson: University of Arizona Press, 1977), 101-2. Norris Hundley, Jr., "The West Against Itself," in Gary D. Weatherford and F. Lee Brown, editors, *New Courses for the Colorado River: Major Issues for the Next Century* (Albuquerque: University of New Mexico Press, 1986), 29.
66. Richardson, *Dams, Parks & Politics*, 69.
67. Harvey, "Echo Park Dam . . .," 10. See also Richardson, *Dams, Parks & Politics*, 135, 142, 151.
68. Fred M. Packard, "Echo Park Dam? Not By a Damsite!" *National Parks Magazine* 29 (July-September 1955): 99.
69. See quote on page 24 of "Echo Park Controversy Resolved," *Living Wilderness* 20 (Winter-Spring 1955–56), 23-5. Harvey, *A Symbol of Wilderness*, 181-205. Harvey, "Echo Park Dam . . .," 12. Hundley, *The Great Thirst*, 307. Cohen, *The History of the Sierra Club*, 154, 157, 165, 172. See also Russell Martin, *A Story that Stands Like a Dam*, 50-74.
70. Harvey, "Echo Park. . .," 46-7, 52, 58-9.
71. Robinson, *Water for the West*, 93. Cohen, *The History of the Sierra Club*, 161.
72. Eliot Porter, *Glen Canyon on the Colorado*, edited by David Brower (Salt Lake City: Peregrine Smith Books, 1988), 8. Cohen, *The History of the Sierra Club*, 177-9. Harvey, "Echo Park . . .," 65-7. Hundley, "The West Against Itself," 29. Hundley, *The Great Thirst*, 307. See also Martin, *A Story That Stands Like a Dam*, 320-32.
73. Wallace Stegner, "Myths of the Western Dam," *Saturday Review* 48 (October 23, 1965), 29.
74. Reuss and Hendricks, *U.S. Army Corps of Engineers*, 20 (however, pages are unnumbered).
75. Elizabeth B. Drew, "Dam Outrage: The Story of the Army Engineers," *Atlantic* 225 (April 1970), 51-2. Robert G. Sherill, "Corps of Engineers: The Pork-Barrel Soldiers," *Nation* (February 14, 1966), 180-3. See also Berkman and Viscusi, *Damming the West*, 73-7. Robinson, "The United States Army Corps of Engineers and the Conservation Community," 84-8. Luther J. Carter, "Dams and Wild Rivers: Looking Beyond the Pork Barrel," *Science* 158 (October 13, 1967), 237-42. Martin Heuvelmans, *The River Killers* (Harrisburg, Pennsylvania: Stackpole Books, 1974), 15. "The Dam Shame: It's Still With Us," *Outdoor Life*, 163 (June 1979), 84-9. Tim Palmer, "Saving the Stanislaus: Must We Wear Chains to Keep Rivers Free?" *Sierra Club Bulletin* 64 (September/October 1979), 10.
76. Quotation in Melosi, *Coping with Abundance*, 297. Gottlieb, *Forcing the Spring*, 105-14. See also Otis L. Graham, Jr., *A Limited Bounty: The United States Since World War II* (New York: McGraw-Hill Company, 1996), 164.
77. Melosi, *Coping with Abundance*, 297-8. Gottlieb, *Forcing the Spring*, 109-10.
78. For an example of ecological effects of dams and endangered species, see Ferrell, *Big Dam Era*, 161-6.

79. Wallis E. McClain, Jr., editor,, *U.S. Environmental Laws: 1994 Edition* (Washington, D.C.: Bureau of National Affairs, Inc., 1994), 9-1. Gottlieb, *Forcing the Spring*, 124-5. Melosi, *Coping with Abundance*, 298. See Alfred S. Harrison, *Water Resources: Hydraulics and Hydrology: Interview with Alfred S. Harrison*, oral history interview conducted by John T. Greenwood, (Alexandria, Virginia: Office of History, Headquarters, U.S. Army Corps of Engineers, 1997), 92-4. See also Valerie M. Fogelman, *Guide to the National Environmental Policy Act* (New York: Quorum Books, 1990), 1-2. Joseph Petulla, *Environmental Protection in the United States* (San Francisco: San Francisco Study Center, 1987), 47-8. Daniel H. Henning and William R. Mangun, *Managing the Environmental Crisis: Incorporating Competing Values in Natural Resource Administration* (Durham, North Carolina: Duke University Press, 1989), 19-20. Dinah Bear, "National Environmental Policy Act of 1969," in Ruth A. Eblen and William R. Eblen, editors, *The Encyclopedia of the Environment* (Boston: Houghton Mifflin Company, 1994), 463-5. Lynton K. Caldwell, "National Environmental Policy Act (U.S.)," in Robert Paehlke, editor,, *Conservation and Environmentalism: An Encyclopedia* (New York: Garland Publishing, 1995), 449-51. Robert V. Bartlett, "Environmental Impact Assessment," in Paehlke, editor,, *Conservation and Environmentalism*, 248-50. Daniel Mazmanian and Mordecai Lee, "Tradition Be Damned! The Army Corps of Engineers Is Changing," *Public Administration* 35 (March-April 1975), 169. William L. Kahrl, "Paradise Reclaimed: The Corps of Engineers and the Battle for the Central Valley," (manuscript, undated), 128-9, 142-3 Files, Office of History, Headquarters, U.S. Army Corps of Engineers, Alexandria, Virginia.

80. Mazmanian and Lee, "Tradition Be Damned! The Army Corps of Engineers Is Changing," 168-9. See also Holmes, *History of Federal Water Resources Programs*, 112-4. T. Michael Ruddy, *Damming the Dam: The St Louis Corps of Engineers and the Controversy Over the Meramec Basin Project from Its Inception to Its Deauthorization* (St. Louis, Missouri: U.S. Army Corps of Engineers, 1992), 68-72. Jacob L. Douma, *Water Resources: Hydraulics and Hydrology: Interview with Jacob H. Douma* oral history interview by John T. Greenwood, (Alexandria, Virginia: Office of History, Headquarters, U.S. Army Corps of Engineers, 1991), 91-2.

81. Melosi, *Coping with Abundance*, 298. McClain Jr., editor, *U.S. Environmental Laws: 1994 Edition*, 9-1. Gottlieb, *Forcing the Spring*, 128-9.

82. Richard N. L. Andrews, "Environmental Protection Agency," in *Conservation and Environmentism: An Encyclopedia,* edited by Robert Paehlke (New York and London: Garland Publishing, 1995). 256. Gottlieb, *Forcing the Spring*, 129. Petulla, *Environmental Protection in the United States*, 48-9. Melosi, *Coping with Abundance*, 298. See also Terrie Davies, "Environmental Protection Agency," in Eblen and Eblen, editors, *The Encyclopedia of the Environment* (Boston: Houghton Mifflin Company, 1994), 221-2. Richard Stren, Rodney White, and Joseph Whitney, editors, *Sustainable Cities: Urbanization and the Environment in International Perspective* (Boulder, Colorado: Westview Press, 1992), 192. Marc K. Landy, Marc J. Roberts, and Stephen Thomas, *The Environmental Protection Agency: Asking the Wrong Questions from Nixon to Clinton* (New York: Oxford University Press, 1990), 22-45. Alfred Marcus, "Environmental Protection Agency," in James Q. Wilson, editor, *The Politics of Regulation* (New York: Basic Books, Inc., 1980), 267-303. Edmund P. Russell III, "Lost Among the Parts Per Billion: Ecological Protection at the United States Environmental Protection Agency, 1970-1993," *Environmental History* 2 (January 1997), 29-51.

83. Andrews, *Managing the Environment, Managing Ourselves*, 218-26. See also Martin V. Melosi, "Lyndon Johnson and Environmental Policy," in Robert A. Divine, editor, *The Johnson Years, Volume Two: Vietnam, the Environment, and Science* (Lawrence: University Press of Kansas, 1987), 113-7.

84. Gottlieb, *Forcing the Spring*, 126, 129. Andrews, *Managing the Environment, Managing Ourselves*, 294-316. Jeffrey Kim Stine, "Environmental Politics and Water Resources Development: The Case of the Army Corps of Engineers during the 1970s," (Ph.D. dissertation, University of California, Santa Barbara, 1984), 21-2, 34-5. Holmes, *History of Federal Water Resources Programs*, 9-10, 13, 17, 111.

85. Mazmanian and Lee, "Tradition Be Damned! The Army Corps of Engineers Is Changing," 166-71. See also Holmes, *History of Federal Water Resources Programs*, 116-7. "The Corps of Engineers, Water, and Ecology," *Audubon Magazine* 72 (July 1970), 102.

86. Martin Reuss, *Shaping Environmental Awareness: The United States Army Corps of Engineers Environmental Advisory Board, 1970-1980* (Washington, D.C.: Historical Division, Office of Administrative Services, Office of the Chief of Engineers, 1980), 67-8. See also Stine, "Environmental Politics and Water Resources Development," 46, 100.

87. Stine, "Environmental Politics and Water Resources Development," 197. See also 94, 97-8, 101, 194-8. Carter, "Dams and Wild Rivers: Looking Beyond the Pork Barrel," 233.

88. Holmes, *History of Federal Water Resources Programs*, 140.

89. Reuss, "Coping with Uncertainty," 129-31. Brent Blackwelder, "In Lieu of Dams," *Water Spectrum* (Fall 1977), 41. Samuel Stafford, "Big Problem: Damming up the Flood of Antipathy," *Government Executive* 1 (December 1969), 49. See also Susan M. Stacy, *When the River Rises: Flood Control on the Boise River, 1943-1985* (Boulder, Colorado: Institute of Behavioral Science, University of Colorado and College of Social Sciences and Public Affairs, Boise State University, 1993), 7-8, 36, 56, 80-106.

90. John D. Echeverria, Pope Barrow, and Richard Roos-Collins, *Rivers at Risk: The Concerned Citizen's Guide to Hydropower* (Washington, D.C.: Island Press, 1989), 4.

91. Pisani, "Federal Water Policy and the Rural West," 139. See also Reisner, *Cadillac Desert*, 48-95.

92. For additional information on water quality and dams see the Environmental Protection Agency's 1989 Report to Congress: *Dam Water Quality Study* (EPA 506/2-89/002). This report was prepared in accordance with the requirement of Section 524 of the Water Quality Act of 1987.

93. Goldsmith and Hildyard, *The Social and Environmental Effects of Large Dams*, 13.

94. Devine, "The Trouble with Dams," 72-3. Committee on Environmental Effects of the United States Committee on Large Dams, *Environmental Effects of Large Dams* (New York: American Society of Civil Engineers, 1978), 5. Echeverria, Barrow, and Roos-Collins, *Rivers at Risk*, 4. Worster, *Rivers of Empire*, 310. Palmer, *The Snake River*, 67. Palmer, *Endangered Rivers and the Conservation Movement*, 1-2. Goldsmith and Hildyard, *The Social and Environmental Effects of Large Dams*, 52, 63. Collier, Webb, and Schmidt, *Dams and Rivers*, 83-4. See also John A. Dixon, Lee M. Talbot, and Guy J. M. Le Moigne, *Dams and the Environment: Considerations in World Bank Projects* (Washington, D.C.: World Bank, 1989), 2.

95. Devine, "The Trouble with Dams," 71-2. Palmer, *The Snake River*, 20-1. Echeverria, Barrow, and Roos-Collins, *Rivers at Risk*, 6.

96. Worster, *Rivers of Empire*, 153-4, 240, 319-24. See also Goldsmith and Hildyard, *The Social and Environmental Effects of Large Dams*, 134-63.

97. Hundley, "The West Against Itself," 37-9: Warne, *The Bureau of Reclamation*, 117-21. See also Berkman and Viscusi, *Damming the West*, 34-41, 46-51. Cullen, *Rivers in Harness*, 120-3. For another significant case of water degradation, see Gene Rose, *San Joaquin: A River Betrayed* (Fresno: Linrose Publishing Company, 1992), 123, 125-30. For a general description of the impact of dams and reservoirs on water quality, see McCully, *Silenced Rivers*, 36-38, 40-1.

98. William Graf, "Landscapes, Commodities, and Ecosystems . . .," 18-20. Joseph L. Sax, "Parks, Wilderness, and Recreation," in Michael J. Lacey, *Government and Environmental Politics* (Washington, D.C.: Woodrow Wilson Center Press, 1989), 124. See also Palmer, *Endangered Rivers and the Conservation Movement*, 2.

99. LeMoigne, Barghouti, and Plusquellec, editors, *Dam Safety and the Environment*, 1. Shaw, " The Search for Dangerous Dams–A Program to Head Off Disaster," 36, 38-39. Kollgaard and Chadwick, editors, *Development of Dam Engineering in the United States*, 1040-41. Worster, *Rivers of Empire*, 309. See also "Silt, Cracks, Floods, and Other Dam Foolishness," *Audubon Magazine* 77 (September 1975), 107-14.

100. The 1972 Act had authorized the Secretary of the Army, acting through the Chief of Engineers, to initiate a national program of dam inspections. See Jansen, *Dams and Public Safety*, 96.

101. See: http://crunch.tec.army.mil/nid/webpages/nid.cfm

102. National Research Council, *Safety of Existing Dams: Evaluation and Improvement* (Washington, D.C.: National Academic Press, 1983), 4-5. For more on the National Inventory of Dams, see the following website: http://crunch.tec.army.mil/webpages/nidwelcome.cfm.

103. Kollgaard and Chadwick, editors, *Development of Dam Engineering in the United States*, 1042. See also Shaw, " The Search for Dangerous Dams–A Program to Head Off Disaster," 44. Vernon K. Hagen, *Water Resources: Hydraulics and Hydrology: Interview with Vernon K. Hagen*, oral history interview by John T. Greenwood (Alexandria, Virginia: Office of History, Headquarters, U.S. Army Corps of Engineers, 1997), 152-53.

104. Mighetto, "Salmon, Science, and Politics," 22-5. See also Mighetto and Ebel, *Saving the Salmon*, 84ff. Darrell J. Turner, "Dams and Ecology: Can They Be Made Compatible?" *Civil Engineering* 41 (September 1971), 78. See also Keith C. Petersen, *River of Life, Channel of Death: Fish and Dams on the Lower Snake* (Lewiston, Idaho: Confluence Press, 1995). Keith C. Petersen, "Battle for Ice Harbor Dam," *Pacific Northwest Quarterly* 86 (Fall, 1995), 178-88.

105. Mighetto, "Salmon, Science, and Politics," 24-6. Mighetto and Ebel, *Saving the Salmon*, 174-6. See also "Fish v. Dams," *Time,* 71 (February 17, 1958), 89. Joel W. Hedgpeth, "The Passing of the Salmon," *Scientific Monthly* 59 (November 1954), 378. Thorson, *River of Promise, River of Peril*, 182-4. Jeffrey F. Mount, *California Rivers and Streams: The Conflict Between Fluvial Process and Land Use* (Berkeley: University of California Press, 1995), 326-9. Senator Frank E. Moss, *The Water Crisis* (New York: Frederick A. Praeger, Publisher, 1967), 137. McCully, *Silenced Rivers*, 41-43, 50-3. Reisner, *Cadillac Desert*, 187-8.

106. Robert K. Davis, "Lessons in Politics and Economics from the Snail Darter," in V. Kerry Smith, editor, *Environmental Resources and Applied Welfare Economics: Essays in Honor of John V. Krutilla* (Washington, D.C.: Resources for the Future, 1988), 213-8. See also William Bruce Wheeler and Michael J. McDonald, *TVA and the Tellico Dam, 1936-1979: A Bureaucratic Crisis in Post-Industrial America* (Knoxville: University of Tennessee Press, 1986). Daniel Deudney, "Rivers of Energy: The Hydropower Potential" (Worldwatch Paper 44, June 1981), 19. Ann Shalowitz, "Endangered Darters 'Endanger' Tellico Dam," *Conservation News* 40 (March 15, 1975), 7-8. Mighetto and Ebel, *Saving the Salmon*, 179.

107. Reisner, *Cadillac Desert*, 295.

108. Robinson, *Water for the West*, 93. "Grand Canyon Dams?" *Audubon Magazine* 67 (May 1965), 181. Arnold Hano, "The Battle of the Grand Canyon," *New York Times Magazine* (December 12, 1965), 56. "Canyon Dams: Dissents from Arizona Scientists," *Science* 157 (July 7, 1967), 46. "Damming the Grand Canyon for a Thirsty Southwest," *New Republic* 154 (April 30, 1966), 9. Cohen, *The History of the Sierra Club*, 314-8, 357-64. See also Warne, *The Bureau of Reclamation*, 99-103. Fradkin, *A River No More*, 194-5, 228-34. Robert Dean, "'Dam Building Still Had Some Magic Then:' Stewart Udall, The Central Arizona Project, and the Evolution of the Pacific Southwest Water Plan, 1963-1968," *Pacific Historical Review* 66 (February 1997), 92. "Water and Power for the Southwest," *National Parks Magazine* 39-40 (September 1965), 2. "'No Compromise' on Grand Canyon Dams: Sierra Club's Reply to Goldwater Plan," *U.S. News and World Report* 61 (December 12, 1966), 60-1. "Dam the Grand Canyon?" *Audubon Magazine* 68 (September 1966), 308-11. "Grand Canyon: Colorado Dams Debated," *Science* 152 (June 17, 1966), 1600-5. "Canyon Controversy: Second Round," *Science News* 91 (April 1967), 302-3. John Ludwigson, "Dams and the Colorado," *Science News* 91 (February 1967), 167. "Grand Canyon Dams Go," *Science News* 91 (February 11, 1967), 135. "The Grand Canyon: Dam It or Not?" *Senior Scholastic* 90 (February 3, 1967), 6-7. "Grand Canyon Dams Blocked," *Audubon Magazine* 68 (November 1966), 462. "Good News on the Grand Canyon," *National Parks Magazine* 41 (March 1967), 289. Stephen Raushenbush, "A Bridge Canyon Dam is Not Necessary," *National Parks Magazine* 38 (April 1964), 4-8.

109. See "Building the Bigger Dam," *Natural History* 72 (December 1963), 66. A Starker Leopold and Justin W. Leonard, "Alaska Dam Could Be Resources Disaster," *Audubon Magazine* 68 (May-June 1966), 176-8. Stephen H. Spurr, "Rampart Dam: A Costly Gamble," *Audubon Magazine* 68 (May-June 1966), 172-5, 179. See also "Fish and Wildlife Service Issues Unfavorable Report on Rampart Dam," *Izaak Walton Magazine* 29 (August-September 1964), 13. "Rampart Dam—Alaska Study," *Izaak Walton Magazine* 31 (May 1966), 12-3. "Rampart Dam," *National Parks Magazine* 41 (August 1967), 20. "Rampart Dam and the Perpetual Engineers," *Field and Stream* 71 (June 1966), 34, 36.

110. "Daniel Boone's Wilderness May Be Tamed by a Lake," *Smithsonian* 6 (September 1975), 56-8.

111. Richard C. Albert, *Damming the Delaware: The Rise and Fall of Tocks Island Dam* (University Park, Pennsylvania: Pennsylvania State University Press, 1987), xv-xvii, 1-4. See also Pete duPont, "The Tocks Island Dam Fight," *Sierra Club Bulletin* 57 (July-August 1972), 13-8.

112. "Engineers' Dam is Coup de Grace for Meramec," *Audubon Magazine* 75 (March 1973), 121-2.

113. Thorson, *River of Promise*. Ferrell, *Big Dam Era*.

APPENDIX A: GUIDELINES FOR APPLYING THE NATIONAL HISTORIC LANDMARKS CRITERIA TO STORAGE DAMS[1]

INTRODUCTION

The following guidelines for applying the National Historic Landmark (NHL) criteria are designed to assist in the evaluation of dams constructed in the United States for the purposes of water storage and distribution (irrigation and domestic supply), hydroelectric power generation, navigation, or flood control.[2]

A singular definition unites these varied purposes: dams "provide active storage for the management of water . . . for striking a balance between natural flow regimes on the one hand and patterns of demand on the other."[3] This general classification encompasses an astonishing number of resources. In 1975, the United States Army Corps of Engineers identified more than 65,000 American dams over twenty-five feet tall or with a reservoir storage capacity of at least fifty acre-feet.[4] By virtue of the centrality of water to human settlement, population growth, and agricultural and industrial endeavor, each of these 65,000 dams had a significant impact on local development. Many impacted a larger geographic region. A few changed the course of American and international technological history, reverberated through the national economy, and marked significant transitions in American politics and culture.

The following discussion of NHL criteria and of general guidelines for applying these criteria to both federal and privately constructed dams should be used in conjunction with the NHL Theme Study: *The History of Large Federal Dams: Planning, Design, and Construction.*[5] This document presents a basic framework of general themes and trends central to the evaluation of federal dams.

GUIDELINES[6]

NHLs are those historic properties that:

Criterion 1: are associated with events that have made a significant contribution to, and are identified with, or that outstandingly represent, the broad national patterns of United States history and from which an understanding and appreciation of those patterns may be gained; and/or

Criterion 2: are associated importantly with the lives of persons nationally significant in the history of the United States; and/or

Criterion 3: represent some great idea or ideal of the American people; and/or

Criterion 4: embody the distinguishing characteristics of an architectural type specimen exceptionally valuable for a study of a period, style or method of construction, or that represent a significant, distinctive and exceptional entity whose components may lack individual distinction; and/or

Criterion 5: are composed of integral parts of the environment not sufficiently significant by reason of historical association or artistic merit to warrant individual recognition but collectively compose an entity of exceptional historical or artistic significance, or outstandingly commemorate or illustrate a way of life or culture; and/or

Criterion 6: have yielded or may yield information of major scientific importance by revealing new cultures, or by shedding light upon periods of occupation over large areas of the United States. Such sites are those which have yielded, or which may reasonably be expected to yield, data affecting theories, concepts, and ideas to a major degree.

The above criteria have rough counterparts in National Register of Historic Places criteria A, B, C, and D. However, NHL-eligible properties must be significant at the national level, versus the local or state level established for National Register properties. Moreover, to be classified as an NHL, nationally significant resources must retain a high degree of historical integrity and should *best represent* their associated class of resources; the determination of "best" representation is based not only on the extent of historical integrity but also on the degree of historical association.

Applying the broad NHL criteria to individual dams or hydraulic systems requires a clear understanding of the patterns, themes, and historic trends with which the dam is associated: a historic context. This context is provided in the associated theme study: *The History of Large Federal Dams: Planning, Design, and Construction*. It also requires an analytical framework in which to evaluate historical integrity and levels and areas of significance. This analytical framework is provided below.

Criterion 1:
Properties that are associated with events that have made a significant contribution to, and are identified with, or that outstandingly represent, the broad national patterns of United States history and from which an understanding and appreciation of those patterns may be gained.

Large dams have dramatically altered the physical landscape. Less obviously, but as importantly, they also defined the placement and density of settlement and commerce, and infused the nineteenth- and twentieth-century political, social, cultural, and environmental debates related to land use, population growth, cultural expansion, and conservation. A level of national significance is therefore sometimes assumed to be an inherent component of large dams and appurtenant resources.

It is particularly beguiling to assign national significance to many of the dams constructed by America's primary federal dam-building bureaus and agencies: the United States Army Corps of Engineers (Corps); the United States Bureau of Reclamation (Reclamation); and the Tennessee Valley Authority (TVA). These organizations were engendered, and have evolved, in response to important national legislation related to the use, distribution, and control of America's rivers. The largest and most expensive of America's dams have been built by Reclamation, the Corps, and the TVA: "Only a national organization. . .could have conceived and built the magnificent dams which brought the United States to the forefront of world developments."[7]

NHL Criterion 1, however, requires *that a resource not only be associated with the broad national patterns of our national history, but that it make "significant contribution to" or "outstandingly represent" those patterns.* In other words, individual dams are not of national significance simply because dam construction is a significant facet of our nation's history. A NHL nomination must also establish that the dam in question is an important illustration of the historic context and that the property possesses the physical characteristics necessary to convey this association (see Integrity, below).

When evaluating dams' national significance, errors of assumption are frequently made in assigning national significance to local and regional events; and in assigning "outstanding" representational value to commonplace sites.

One of the most compelling justifications for national significance relates to impact or influence. Large dams[8] often require more man-hours and more material to build, impacting the local economy through an influx of construction workers and long-lasting economic development. They often create the largest reservoirs, resulting in the most dramatic restructuring of natural river systems; dictating settlement patterns; inspiring long-lasting and potentially significant political battles over water use by conflicting interests;[9] and otherwise altering the social and physical landscape. However, size is not a virtue in its own right: it must heighten and define substantive technological, social, political, cultural, and environmental consequences.

A dam's size is also often directly related to the economic benefits of dam construction: the power generated, the ratio of reservoir storage to annual runoff, and the acre-feet of water made available for irrigation, municipal, and navigational use. While these benefits may have important policy implications at the national and international level, the direct benefits of power, water, and flood control clearly relate most directly to a dam's local and state significance and do not necessarily contribute to the dam's national significance.

Precedence – "the first" – may also relate directly to a property's historical significance. For example, Garrison Dam (North Dakota), the first Corps dam constructed under the terms of the Pick-Sloan Missouri Basin Program, may most effectively represent the dams that followed – not because it is the most important dam, but because it was the first, setting the stage for a larger system. To be nationally significant, the "first" in a series of dams must provide a meaningful foundation for the dams that followed, either in terms of technological advances (see criterion 4, below) or in terms of policy initiatives. It cannot simply be the first in a contemporaneous cluster of resources.

Care must also be taken not to assign national significance to local or regional events. While the cumulative impacts of, for example, a vibrant Pacific Northwest economy may be felt at the national level, this impact is not necessarily "important" as required NHL criteria; examples of the national scope of the American economy are endless.[10] Similarly, New Deal wages of 50 cents an hour, paid to the more than 10,000 men and women who worked on Fort Peck Dam (Montana), were profoundly significant to individual families and communities. The seep of cash outward touched the far corners of the nation. The potential national economic and social significance of the dam, however, relates not to this direct but ultimately short-term economic impact, but to the larger economic policy behind federal economic intervention of the scale initiated with Fort Peck Dam construction

There will be exceptions, where the direct impact of power production, flood control, irrigation, or associated population growth can be demonstrated to have greatly influenced the direction of American history. For example, California's Central Valley (Reclamation) Project supports the cultivation of over 80% of the nation's fruits and vegetables and altered regional and national distribution patterns and the size of economically viable farms. The hydroelectric generating capability of Bonneville (Washington) and Grand Coulee (Washington) Dams was central to America's World War II industrial war effort, allowing regional concentration of aluminum refining, aircraft manufacture, and liberty ship manufacture. The argument for national significance, however, must be carefully documented and care must be taken, first, to distinguish between association with regional versus national trends and, second, to assess whether the association was important.[11]

Finally, "outstanding" representational value should not be assigned to commonplace sties. Characteristics of the Bureau of Reclamation's Belle Fourche Project include federal intervention in a private reclamation effort; social engineering based on turn-of-the-century pastoral ideals; Reclamation misunderstanding of project benefits; and social and political upheaval associated with repayment schedules. These characteristics, however, are represented at many early Reclamation projects, including the North Platte Project with Pathfinder Dam (Wyoming), the Shoshone Project with Buffalo Bill Dam (Wyoming), and the Boise Project with Arrowrock Dam (Idaho). The same is true for the large Depression-era Public Works Administration-funded dams, including Grand Coulee, Fort Peck, Hoover (Arizona and Nevada), Owyhee (Oregon), and Norris (Tennessee) – all of which provided employment and settlement opportunities to vast numbers of farmers and other Americans displaced by the Dust Bowl and Depression and all of which could stake a claim to national significance on that social and economic foundation. While these dams may be eligible to the National Register of Historic Places, any NHL nomination based solely on their ability to "outstandingly represent the broad national patterns of United States history" must meet a much higher standard. The nomination must provide a comparative framework in which the significance and the historical integrity of the dam is contrasted with that of similar, historically associated resources.

The historic context in which a dam is evaluated for national significance might not be directly related to dam construction and the water-related improvements of flood control, navigation, power generation, or irrigation. For instance, at Oahe (South Dakota) and Garrison Dams (North Dakota) on the Missouri River, the inundation of Native American traditional and treaty lands during the 1940s and 1950s resulted in significant federal legislation and policy initiatives. Environmental impacts associated with Glen Canyon Dam and Lake Powell (Arizona and Utah) construction helped redefine and redirect the national environmental debate in the 1960s. Court battles associated with repayment schedules at Belle Fourche Dam and project (South Dakota) resulted in passage of the Reclamation Extension Act of 1914. New Mexico's Elephant Butte Dam, on the Rio Grande Project, was funded not only as a means of reclaiming 178,000 acres of the Rio Grande drainage but also in order to meet Mexico's water rights, as established in international treaty.[12]

Associations with historical events such as these *might* be sufficient to support a claim of NHL eligibility if it can be shown that the historical event is nationally significant and that the selected dam is importantly associated with that nationally significant theme. This "important association" cannot be safely assumed in any of the above examples: the national significance of the Reclamation Extension Act might prove difficult to establish, despite the Act's importance to Bureau of Reclamation history and to farmers in the West. The treaty obligations met by construction of Elephant Butte may not have had ramifications beyond the limited watershed of the Rio Grande (level of significance: state or local). While the Native American civil rights movements is of national significance, legislation growing out of Missouri River dam construction must be shown to have played an important part in that movement. While the modern environmental movement is arguably of national significance, Glen Canyon Dam may be a less important example than O'Shaughnessy (Hetch Hetchy) Dam (California) or the proposed Echo Park Dam site (Utah).

The historical impact of private and municipal dams can also extend to the national level. Historian Donald Jackson writes "the tentacles of seemingly 'local' water development projects often extend beyond the bounds of where the water is actually used, into far-flung regional and national networks of engineers, businessmen, financiers and bureaucrats."[13] Yet as with federally constructed dams, once these tentacles have been identified, the argument must be effectively made that the impact was not only national, but also significant. For example, debate surrounding the 1915 construction of the O'Shaughnessy Dam in the Hetch Hetchy Valley of Yosemite National Park marked a critical juncture in the modern conservation movement, introducing an element of environmental and social concern to the question of dam construction and intensifying the philosophical debate surrounding man's relationship to nature. Hetch Hetchy has become part of our vernacular lexicon – a shorthand reference to a wide range of conservation, preservation, and wise use issues. In contrast, Pardee Dam, the kingpin of East Bay Municipal Utility District's (California) municipal water system, "was completed in about five years from initial studies to water deliveries and with essentially no controversy." While this equanimity may be "remarkable in the larger context of public works," as argued in the Pardee Dam National Register nomination, it is not necessarily of national significance.[14]

Criterion 2:
Properties with an important association with the lives of persons nationally significant in the history of the United States.

Dams evaluated under Criterion 2 must follow the general guidelines established by the National Historic Landmarks Survey for *any* resource associated with significant people. The dam must be associated with the individual's productive life, during which he or she achieved significance; the property must be the resource determined to *best* represent the person's historic contributions; and the person associated with the property must be of national significance. Under this criterion, the determination of "best" will relate directly to the nature of the individual's contribution to dam construction or to dam design. There may be a few designers/engineers that are nationally significant, e.g. John L. Savage, and a dam might be an appropriate illustration of their important contribution. It is critical to confirm that the engineer's role in design was an important one, rather than titular (many people are involved in design and construction). It is unlikely that political/administrative figures would be directly associated with a dam's design or construction.

Resources eligible under Criterion 2 are to be *illustrative* rather than *commemorative* of a person's contributions. Thus, for example, Lake Powell behind Glen Canyon Dam would not be eligible for its association with namesake John Wesley Powell, nor was Roosevelt Dam (Arizona) designated a NHL for its association with Theodore Roosevelt.

Criterion 3:
Properties represent some great idea or ideal of the American people.

Criterion 3 is rarely, if ever, used alone. Dams, however, have been defined as: "one of the most essential aspects of man's attempt to harness, control and improve his environment;" the "Useful Pyramids;" modern wonders of the world.[15] Large public expenditure and (historically) rarely tempered support for the construction of such powerful structures make dams an outstanding representation of American cultural keystones: growth of empire and economic expansion. Moreover, the largest of our massive dams bolster our cultural self-definition of a people resilient in the face of adversity, whether economically or environmentally imposed. In 1936, a partial image of the spillways at Fort Peck Dam adorned the inaugural issue of *Life*. Hoover Dam, sixty percent taller than any dam built before it, "became the symbol of price and accomplishment during a period of national despair."[16] American icon and the voice of an era, Woody Guthrie, celebrated Grand Coulee Dam:

> Mightiest thing ever built by a man.
> To run the great factories and water the land.
> It's roll on, Columbia, roll on.[17]

And therein lies the reason that NHL Criterion 3 is rarely used alone. The *Life* image may have been the most outstanding physical representation of American social and political life in the 1930s, a portent for a newly defined public-private political, economic, and social contract. It may also have been simply the most striking graphic image and the Fort Peck story, with bold stanzas of sex and alcohol and the Wild West of American nostalgia, may have been the most titillating. Guthrie may have summarized the collective dreams and visions and aspirations of a nation burdened by economic depression. He was also paid by the federal government as part of a mighty inspirational campaign. Cultural significance, or popular affection, is central to the evaluation or our national icons and must be addressed. Yet it is intellectually dangerous to base any argument for national significance solely on popularity, on, for example, the percentage of Americans who recognize the Golden Gate Bridge, or who have visited the Liberty Bell, or who can recite the Statue of Liberty's welcome. Popular affection assumes the greatest weight when traced to its roots in the more objective realms of American social, economic, technological, or political history: when NHL Criterion 3 is paired persuasively with NHL Criteria 1, 4, or 6.

Criterion 4:
Properties that embody the distinguishing characteristics of an architectural type specimen exceptionally valuable for a study of a period, style or method of construction, or that represent a significant, distinctive and exceptional entity whose components may lack individual distinction.

Criterion 4 encompasses the historical significance associated with design or construction merit, with technological innovation, and with artistry. National significance in any of these areas relates to the property's ability to outstandingly represent the defining characteristics of a design/construction "type" or to outstandingly represent the *transition* between styles of resources (most often as part of a chronological chain of design evolution or technological innovation).

"Unique" or "transitional" properties will be determined nationally significant *only* if it can be demonstrated that the variation represents a significant technological or design step rather than an alteration that had no subsequent impact. In regard to dams and other large civil engineering works, for example, variations on standard design are often demanded by the unique geological characteristics of the construction site and of the river; only rarely can these variations be defined as innovations that altered subsequent technological theory and practice.

It is equally rare that the variations suggest evaluation of the associated dam as an "outstanding representation" of a distinct and definable dam type. Despite variations dictated by site or river characteristics, all dams can be classified as one of two dam types: either "massive" or "structural." The massive tradition includes all *gravity dams* wherein stability is provided solely by the weight of the material used; civil engineering professor Jerome Raphael (University of California-Berkley) argues that "this definition was as applicable a hundred years ago as it is today."[18] Variations in gravity dam design – e.g., a straight, curved, arched, or crooked crest – do not alter the basic design principle. Engineers further categorize massive dams by material – *concrete, masonry, earthfill, or rockfill* – in recognition of the divergent construction methods demanded by each material type.

The "structural tradition" encompasses those dams where stability is provided not by gravity but by application of "progress in science and engineering" to concerns of "need and demand," namely limited funds, limited manpower, and limited material. Important variations within the two principal structural dam types (*arch dams and buttress dams*) relate to evolution in three general characteristics: shape, height, and geometric properties. These variations have resulted from four primary factors: (1) advances in analytical methods including computer application, (2) measurements of actual behavior (improved field testing and instrumentation), (3) improvements of concrete and other construction materials, and (4) improved construction methods and utilization of equipment.[19]

In reference to structural dams, "large scale" refers most meaningfully to height or length, and engineering importance results primarily from the inverse proportion of massing to height and to length. For example, engineer Jan A. Veltrop notes that the engineering fraternity described California's Bear Valley Dam (1884) as "the eighth wonder of the world," in recognition not of its size (its 64-foot height was not particularly remarkable) but of its remarkable economy of materials: the bottom thickness of the dam measured only twenty-two to twenty four feet, the top three feet, and the crest length exceeded 450 feet.[20] In contrast, the size of the world's most massive concrete dams may reflect only the large-scale application of the central (and essentially unchanged) design principle of gravity construction and may not be an indicator of design significance.

As suggested by the above discussion, evaluation of national significance under Criterion 4 depends in part on an understanding of when the characteristics of size and precedent might be important. An evaluation of the importance of these two characteristics requires an understanding of the engineering process and of the important distinction between design and construction. The following guidelines, related to design and construction innovation, are intended to provide guidance on when the superlatives "biggest," "highest," "longest," "first" are important as defining characteristics of a dam's NHL significance under Criterion 4. The negative guidance condenses to a single caveat: size and statistical glory are not important in and of themselves and dams defined solely in those terms will not meet NHL criteria. However, there may be instances in which size (whether great or representative of economy of material and leaps in scientific ability) is an important and demonstrable component of a resource's national significance.

Design Significance

In a 1935 editorial focused on Public Works Administration-funded dam construction, the editors of *Compressed Air Magazine* "conceded" that:

> there are no considerable physical or technical problems that would prevent the erection of a skyscraper having twice the height of the Empire State Building. There are economic limitations, of course, and these in the long run, are the factors that control the size of structures.[21]

From 1932 until completion of Hoover Dam in 1936, southeastern Oregon's Owyhee Dam (417 feet) stood as the tallest dam in the world, sixty-six feet taller than the previous record holder, Arrowrock Dam in southwestern Idaho.[22] Yet construction of Owyhee *to that height* involved no design innovation and nomination of Owyhee to the NHL on the basis of fleeting statistical glory would inevitably lead to confusion over the significance of the numerous dams that at some point, and often only briefly, earned the accolade of the world's biggest. The central NHL caveat for any resource evaluated under criterion 4 applies: the dam must be "exceptionally valuable" for the *study* of the design and construction techniques rather than simply an "embodiment of characteristics of the type" – in this case, simply a large embodiment of the type.

Construction of Hoover Dam entailed a concrete placing effort larger than the aggregate of Bureau of Reclamation dams since its establishment in 1902. Preliminary trial load analysis revealed that the dam was being overbuilt by half.[23] The original design of Grand Coulee Dam[24] was rejected as "well beyond the range of recognized engineering procedure. While technical skill may overcome them . . . the hazard of the operation is one that should not needlessly be assumed."[25] The financial and human risk associated with the dam failure is extreme.[26] The size of our largest dams and spillways therefore may represent a conservative precaution that does not represent the extent of American engineering ability at the time of construction.[27]

Extreme size may also more accurately reflect the economic and psychological health of the nation (see Criterion 1) than the skill of our civil engineers. Of the massive dams constructed during the New Deal Era, historian Donald Jackson writes:

> the "celebration of mass" became the dominant ideology associated with dam construction: the more material a dam required, the more acclaim and adulation it received. In an era of limits and diminished expectations, American culture apparently

> derived psychological satisfaction from creating something big in the face of adversity. Thus the Grand Coulee Dam drew praise for being the first masonry structure in more than three millennia to use more material than the largest Egyptian pyramid. Similarly, no one complained that the Hoover Dam would have been grossly overbuilt even without its pronounced upstream curve . . . dam financing became as much a vehicle for distributing federal funds as it did a fiscal basis for undertaking a water storage project.[28]

Technological innovations are often realized not in dramatic leaps in height and massing but in lab testing; in small-scale field applications incorporated within tried and trusted designs; and, most importantly, are usually realized in increments, in the new application or unprecedented combination of exiting technologies. Scale of application is relevant to a discussion of technological merit *only* if the sheer mass of the property introduces technological obstacles that require innovative combinations and variations on existing technology. Hoover Dam again serves as an example: the extreme volume of concrete used demanded innovative solutions to problems of heat generated by chemical reaction during the concrete-cooling process; Owyhee Dam, already under construction, was selected as a field-test laboratory for cooling pipes later incorporated within Hoover's design. This innovation made possible subsequent construction of the world's largest dams, including not only Hoover but also Grand Coulee and Shasta.[29]

Pathfinder and Buffalo Bill (Shoshone) Dams in Wyoming also serve as examples of high dams made possible by technological innovation. Following the example of Frank E. Brown and John Eastwood's thin arch and buttress designs, most notably California's Bear Valley and Sweetwater Dams, Bureau of Reclamation engineers,

> introduced the concept of a co-existing system of horizontal arches and vertical cantilevers. . . . The so-called 'trial load method' put arch dam design on a much sounder footing, particularly if a series rather than one cantilever was used, and especially if complete compatibility of movements at all points was achieved. The technique has remained fundamental to the American school of arch dam design to this day and has been the method used to design many large and famous dams: the Arrowrock dam [sic.] (400 feet high) . . . the Gibson dam [sic.] (200 feet high) . . . the mighty Hoover dam [sic.] (727 feet high) . . . and the Hungry Horse dam [sic.] (564 feet high). . . .[30]

The language here is found only infrequently in reasoned discussions of dam design and clearly introduces the possibility of technological innovation warranting evaluation under NHL Criterion 4: "*introduced* the concept," "*remained fundamental* to the American school of arch dam design." Trial load analysis, or "Arch and Crown-Cantilever Method" as it was first defined, was both *innovative* and, equally important, of *impact*. These two standards are central to the evaluation of national significance under NHL Criterion 4. Either the first use of the new technology (Pathfinder Dam, 1907) or the first large-scale use (Buffalo Bill, 1910) might prove to be nationally significant manifestations of the innovation.

The significance of the largest-scale use of technology is analogous, in many cases synonymous, to the significance of the first use. The small and unassuming Boyd's Corner Dam in New York (1870), the first dam constructed with a concrete core since Roman times, is a potential NHL.[31] (This despite the fact that is serves a strictly parochial function [municipal water supply].) The more difficult task, however, involved determining which "first" best represents the critical innovation: San Francisco's San Mateo Dam, constructed entirely of concrete between 1887-1889, may be a better candidate than Boyd's Corner.[32] And yet, concrete's use in construction of massive dams remained restricted until completion of Owyhee Dam (1932). However, only panels 3, 4, and 8 at Owyhee incorporated the pipe cooling system; the system was an incidental afterthought in Owyhee's design and construction. The success of the field tests, however, allowed for the completion of Hoover Dam, where the use of previously realized technological innovation made possible construction of a dam of unprecedented size.[33]

All three firsts, Boyd's Corner, San Mateo, Owyhee, led logically and chronologically to construction of the massive concrete-gravity dams during the period 1930 to 1980. Additional examples of sequential development leading to a confused string of first and biggest abound (with advances in trial load analysis and model studies the most often cited). Robert Vogel defines the quandary:

> The Brooklyn Bridge is a perfect example of a structure whose design is the result of logical evolution and proven construction methods, yet also embracing innovative technologies and materials. But even in the case of the new departure, nothing was adopted by either John or Washington Roebling that had not been shown to be effective and safe in prior undertakings.[34]

Engineer Lawrence Sowles reiterates:

> The design of [Hoover Dam] is based upon researches which have been carried on almost since the Bureau of Reclamation was created in 1902, and upon the accumulated experience obtained in building more than 50 concrete dams during that period. *Incidentally*, [Hoover] Dam will contain more concrete than the preceding 50, a circumstance that warranted special intensive study to determine beforehand how concrete would behave when poured in a mass of this size and height and to make it possible to specify materials, methods, and practices that would serve to best advantage.[35]

In sum, engineering significance is most often defined as the use of existing technology in new ways and in new combinations; the field conditions and design characteristics that demand these innovations might relate to size. There is therefore merit in looking to size as a *potential indicator* of national design significance. The subsequent analysis, however, must include both an evaluation of design innovations and impacts demanded or defined by great size and also an evaluation of the most significant "first" – what resource, regardless of size, most effectively represents the critical innovation?

If civil engineering is in fact a process of small-scale steps, marked by few easily defined "leaps" or engineering breakthroughs, then it is possible that small-scale auxiliary resources will define a dam's engineering significance. Examples might include innovative gate or spillway design. Again, the evaluation of the importance of the innovation must include an analysis of impact: was the innovation demanded only by unique site conditions? Was it used subsequently? Did it allow construction of dams in previously untenable sites, under previously untenable conditions?

A dam's aesthetic appeal may also suggest design significance. Civil engineer David P. Billington has defined "structural art" as the deliberate manipulation of dams and spillway design to create functional sculpture, resources of exceptional beauty that transcend the vagaries of style. Examples include California's East Park Dam, where the spillway undulates along the natural contour lines, not only breaking the water force and potential for erosion but creating striking man-made waterfalls. Similar hydraulic characteristics of the spillway were realized, for example, at Fort Peck Dam (where concrete blocks impede the force of the water) with none of the aesthetic benefits. Manipulation of functional resources for aesthetic purposes might be central to a dam's design, in which case it should be addressed in any analysis of the resource's overall design and construction significance. However, it is unlikely that a dam will be determined a NHL strictly on its quality as sculpture: that aesthetic appeal is too subjective and its impact on the history of the nation too nebulous.[36] Justifying a determination of national significance solely on the basis of aesthetic appeal requires a resource of exceptional merit.[37]

Construction Significance

Engineers not only decide "what work to do" – problems of "choice and character of structure" – but also "how to do work" – how to place enormous volumes of earth, rock, and concrete.[38] Variations in material used within the massive tradition (concrete, masonry, earthfill, or rockfill) are most relevant within this context; large-scale application of each of these primary materials poses a unique set of problems that may demand distinct innovations. Fort Peck Dam, where "[material volume] figures difficult to grasp confront the construction engineer," was made possible "only by bringing into action all the resources of invention, of business organization and of technical skill that *engineering and contracting and the equipment manufacturers* have created in a generation of great public and private construction."[39] Engineer Lawrence P. Sowles described the construction of Hoover as

> primarily interesting because of the enormity of the work involved. The building of a big concrete dam consists essentially of placing a large volume of concrete in one mass in accordance with a specific design, and of providing the necessary appurtenances and accessories. Except for their scale of application, the operations in the case of Hoover Dam are much the same as those that have been performed in rearing many other dams of more or less consequence.[40]

"Except for their scale of application." The exception is important, establishing that scale and size introduce a degree of construction difficulty that *may* in turn demand important innovations in construction methods.

However, while the level of difficulty and of construction achievement is an element of significance, it is important to recognize that it is shared by all large and/or isolated construction projects undertaken in inhospitable environments. The ability of thousands of men to stack concrete or to pile dirt higher than it has ever been piled before, in a more rugged environment, at more extreme temperatures, is not unusual and is not in and of itself significant. Are the "mechanical shortcuts devised" or the "startling departures that set new standards for speed and economy of performance" important to the nation? Do they have an important influence or impact on future, significant projects? If yes, what resource (if there is more than one) best represents the innovation and what components relate most directly to historical significance?[41]

Construction innovation and ingenuity are frequently demanded not only by the mass of material applied but also by the topographic and geological characteristics of a dam site. At Fort Peck Dam, engineer Harold O'Connell noted that "size, in itself, would not make the construction of [Fort Peck Dam] the engineering feat that it is. Difficulties imposed by the nature of the topography immediately surrounding the site [is a] contributing factor."[42] Most notably, the prevalent Bearpaw Shale, virtually impervious in its undisturbed state, yet prone to disintegration upon exposure to the atmosphere, required the application of bituminous material during tunneling and was cut with rock saws, a technology borrowed from coal mines. Although this methodology was "unprecedented" and "unique" to Fort Peck, an effective analysis of its national significance under Criterion 4 would have to include a discussion of the variations' significant, subsequent use: did the variations represent an important step in hydraulic-fill dam design and construction or was their impact merely local? The same critical evaluation is demanded, for example, by Arrowrock Dam in southern Idaho where an increase in the water content of the concrete slurry (simplifying placement in the concrete forms) proved to be "unfortunate," resulting "in an insufficiently durable concrete, especially under severe frost."[43] Although innovative, and important to implementation of the Arrowrock design, the innovation was not repeated and is not significant as defined by NHL Criterion 4.

Ironically, failed dams may also meet NHL criteria. Within the United States, the failure of the South Fork Dam, Johnstown (Pennsylvania), St. Francis (California), and Teton (Idaho) Dams, and the massive slide at Fort Peck, led to technological modifications (Criterion 4) and policy initiatives (Criterion 1) that *might* rank as nationally significant. Jerome Raphael reports that the March 12, 1928, failure of St. Francis Dam "changed forever the thinking about foundation engineering for dams. . . . The great lesson of the disaster was the necessity of thorough investigation of the geology of a proposed damsite and reservoir."[44] As in every instance, the critical question remains: are the modifications and initiatives of national significance?

Obvious threats to integrity of design and workmanship associated with the dam failure lie at the center of a failed dam's significance, and thus might not adversely affect NHL eligibility (see *integrity*, below).[45]

**Criterion 5
Resources composed of integral parts of the environment
not sufficiently significant by reason of historical association
or artistic merit to warrant individual recognition but**

collectively compose an entity of exceptional historical or artistic significance, or outstandingly commemorate or illustrate a way of life or culture.

Criterion 5 facilitates the nomination of resources that individually might not be of national significance but that when combined compose a historically or artistically significant entity. This criterion is most often applied to architectural districts. In the case of dams, elements of the "whole" must be historically interconnected, as in the large-scale federally sponsored river-basin flood-control and multiple-use projects of the mid to late twentieth century. To meet Criterion 5, these dams must be part of a *planned* system of national significance. It is not enough that the dam is operated in conjunction with other dams along a river system (most dams are), but that the system was envisioned from the beginning. For instance, Shasta Dam in California's Central Valley Project may not be of national significance. However, if it met the 50-year rule and other requirements, the Central Valley Project might be eligible even if none of its elements, such as Shasta, San Luis, Trinity, Folsom, Friant, and other dams, proved individually significant. The system defines the water exchange that makes the Central Valley Project a path-breaking innovation in water resource management. Similarly, the Army Corps of Engineers' navigational efforts on the Mississippi and the Ohio Rivers or flood control efforts in the lower Mississippi Valley may be nationally significant only in the context of a larger system designed with multiple dams working in tandem.

It is important to note that Criterion 5 relates not simply to boundaries (the identification of those associated resources possessing historical integrity, such as a dam and supply road), but rather to the identification and documentation of those contiguous and discontiguous components critical to a system's national significance. It is especially relevant when evaluating the dams of the middle to late twentieth century, constructed after both Reclamation and the Corps had embraced the concept of unified river basin development.

**Criterion 6:
Resources that have yielded or may yield information of major scientific importance by revealing new cultures, or by shedding light upon periods of occupation over large areas of the United States. Such sites are those which have yielded, or which may reasonably be expected to yield, data affecting theories, concepts, and ideas to a major degree.**

Dam ruins are most likely to be evaluated under Criterion 1 and/or 4. When evaluation under Criterion 6 is deemed appropriate, the evaluation must focus on the importance of the remains and how they inform theories, concepts, and ideas. In those instances, when there are insufficient remains to yield significant data, the dam site might be evaluated for its associative value, under the terms of the NHL exclusion pertaining to the site of a structure no longer standing (see Exclusions, below).

National Historic Landmark Exclusions

Cemeteries, birthplaces, or graves of historical figures; properties owned by religious institutions or used for religious purposes; a site of a building or structure no longer standing; commemorative properties; structures that have been moved from their original locations; reconstructed historic buildings; and properties that have achieved significance within the past fifty years are generally not eligible for National Historic Landmark designation. Of these exclusions, restrictions against commemorative, non-extant, and modern resources may prove relevant to the evaluation of dams.

Dams and associated infrastructure named in commemoration of an individual (Hoover Dam, Roosevelt Dam, Lake Powell, Bonneville Dam, Barkley Dam, Richard Russell Dam, Melvin Price Lock and Dam, to name seven of the many) are not nationally significant on the basis of this commemorative association.

The exclusion of commemorative properties is closely related to the exclusion of sites where the significant building or structure (in this context, the dam) is no longer standing. Such sites may be nationally significant only if "the persons or events with which they are associated are of *transcendent* national significance *and* the association is significant." An example might include the site of the St. Francis Dam failure, if the failure of the dam can be shown to have been of transcendent importance in dam engineering. (By the nature of dams and other civil engineering works, the association between the site and the "event" of construction or failure is significant. If, however, St. Francis Dam designer William Mulholland is identified as an individual of transcendent national importance, then the site's important association with Mulholland would also have to be established.)

Resources achieving national significance within the past fifty years qualify for NHL consideration if they are of "extraordinary national importance." The evaluation process can be complicated by the absence of time in which to view the properties in a more objective and informed context, of an established historic context, and of secondary sources by which to judge a resource's impact. However, the impact of exceptional resources will most often be of such immediate magnitude that historic contexts will have been developed, the immediate impacts documented, and hypothesis for future reverberations clearly articulated. Examples of designated NHLs less than fifty years old include the most important resources associated with such cataclysmic events as the Civil Rights Movement, Space Exploration, and the Cold War. It is not anticipated that any U.S. dam will meet this burden of exceptional national significance.

Note that the impact of a dam, whether related to design, construction, or associative value, may date to the design or construction period. If warranted, the period of significance may begin with the initiation of formal design work or the start, rather than the completion, of construction.

Integrity

National Historic Landmarks must possess a higher degree of historical integrity than that required for National Register listing and must retain all seven aspects of integrity – setting, location, design, material, workmanship, feeling, and association. Integrity of important interior spaces and interior mechanization must be factored into the evaluation. Those aspects of integrity that relate most directly to a property's character-defining features must be evaluated the most carefully.

In general, threats to integrity, whether of design or association, are cumulative: there is a "critical mass" at which the resource is unable to absorb either the volume or the visual impact of modifications. (In terms of visual and design impact, the loss of historical material is often more damaging than the addition of new materials.) By virtue of function and the constraints of geographic setting, dams will generally possess integrity of setting and will always possess integrity of location. Potential threats to integrity of design and material include raising the crest height; refacing a dam face with a more stable material; modifying the spillway; and constructing or reconstructing auxiliary structures, such as powerhouses, fish ladders, or temperature control devices.

Because of their massive size, dams are able to absorb a degree of new material and auxiliary infrastructure without losing high integrity. For example, a 250-foot-wide by 300-foot-high Temperature Control Device (TCD) has recently been appended to the upstream face of 602-foot-high Shasta Dam (California); the TCD's seventeen gates, each of which weighs fifty-eight tons, allow Reclamation to discharge water from varied levels of the reservoir, thereby maintaining the colder water temperatures needed by spawning chinook salmon.[46] Not only does the reservoir mask much of the TCD, but the scale of the downstream dam face dwarfs the exposed TCD components; the TCD's minor visual impact is equal to its overall technological impact – it is clearly an auxiliary component with little bearing, and little adverse physical impact, on the overall size and design of the dam itself.

In contrast, Stoney Gorge Dam (California) approaches the limit of absorption. Although the original structure remains intact, the adjacent spillway has been modified, a section of the dam crest raised, and a powerhouse constructed. While these additions do not mask the original design characteristics, particularly the flat slabs and buttresses, additional modification of the buttresses, original overflow-type gates, and other integral components of the design would likely render the property ineligible as a NHL.

Boundary Justification/Identification of Contributing and Noncontributing Resources

The boundaries of a proposed NHL should be given careful consideration and should be based upon the established national area of significance, period of significance, and evaluation of integrity. For all potential landmarks, the following guidelines apply:[47]

1. Carefully select boundaries to encompass, but not to exceed, the full extent of the nationally significant resource.

2. The area to be nominated should be large enough to include all nationally significant features of the property, but should not include "buffer zones" or acreage not directly contributing to the significance of the property.

3. Leave out peripheral areas of the property that no longer retain a high degree of integrity, due to subdivision, development, or other changes.

The determination of boundaries and the identification of contributing and noncontributing features are closely related. Both are based upon an assessment of the nominated property's area and period of significance, as well as the historical integrity of individual components. (For example, while transformer yards may be included within site boundaries, and may be recognized as historically significant, they often do not retain sufficient integrity, as defined by NHL criteria, to be identified as contributing components.) In certain circumstances, including those in which visual continuity is not a factor of historic significance; when resources are geographically separate; and when the intervening space lacks significance, contributing resources will be defined as discontiguous elements thus excluding the nonsignificant landbase and noncontributing resources from the NHL boundaries.

Dam construction, particularly in the mountainous stretches of the unsettled West, involved construction of a wide array of support infrastructure (both temporary and permanent), including river diversion tunnels, powerhouses, transformer yards, spillways and penstocks, wagon roads and railroads to haul equipment and supplies, lumber mills, cement production plants, rock quarries, borrow areas, telephone lines, and townsites.[48] A description of Fort Peck Dam in 1935 gives some idea of the magnitude and diversity of these resources. The project included the following: a detached spillway; a thirteen-mile branch line from the Great Northern tracks at Wiota; a 300-foot, three-span bridge over the Missouri River; a 13,000 foot truss-and-trestle rail and auto bridge along the downstream face of the dam, 238 miles of powerline and two substations, and the town of Fort Peck. The town contained nine blocks of bunkhouses, nine mess halls, eighteen shower houses, three blocks of foremen's dormitories, blocks of homes for married staff and executives, a central administration building, a shopping center, a cold-storage plant, a commissary warehouse, a hospital, a garage and maintenance yard, and a community water treatment and sewage plant.[49]

National Register evaluation of a dam at the local or state level of significance demands an evaluation of those extant auxiliary structures and buildings associated with dam construction and use and therefore associated with the dam's impact on the local community. However, in most cases, these resources will not relate directly to a dam's *national* significance and will be excluded from the landmark nominations. (Frequently, landmark boundaries may be drawn within a larger National Register Historic District boundary.)

Integral components of a dam may be grouped in two ways: (1) those resources central to the safe and effective operation of the dam and (2) those central to product (water and power) delivery to an outlying distribution system. The first group of resources might include diversion tunnels, outlets, inlets, stilling basins, guide walls, gates, spillways, trash racks, and (possibly) penstocks; the second includes transformer yards, pumping stations, powerhouses, penstocks, turbines, fish ladders, and temperature control devices. *If* they possess high integrity and date to the defined period of national significance, both classes of resources will most often be included within the NHL boundaries.

There will be exceptions and, again, the determination of boundaries and of contributing versus noncontributing resources is based upon the property's area and period significance. Fort Peck Dam was authorized and constructed to assure an 8' to 9' navigation channel below Sioux City, Iowa; to provide minimal flood control and erosion protection; and "to relieve serious local unemployment and distress." Neither power production nor irrigation was part of the original design plan. Unless the dam's national significance is based at least in part upon post-construction, multiple-purpose development, the powerhouses added to the dam after the initial construction are not likely to be included within the district boundaries. In contrast, the powerhouses at Bonneville Dam were central to the original design and intent of dam construction and have been defined as contributing components of the NHL property. Similarly, the pumping station at Grand Coulee Dam, components of which extend upward beyond the immediate dam site to the canyon rim where they feed the primary distribution canals, would likely be included within an NHL nomination as resources integral to the effective operation of the dam.

Non-integral components may also be grouped in two ways: (1) residential, recreational, and administrative facilities and (2) reservoirs and resources related to the distribution system, such as powerlines, canals, and feeder ditches.

"Company Towns," administrative headquarters, and boating facilities associated with large federal dam construction projects may be included within the landmark boundary. This inclusion, however, is dependent upon the resources' important association with nationally significant economic, labor, and social themes in American history and upon their historical integrity.

Statements of national significance, whether related to a dam's associative or design significance, will inevitably include a discussion of the importance of reservoirs and auxiliary resources related to the distribution system. Dams are only one component of a tightly interrelated *technological system*: canals are central to the function of irrigation dams, as water lines are to municipal water-supply dams, as transmission lines are to power dams, as reservoirs are to flood control and navigation dams. However, historically associated and historically significant infrastructure located outside the immediate boundaries of a dam site – that boundary imposed by the dam itself and its *structurally integral* infrastructure – will most often be excluded from the NHL boundary.

Arguably, it is the associated infrastructure that best defines a dam's significance. The Bureau of Reclamation's impact on the West is conveyed by the myriad of canals and disitribution ditches that carry the water to previously arid land. The U. S. Army Corps of Engineers' impact on the nation may be best represented by merchant ships that ply downstream shipping lanes. Yet the National Historic Landmark program is in part one of the properties that "possess exceptional value or quality in *illustrating* or interpreting the heritage of the United States." Exclusion of reservoirs, canals, water supply lines, power lines, and irrigated fields from NHL dam boundaries is a practical concession, acknowledgment of the difficulty of defining, administering, and describing a property of such proportions. It is also valid recognition of the power of contiguous visual imagery, of a dam's remarkable ability to represent the larger technological system with which it is associated.

Endnotes

1. Guidelines developed by the National Park Service for evaluating a property's historical significance and physical integrity, outlined in U.S. Department of the Interior, National Park Service, *National Register Bulletin 15: How to Apply the National Register Criteria for Evaluation* (Washington D.C.: Government Printing Office, 1991), apply generally to the evaluation of potential NHLs. The categories of historic properties are defined the same way; historic contexts are identified similarly; and comparative evaluation is carried out on the same principles. This information should be used as a general reference when evaluation American dams for NHL designation.
2. While these guidelines were prepared as part of a review of large multiple-purpose dams, they have broader applicability to the evaluation of all dams. As part of this study, four NHL nominations of Corps of Engineers and Reclamation dams were prepared. These four were selected from a larger list of potential NHLs.
3. Schnitter, *A History of Dams*, ix. The National Register defines dams as *structures* although the boundaries of a nominated property may include a variety of *buildings, objects,* and *sites* united in a *district* (see boundary justifications, below).
4. U.S. Army Corps of Engineers Survey cited in Kollgaard and Chadwick, editors, *Development of Dam Engineering*, xii.
5. As specified in the Scope of Work associated with this project, a context has been developed specifically for federal dams (although their private precursors are references when appropriate). However, the basic principles for evaluation of public and private dams are similar. This larger class of resources is covered by the guidelines.
6. The Secretary of the Interior designates NHLs based on an evaluation of formally nominated properties.
7. Smith, *A History of Dams*, 228.
8. Defined for the purpose of this study as those in excess of fifty feet high.
9. For example, flood control requires a large storage area to accommodate swollen rivers and heavy precipitation, while hydropower interests prefer that the reservoirs remain as full as possible; irrigation takes water away, while recreational champions desire high lake levels; cities want their reservoirs full for water supply, but navigation interests want to use the reservoir water downstream to ensure adequate navigation depths. When these conflicts, or their resolutions, have an important and substantive impact on the nation as a whole, they may demonstrate national significance.
10. C. H. Vivian, "Developing the Mighty Columbia," *Compressed Air Magazine* (September 1935), 4819.
11. Any analysis of the degree of historical impact should be based on the actual development not on preconstruction projections. Grand Coulee Dam's potential, for example, to irrigate 1,200,000 acres of "fertile but arid" land is less important than the acreage ultimately irrigated and farmed.
12. Kollgaard and Chadwick, editors, *Development of Dam Engineering*, 74.
13. Jackson, *Building the Ultimate Dam*, 4.
14. Stephen D. Mikesell, "Pardee Dam: Evaluation of National Register Eligibility," report prepared for John Holson, BioSystems Analysis, In., 1994, 13.
15. Norman Smith, as quoted in Schnitter, *A History of Dams*, xi.
16. Robinson, *Water for the West*, 51.
17. Pitzer, *Grand Coulee*, 265.
18. Kollgaard and Chadwick, editors, *Development of Dam Engineering*, 13.
19. Kollgaard and Chadwick, editors, *Development of Dam Engineering*, 226.
20. Kollgaard and Chadwick, editors, *Development of Dam Engineering*, 74.
21. "Editorial: Construction Progress," *Compressed Air Magazine* (October 1935), 4864.
22. Frederick L. Quivik and Amy Slaton, Historic American Engineering Records (HAER No. OR-17), Prints and Photographs Division, Library of Congress. Owyhee Dam, Nyssa Vicinity, Malheur County, Oregon [hereinafter "Quivik and Slaton, HAER No. OR-17]
23. Schnitter, *A History of Dams*, 177.
24. Requiring construction of a low dam for power generation purposed and subsequent incorporation of this low dam in a high dam design for irrigation purposes.
25. "A Mistake that Should be Corrected," *Engineering News-Record* (January 3, 1935), 23.
26. Smith, *A History of Dams*, 214.
27. Robinson, *Water for the West*, 52.
28. Jackson, *Building the Ultimate Dam*, 246-51.

29. Similarly, the subaqueous St. Clair River Tunnel (Minnesota) has been designated an NHL under criterion 4 because of the successful combination of existing technologies (the shield method of excavation, the cast iron tunnel lining, and excavation in a compressed air environment) in a setting, and on a scale, suitable to full-size transportation needs (Robie Lange, "National Historic Landmark Nomination: St. Clair River Tunnel," February 1993, 11).
30. Smith, *Man and Water,* 63-4
31. Schnitter, *A History of Dams*, 174.
32. Smith, *A History of Dams*, 220.
33. Quivik and Slaton, HAER No. OR-17, 50.
34. Robert Vogel, quoted in Donald C. Jackson with foreword by David McCullough, *Great American Bridges and Dams*, Great American Places Series (Washington, D.C.: The Preservation Press, National Trust for Historic Preservation, 1988), p. 131.
35. Lawrence P. Sowles, "Construction of Boulder Dam: Description of Methods of Pouring the Concrete," *Compressed Air Magazine* (April 1934), 4385
36. David P. Billington referenced in Jackson, *Great American Bridges and Dams*, 243.
37. Dam cladding, decorative treatment disconnected from all function, may also possess artistic merit; the Moderne facade of TVA's Hiwassee Dam is an often-cited example. This cladding, however, is irrelevant to the functioning design of the dam itself and is evaluated as all works of art are evaluated to NHL significance: "considered in the context of history's judgement" does it "so fully articulate a particular concept of design that it expresses an aesthetic ideal . . ." See National Park Service, *National Register Bulletin 1*; 20, 51).
38. "This Issue," *Engineering News-Record* (August 1935), 279.
39. "Men and Mechanization," 684, emphasis added.
40. Sowles, "Construction of the Boulder Dam. . .," 4385.
41. "Editorial: Construction Progress," *Compressed Air Magazine* (October 1935), 4864.
42. Harold O'Connell, "Fort Peck Dam: Part 1 – The Project and Its Scope," *Compressed Air Magazine* (March 1935), 4672.
43. Schnitter, *A History of Dams*, 177.
44. Kollgaard and Chadwick, editors, *Development of Dam Engineering. . .*, 74.
45. This modified integrity standard obviously does not apply if evidence of the ruin itself, or of the failed component, has been removed.
46. "Shasta Temperature Control Device, Planned, Designed and Implemented by the Bureau of Reclamation," no date, on file at the Bureau of Reclamation Northern California Area Office, Shasta lake, California.
47. U. S. Department of the Interior, National Park Service, National Register, History and Education Division, *How to Prepare National Historic Landmark Nominations* (Washington, D.C.: Government Printing Office, 1999), *passim*.
48. Robinson, *Water for the West*, 20.
49. O'Connell, "Fort Peck Dam: Part I. . .," 4672-3.

APPENDIX B: RESEARCH METHODOLOGY AND LIST OF POTENTIAL NATIONAL HISTORIC LANDMARK MULTIPLE-PURPOSE DAMS

Historical Research Associates, Inc. (HRA)

The following methodology details the means by which Historical Research Associates, Inc. arrived at a historic context and list of potential multiple-purpose National Historic Landmark (NHL) dams, constructed by the U.S. Army Corps of Engineers or the Bureau of Reclamation within the historic period (up to 1950).

I. Consultation with Project Historians

Many of the dams on the list of potential NHL dams that follow were identified in the course of extensive discussion with project sponsors and with project historians (the subject experts) as significant components of the watersheds, policy decisions, and construction innovations around which the associated NHL theme study *Federal Multipurpose Dam* is organized. For example, Elephant Butte will figure prominently in the chapter on federal water law and policy; Garrison and Fort Peck Dams will figure prominently in the discussion of Missouri River development and the Pick-Sloan plan; Grand Coulee Dam will figure prominently in the discussion of hydropower development. Similarly, Pathfinder, Buffalo Bill (Shoshone), and Gibson Dams have been identified by David P. Billington and others as the dams most directly associated with significant engineering advances. The subject experts have therefore concluded that the dams on the attached list are those that best represent the key characteristics and areas of significance (whether technological, social, political) defined in the historic context and may be worthy of NHL nomination.

II. Secondary Source Review

In the course of developing the criteria for National Historic Landmark eligibility and of developing a list of potential NHL dams, HRA reviewed the following sources:

Arnold, Joseph L. *The Evolution of the 1936 Flood Control Act.* Fort Belvoir, Virginia: Office of History United States Army Corps of Engineers, 1988.

Barrows, H. K. *Floods. Their Hydrology and Control.* New York: McGraw-Hill Book Company, Inc., 1948.

Chambers, John Whiteclay II. *The North Atlantic Engineers. A History of the North Atlantic Division and its Predecessors in the U.S. Army Corps of*

Engineers, 1775-1975. Report prepared for the North Atlantic Division, U.S. Army Corps of Engineers, 1980.

Gates, Paul W. *History of Public Land Law Development.* Washington, D.C.: Zenger Publishing Company Inc, 1968.

Goldsmith, E., and N. Hildyard, *The Social and Environmental Effects of Large Dams, Volume I, Overview.* A report to the European Ecological Action Group (ECOROPA), 1984.

Graf, William L. "Landscapes, Commodities, and Ecosystems: The Relationship Between Policy and Science for American Rivers," *Sustaining Our Water Resources.* Washington, D.C.: National Academy Press, 1993.

Hays, Samuel. *Conservation and the Gospel of Efficiency: The Progressive Conservation Movement 1890-1920.* New York: Atheneum Press, 1972 (first published by Harvard University Press, 1959).

Hundley, Norris. *Water and the West: The Colorado River Compact and the Politics of Water in the American West.* Berkeley, Los Angeles, London: University of California Press, 1975.

Jackson, Donald C. *Building the Ultimate Dam: John S. Eastwood and the Control of Water in the West.* University Press of Kansas, 1995.

Johnson, Leland R. *The Headwaters District: A History of the Pittsburgh District, U.S. Army Corps of Engineers.* Report prepared for the U.S. Army Corps of Engineers, n.d.

Kollgaard, Eric B. and Wallace L. Chadwick. *Development of Dam Engineering in the United States.* New York: Pergamon Press, 1988.

Maass, Arthur. *Muddy Waters. The Army Engineers and the Nation's Rivers.* Cambridge, Massachusetts: Harvard University Press, 1951.

Martin, Russell. *A Story that Stands Like a Dam: Glen Canyon and the Struggle for the Soul of the West.* New York: Henry Holt and Company, 1989.

Pitzer, Paul. *Grand Coulee: Harnessing a Dream.* Pullman, Washington: Washington State University Press, 1994.

Robinson, Michael C. *Water for the West: The Bureau of Reclamation 1902-1977.* Chicago, Illinois: Public Works Historical Society, 1979.

Schnitter, Nicholas J. *A History of Dams: The Useful Pyramids.* Rotterdam and Brookfield: A. A. Balkema, 1994.

Settle, William A. Jr. *The Dawning. A New Day for the Southwest: A History of the Tulsa District, Corps of Engineers, 1939-1971.* Tulsa Oklahoma: U.S. Army Corps of Engineers, Tulsa District, 1975.

Shallat, Todd. *Structures in the Stream: Water, Science, and the Rise of the U.S. Army Corps of Engineers.* Austin: University of Texas Press, 1994.

Smith, Norman. *Man and Water: A History of Hydro-Technology.* Charles Scribner's Sons, 1975.

_____. *A History of Dams.* London: Peter Davies, 1971.

Stegner, Wallace. *Beyond the Hundredth Meridian: John Wesley Powell and the Second Opening of the West.* Lincoln and London: University of Nebraska Press, 1953.

Welsh, Michael. *A Mission in the Desert: Albuquerque District, 1935-1985.* Report prepared for the U.S. Army Corps of Engineers, n.d.

Williams, Albert N. *The Water and the Power: Development of the Five Great Rivers of the West.* New York: Duell, Sloan, and Pearce, 1951.

Worster, Donald. *Rivers of Empire: Water, Aridity, and the Growth of the American West.* New York: Pantheon Books, 1985.

III. Comparative Analysis

Phases I and II of the research effort successfully identified obviously noteworthy dams, significant by dint of their *unique* impact upon construction and engineering technology or *unique* association with national water policy. Once identified, these dams were preliminarily assessed for national significance using the criteria for evaluation. The research methodology also identified those nationally significant initiatives that dramatically and significantly changed the federal government's role in development of the nation's water resources. These include the genesis of the Reclamation Service (1902); passage of the Flood Control Act of 1936; and authorization of the Pick-Sloan plan of multiple-agency, integrated multiple-purpose, basin-wide development (1944).

The methodology was less successful at providing a systematic means of comparative analysis of the merits of those *representative* dams best able to symbolize the significance of these policy initiatives. For the purposes of comparative analysis of Bureau of Reclamation dams associated with (representative of) genesis of a national reclamation policy and the first application of multiple-purpose irrigation/hydropower development, HRA relied upon the Bureau of Reclamation's *Project Data Book*. Dams identified in this comprehensive database were evaluated under the evaluation criteria

In the absence of a comprehensive project directory, evaluation of the most appropriate representatives of the U.S. Army Corps of Engineers multiple-purpose dam development required a systematic comparative analysis of all Corps multiple-purpose dams. At the suggestion of COE Senior Historian Martin Reuss, HRA initiated this analysis with review of the U.S. Army Corps of Engineers Data Base, searching for multiple-purpose dams constructed in the historic period (prior to 1950) and currently owned by the U.S. Army Corps of Engineers. Single-purpose dams generating multiple use incidental to their construction–recreation; fire protection; fish and wildlife habitat–were not defined as multiple–purpose. This criteria eliminated many of the regulatory dams designed principally for navigation purposes.

Guided by this list, HRA reviewed the U.S. Army Corps of Engineers Annual Reports and Division and District histories for additional information related to the multiple-purpose dams and then assessed the significance, integrity, and eligibility of these dams against the criteria for evaluation. The results of this document search/comparative analysis are incorporated within the list of potential NHL dams.

Historic Context

Three thousand years before the birth of Christ, Mesopotamian civilizations constructed crude gravity dams, initiating "one of the most essential aspects of man's attempt to harness, control and improve his environment."[1] Hydraulic development in America boasts only slightly more recent antecedents, initiated by the Hohokam and Anasazi people of the arid Southwest and continued thousands of years later by Euro-American settlers who constructed dams to power their mills, to provide domestic water, to water their fields, to improve navigation, and to protect their fields and homes from flood.

Until the twentieth century, private enterprise and local, territorial, or state government assumed sole control for the construction of American irrigation and power dams, a response to the young Republic's faith in states' rights and the power of capitalism. The history of navigational structures, however, begins with federal authority, when nascent communities turned to the U.S. Army

Corps of Engineers for the construction of canals, dams, locks, and harbor improvements.[2]

Corps

On March 16, 1802, Congress granted the president authority to organize and establish an engineering and construction agency responsible not only for military construction, supply, and training, but also for the design, construction, operation, and maintenance of internal improvements and related works: the U.S. Army Corps of Engineers.[3]

Debate over the power and accountability of the nascent Corps symbolized the larger debate over the character of the new nation. Throughout the nineteenth century, states rights advocates denounced the Corps as a tool of centralization and a threat to the constitutional prerogatives of the states. Federalist Henry Clay, author of the "American Plan," countered that Congress had a "great national duty" to open the veins of commerce that bound the East to the West."[4] In 1847, Whig politician John C. Spencer defined the United States as a boundless hydrographical system, in which the rivers and lakes formed a five-thousand mile conduit of navigation, linking New England to the West and the Gulf of Mexico, a liquid link between the states, creating "one government, ... a union." A substantial and persistent investment in river and harbor construction would protect, preserve, and strengthen that union.[5]

The nation's investment was substantial, involving watershed surveys, canal, dam, and wharf construction at a cost of over $43 million federal dollars (1802–1861).[6]

This national investment was guided in theory by an engineering elite committed to scientific principles, schooled in government service, and dedicated to the good of the many rather than the profit of the few. Yet Corps projects waited on Congressional appropriations and by the turn of the century an emerging class of "conservationists" scorned the Corps' efforts as evidence of pork-barrel politics and of inefficiency spurred by political concerns rather than the impartial laws of science.[7]

Reclamation

With the Louisiana Purchase (1802) and victory in the Mexican-American War (1848), the United States acquired extensive semi-arid and arid lands that would sustain families, temper urban growth and class conflict, generate taxes, and contribute to the wealth of the nation only if irrigated. And, Major John Wesley Powell of the United States Geological Survey argued in 1878, only if federal land policy recognized the primacy of water in land with little of it, recognized the limits to western settlement, and undertook to study, define, manage, and distribute western land on the basis of hydrographic rather than political boundaries.[8]

By the turn of the century, Powell's admonitions were echoed by those men and women who had failed to bring water to their Desert Land Act claims or who had realized too late the limits to 160 (or 320) semi-arid homestead acres. Settlers, irrigation districts, corporations, speculators, and municipal and state agencies that had been unable to muster the technological expertise and the financial resources to harness the major rivers of the West clamored for a revised federal reclamation policy. An emerging class of "conservationists" committed to "putting rivers. . .to work in the most efficient way possible for the purpose of maximizing production and wealth" both heeded and inspired the popular cry.[9] Historian Donald Worster writes: "The West had gone as far as it could on its own hook. It had tried partnerships, theocracy, foreign and local capital and still most of the rivers ran on freely to the sea. . .. So they raised their voices in one loud, sustained chant. . .. 'We need the state!' And the federal government responded by passing the National Reclamation Act in 1902.[10]

The Reclamation Act (also known as the Newlands Act in recognition of its sponsor Francis R. Newlands [Nevada]) established that the federal government would finance irrigation projects through a Reclamation Fund composed of proceeds from the sale of project lands. The Reclamation Service within the U.S. Geological Survey, a federal bureaucracy independent from project-specific Congressional appropriations, would decide the course of water resource development and would determine the goals and methods of water programs. In theory, these decisions were to be made by engineers, on the basis of field data and scientific principles, rather than by politicians, in response to partisan politics and the competing, parochial interests of water users."[11]

Although the Service's focus was on the technological components of dam and canal construction, a measure of social engineering was included within their mission. Water would be distributed only to farms of 160 acres or less, allowing equitable distribution to the anticipated hordes of landless immigrants that clogged American cities (and assuring that western lands would effectively relieve labor unrest and class conflict thereby contributing to the growth and prosperity of the nation). Those with larger existing holdings within project boundaries were to sell their excess land at pre-project prices, or surrender their claim to government water.[12]

Between 1902 and 1920, the U.S. Reclamation Service (separated from the Geological Survey and renamed the Bureau of Reclamation in 1923) constructed the Truckee and the Carson River projects in Nevada; Salt River and Yuma projects in Arizona; North Platte in Nebraska-Wyoming; Shoshone in Wyoming; Milk River, Huntley, and Lower Yellowstone in Montana; Belle Fourche in South Dakota; Payette-Boise River in Idaho; Yakima and Okanagan in Washington; Klamath and Umatilla in Oregon; Strawberry Valley in Utah; Uncompahgre in Colorado; and the Rio Grande in New Mexico.[13]

By virtually all accounts, these early years were difficult. Enforcement of the 160-acre land limit was lax, allowing a government windfall to established farmers and fueling land speculation where absentee landowners leased their project land to a new class of western sharecroppers. In their focus on dam construction, Reclamation paid insufficient attention to soil condition, crop suitability, and the need to provide training in new methods of crop selection, rotation, and cultivation. Construction costs had been greatly underestimated, increasing from $9 per watered acre before 1910 to $50 per watered acre by 1915. Settlers' land costs escalated accordingly, adding additional burden to the already onerous task of growing redundant crops on previously undeveloped land in undeveloped communities. By 1923, only 1.2 million acres of the 20 million acres of irrigated land in the West were on federal projects. Two-thirds of all project farms predated the reclamation program.[14]

Multiple-Purpose River Development

The debate over the cost-benefit ratio and payback schedule of federally funded reclamation ultimately merged with on-going debate over the cost-benefit ratio of federally funded navigation projects. By the late 1890s rail rates had escalated, prompting American shippers to promote inland navigation. Merchants and manufacturers were joined by local boosters who based future hopes for community growth and prosperity on the availability of cheap transportation. Proposals included development of an intercoastal waterway from Boston to the Rio Grande, by way of a man-made canal across Florida, and of a "Lakes-to-the-Gulf" channel connecting Chicago and the Gulf of Mexico, by way of the Illinois and Mississippi Rivers. Less ambitious plans included deepening the Missouri and upper Mississippi. Both the U.S. Army Corps of Engineers and the House Rivers and Harbors Committee opposed the inland navigation projects, arguing that, in contrast to coastal river improvements and harbor development, inland waterway improvements and canal construction were not worth their extreme cost.[15]

And as the reclamation and navigation debates raged, so did the unharnessed rivers, destroying property in seasonal flooding and prompting continued debate over the most effective means of flood control.[16]

If American rivers were to sustain and unite a growing empire, then federal water development had to proceed on a rational, effective, and financially sustainable basis. In 1907, President Roosevelt created the Inland Waterways Commission, charging its members with development of a new approach to river development.

> Works designed to control our waterways have thus far usually been undertaken for a single purpose, such as the improvement of navigation, the development of power, the irrigation of arid lands, the protection of lowlands from floods, or to supply water for domestic and manufacturing purposes.... The time has come for merging local projects and uses of the inland waters in a comprehensive plan designed for the benefit of the entire country.[17]

The commission defined that basis as "multiple-purpose" development, a rational planning process in which rivers were developed as interrelated systems, where "the several parts are so closely interdependent that no section can be brought under control without at least partial control of all other portions."[18] The commission recommended that hereafter, plans for the improvement of navigation in inland waterways take into account "the purification of the waters through protection of watershed, the development of power, the control of floods, the reclamation of land by irrigation and drainage, and all other uses of the waters or benefits to be derived from their control." In addition, it called for creation of a permanent commission to oversee and direct the regional and national development of the nation's river systems, immune from interagency competition and jealousies.[19]

Only commission member Brigadier-General Alexander Mackenzie, Chief of the U.S. Army Corps of Engineers, dissented. The Army Corps maintaining that hydro-electric and navigational development were incompatible, that reservoirs were an ineffective means of controlling flood waters, that radical waterways improvements were both vulnerable to Congressional logrolling and an improvident use of federal funds, and that creation of the permanent commission represented an unconstitutional expansion of executive authority. Many in Congress agreed.[20]

In December 1907, Senator Francis G. Newlands presented Congress with a bill to carry out the recommendations of the waterways commission. Multiple-purpose projects would be chosen by a non-partisan commission, would be financed by an initial Congressional appropriation of $50,000,000 and would be sustained by the sale of hydroelectric power associated with river-basin development. The Newlands waterways bill passed the house by a wide margin but was defeated in the Senate.[21] After much political wrangling, Newlands' Waterways Commission was finally approved in 1917. President Wilson, however, made no appointments to the commission and, with the 1920 creation of the Federal Power Commission, the "whole Waterways Commission idea was shelved."[22]

Multiple use river basin planning would not be realized until the Depression Era, when extravagant use of federal funds became central to American employment and when the Executive Branch asserted unprecedented powers in response to the economic emergency. Public power became the "paying partner" of irrigation projects and the Bureau of Reclamation and the U.S. Army Corps of Engineers spearheaded multimillion-dollar New Deal projects spanning entire river basins.[23]

Engineering Considerations - Prepared by the Institute for the History of Technology and Industrial Archaeology, 1994

The pace, nature, and direction of American hydraulic development were dictated not only by federal policy but by the limits of existing dam technology. Regardless of the construction materials employed, ancient dams were inevitably of the gravity type, that is, a structure that resists the pressures of the impounded water by its sheer mass. Until the late nineteenth century, gravity dams were constructed of stone masonry, or soil and/or rockfill. Rarely was brick used. Advances in the analysis of gravity dams, the advent of Portland cement concrete, and construction equipment and techniques allowed dams of unprecedented size to be built during the past century.

The modern era of gravity dam design begins with work published by J. A. T. de Sazilly in 1853. He provided a means of calculating the stress at every section of a dam and stated that all stresses must not exceed safe limits. This work was enhanced by Delocre in 1858 and later by the well known mathematician and engineer W. J. M. Rankine in 1872. (A century earlier, in 1750, B. F. Belidor established the basis for determining the stability of gravity dams against sliding and overturning but not of calculating internal stresses.) By the beginning of the nineteenth century, the design of masonry and concrete gravity dams provided engineers with a sound approach to dam design that persists to the present.

This approach to dam design was challenged early in the new century by Atcherley and Pearson at University College in London, who claimed that the most critical region of tensile stresses was across vertical, rather than horizontal, sections of the dam. In the debate that followed, the traditional approach was validated, in part, because the experimental results on rubber models of dams agreed well with established analytical methods. Subsequently, other researchers attempted to use the theory of elasticity to obtain mathematical solutions for stresses in gravity dams. This concern for a mathematical solution for gravity dams persisted over a number of decades but apparently had little influence on the actual design of such dams. Richard Southwell developed an iterative solution for the well known R101 rigid Zeppelin and the method was used in 1947 before the age of computers to solve for the stresses in gravity dams.

Waterproof concrete had been used by the Greeks and later the Romans for a wide variety of hydraulic structures. During the nineteenth century, its direct descendent, natural concrete, was essential to the construction of canals and widely used in building construction. It was not, however, until the introduction of Portland cement in America during the last decade of the nineteenth century that concrete was used in construction of large dams.

Mass concrete is essentially an artificial stone without any reinforcement to overcome its inherent weakness in tension. Thus, it was applied in gravity and arch dams where tensile stresses could be kept to a minimum. Concrete gravity dams are amongst the most massive structures ever constructed. Even a typical large multi-purpose dam such as the Tygart Dam in West Virginia, completed in 1938, required over one million cubic yards of concrete.

The embankment dam represents the other category of gravity dam construction, where the material used is either earth or rock. It, too, is of ancient origin and, until this century, these dams were built without knowledge of how these materials behaved in service. This empirical approach limited the height that could safely be reached while using these materials. The other factor limiting the development of embankment dams was the necessity of providing earth moving machinery to efficiently place material.

The field of soil mechanics is of quite recent origin. Beginnings of this discipline trace back to the pioneering work of the eighteenth-century French engineer Coulomb who was concerned with earth pressures against walls and the stability of embankments. The late nineteenth century saw the field expand as the result of work by Poncelet, Rankine, Culmann, Ritter, and others. This research was concerned with the further development of the classical theory of earth thrust as a function of cohesion and internal friction of the embankment material. Modern soil mechanics, as developed in the twentieth century, is concerned not only with earth pressures, but also with settlement, permeability, stability, and compaction of earth works. The discipline has been advanced by both mathematical analysis and experimental work. Karl Terzaghi, who is considered by many to be the "father" of modern soil mechanics, began his work before the First World War. By the 1950s, engineers could undertake the design of large embankment dams with confidence rooted in knowledge developed in the field of soil mechanics.

Large embankment dams requiring more than a million cubic yards of material were not economically feasible without the application of modern earth moving equipment. America has led the world in the development of a wide range of earth moving machines capable of shifting very large amounts of material expeditiously and economically. The combination of soil mechanics for design and massive machines for construction has resulted in the building of large earth embankment dams.

An alternative method of constructing an embankment dam is to use standard dredge technology and to apply soil in an embankment by hydraulic placement. The greatest example of the use of this particular technology is the Fort Peck Dam on the upper Missouri River. It represents the end of an era since no further hydraulic-fill embankment dams of the magnitude of the Fort Peck have been constructed subsequently.

Both stone and concrete are strong in compression but operatively weak in tension. As a result, both materials should be applied to structural forms that ensure only negligible tensile stresses will occur under any loading conditions. As the ancients discovered, the two principle structural forms are column and arch structures. American use of these forms occurred first in 1884 when Frank E. Brown designed and constructed a daringly thin arch dam in the San Fernando Valley, Bear Valley Dam, to impound irrigation water.[24] The dam stood 65 feet high with a radius of 335 feet. The most notable feature of the dam was the thinness of the cross section that varied from 22-24 feet near the foundation to a mere 3 feet at the crest. Its success provided a symbol and guide for the building of subsequent arch dams.

The Pathfinder Dam, completed in 1907, and the Shoshone (a.k.a. Buffalo Bill) Dam, also of 1907, were America's first ventures into the construction of high arch dams. It should be noted that at a height of 325 feet the Shoshone Dam was the highest in the world when completed and was a great leap forward compared to the Bear Valley Dam. These two latter dams represent the origins of the "trial load" method of analysis. This theory, with modifications, remained the American analytical approach to arch dam designs until recently. In concept, the method envisages an arch dam resisting impounded water through horizontal arch action of a series of horizontal arches and vertical cantilevers fixed at the foundation level. In order to obtain compatibility of deflections it was necessary to ensure that the arch deflections at the crown of the dam matched the cantilever deflections at the same point. The method was greatly refined in 1929 by using elastic theory to determine arch deflections at a series of points and not just at the crown. The method requires that a series of cantilevers match the deflections of a series of horizontal arches to achieve compatibility. Later, the "trial load" method was further modified to account for twisting and tangential movements. It was the "trial load" method that served as the basis for the design of the 727 foot high Hoover Dam.

European engineers, on the other hand, did not subscribe to the "trial load" method because they believed the calculations required an inordinate amount of effort that produced designs with unnecessarily massive cross sections. It was felt that there was too much reliance on the gravity action of the dam which was believed to resist the water pressure largely by arch action. The "trial load" method produces conservative results but it should be mentioned that it has resulted in such dams being the safest of all the American types, with embankment dams giving the most trouble in service.

The arch and gravity dams employed concrete without reinforcement. The introduction of Portland cement to replace natural cement at the end of the nineteenth century was accompanied by revolutionary developments that overcame the inherent weakness of concrete in tension by the addition of steel reinforcement. The resulting structural material was reinforced concrete that could resist tensile as well as compressive stresses effectively. Thus, it could be used in components subjected to bending as well as tension. This material allowed concrete to be applied not only for beams and columns, but also for slabs and shell members. The application to dams came early with the introduction of reinforced concrete buttress dams. These structures are associated with the Ambursen System, which was patented, and resulted in a series of buttress supporting reinforced concrete slabs as early as 1903. By sloping the water face, the resultant gravity load of the water assures that loads pass through the middle third of the dam which, in turn, prevents overturning. The series of slabs also permit articulation of the structure if uneven settlements of the foundation, due to either static or dynamic loads, happens to occur. From the construction point of view, a buttress dam requires far less material than corresponding gravity structures. In addition, because the structure is essentially hollow behind the face, uplift pressure that might develop if the foundation is permeable is completely avoided. The vaulted buttress dam was invented by John Eastwood and first applied to Hume Lake Dam near Fresno, California in 1909. Instead of using flat slabs, a number of vaults spanned between the buttresses reducing the amount of reinforcement required and causing a further reduction in the amount of concrete needed for a given design.

Consideration of Hydraulic Designs

The hydraulic design of dams is as important as the structural forms discussed above. The design involves the provision of spillways and penstocks to pass water through or over the dam, to control water levels downstream, and in the case of floods, to pass water safely past the dam without damaging the structure. Penstocks are used for hydroelectric generation and for diverting water for irrigation purposes.

The design of emergency spillways, which function to pass water over or around the dam, must be proportioned so that the water moving over the spillway crest in a horizontal direction is turned to run down the face of the spillway in a smooth and essentially vertical direction and then diverted back to the horizontal direction or even a reverse gradient at the bottom. In order to protect the face of the dam it is essential that the discharge water flows smoothly down the face of the spillway and does not create vacuum pockets that cause cavitation to occur with attendant erosion of the face. Primary spillways, on the other hand, are used for water release as part of the normal operation of a dam. These releases may be necessary for minimum flow requirements for low water augmentation and controlled releases during flood stages.

To avoid erosion at the toe of the dam, the energy of the water at the bottom must be dissipated. This is usually accomplished by designing a stilling basin or "cushion pool" that absorbs the energy of the water before it is released into the natural streambed. The resulting shape is the ogival curve that was first used in America by John Jervis in his Croton Dam completed in 1842.

Although studies of weirs and pipe flow were amenable to analytic solutions, the design of spillways and stilling basins remains a complex problem that is most often either solved on the basis of theoretical assumptions and confirmed by model analysis. The establishment of hydraulic laboratories by the U.S. Army Corps of Engineers, the Bureau of Reclamation, and selected universities is an important chapter in the history of dam design and makes possible the erection of dams that function safely and efficiently in controlling water passing across the dam.

Most of the spillways that pass water over the top of the dam are what are called "fixed crest" spillways. However, there are a significant number of dams that have control gates at the crest or slide gates in the face of the dam. Both roller gates, a design that was developed in Germany, and tainter gates, which are curved gates supported on radial struts, enable engineers to control the flow over the top of the dam and are important features that should be noted in an evaluation of the design of a historic dam.

No design could be assumed to be successful unless it was based upon the knowledge of the hydrology of the watershed served by a given dam. The determination of rainfall and runoff data are essential in designing any hydraulic structure connected with a dam. The first application of such data gleaned from rainfall gauges and streamflow records was at summit level reservoir on the Chenango Canal, which was a feeder to the Erie Canal in upstate New York. It was completed to a design of John Jervis in 1836. This pioneering effort was entirely successful and led the way for the application of hydrological information to the design of large multiple-purpose dams.

Large dams can serve a number of purposes. One of the most important is hydroelectric power generation. Although current and static electricity had been studied as early as the eighteenth century, this new form of power did not become a public utility until the opening of Edison's Pearl Street Station in New York in 1882. In the same year, halfway across the country, in Appleton, Wisconsin, a diminutive hydroelectric station was put online. Through this structure, electric power was generated by a vertical shaft turbine driving a generator mounted on top. By the end of the decade, the first hydroelectric dam was completed at Oregon City, Oregon. This humble beginning led the way to later great hydroelectric installations. As developed, these facilities added new features to conventional dam design. These included penstocks with control devices for supplying the turbines in the power house, and discharge conduits that reduced downstream scouring. Power stations of immense size with a series of turbine generators are an integral part of a hydroelectric power dam installation. Although often removed from the dam itself, the switch yard serves as the point of distributing electrical energy into the power grid. Clearly the development of generating and distribution systems comprises an important part of the history of electrical engineering, and of the power industry. Although hidden from view, the turbines in a hydroelectric power-generating dam are an essential feature in the design of such a facility. Through a series of advances in engineering, turbine-generated power is amongst the most efficient machines known.

List of Potential National Historic Landmark Dams

Anderson Ranch Dam (1940 - 1950) Idaho, South Fork Boise River, earthfill (rolled-earth and rockfill) (Reclamation)

Completion of Anderson Ranch Dam, the highest earthfill dam in the world at the time of construction, "posed unprecedented problems for both the designers and the constructors" (Kollgaard and Chadwick 1988, 803). If further research reveals that the solutions to these problems impacted future dam design and construction, Anderson Ranch Dam may be nationally significant under NHL Criterion 4.

Arrowrock Dam (constructed 1911-1915; crest raised 1935-1937; concrete slab applied to downstream face and spillway, 1935 - 1937) Idaho, Boise River, curved concrete gravity (Reclamation)

Construction of Arrowrock Dam involved innovative integration of sand and Portland cement, in efforts to control temperature rise (during concrete curing) associated with the dam's mass. (This innovation resulted in a 30 degree F temperature rise, significantly less than would have been realized with straight Portland cement mix.) Kollgaard and Chadwick suggest that this innovation was significant, with impact on future dam design. Dam also "a pioneer in instrumentation" (Kollgaard and Chadwick 1988, 18).

The raised crest, although a modification to the dam's original design, is not likely to be determined an unacceptable impact to integrity, particularly if significance under NHL Criterion 4 is derived from the pioneer use of instrumentation and the integration of sand/Portland cement. In contrast, the concrete slab applied to the downstream face and spillway was designed to stop the disintegration of the excessively porous exposed sand-cement concrete; impact to integrity of design and materials may prove to be severe.

Bartlett Dam (1939) Arizona, Verde River, buttress dam (multiple arch) (Reclamation)

Private agencies are responsible for design and construction of the vast majority of buttress dams, which are significantly less expensive to build than traditional arch or gravity dams. An evaluation of the national engineering significance of federally constructed buttress dams must include a comparative analysis of privately constructed dams that may possess greater significance and greater integrity. That said, Bartlett retains integrity, and is directly associated with important innovations in buttress-dam engineering, including 1) design of the arches as cylindrical, full half-circles analyzed by the trial-load method and 2) curvature, in plan, of the dam upstream to better fit topographic conditions (Kollgaard and Chadwick 1988, 550).

Buffalo Bill Dam (Shoshone) Wyoming, concrete arch (Reclamation)

First of the Bureau of Reclamation's multiple-purpose projects (Criterion 1); also the first dam to be built of mass concrete with a large percentage of plum rock, and one of the first to be analyzed by the trial-load method, using arch and crown cantilever radial adjustments. The Bureau of Reclamation, however, reports recent and significant threats to physical integrity, including construction of a new visitor center and raising of the crest height. The dam may not meet the NHL requirement of a high degree of integrity.

Central Valley Project - Shasta Dam and Friant Dam (Reclamation) (NHL Criterion 5)

The CVP may be nationally significant for its impact on national settlement and agricultural patterns: "each year, between 3 and 4 million ac-ft of water are delivered through the Central Valley Project for irrigation use on nearly 2 million acres of fertile land. This land produces more than $1 billion in crops annually." The project also provides 320,000 acre feet of municipal water and in excess of 5.5 billion kW of hydroelectric power annually. (Kollgaard and Chadwick 1988, 132). While much of the system is less than fifty years old, the two principal components–Shasta and Friant–both date to the historical period and might be evaluated under Criterion 5. Difficulties with the nomination will include using only these two historic components to define a largely modern project. Each dam may also be nationally significant on (individual) technological merit.

Shasta Dam (1945) California, Sacramento River, curved concrete gravity (Reclamation)

Partner to Friant Dam, below. Capitalizing on experience gained at Grand Coulee, materials were conveyed by an 11-mile conveyor belt between the screening plant and stockpiles at the receiving plant. Another "outstanding feature of construction was the immense cableway head tower" (Kollgaard and Chadwick 1988, 33).

The pipe cooling system resembles that of Hoover Dam but "the operating techniques were vastly improved." The instrumentation installed was "one of the most complete. . . . Analysis of all these data eventually led to the reconsideration of methods of stress analysis for concrete gravity dams, and became background material for the formulation of the finite element analyses of dams" (Kollgaard and Chadwick 1988, 34).

Shasta may be nationally significant on the basis of this contribution to engineers' understanding of the behavior of large dams (Criterion 4). In terms of historic association (Criterion 1), the dam is most effectively evaluated in conjunction with Friant Dam.

Friant Dam (1942) California, San Joaquin River, concrete gravity (Reclamation)

One of two storage dams of the Central Valley Project (see Shasta, above, re: Criterion 1). Friant also represents Reclamation's first extensive use of pumicit pozzolan admixture to reduce cement content and thus heat of generation (Kollgaard and Chadwick 1988, 131).

Elephant Butte Dam (1911–1916) New Mexico, Rio Grande, concrete gravity (Reclamation)

Possibly nationally significant due to 1906 treaty obligations that require the U.S. to provide Mexico with 60,000 acre-feet of water from the Rio Grande. These treaty obligations are met by Elephant Butte. (The 1906 treaty followed continued water shortages for Mexican farmers along the Rio Grande and an 1894 challenge by the Mexican minister in Washington for monetary damages, as allowed by the right of prior appropriation under international law.) This treaty was precedent setting: historian David A. Philips, in his National Register nomination of Elephant Butte Dam, argues that prior to the Elephant Butte controversy, the federal government had all but surrendered control of U.S. waterways to the states. Faced with the need to resolve squabbles over the Rio Grande, the federal government found legal mechanisms by which it could take back a great deal of its control over those waters - and it uses those same legal mechanisms to this day. Thus, Elephant Butte Dam is a symbol of an important moment in the development of national water law and in the changing balance of power between the federal government and the states.

Like Arrowrock Dam (see above), Elephant Butte is an early example of sand-cement use: "Five percent of the cost of the dam was saved by the use of sand-cement instead of normal Portland cement, but the real advantage was in the decreased heat development." The dam was also the site of potentially significant construction innovations: as at Arrowrock, the expansive construction plant produced 2,651 cubic yards of concrete in 16 hours, 380,000 cubic yards in 1 year . . . "efficiency demanded a single, smoothly functioning material flow system. With this basic need recognized, the masonry dam era ended" (Kollgaard and Chadwick 1988, 19). Elephant Butte, however, appears to possess better integrity and may prove to be a more appropriate representative of this construction innovation.

Fort Peck Dam (1940) Montana, Missouri River, earthfill (hydraulic fill) (Corps)

Construction of this massive dam (126 million cubic yards of earthfill) -- the largest in the world for almost 30 years -- was made economically feasible by significant innovations in soil mechanics, including Corps flownet and filter test research. The 5 million cubic yard slide at Fort Peck, as the dam neared completion, also significantly impacted future earth-dam construction techniques.

Garrison Dam (1947-54) North Dakota, Missouri River, earthfill (rolled earth embankment) (Corps)

The Pick-Sloan agreement (by which Corps and Bureau Missouri River Basin Programs were coordinated) was authorized by the Flood Control Act of 1944 and stands as an early and outstanding example of coordinated river-basin development. More than 100 reservoirs throughout the Missouri Basin were authorized under this agreement, including five, multiple-purpose, mainstem dams. Garrison is the first of these main-stem dams and an integral component of the largest reservoir system in the United States, a system with a storage capacity of more than seventy-four million acre feet and a surface area of over one million acres (Farrell, x-xii). This size translates very directly to impact–flood control, navigation, recreation, and commensurate economic and social changes–over a vast region of the United States. Further research may establish that this five dam main-stem system is of national significance. If not, the first dam, Garrison effectively represents the nationally significant compromise.

Gatun Dam (1914) Panama, Chagres River, earthfill (Corps)

On the route of the Panama Canal, Gatun Dam is one of the first Corps flood control dams, but its primary significance is that it supports navigation, as the large reservoir it creates forms twenty-three miles of the canal. A tertiary purpose is power generation. The dam provided the only feasible engineering method of building a canal across Panama, which reduces the sea route from the eastern U.S. to the West Coast by two-thirds and avoids the dangerous passage around Cape Horn. The canal, along with the interior placement of the locks by the dam, away from the coast, was driven by important military considerations for a young, imperial America. The original plan to use a dam instead of a sea-level route was proposed by French engineer Godin de Lepinay in the nineteenth century, but rejected by the French construction company; Gatun Dam was designed along the de Lepinay plan by Joseph Ripley and Alfred Noble of Soo (Sault Ste Marie) Canal fame. With the recent transfer of the Panama Canal, Gatun Dam is no longer under U.S. jurisdiction.

Gibson Dam (1926-1929) Montana, Sun River, concrete arch dam (Reclamation)

Civil Engineer David P. Billington reports that Gibson Dam was the first U.S. dam designed and constructed in response to trial-load analysis, as presented by Reclamation engineers Howell and Jaquith in January of 1925: "[at] Gibson Dam on the Sun River in Montana...the trial load method was used not just for analysis but for design, resulting in a savings of more than 41,000 cubic yards of concrete over the gravity design." Gibson Dam is potentially nationally significant as the first application of this evolved mathematical process of analysis. See also Pathfinder Dam, below.

Glen Canyon Dam (1957-1964) Utah and Arizona, Colorado River, concrete arch (Reclamation)

Although not yet 50 years old, Glen Canyon Dam may be nationally significant for its association with the modern conservation movement (Criterion 1). Despite some engineering innovations -- winner of the American Society of Civil Engineers' annual Outstanding Civil Engineering Award for the "project that demonstrated the greatest engineering skills . . . " (including use of new computer-run trial-load analysis [Kollgaard and Chadwick 1988, 456]). Extraordinary national significance under Criterion 4 (as demanded of resources less than 50 years old) will be difficult to establish.

Grand Coulee Dam (1938) Washington, Columbia River, concrete gravity (Reclamation)

Construction of massive Grand Coulee Dam required significant construction innovations. These include diversion of the large flow of the Columbia River during construction; special grouting adaptations to address adverse geologic conditions of the site, and unique conveyor systems designed for efficient transport of large quantities of material (Kollgaard and Chadwick 1988, 31). The revolutionary water storage system for irrigation, which utilizes a secondary, natural reservoir 280 feet above the dam, also represents a technological innovation. (This innovation, however, relates strictly to unique site conditions and is not thought to have had a significant impact, as required in the evaluation guidelines, on subsequent designs).

In terms of NHL Criterion 1, power from the dam was important in WWII manufacturing and atomic bomb development at the Hanford Reservation in Washington.

Owyhee Dam (1932) Oregon, Owyhee River, curved concrete gravity (Reclamation)

Kollgaard and Chadwick describe Owyhee Dam as a major precursor to Hoover Dam in terms of design and project management: "Owyhee Dam became a proving ground for theories being developed to assist with the design and construction of Hoover Dam."

The spillway at Owyhee - a morning glory design - had never before been used. An assessment of the national significance of this innovation would require an analysis of subsequent use and the degree to which unique site conditions dictated use of the morning glory design.

Pathfinder Dam (1909) Wyoming, North Platte River, masonry arch dam (Reclamation)

First arch dam to be analyzed by Wisner and Wheeler's arch-and-crown cantilever method (predecessor to the trial-load method). As described by David P. Billington, the arch-and-crown cantilever method was one of a series of significant treatises on gravity dam theory, beginning with the 1850s publication by J. Augustin de Sazilly of a profile of equal resistance. Billington reports that analysis of Pathfinder Dam design, using the arch-and-crown cantilever method, is significant in that "for the first time the engineers [were] seeking to take advantage of the true behavior of a curved dam in a narrow canyon to the end of making it more economical through saving material." However, although the arch-and-cantilever method of mathematical analysis confirmed the safety of Pathfinder Dam, it did not inform the actual design. Not until construction of Gibson Dam, Sun River, Montana, was a dam designed according to specifications suggested by the arch-and-crown cantilever method (refined and redefined in the 1920s as the trial-load method).

Pine Flat Dam (1954) California, Kings River, concrete gravity (Corps)

Kollgaard and Chadwick write "While the design of the dam and appurtenant works resembles that of most gravity dams of the period, the details of foundation treatment, concrete production, temperature control, and seismic analysis all are innovative" (Kollgaard and Chadwick 1988, 35).

Sardis Dam (1940), Mississippi, Little Tallahatchie River, earthfill (hydraulic fill) (Corps)

Kollgaard and Chadwick write "After the dam was placed in operation, seepage was noted on the downstream slope. . . . In 1942, a system of forty-eight experimental wells was installed along the line of the first well system to study the efficiency and length of life of various commercial well screens and to analyze the effects of well diameter, spacing and penetration. Many of the analytical procedures used by the [Corps] today stem from this study," (Kollgaard and

Chadwick 1988, 797). Further research may reveal that this impact on analytical procedures is nationally significant under NHL Criterion 4.

Troy Dam #1 (1915) New York, Hudson River, buttress dam (flat slab) (Corps)

Troy Dam is one of a series of structures built by the COE on the Hudson River between Waterford and Hudson, to increase stream depth and allow navigation. The Hudson River had been under improvement since 1789, when the state of New York first constructed dikes and dredged the channel. The original federal project for improvement was adopted in 1834 (in cooperation with the state) and modified in 1852 and 1866. The second project was adopted in 1892 and modified in 1899. Construction of Troy Dam and associated structures —constituting a third-phase of river improvement—was first authorized by the River and Harbor Act of June 25, 1910. In addition to removal of an existing dam at Troy and construction of the new lock, dam, and powerhouse, the Corps widened the river channel to form harbors, constructed longitudinal dikes between Troy and Stuyvesant, and - in cooperation with the state of New York - established canal terminals in Troy and Albany. Completion of this comprehensive improvement program resulted in "considerable increased navigation facilities." The U.S. Army Corps of Engineers reported that "the through waterway formed by the Hudson River and the New York State canals determines through freight rates from the Middle West to the Atlantic Ocean" (Chief of Engineers 1915, 225).

Completion of Hudson River navigation improvements, and the transition from a state, to a cooperative, to a federal effort, may be nationally significant. The Hudson River improvements may also significantly represent U.S. Army Corps of Engineers early navigation-improvement efforts. The incorporation of (modest) hydro-electric power facilities in the modified system, may also be a significant example of early Corps efforts at multiple-purpose development.

Comprehensive Coordinated River-Basin Flood-Control Systems (U.S. Army Corps of Engineers)

Annual Reports of the Chief of Engineers, U.S. Army, define the Flood Control Act of 1936 as a critical juncture in the Corps' history and in federal policy. The act established a definitive policy for federal participation in the construction of flood-control projects throughout the nation and, like the Pick-Sloan agreement (see Garrison Dam, above), expanded the Corps' legal authority (see, for example, Chief of Engineers 1938, 7). (Note that the Flood Control Act *did not* authorize multiple-purpose dams, and the significance of the act lies in the expansion of Corps authority and in the impact of the flood control improvements not within the context of multiple-purpose development. Dams authorized under the 1936 act were authorized for flood-control alone, with the stipulation that penstocks could be constructed in anticipation of hydropower development. Navigation related purposes, although of real benefit, were incidental to authorization.)

With passage of the act, Congress authorized approximately 270 flood-control projects, based upon a series of studies completed by the Corps over the preceding twenty years. An additional 240 preliminary examinations and surveys were also ordered. By the end of the 1942 fiscal year, authorizations totaling $930,400,000 had been provided by Congress for the construction of 485 multi-unit projects widely dispersed throughout the United States. Of this massive number of authorized improvements, however, the number of dams constructed—in contrast to levees, drains, or channel improvements—was relatively low, particularly when the scope of inquiry is limited to the historic period, 1936-1949 (Chief of Engineers 1942, 4-7).[25]

Of these dams, all were constructed as part of comprehensive coordinated river-basin systems. This coordinated role poses certain difficulties when evaluating the NHL eligibility of *individual dams* within each of the basin systems. As stipulated in the criteria for evaluation, individual properties without unique and distinguishing features may be nationally significant if they *best represent* a larger class *of important* properties. The ability to best represent that larger class may be based on standing as the first constructed, on standing as the pivotal structure within a significant system, or on symbolic value. This symbolic value might, for example, relate to geography–a flood-control dam built to protect Pittsburgh from flooding might best symbolize the importance of the Ohio River floods of 1927 and 1936 to passage of the 1936 Flood Control Act. Large kingpin dams, integral by virtue of size–and therefore function–to the system as a whole, are also appropriate symbols of the larger system. An entire comprehensive system might also qualify as an NHL, under the conditions of NHL Criterion 5. This approach, however, creates difficulty in resource management and in the delineation of appropriate boundaries.

Ohio River Basin
Allegheny and Monongahela River System

Kollgaard and Chadwick write that "the concept of a project for flood control of the Pittsburgh area dates to shortly after the disastrous flood of 1907. . . . The federal government [COE], with an interest in water supply for navigation as well as flood control, in 1912 authorized a report on the practicability of establishing a system of reservoirs on the headwaters of the Allegheny, Monongahela, and Ohio Rivers, and their tributaries. Through the ensuing years, various reports were written until, under the provisions of the National Industrial Recovery Act of 1933, the construction of the Tygart River Reservoir was authorized as Federal Project No. 44 and funds were appropriated for its construction" (Kollgaard and Chadwick 1988, 116).

Despite early identification of Tygart as the central component of an integrated system of flood-control dams within the Allegheny and Monongahela [upper Ohio] watersheds, its construction was approved, and PWA funding provided, on the basis of its important role in maintaining adequate flow for navigation on the Monongahela River. Proposed flood-control dams—on Loyalhanna, Tionesta, French, Redbank, and Crooked Creeks of the Allegheny River, and on the West Fork of the Monongahela River—were not approved for construction, a complex decision related to disagreement over cost-benefit equations, cooperative funding, and restrictions on Corps authority (Johnson n.d., 200).

In December of 1936, armed with the authority of the Flood Control Act of 1936, the Corps began the process of "planning and building an unbreakable chain of dams upstream of Pittsburgh." Upon completion in 1943, this chain of dams included Tygart, rolled earth embankment dams on Tionesta and Crooked creeks, concrete-gravity dams on Mahoning and Loyalhanna creeks, and a rolled earth/concrete dam on the Youghiogheny River (Johnson n.d., 211).

Tygart Dam was completed in February 1938. Engineer H. K. Barrows noted ten years later that this *completion marked the beginning of federal flood-control measures* (Barrows 1948, n.p.). The concrete-gravity dam, 1,880 feet in length at the top and rising 209 feet above the river bed at the spillway section, provides for the storage of 100,000 acre-feet to make up the deficiency of water in the Monongahela River during low rainfall periods and for the additional controlled storage of over 178,000 acre-feet for flood prevention in the Monongahela and Ohio valleys. The dam first prevented major flood damage in the spring of 1939, one month after closure. In the fall drought of that same year it prevented interruption of navigation on the Monongahela. "It is believed," said Tom Reilly [future Hydrology Branch Chief], "that the results achieved thus far by the Tygart Dam . . . amply justify the funds expended for its construction and operation and are the most conclusive proof that can be offered for the extension of the flood control system of the Upper Ohio" (quoted in Johnson n.d., 203).

Construction of **Youghiogheny Dam** - the last and largest of the upper Ohio basin dams proceeded during the war years as a construction effort directly related to the war effort: specifically, to the protection of manufacture plants from destruction by flood and prevention of interruption of the river-corridor transport system. As described by the Corps, the integrated Ohio Basin flood-control system was not deemed fully functional without Youghiogheny Dam (Johnson n.d., 211).

Either Tygart Dam, symbol of the Corps' expanding mission to flood-control dam construction upon passage of the Flood Control Act of 1936, or Youghiogheny Dam, the first kingpin dam constructed in the Ohio Basin solely under the authority of the Flood Control Act of 1936 (as amended in 1938 and 1941) may be nationally significant. The symbolic value of both dams is dramatically enhanced by their placement within the Ohio River drainage, site of the most devastating floods of 1936 and the site of considerable debate and study—beginning in the nineteenth century—on the proper means of flood control, and on an appropriate federal role in flood-control efforts.

North Atlantic Division:

John Chambers, a Corps contract historian, wrote:

like a recurring plague, heavy rains came summer after summer between 1934 and 1936. In midsummer 1935, torrential rains sent terror-producing floods across a ten-county area of south-central New York State.... The deluge returned with even greater vengeance in the spring of 1936: Hartford, Connecticut and Pittsburgh, Pennsylvania were the worst hit. In the Connecticut River Valley people called it the worst major disaster in memory.... The industrial life of the region slopped to a halt. Rivers roared out of control - the Susquehanna, Monongahela, Penobscot, Housatonic, Allegheny . . . [and] Congress passed the Flood Control Act of 1936. In south-central New York, the Flood Control Act provided for 13 local protection projects and seven dams and reservoirs. In 1943, Congress authorized $46 million for flood protection in the Connecticut River Valley. This included some 20 dams and retaining reservoirs (Chambers 1980, 45-47).

Flood control systems in New England, as authorized by the Flood Control Act of 1936, are described below:

Connecticut River Basin, Vermont, New Hampshire, Massachusetts, and Connecticut

The Flood Control Act of 1936 authorized construction of twenty dams and reservoirs and seven dike projects within the Connecticut River Valley, "[all]

necessary parts of an adequate, balanced regional plan of flood control." These dams and reservoirs include those at Knightville, West Brookfield, Barre Falls, Tully, Birch Hill, Honey Hill, Surry Mountain, Claremont, West Canaan, Sugar Hill, Williamsville, Cambridgeport, North Springfield, Ludlow, North Hartland, South Tunbridge, Gaysville, Union Village, and Victory (Chief of Engineers 1942, 134-147). The Corps describes six of these dams as "major" components:

Claremont: rolled-fill earth dam, 2,300 feet long at the crest line, rising 122.5 feet above the stream bed and providing a total storage capacity of 78,400 acre-feet.

Union Village: hydraulic-fill earth dam, 1,100 feet long rising 169 feet above the stream bed and providing a total storage capacity of 30,200 acre-feet.

Surry Mountain Reservoir: rolled-fill earth dam 1,670 feet long at its crest, rising 85 feet above the stream bed, and providing a total storage capacity of 32,500 acre-feet.

Birch Hill: earth dam of rolled-fill construction, 1,400 feet long and rises 56 feet above the river bed. Storage capacity of 49,400 acre-feet at spillway crest.

Knightville Reservoir: hydraulic-fill earth dam, 1,600 feet long at the crest line, rising 160 feet above the stream bed and provides for a total storage capacity of 39,300 acre-feet. **Tully Reservoir:** rolled earth dam, 1,500 feet at its crest, rising 62 feet above the stream bed, providing a total storage capacity of 22,150 acre-feet.

Individually, none of these dams plays a defining role in the effectiveness of the system as a whole. The integrated Connecticut River Basin system, however, may be a National Historic Landmark, under Criterion 5, eligible for its early and direct association with the Flood Control Act of 1936. As with the Monongahela system, the importance of the Connecticut River flooding and destruction of Hartford to Congressional support of the act heightens the representative value of the system.

Los Angeles County Flood Control and Water Conservation District, California

The Flood Control Acts of 1936, as amended May 15, 1937, authorized the construction of reservoirs and channel improvements in Los Angeles and San Gabriel Rivers and Ballona Creek and tributaries thereof. In addition to channel improvements, cut-offs, and storm drains, the system included small flood-control dams in Sepulveda, Hansen, and Lopez Canyons (Chief of Engineers 1942,

1582-1601). Further research may determine that this network of resources might be nationally significant for its impact on the history and development of one of the nation's most important metropolitan areas.

Red River, Texas
Denison Dam (1944) Texas, Red River, rolled-earth (Corps)

Study of a flood control plan for the Red River and its tributaries was first authorized by the Flood Control Act of 1928; the study report of nine tributary dams and a main-stem dam (Denison) found no economic justification for navigation, flood control reservoirs, and hydroelectric power development. The Flood Control Act of June 22, 1936 authorized another study of the Red River system. This study recommended that the proposed Denison Reservoir, "developed either for flood control and power or flood control only, be classed as a project with benefits and charges approaching an economic balance." Congress authorized construction of the multiple-purpose dam on June 28, 1938 as part of a comprehensive plan for control of the Mississippi River and its tributaries (Settle 1975, 60). Construction was expedited to meet projected wartime-industry power needs: closure of the dam in July 1942 placed it in operation for flood control purposes; by March 1945 the first generating unit began power production (Settle 1975, 63). The dam is a rolled earthfill structure, 15,200 feet in length and with a maximum height of 165 feet above the riverbed. Subsequent tributary dams have been constructed in the modern period; Denison, however, remains the primary component of the Red River flood control system.

Opponents to the dam sought court injunctions to halt construction, arguing that "the Denison project could not be sustained under either the interstate power or the general welfare clause" (Settle 1975, 63). The U.S. Supreme Court ruled that "the Denison Reservoir, as part of a comprehensive scheme to control the floods of the Mississippi River and its tributaries, was a valid exercise of the commerce clause" (Settle 1975, 63). Settle has argued that this Supreme Court decision "had the effect of making clear the extent of the power of the federal government in developing the non-navigable tributaries and nonnavigable portions of navigable streams. . . . And it stated firmly the doctrine of Federal supremacy in the instances of conflict between legitimate Federal and state projects. It can well be called a landmark case" (Settle 1975, 63). Further research may reveal that this court case was nationally significant as required by NHL Criterion 1.

The "Denison double-tube core barrel," used subsequently in drilling mudded holes in earthen dams, was developed during the course of investigating the Denison Dam site (Kollgaard and Chadwick, 692). Corps Senior Historian Marty Reuss also reports that the dam is associated with significant innovations in rolled-earth fill techniques. These innovations may prove to be nationally significant under NHL Criterion 4.

Endnotes

1. Norman Smith, quoted in Schnitter, *A History of Dams: The Useful Pyramids,* xi.
2. Donald Worster, *Rivers of Empire: Water, Aridity, and the Growth of the American West* (New York: Pantheon Books, 1985), 130.
3. Arthur Maass, *Muddy Waters: The Army Engineers and the Nation's Rivers* (Cambridge, Massachusetts: Harvard University Press, 1951), 20-1. Flood control work for the nation as a whole was more definitely assigned to the Army Engineers by the provisions of the Flood Control Act of 1936.
4. Henry Clay, quoted in Todd Shallat, *Structures in the Stream* (Austin: University of Texas Press, 1994), 125.
5. Quoted in Shallat, *Structures in the Stream,* 1.
6. Ibid., 118.
7. Ibid., 4-5.
8. Wallace Stegner, *Beyond the Hundredth Meridian John Wesley Powell and the Second Opening of the West* (Lincoln and London: University of Nebraska Press, 1953), passim.
9. Worster, *Rivers of Empire,* 155.
10. Ibid., 130.
11. Samuel P. Hays, *Conservation and the Gospel of Efficiency: The Progressive Conservation Movement 1890-1920* (New York: Atheneum Press, 1972 [first published by Harvard University Press, 1959]), 12-5.
12. Worster, *Rivers of Empire,* 167-9.
13. Worster, *Rivers of Empire,* 171.
14. *Ibid.,* 176-8.
15. Hays, *Conservation and the Gospel of Efficiency,* 92-4.
16. *Ibid.,* 223-39.
17. Roosevelt to members of the Inland Waterways Commission, March 14, 1907, quoted in Hays, *Conservation and the Gospel of Efficiency,* 106.
18. W. J. McGee, Inland Waterways Commission, quoted in Hays, *Conservation and the Gospel of Efficiency,* 104.
19. Commission, quoted in Robinson, *Water for the West,* 26.
20. Hays, *Conservation and the Gospel of Efficiency,* 203-4. Robinson, *Water for the West,* 27.
21. Hays, *Conservation and the Gospel of Efficiency,* 114.
22. Martin Reuss, U.S. Army Corps of Engineers Senior Historian, to Ann Hubber, Historical Research Associates, February 18, 1999.
23. Robinson, *Water for the West,* 49-51.
24. This dam was privately, rather than federally, owned and – like Bear Valley Dam – has not been evaluated in the course of this current project.
25. For example, the Sacramento River Valley project, often cited as a significant result of the Flood Control Act of 1936, consisted of enlargement of river channels, excavation of cut-offs, and construction of levees, weirs, bank protection, and pumping facilities.

BIBLIOGRAPHY

Abbott, Carl. *The Metropolitan Frontier: Cities in the Modern American West.* Tucson: University of Arizona Press, 1998.

———. *Urban America in the Modern Age: 1920 to the Present.* Arlington Heights, Illinois: Harlan Davidson, 1987.

Albert, Richard C. *Damming the Delaware: The Rise and Fall of Tocks Island Dam.* University Park, Pennsylvania: Pennsylvania State University Press, 1987.

Alexander, Thomas G. "The Powell Irrigation Survey and the People of the Mountain West." *Journal of the West* 7 (January 1968).

Allen, Marion V., editor. *Redding and the Three Shastas.* Shingletown, California: published by the author, 1989.

Allin, Craig W. *The Politics of Wilderness Preservation.* Westport, Connecticut: Greenwood Press, 1982.

Ambrose, Stephen E. *Duty, Honor, Country: A History of West Point.* Baltimore: Johns Hopkins Press, 1966.

"Amended Height of Garrison Dam Attacked on Three Fronts." *Bismarck Tribune* (December 6, 1945).

Andrews, Richard N. L. "Environmental Protection Agency." In *Conservation and Environmentalism: An Encyclopedia,* edited by Robert Paehlke. New York and London: Garland Publishing, 1995.

———. *Managing the Environment, Managing Ourselves: A History of American Environmental Policy.* New Haven, Connecticut: Yale University Press, 1999.

Anfinson, John O. "Commerce and Conservation on the Upper Mississippi River." *The Annals of Iowa* 52 (Fall 1993).

Armstrong, Ellis L., Michael C. Robinson, and Suellen M. Hoy, editors. *History of Public Works in the United States, 1776-1976.* Chicago: American Public Works Association, 1976.

"Army General Blasts At Lower Pool Level." *The Bismarck Tribune* (August 25, 1949).

"Army Plans to Begin Building Town in Spring." *Bismarck Tribune* (December 21, 1945).

Arnold, Joseph L. *The Evolution of the 1936 Flood Control Act.* Fort Belvoir, Virginia: Office of History, U.S. Army Corps of Engineers, 1988.

Arrington, Leonard J. *Great Basin Kingdom: Economics History of the Latter-Day Saints.* Lincoln: University of Nebraska Press, 1958.

―――― and Dean May. "A Different Mode of Life: Irrigation and Society in Nineteenth Century Utah." *Agricultural History* 49 (January 1975).

"Arthur Powell Davis." *Transactions of the American Society of Civil Engineers* 100 (1935): 1582-91.

Austin, Mary. "The Colorado River Controversy." *Nation* 125 (November 9, 1927).

Bailey, Paul. *The Development of the Upper Sacramento River.* Sacramento: California State Printing Office, 1928.

Banks, F. A. "Columbia Basin Project is Described by Construction Engineer." *Southwest Builder and Contractor* (November 23, 1934).

Barry, John M. *Rising Tide, the Great Mississippi Flood of 1927 and How it Changed America.* New York: Simon and Schuster, 1995.

Bartlett, Robert V. "Environmental Impact Assessment." *In Conservation and Environmentalism: An Encyclopedia* edited by Robert Paehlke. New York: Garland Publishing, 1995.

Bates, Sarah F., Lawrence J. Getches, Lawrence J. MacDonnell, and Charles F. Wilkinson. *Searching Out the Headwaters: Change and Rediscovery in Western Water Policy.* Covelo, California, Washington, D.C.: Island Press, 1993.

Baumhoff, Richard G. "Huge Missouri-Souris Diversion Project Apparently to be Dropped." *St. Louis Post-Dispatch* (October 12, 1950).

———. "Record Flood Damage This Year Emphasizes Need for Action Now." *St. Louis Post Dispatch* (July 20, 1947).

Bear, Dinah. "National Environmental Policy Act of 1969." In *The Encyclopedia of the Environment* edited by Ruth A. Eblen and William R. Eblen. Boston: Houghton Mifflin Company, 1994.

Berkey, Charles P. "Foundation Conditions for Grand Coulee and Bonneville Projects." *Civil Engineering* 5, No. 2 (February 1935).

Berkman Richard L., and W. Kip Viscusi. *Damming the West*. New York: Grossman Publishers, 1973.

"Berthold Tribe Did Not Always Object to New Location." *The Fargo Forum* (December 15, 1946).

"Big Equipment Being Used at Garrison Dam Site." *Minot Daily News* (April 29, 1948).

"Big Mattresses Built For River Diversion." *Minot Daily News* (September 15, 1952).

Bilharz, Joy A. *The Allegheny Senecas and Kinzua Dam: Forced Relocation Through Two Generations*. Lincoln: University of Nebraska Press, 1998.

Billington, David P. "History and Esthetics in Suspension Bridges." *Journal of the Structural Division*, ASCE, 103, ST 8, August 1977, with Discussions and Closure, March 1979.

Binder, R. C. *Fluid Mechanics*. New York: Prentice-Hall Inc., 1943.

Bissell, Charles A., compiler and editor. *The Metropolitan Water District of Southern California: History and First Annual Report*. Los Angeles, 1939.

Bissell, Charles H., and F. E. Weymouth. "Memoir for Arthur Powell Davis." *Transactions of the American Society of Civil Engineers* 100 (1935).

Biswas, Asit K., Mohammed Jellali, and Glenn E. Stou, editors. *Water for Sustainable Development in the Twenty-first Century*. Delhi: Oxford University Press, 1993.

Blackwelder, Brent. "In Lieu of Dams." *Water Spectrum* (Fall, 1977).

Bohn, Frank, editor. *Boulder Dam: From the Origin of the Idea to the Swing-Johnson Bill.* 1927.

Bollens, John C., and Henry J. Schmandt. *The Metropolis: Its People, Politics, and Economic Life,* 2nd edition. New York: Harper and Row, Publisher, 1970.

Boulder Dam Association. "Colorado River Development: Statements by Congressman Addison T. Smith of Idaho and Mayor George E. Cryer of Los Angeles." Los Angeles: Boulder Dam Association, 1925.

———. "The Federal Government's Colorado River Project." Los Angeles: Boulder Dam Association, September 1927.

———. "The Story of a Great Government Project for the Conquest of the Colorado River." Los Angeles: Boulder Dam Association, 1927.

Bowie, A. J. *A Practical Treatise on Mining in California.* New York: D. Van Nostrand, 1905.

Branyan, Robert L. *Taming the Mighty Missouri: A History of the Kansas City District, Corps of Engineers: 1907-1971.* Kansas City, Missouri: U.S. Army Corps of Engineers, 1974.

Briggs, Peter. *Rampage: The Story of Disastrous Floods, Broken Dams, and Human Fallibility.* New York: David McKay Company, Inc., 1973.

Brigham, Jay. *Empowering the West: Electrical Politics Before FDR.* Lawrence: University Press of Kansas, 1998.

Brooks, Charles E. *Frontier Settlement and Market Revolution: The Holland Land Purchase.* Ithaca, New York: Cornell University Press, 1996.

Brough, Charles H. *Irrigation in Utah.* Baltimore: The Johns Hopkins University Press, 1898.

Brown, F. E. "The Bear Valley Dam." *Engineering News* 19 (June 23, 1888).

Brown, F. Lee, and Helen M. Ingram. *Water and Poverty in the Southwest.* Tucson: University of Arizona Press, 1987.

Buchanan, Spencer J. "Levees in the Lower Mississippi Valley." *Proceedings of the ASCE* 63, No. 7, (September 1937).

"Building the Bigger Dam." *Natural History* 72 (December 1963).

Burdick, Usher L. "The Truth About the Army Engineers and Garrison Dam." *The Williston Daily Herald* (August 22, 1949).

Butler, John S.. "Comprehensive Study by Army Engineers." *Civil Engineering* 1, No. 12 (September 1931).

Cabot, Philip. "Danger in the Boulder Canyon Project." In"Boulder Dam: Complete Bibliography, References, Engineers' Charts, Studies and Reports, the Swing-Johnson Bill, Minority Reports and General Comments," Frank Bohn, editor, (n.p. circa 1928).

Cain, William. "The Circular Arch Under Normal Loads." *Transaction of the American Society of Civil Engineers* 85 (1922), 233-48.

Caldwell, Lynton K. "National Environmental Policy Act (U.S.)." In *Conservation and Environmentalism: An Encyclopedia*, edited by Robert Paehlke. New York: Garland Publishing, 1995.

Cannon, Brian Q. "'We Are Now Entering a New Era': Federal Reclamation and the Fact Finding Commission of 1923-24." *Pacific Historical Review* 66 (May 1997).

"Canyon Controversy: Second Round." *Science News* 91 (April 1967).

"Canyon Dams: Dissents from Arizona Scientists." *Science* 157 (July 7, 1967).

Carroll, Jane Lamm. "Gull Lake Reservoir Dam." (*Historic American Engineering Record*, Report No. MN-70). Prints and Photographs Division, Library of Congress.

———. "Lake Pokegama Reservoir Dam." (*Historic American Engineering Record*, Report No. MN-66). Prints and Photographs Division, Library of Congress.

———. "Lake Winnibigoshish Reservoir Dam." (*Historic American Engineering Record*, Report No. MN-65). Prints and Photographs Division, Library of Congress.

———. "Leech Lake Reservoir Dam." (*Historic American Engineering Record*, Report No. MN-67). Prints and Photographs Division, Library of Congress.

———. "Mississippi River Headwaters Reservoirs." (*Historic American Engineering Record*, Report No. MN-6). Prints and Photographs Division, Library of Congress.

———. "Pine River Reservoir Dam." (*Historic American Engineering Record*, Report No. MN-68). Prints and Photographs Division, Library of Congress.

———, "Sandy Lake Reservoir Dam and Lock" (*Historic American Engineering Record*, Report No. MN-69). Prints and Photographs Division, Library of Congress.

Carter, Charles H. "Change in Plan for Grand Coulee Dam Explained by Engineer." *Southwest Builder and Contractor* (August 23, 1935).

Carter, Luther J. "Dams and Wild Rivers: Looking Beyond the Pork Barrel." *Science* 158 (October 13, 1967).

Casagrande, Arthur. "Discussion." Of Glennon Gilboy article, "Mechanics of Hydraulic Fill Dams." *Journal of the Boston Society of Civil Engineers* 21, No. 3 (July 1934).

———. "Seepage through Dams." Contributors to Soil Mechanics 1925-1940, *Boston Society of Civil Engineers*, 1940.

Case, Robert Ormond. "The Eighth World Wonder." *The Saturday Evening Post* (July 13, 1935).

"Charges Army Influenced by Private Power." *The Fargo Forum* (August 5, 1951).

Chittenden, Hiram Martin. "Preliminary Examination of Reservoir Sites in Wyoming and Colorado." H. Doc. 141, 55th Congress, 2d. Session, Serial 3666, 1897.

Chorpening, C. H. "Fort Peck Dam—Progress of Construction." *The Military Engineer* 27, No. 151 (January-February 1935).

Christie, Jean. "Morris L. Cooke and Energy for America." In *Technology in America: A History of Individuals and Ideas*, Carroll W. Purcell, Jr., editor, 2nd edition. Cambridge, Massachusetts: MIT Press, 1991.

Chudacoff, Howard P., and Judith E. Smith. *The Evolution of American Urban Society*, 4th edition. Englewood Cliffs, New Jersey: Prentice-Hall, 1994.

Clarke, R. R., and H. E. Brown, Jr. "Portland-Puzzolan Cement as used in the Bonneville Spillway Dam." *Journal of the American Concrete Institute* (January - February 1937).

Coffey, Max. "Engineers' Lay-Off Curbs River Job." *Rocky Mountain Empire Magazine* (November 10, 1946).

Cohen, Michael P. *The History of the Sierra Club, 1892-1970*. San Francisco: Sierra Club Books, 1988.

———. *The Pathless Way: John Muir and American Wilderness*. Madison: University of Wisconsin Press, 1984.

Collier, Michael, Robert H. Webb, and John C. Schmidt. *Dams and Rivers: A Primer on the Downstream Effects of Dams*. Tucson: U.S. Geological Survey, Circular 1126, June 1996.

Committee on Environmental Effects of the United States. Committee on Large Dams. *Environmental Effects of Large Dams*. New York: American Society of Civil Engineers, 1978.

"Concrete Flows as Big Year Begins at Garrison." *The Minot Daily News* (April 9, 1951).

"Concrete Placing at Grand Coulee: From Gravel Pit to Forms." *Western Construction News* (June 1939).

"Concrete Structures for Spillway." *Engineering News-Record* (August 29, 1935).

"Conflict in House Reports on Col. Cooper's Muscle Shoals Work." *Engineering News-Record* 84, No. 22 (1920).

"Constructing the First Cofferdam." *Engineering News-Record* (August 1, 1935).

Cooper, Erwin. *Aqueduct Empire: A Guide to Water in California, Its Turbulent History and its Management Today.* Glendale California: Arthur H. Clark Company, 1968.

Cooper, Hugh L., and Co. "Electric Power for the Tennessee River, Muscle Shoals, Alabama." *Interim Construction Progress Bulletin* (November 1, 1923).

———. "The Water Power Development of the Mississippi River Power Company at Keokuk Iowa." *Journal of the Western Society of Engineers* 17 (January to December 1912).

"The Corps of Engineers, Water, and Ecology." *Audubon Magazine* 72 (July 1970).

Cosco, Jon M. *Echo Park: Struggle for Preservation.* Boulder: Johnson Books, 1995.

Covell, W. E. R. "Flood Control in the Pittsburgh District." *The Military Engineer* (May-June 1939).

Cowdry, Albert E. *A City for the Nation: The Army Engineers and the Building of Washington D.C., 1790-1967.* Washington D.C.: GovernmentPrinting Office,1978.

Cross, Hardy. "Discussion of Design of Symmetrical Concrete Arches by C. S. Whitney." *Transactions of the American Society of Civil Engineers* 88 (1925).

Cullen, Allan H. *Rivers in Harness: The Story of Dams.* Philadelphia: Chilton Company, 1962.

"Dakota Floods Share Nation's Top Picture News Spotlight." *The Fargo Forum* (April 13, 1952).

Daly, Edmund L. "The Mastery of the Ohio River." *The Military Engineer* 19 (May-June 1927).

"Dam the Grand Canyon?" *Audubon Magazine* 68 (September 1966).

"Damming the Grand Canyon for a Thirsty Southwest." *New Republic* 154 (April 30, 1966).

"Dams and Destruction." *Nature Magazine* 31 (October 1938).

"Dams and Wild Rivers: Looking Beyond the Pork Barrel." *Science* 158 (October 13, 1967).

"Dams or Wilderness Areas?" *Audubon Magazine* 52 (September-October 1950).

"The Dam Shame: It's Still With Us." *Outdoor Life* 163 (June 1979).

"Dams Threaten West Coast Fisheries Industry." *Oregon Business Review* 6 (June 1947).

"Daniel Boone's Wilderness May Be Tamed by a Lake." *Smithsonian* 6 (September 1975).

Darcy, Henry. *Les Fontaines publiques de la ville de Dijon: Exposition et application des principes à suivre et des formules à employer dans les questions de distribution d'eau.* Paris: Victor Dalmont, 1856.

Darland, Alvin F. "The Columbia Basin Project." *Electrical Engineering* 56, No. 11 (November 1937).

Davies, Terrie. "Environmental Protection Agency." In *The Encyclopedia of the Environment*, edited by Ruth A. Eblen and William R. Eblen. Boston: Houghton Mifflin Company, 1994.

Davis, Arthur P. *Irrigation Near Phoenix, Arizona.* U.S. Geological Survey Water Supply Paper No. 2. Washington, D.C.: Government Printing Office, 1897.

———. *Irrigation Works Constructed by the U.S. Government.* New York: John Wiley and Son, 1917.

Davis, John P. "Locks and Mechanical Lifts: State of the Art." Paper prepared for the National Waterways Roundtable, Norfolk, Virginia, April 22-24, 1980.

Davis, Robert K. "Lessons in Politics and Economics from the Snail Darter." In *Environmental Resources and Applied Welfare Economics: Essays in Honor of John V. Krutilla*, edited by V. Kerry Smith. Washington, D.C.: Resources for the Future, 1988.

Davison, Stanley. *The Leadership of the Reclamation Movement: 1875-1902.* New York: Arno Press, 1979.

Dean, Robert. "'Dam Building Still Had Some Magic Then': Stewart Udall, the Central Arizona Project, and the Evolution of the Pacific Southwest Water Plan, 1963-1968." *Pacific Historical Review* 66 (February 1997).

"Deep Sheetpile Cutoff Well for Fort Peck Dam." *Engineering News Record* (January 10, 1935).

Delocre, M. *"Memoire sur la forme du profil a adopter pour les grande barrs ages en maconnerie des reservoirs." Annales des Ponts et Chaussee, Memoires et Documents* (1866).

Department of the Army, U.S. Army Corps of Engineers. "Analysis of Design: Spillway Structures and Gates." *Missouri River Garrison Reservoir*, North Dakota, Office of the District Engineers, Garrison District, Bismarck, North Dakota, April 1952.

———. *The Federal Engineer: Damsites to Missile Sites: A History of the Omaha District* Omaha: Government Printing Office, 1984.

———. *Garrison Dam and Reservoir: Analysis of Design, Excavation and Main Embankment.* Garrison District, (May 1948).

———. *The History of the U.S. Army Corps of Engineers*, 2nd edition. Alexandria, Virginia: 1998.

———. *Oahe Reservoir: Definitive Project Report.* Omaha Office of the District Engineer, February, 1948.

———. *Ohio River Navigation.* Cincinnati, 1979.

de Sazilly, M. *"Sur un type de profil d'egale resistance propose pour les murs des reservoirs d'eau." Annales des Ponts et Chaussees* (1853).

de Stanley, Mildred. *The Salton Sea: Yesterday and Today.* Los Angeles: Triumph Press, Inc., 1966.

Deudney, Daniel. "Rivers of Energy: The Hydropower Potential." Worldwatch Paper 44, June 1981.

Devine, Robert S. "The Trouble with Dams." *Atlantic Monthly* 276 (August 1995).

Dick, Wesley Arden. "When Dams Weren't Damned: The Public Power Crusade and Visions of the Good Life in the Pacific Northwest in the 1930s." *Environmental Review* 13 (Fall/Winter 1989).

Dill, Clarence C. *Where Water Falls*. Spokane, Washington: C. W. Hill Printers, 1970.

"Diversion-Tunnel Driving." *Engineering News-Record* (August 29, 1935).

Divine, Robert A., editor. *The Johnson Years, Volume Two: Vietnam, the Environment, and Science.* Lawrence: University Press of Kansas, 1987.

Dixon, John A., Lee M. Talbot, and Guy J. M. Le Moigne. *Dams and the Environment: Considerations in World Bank Projects*. Washington, D.C.: World Bank, 1989.

Dodds, Gordon B. *Hiram Martin Chittenden: His Public Career*. Lexington: University Press of Kentucky, 1973.

———. "The Stream-Flow Controversy: A Conservation Turning Point." *Journal of American History* 56 (June 1969).

Dorn, Harold. "Hugh Lincoln Cooper and the First Détente." *Technology and Culture* 20, No. 2 (April 1979).

Douglas, C. T. "Aggregate Production for Friant Dam." *Civil Engineering* 12 (March 1942).

———. "Concrete Mixing Plant for Friant Dam." *Civil Engineering* 12 (April 1942).

Douma, Jacob L. *Water Resources: Hydraulics and Hydrology: Interview with Jacob H. Douma*. Oral history interview by John T. Greenwood. Alexandria, Virginia: Office of History, Headquarters, U.S. Army Corps of Engineers, 1991.

Downs, L. Vaughn. *The Mightiest of Them All: Memories of Grand Coulee Dam*. Fairfield, Washington: Ye Galleon Press, 1986.

Drew, Elizabeth B. "Dam Outrage: The Story of the Army Engineers." *Atlantic* 225 (April 1970), 51-62.

Dunbar, Robert G. *Forging New Rights in Western Waters*. Lincoln: University of Nebraska Press, 1983.

duPont, Pete. "The Tocks Island Dam Fight." *Sierra Club Bulletin* 57 (July-August 1972).

Eastwood, John S. "An Arch Dam Design for the Site of the Shoshone Dam." *Engineering News* 63 (June 9, 1910).

Eblen, Ruth A., and William R. Eblen, editors. *The Encyclopedia of the Environment*. Boston: Houghton Mifflin Company, 1994.

Echeverria, John D., Pope Barrow, and Richard Roos-Collins. *Rivers at Risk: The Concerned Citizen's Guide to Hydropower*. Washington, D.C.: Island Press, 1989.

"Echo Park Controversy Resolved." *Living Wilderness* 20 (Winter-Spring 1955–56).

"Editorial: Construction Progress," *Compressed Air Magazine* (October 1935).

Edsom, Peter. "Army-Interior Duplication Doomed." *St. Paul Pioneer Press* (December 9, 1948).

Egan, Timothy. "A G.O.P. Attack Hits Bit Too Close to Home." *New York Times* (Friday, March 3, 1995).

Ekirch, Arthur A., Jr. *Man and Nature in America*. New York: Columbia University Press, 1963.

"Electric Brains to Guide Basin Reservoir Operation." *Minot Daily News* (December 3, 1953).

Emperger, Friedrich Ignaz Edler von, editor. *Handbuch für Eisenbetonbau*. Berlin: Wilhelm Ernst & Sohn, 1911 edition.

"Engineers' Dam is Coup de Grace for Meramec." *Audubon Magazine* 75 (March 1973).

Environmental Protection Agency. (1989 Report to Congress) *Dam Water Quality Study* (EPA 506/2-89/002).

"Erosion Losses." *St. Louis Globe Democrat* (July 17, 1947).

Evans, Gail E. "Storm Over Niagara: A Catalyst in Co-shaping Government in the United States and Canada During the Progressive Era." *Natural Resources Journal* 32, No. 1 (Winter 1992).

Ferrell, John R. *Big Dam Era: A Legislative and Institutional History of the Pick-Sloan Missouri Basin Program.* Omaha: U.S. Army Corps of Engineers, 1993.

———. "From Single- to Multi-Purpose Planning: The Role of the Army Engineers in River Development, 1824-1930." Draft manuscript, Historical Division, Office of the Chief of Engineers, February 1976.

Fink, Leon, editor. *Major Problems in the Gilded Age and the Progressive Era.* Lexington, Massachusetts: D.C.: Heath and Company, 1993.

Finley, William L. "Salmon or Kilowatts: Columbia River Dams Threaten Great Natural Resource," *Nature Magazine* 26 (August 1935).

"Fish and Wildlife Service Issues Unfavorable Report on Rampart Dam." *Izaak Walton Magazine* 29 (August-September 1964).

"Fish Stairways." *Literary Digest* 121 (May 30, 1936).

"Fish v. Dams." *Time* 71 (February 17, 1958).

Fiske, Harold C. "Preliminary Examination of Tennessee River and Tributaries." January 15, 1921, 9, in *Tennessee River and Tributaries North Carolina, Tennessee, Alabama, and Kentucky,* House of Representatives, 67th Congress, 2nd Session, Document No. 319, May 19, 1922.

"Flood Control? Not by a Dam Site!" *Outdoor America* 16 (March-April 1951).

Fogelman, Valerie M. *Guide to the National Environmental Policy Act.* New York: Quorum Books, 1990.

Forchheimer, P. "Über die Ergiebigkeit von Brunnenanlagen und Sickerschlitzen." *Zeitschrift, Architekten-und Ingenieur-Verein.* Hanover, 32, No. 7 (1886).

"Fort Peck Dam." *Engineering News-Record* (November 29, 1934).

"Fort Peck Diversion Tunnels Redesigned for Faster Driving." *Engineering News-Record* (March 26, 1936).

Fox, Stephen. *The American Conservation Movement: John Muir and His Legacy.* Madison: University of Wisconsin Press, 1981.

Frankin, Philip. *A River No More: The Colorado and the West.* Tucson: University Press of Arizona, 1981.

"Frederick Haynes Newell." *Transactions of the American Society of Civil Engineers* 98 (1933).

Freeman, John R. John R. Freeman Papers. Massachusetts Institutes of Technology Archives and Special Collections. Cambridge, Massachusetts.

Freeman, Peter H. *Large Dams and the Environment: Recommendations for Development Planning.* A Report prepared for the United Nations Water Conference, Mar Del Plata, Argentina, March 1977.

"Ft. Peck Power Flows to Dam." *The Williston Daily Herald* (May 17, 1951).

Fox, Stephen. *The American Conservation Movement: John Muir and His Legacy.* Madison: University of Wisconsin Press, 1981.

Gage, John. *Turner: Rain, Steam and Speed.* New York: Allen Lane, 1972.

Galbraith, C. G. "Kaplan Turbines for Bonneville." *Engineering News-Record* (May 27, 1937).

Ganoe, John T. "The Beginnings of Irrigation in the United States." *Mississippi Valley Historical Review* 25 (1938).

———. "The Origins of a National Reclamation Policy." *Mississippi Valley Historical Review* 18 (June 1931).

Garcia-Diego, J. A. "The Chapter on Weirs in the Codex of Juanelo Turriano: A Question of Authorship." *Technology and Culture* 17 (April 1976), 217-34.

———. "Old Dams in Extramadura." *History of Technology* 2 (1977).

"Garrison Dam Activity Centered in Powerhouse Site." *The Minot Daily News* (April 28, 1951).

"Garrison Dam Closed." *The Bismarck Tribune* (April 16, 1953).

"Garrison Dam Contractors Featured in *Time* Article." *The Bismarck Tribune* (August 23, 1952).

"Garrison Dam Featured by Contractors Magazine." *The Bismarck Tribune* (August 1, 1950).

"Garrison Dam Pool Limit Rejected." *St. Paul Pioneer* (August 23, 1949).

Garrison Dam Takes Shape For the Casual Observer." *The Fargo Forum* (October 9, 1949).

"The Garrison Story." In *The Story of Garrison Dam: Taming the Big Muddy*, Sheila C. Robinson, compiler. Garrison, North Dakota: BHG, Inc., 1997, 55-66.

"Garrison Tunnels Under Construction." *McLean County Independent* (October 6, 1949).

Gates, Paul W. *History of Public Land Law Development*. Washington, D.C.: Zenger Publishing Co., Inc., 1968.

Gault Homer J., and W. F. McClure. *Report on Iron Canyon Project California*. Washington D.C.: Government Printing Office, 1921.

"General Says Pick-Sloan Plan Would have Averted '52 Flood." *The Fargo Forum* (May 7, 1952).

"Giant Embankment Now Taking Shape." *Grand Forks Herald*. (July 4, 1948.)

Gilboy, Glennon. "Mechanics of Hydraulic Fill Dams." *Journal of the Boston Society of Civil Engineers* 21, No. 3 (July 1934).

Goldmark, Henry. "The Power Plant, Pipe Line and Dam of the Pioneer Electric Power Company at Ogden, Utah," with Discussion and Correspondence. *Transactions of the American Society of Civil Engineers* 38 (December 1897), 246-314.

Goldsmith, Edward, and Nicholas Hildyard. *The Social and Environmental Effects of Large Dams*. San Francisco: Sierra Club Books, 1986.

"Good News on the Grand Canyon." *National Parks Magazine* 41 (March 1967).

Goodman, Joseph C. "Build a Dam, Save a Stream." *American Game* 20 (March-April 1931).

Goodman, Richard E. *Karl Terzaghi: The Engineer as Artist*. Reston, Virginia: ASCE Press, 1998.

Gordon, J. E. *Structures: Or Why Things Don't Fall Down*. New York: Penguin Books, 1978.

Gorlinski, J. S. "The Bonneville Dam." *The Military Engineer* 27, No. 153, (May-June 1935).

Gottlieb, Robert. *Forcing the Spring: The Transformation of the American Environmental Movement*. Washington, D.C.: Island Press, 1993.

Gould, E. Sherman. *High Masonry Dams*. New York: D. Van Nostrand, 1897.

Graf, William L. "Landscapes, Commodities, and Ecosystems: The Relationship Between Policy and Science for American Rivers." In Water Science and Technology Board. National Research Council. National Academy of Sciences. *Sustaining Our Water Resources*. Washington, D.C.: National Academy Press, 1993.

Graham, Otis L., Jr. *A Limited Bounty: The United States Since World War II*. New York: McGraw-Hill Company, 1996.

"Grand Canyon: Colorado Dams Debated." *Science* 152 (June 17, 1966).

"The Grand Canyon: Dam It or Not?" *Senior Scholastic* 90 (February 3, 1967).

"Grand Canyon Dams?" *Audubon Magazine* 67 (May 1965).

"Grand Canyon Dams Blocked." *Audubon Magazine* 68 (November 1966).

"Grand Canyon Dams Go." *Science News* 91 (February 11, 1967).

"Grand Coulee Revised." *Engineering News-Record* (June 20, 1935).

Grant, U. S., III. "The Dinosaur Dam Sites Are Not Needed." *Living Wilderness* 15 (Autumn, 1950).

"Great Depression." *Encyclopedia Britannica*, 1997.

Gregg, Frank. "Public Land Policy: Controversial Beginnings for the Third Century." In *Government and Environmental Politics*, edited by Michael J. Lacey. Washington, D.C.: Woodrow Wilson Center Press, 1989.

Griffith, Ernest S., and Charles R. Adrian. *A History of American City Government, 1775-1870*. Washington, D.C.: University Press of America, 1983.

Hagen, Vernon K. *Water Resources: Hydraulics and Hydrology: Interview with Vernon K. Hagen,* oral history interview by John T. Greenwood. Alexandria, Virginia: Office of History, Headquarters, U.S. Army Corps of Engineers, 1997.

Hagwood, Joseph J., Jr. *The California Debris Commission: A History*. Sacramento: U.S. Army Corps of Engineers, 1981.

Hall, John W. "The Control of Mixtures and Testing of Wilson Dam Concrete." Paper delivered at a meeting of the American Concrete Institute, Chicago, February 24, 1926.

Halloran, James J. "Earth Movement at Fort Peck Dam." *The Military Engineer*, 31.

Hano, Arnold. "The Battle of the Grand Canyon." *New York Times Magazine* (December 12, 1965).

Hardin, John R. "Fort Peck Dam Spillway." *The Military Engineer* 29, No. 163 (January-February 1937).

Harding, S. T. *Water Rights for Irrigation: Principles and Procedures for Engineers*. Palo Alto, California: Stanford University Press, 1936.

Harrison, Alfred S. *Water Resources: Hydraulics and Hydrology: Interview with Alfred S. Harrison*, oral history interview conducted by John T. Greenwood. Alexandria, Virginia: Office of History, Headquarters, U.S. Army Corps of Engineers, 1997.

Harrison, C. L. "Provision for Uplift and Ice Pressure in Designing Masonry Dams." *Transactions of the American Society of Civil Engineers* 75 (1912).

Harrison, Robert W. *History of the Commercial Waterways & Ports of the United States, Volume I.* Fort Belvoir, Virginia: U.S. Army Engineer Water Resources Support Center, 1979.

Harrold, J. C. "Outline of the General Principles Used in the Design of Dams on the Upper Mississippi River." *Engineering As Applied to the Canalization of a River.* St. Louis: September 1935.

Hart, Henry C. *The Dark Missouri.* Madison: University of Wisconsin Press, 1957.

Harvey, Mark W. T. *A Symbol of Wilderness: Echo Park and the American Conservation Movement.* Albuquerque: University of New Mexico Press, 1994.

———. "Echo Park Dam: An Old Problem of Federalism." *Annals of Wyoming* 55 (Fall 1983).

———. "Echo Park, Glen Canyon, and the Postwar Wilderness Movement." *Pacific Historical Review* 60 (February 1991).

———. "Symbols from the Big Dam Era in the American West." Unpublished paper delivered at the American Society of Environmental History conference, San Antonio, Texas, 1998.

Harza, L. F. "Uplift and Seepage Under Dams on Sand." *Transactions of the American Society of Civil Engineers* 100 (1935).

Hatch, Harry H. "Tests for Hydraulic-Fill Dams." *Transactions of the American Society of Civil Engineers* 99 (1934).

Hawgood, H. "Huacal Dam, Sonora, Mexico." *Transactions of the American Society of Civil Engineers* 78 (1915), *564-601.*

Hayden, T. A. "Salt River Project, Arizona, Irrigation and Hydroelectric Power Development by the Salt River Water Users' Association-Six Major Dams." *Western Construction News* 5 (June 25, 1930).

Hays, Samuel P. *Conservation and the Gospel of Efficiency: The Progressive Conservation Movement, 1890-1920*. New York: Atheneum, 1972; [originally published 1959].

Hazen, Allen. "Hydraulic-Fill Dams." *Transactions of the American Society of Civil Engineers* 83 (1919-1920).

———. "On Sedimentation." *Transactions of the American Society of Civil Engineers* 53 (December 1904).

Hedgpeth, Joel W. "The Passing of the Salmon." *Scientific Monthly* 59 (November 1954).

Helfman, Elizabeth S. *Rivers and Watersheds in America's Future*. New York: David McKay Company, Inc., 1965.

Helm, S. W. "Jawa, A Fortified Town of the Fourth Millennium B.C." *Archaeology* 27 (1974).

Henning, Daniel H., and William R. Mangun. *Managing the Environmental Crisis: Incorporating Competing Values in Natural Resource Administration*. Durham, North Carolina: Duke University Press, 1989.

Henny, D. C. "Stability of Straight Concrete Gravity Dams." *Transactions of the American Society of Civil Engineers* 99 (1934).

Herzog, Max A. M. *Practical Dam Analysis*. London: Thomas Telford, Ltd.: 1999.

Heuvelmans, Martin. *The River Killers*. Harrisburg, Pennsylvania: Stackpole Books, 1974.

"High Dams in the United States." *Reclamation Era* 26 (January 1936).

Hill, Forest G. *Roads, Rails & Waterways: The Army Engineers and Early Transportation*. Norman: University of Oklahoma Press, 1957.

Hirshleifer, Jack, James C. DeHaven, and Jerome W. Milliman. *Water Supply: Economics, Technology, and Policy*. Chicago: University of Chicago Press, 1960.

Hoffman, Abraham. *Vision or Villainy: Origins of the Owens Valley-Los Angeles Water Controversy*. College Station, Texas: Texas A&M Press, 1981.

Holmes, Beatrice Hort. *A History of Federal Water Resources Programs, 1800-1960*. Washington, D.C.: Economic Research Service, U.S. Department of Agriculture, Miscellaneous Publication No. 1233, June 1972.

Holmes, J. Albert. "Some Investigations and Studies in Hydraulic-Fill Dam Construction." *Transactions of the American Society of Civil Engineers* 84 (1921).

Holt, W. Stull. *The Office of the Chief of Engineers of the Army: Its Non-Military History, Activities, and Organization*. Baltimore: Johns Hopkins Press, 1923.

Horwitz, Morton J. *The Transformation of American Law, 1780-1860*. Cambridge, Massachusetts: Harvard University Press, 1977.

Howell, C. H., and A. C. Jaquith. "Analysis of Arch Dams by the Trial Load Method," with Discussion. *Transactions of the American Society of Civil Engineers* 93 (1929), 1191-1316.

Hubbard, Preston J. *Origins of the TVA: The Muscle Shoals Controversy, 1920-1932*. New York: Norton, 1968 [originally published Nashville: Vanderbilt University Press, 1961].

Hughes, Thomas P. *Networks of Power: Electrification in Western Society, 1880-1930*. Baltimore: The Johns Hopkins University Press, 1983.

Hundley, Norris. *The Great Thirst: Californians and Water, 1770-1990*. Berkeley: University of California Press, 1992.

———. *Water and the West: The Colorado River Compact and the Politics of Water in the American West*. Berkeley: University of California Press, 1975.

———. "The West Against Itself." In *New Courses for the Colorado River: Major Issues for the Next Century*. Gary D. Weatherford and F. Lee Brown, editors. Albuquerque: University of New Mexico Press, 1986.

Hunter, Louis C. *A History of Industrial Power in the United States, 1780-1930 Volume One: Water in the Century of the Steam Engine*. Charlottesville: University Press of Virginia, 1979.

———. *Steamboats on the Western Rivers: An Economic and Technological History.* Cambridge: Harvard University Press, 1949.

Hurt, R. Douglas. *Indian Agriculture in America: Prehistory to the Present.* Lawrence: University Press of Kansas, 1987.

———, editor. *The Rural West Since World War II.* Lawrence: University Press of Kansas, 1998.

Hutton, S. E. "The Grand Coulee Dam and the Columbia Basin Reclamation Project." *Mechanical Engineering* 62 (September 1940).

Ickes, Harold. "The Wide Missouri Gets Wider." *Field and Stream* (March 1954).

Independent Panel to Review Cause of Teton Failure. *Report to the U.S. Department of the Interior and State of Idaho on Failure of Teton Dam.* Washington, D.C.: Government Printing Office, 1976.

"Indians' Fight Over Big Dam at End as Contract is Signed." *Fargo Forum* (May 21, 1948).

"Indians Want Retention of Rights to Garrison Land." *New Town News* (June 4, 1953).

Institute for Government Research. *The U.S. Reclamation Service: Its History, Activities and Organization.* New York: D. Appleton and Company, 1919.

Introcaso, David M. "The Politics of Technology: The 'Unpleasant Truth About Pleasant Dam.'" *Western Historical Quarterly* 26 (Autumn 1995).

"Introduction to the First Issue of Life," *Life* 1, No. 1 (November 23, 1936).

Iverson, Peter. *"We're Still Here:" American Indians in the Twentieth Century.* Wheeling, Illinois: Harlan Davidson, Inc., 1998.

Jackson, Donald C. *Building the Ultimate Dam: John S. Eastwood and the Control of Water in the West.* Lawrence: University Press of Kansas, 1995.

———, with foreword by David McCullough, *Great American Bridges and Dams*, Great American Places Series, (Washington, D.C.: The Preservation Press, National Trust for Historic Preservation, 1988).

———. "History of Stewart Mountain Dam." Historic American Engineering Record Collection, Prints and Photographs Division, Library of Congress, Washington, D.C.

Jackson, Kenneth T. *Crabgrass Frontier: The Suburbanization of the United States*. New York: Oxford University Press, 1985.

Jackson, W. Turrentine, Rand F. Herbert, and Stephen R. Wee, introduction. *Engineers and Irrigation: Report of the Board of Commissioners on the Irrigation of the San Joaquin, Tulare and Sacramento Valleys of the State of California, 1873*. Fort Belvoir, Virginia: Office of History of the U.S. Army Corps of Engineers, 1990 – Reprint published as Engineer Historical Studies Number 5.

Jacobs, J. D. "Grouting the Tunnels." *Engineering News-Record* (August 20, 1936).

Jameson, John R. "Bonneville and Grand Coulee: The Politics of Multipurpose Development on the Columbia." Office of History, U.S. Army Corps of Engineers, Fort Belvoir, Virginia, 64 pages.

Jansen, Robert B. *Dams and Public Safety*. Washington, D.C.: Government Printing Office and U.S. Department of Interior, Water and Power Resources Service, 1980.

Jefferson, T. B. "Longer Life for Dredge Pumps." *Engineering News-Record* (December 2, 1937).

Johnson, Leland R. *The Davis Island Lock and Dam, 1870-1922*. Pittsburgh: U.S. Army Engineer District, 1985.

———. *Engineers on the Twin Rivers*. Nashville: U.S. Army Corps of Engineers, 1978.

———. *The Headwaters District*. Pittsburgh: U.S. Army Corps of Engineers, c.1978.

———. *The Ohio River Division: U.S. Army Corps of Engineers*. Cincinnati: U.S. Army Corps of Engineers, 1992.

Johnson, Leland R., and Jacques S. Minnotte. *The Headwaters District Roundtables: An Eyewitness History of the Pittsburgh District, United States Army Corps of Engineers: 1936-1988*. Pittsburgh: U.S. Army Corps of Engineers, 1989 [Revised edition of *The Headwaters District*, c. 1978].

Johnson, Rich. *The Central Arizona Project, 1918-1968*. Tucson: University of Arizona Press, 1977.

Johnson, W. A. "Report of the District Engineer," in *Ohio River*, October 31, 1933. 74th Congress, 1st Session, August 23, 1935, House Document 306. Washington, D.C.: Government Printing Office.

Jones, Holway R. *John Muir and the Sierra Club: The Battle for Yosemite*. San Francisco: The Sierra Club, 1965.

Jorgensen, Lars. "The Constant-Angle Arch Dam." *Transactions of the American Society of Civil Engineers* 78 (1915).

———. "Improving Arch Action in Dams." *Transactions of the American Society of Civil Engineers* 83 (1919-1920), (New York: 1921)

———. "Memorandum on Arch Dam Developments." *Proceedings American Concrete Institute* 27 (1931).

———. "The Record of 100 Dam Failures." *Journal of Electricity,* San Francisco 44, No. 6, March 15, 1920.

Justin, Joel D. "The Design of Earth Dams." *Transactions of the American Society of Civil. Engineers* 87 (1924).

———. "Discussion [of Middlebrooks]." *Transactions of the American Society of Civil Engineers* 107 (1942).

———. *Earth Dam Projects*. New York: John Wiley, 1932.

Kahrl, William L. "Paradise Reclaimed: The Corps of Engineers and the Battle for the Central Valley." Undated manuscript, 128-9, 142-3. Files, Office of History, Headquarters, U.S. Army Corps of Engineers, Alexandria, Virginia.

———. *Water and Power: The Conflict over Los Angeles' Water Supply in the Owens Valley*. Berkeley, University of California Press, 1982.

Katz, Alfred H., and Jean Spencer Felton, editors. *Health and the Community*. New York: Free Press, 1965.

Kelley, Robert. *Battling the Inland Sea: American Political Culture, Public Policy, and the Sacramento Valley, 1850-1886*. Berkeley: University of California Press, 1989.

———. *Gold vs. Grain: The Mining Debris Controversy*. Glendale, California: Arthur H. Clarke Company, 1959.

Kendall, A. W. "Keeping Up with the 'Big Muddy'." *The Military Engineer* 30 (March-April 1938).

Kennedy, Lorne. "Fund Restored for Projects on Missouri River." *World Herald* (November 30, 1945).

Ketcham, Valentine, and M. C. Tyler. "Hugh Lincoln Cooper Memoir." *Transactions, American Society of Civil Engineers* 103 (1938).

King, Judson. *The Conservation Fight: From Theodore Roosevelt to the Tennessee Valley Authority*. Washington, D.C.: Public Affairs Press, 1959.

Kleinsorge, Paul L. *Boulder Canyon Project: Historical and Economic Aspects*. Palo Alto, California: Stanford University Press, 1941.

Kluger, James R. *Turning on Water with a Shovel: The Career of Elwood Mead*. Albuquerque: University of New Mexico Press, 1992.

Kollgaard Eric B., and Wallace L. Chadwick, editors. *Development of Dam Engineering in the United States*. New York: Pergamon Press, 1988.

Koppes, Clayton R. "Efficiency/Equity/Esthetics: Towards a Reinterpretation of American Conservation." *Environmental Review* 11 (Summer 1987).

Kuentz, Oscar O. "The Lower Columbia River Project." *The Military Engineer* 25, No. 139 (January-February 1933).

"Laboratory Research Applied to the Hydraulic Design of Large Dams." Bulletin No. 32, Waterways Experiment Station, Vicksburg, Mississippi, June 1948.

Lacey, Michael J. *Government and Environmental Politics.* Washington, D.C.: Woodrow Wilson Center Press, 1989.

"Land of the Big Muddy." *Times* (September 1, 1952).

Landy, Marc K., Marc J. Roberts, and Stephen Thomas. *The Environmental Protection Agency: Asking the Wrong Questions from Nixon to Clinton.* New York: Oxford University Press, 1990.

Lane, E. W., and J. R. Ritter. "The Life of Hoover Dam." *Reclamation Era* 34 (April 1948).

"Large Slide in Fort Peck Dam Caused by Foundation Failure." *Engineering News-Record* 122 (May 11, 1939).

Larson, John Lauritz. "A Bridge, A Dam, A River: Liberty and Innovation in the Early Republic." *Journal of the Early Republic* 7 (Winter 1987).

Lawson, Michael L. *Dammed Indians: The Pick-Sloan Plan and the Missouri River Sioux, 1944-1980.* Norman: University of Oklahoma Press, 1994 [originally 1982], paperback edition.

———. "The Oahe Dam and the Standing Rock Sioux." *South Dakota History* 6 (Spring 1976).

Layton, Edwin. "Mirror-Image Twins: The Communities of Science and Technology in 19th-Century America." *Technology and Culture* 12 (October 1971).

Lear, Linda J. "Boulder Dam: A Crossroads in Natural Resource Policy." *Journal of the West* 24 (October 1985).

Lee, Lawrence B. *Reclaiming the American West: An Historiography and Guide.* Santa Barbara, California: ABC-Clio Press, 1980.

———. "Water Resource History: A New Field of Historiography?" *Pacific Historical Review* 57 (November 1988).

Leggett, Robert F. "The Jones Falls Dam on the Rideau Canal." *Transactions of the Newcomen Society* 31 (1957-59).

Legos, James. "Concrete Buttress Dams." In *Development of Dam Engineering in the United States,* edited by Eric B. Kollgaard and Wallace L. Chadwick. New York: Pergamon Press, 1988, 533-670.

LeMoigne, Guy, Shawki Barghoutti, and Herve Plusquellac, editors. *Dam Safety and the Environment*. Washington, D.C.: The World Bank, 1990.

Leopold, A Starker, and Justin W. Leonard. "Alaska Dam Could Be Resources Disaster," *Audubon Magazine* 68 (May-June 1966).

Leopold, Luna B. and Thomas Maddock, Jr. *The Flood Control Controversy: Big Dams, Little Dams, and Land Management*. New York: Ronald Press Company, 1954.

Lewis, Meriwether, and William Clark. *The Journals of Lewis and Clark*, edited by Bernard DeVoto. Boston: Houghton Mifflin Company, 1953.

Limerick, Patricia Nelson. *The Legacy of Conquest: The Unbroken Past of the American West*. New York: W. W. Norton, 1987.

Link, Arthur S. *Wilson: The New Freedom*. Princeton: Princeton University Press, 1956.

Lippincott, Isaac. "A History of River Improvement." *Journal of Political Economy* (1914).

Los Angeles Department of Water and Power. "You Are Needed-Help Build Boulder Dam." Los Angeles: Los Angeles Department of Water and Power, December 1928.

Lowe, John III, "Earthfill Dams." In *Development of Dam Engineering in the United States*. Edited by Eric B. Kollgaard and Wallace L. Chadwick. New York: Pergamon Press, 1988, 671-884.

Lowitt, Richard. *The New Deal and the West*. Bloomington: Indiana University Press, 1984.

Lowrey, Ralph. "Construction Features of Shasta Project Part 1: General Layout and Aggregate Production." *Civil Engineering* 11 (June 1941).

———. "Construction Features of Shasta Project Part 2: Conveying Aggregate, Mixing and Placing Concrete." *Civil Engineering* 11 (July 1941).

Ludwigson, John. "Dams and the Colorado." *Science News* 91 (February 1967).

Lukesh, G. R. "The Columbia River System." *The Military Engineer* 22, No. 124 (July-August 1930).

Maass, Arthur. *Muddy Waters: The Army Corps of Engineers and the Nation's Rivers*. Cambridge, Massachusetts: Harvard University Press, 1951.

Magnusson, Carl Edward. "Hydroelectric Power in Washington." *Bulletin No. 78*, Engineering Experiment Station Series, Seattle, February 1934.

Marcus, Alfred. "Environmental Protection Agency." In *The Politics of Regulation*, James Q. Wilson, editor. New York: Basic Books, Inc., 1980.

Marshall, Jim. "Dam of Doubt." *Colliers* (June 19, 1937).

Martin, Ben Robert. "The Hetch Hetchy Controversy: The Value of Nature in a Technological Society." PhD dissertation, Brandeis University, 1982.

Martin, Russell. *The Story that Stands Like a Dam: Glen Canyon and the Struggle for the American West*. New York: Henry Holt and Company, 1989.

Matthes, Gerard H. "Aerial Photography as an Aid in Map Making, with Special Reference to Water Power Surveys." *Transactions of the American Society of Civil Engineers* 86 (1923).

———. "Discussion." *Transactions of the American Society of Civil Engineers* 100 (1935).

Mawn, Geoffrey P. "Phoenix, Arizona: Central City of the Southwest, 1870-1920," Ph.D. dissertation, Arizona State University, 1979.

Mazmanian, Daniel, and Mordecai Lee. "Tradition Be Damned! The Army Corps of Engineers Is Changing." *Public Administration* 35 March-April 1975).

McAlpine, W. H. "Roller Gates in Navigation Dams." *The Military Engineer* 26, No. 150.

"The McCall Ferry Hydro-Electric Powerplant on the Susquehanna River." *Engineering News* 58, No. 11 (September 12, 1907).

McCarthy, G. Michael. *Hour of Trial: The Conservation Conflict in Colorado and the West, 1891-1907.* Norman: University of Oklahoma Press, 1977.

McClain, Wallis E., Jr., editor. *U.S. Environmental Laws: 1994 Edition.* Washington, D.C.: Bureau of National Affairs, Inc., 1994.

McClintock, James H.. *Arizona*, 3 volumes. Chicago: S. J. Clarke Publishing Company, 1916.

McCully, Patrick. *Silenced Rivers: The Ecology and Politics of Large Dams.* London: Zed Books, 1996.

McGeary, M. Nelson. *Gifford Pinchot: Forester - Politician.* Princeton, New Jersey: Princeton University Press, 1960.

McWilliams, Carey. *Southern California Country: An Island on the Land.* New York: Duell, Sloan, & Pearce, 1946.

Mead, Elwood. *Helping Men Own Farms: A Practical Discussion of Government Aid in Land Settlement.* New York: The Macmillan Company, 1920.

———. *Irrigation Institutions: A Discussion of the Economic and Legal Questions Created by the Growth of Irrigated Agriculture in the West.* New York: The Macmillan Company, 1903.

Melosi, Martin V. *Coping with Abundance: Energy and Environment in Industrial America.* New York: Alfred A. Knopf, 1985.

———. "Lyndon Johnson and Environmental Policy." In *The Johnson Years, Volume Two: Vietnam, the Environment, and Science*, Robert A. Divine, editor. Lawrence: University Press of Kansas, 1987.

———, editor. *Pollution and Reform in American Cities, 1870-1930.* Austin: University of Texas Press, 1980.

"Members of Tribes Losing Land to Participate in Dam Closure Ceremony on June 12." *Williston Daily Herald* (June 3, 1953).

"Men and Mechanization." *Engineering News-Record* 114, 19 (May 9, 1935), 684.

Merchant, Carolyn, editor. *Major Problems in American Environmental History.* Lexington, Massachusetts: D.C. Heath and Company, 1993.

Merritt, Raymond H. *The Corps, the Environment, and the Upper Mississippi River Basin.* Washington D.C.: Historical Division, Office of the Chief of Engineers, 1981.

Meyer, Michael C. *Water in the Hispanic Southwest: A Social and Legal History, 1550-1850.* Tucson: University of Arizona, 1984.

Meyer, Roy W. "Fort Berthold and the Garrison Dam." *North Dakota History* 35 (Summer-Fall 1968).

Miami Conservancy District. *Official Plan Miami Conservancy District.* Dayton, Ohio: Miami Conservancy District, May 1916.

Middlebrooks, T. A. "Fort Peck Slide." *Transactions of the American Society of Civil Engineers* 107 (1942).

———. "Progress in earth-dam design and construction in the United States." *Civil Engineering* (September 1952).

———. Thomas A. Middlebrooks Papers. Office of History, Headquarters, U.S. Army Corps of Engineers, Alezandria, Virginia.

Mighetto, Lisa. "Salmon, Science, and Politics: Writing History for the U.S. Army Corps of Engineers." *Public Historian* 17 (Fall 1995).

Mighetto, Lisa, and Wesley J. Ebel. *Saving the Salmon: A History of the U.S. Army Corps of Engineers' Efforts to Protect Anadromous Fish on the Columbia and Snake Rivers.* Seattle: Historical Research Associates, Inc., 1994.

Mikesell, Stephen D. "Pardee Dam: Evaluation of National Register Eligibility."
A report prepared for John Holson, BioSystems Analysis, Inc., 1994.

Mikkelson, Gordon. "Debate Rages Over Ninth of River Plans." *Minneapolis Morning Tribune* (January 21, 1950).

Miller, M. Catherine. *Flooding the Courtrooms: Law and Water in the Far West*. Lincoln: University of Nebraska Press, 1993.

Miller, Zane L., and Patricia M. Melvin. *The Urbanization of Modern America: A Brief History*. San Diego: Harcourt Brace Jovanovich, 1987 [2nd edition].

"Mississippi Congressman Calls for Missouri Valley Authority." *The Leader* (Bismarck) (July 26, 1951).

"Missouri Basin is Major Problem in Nation's Economy." *The Leader* (Bismarck) (September 7, 1950).

" The Missouri Moves Over." *Life* (May 1953).

" A Mistake That Should Be Corrected." *Engineering News-Record* (January 3, 1935).

Moeller, Beverly Bowen. *Phil Swing and Boulder Dam*. Berkeley: University of California Press, 1971.

Molloy, Peter. "19th Century Hydropower: Design and Construction of Lawrence Dam, 1845-1848." *Winterthur Portfolio* 15 (Winter 1980).

Montgomery, M. R. *Saying Goodbye*. New York: Alfred A. Knopf, 1989.

Moore, George Holmes. "Neglected First Principles of Masonry Dam Design." *Engineering News* 70 (September 4, 1913).

Moreell, Ben. *Our Nation's Water Resources–Policies and Politics*. Chicago: Law School, University of Chicago, 1956.

Morgan, Arthur E. *The Miami Conservancy District*. New York: McGraw Hill, 1951.

Morris, Robert M., and Thomas L. Reilly. "Operation Experiences, Tygart Reservoir." *Transactions of the American Society of Civil Engineers* 107 (1942).

———. *Tygart Dam: Review of Operations 1938-1939*. Pittsburgh: U.S. Army Corps of Engineers, April 1940.

Moss, Senator Frank E. *The Water Crisis*. New York: Frederick A. Praeger, Publisher, 1967.

Mount, Jeffrey F. *California Rivers and Streams: The Conflict Between Fluvial Process and Land Use*. Berkeley: University of California Press, 1995.

Mundt, Karl E. "Not All Dams are Damnable!" *Outdoor America* 8 (December 1943).

Nash, Roderick Frazier, editor. *American Environmentalism: Readings in Conservation History*. New York: McGraw-Hill Publishing Company, 1990.

———. *Wilderness and the American Mind*. New Haven, Connecticut: Yale University Press, 1967.

National Archives and Records Administration, Denver, Colorado, Records of the Bureau of Reclamation, Record Group 115.

———. College Park, Maryland, Records of the U.S. Army Corps of Engineers, Record Group 77.

——— College Park, Maryland, Records of the Secretary of the Interior, Record Group 48.

———. Philadelphia, Records of the U.S. Army Corps of Engineers, Record Group 77.

National Research Council. *Safety of Existing Dams: Evaluation and Improvement*. Washington, D.C.: National Academic Press, 1983.

Newell, Frederick Haynes. Frederick Haynes Newell Papers. Manuscript Division, Library of Congress, Washington D.C.

———. "Irrigation: An Informal Discussion." *Transactions of the American Society of Civil Engineers* 62 (1909).

———. *Irrigation in the United States*. New York: T. Y. Crowell, 1902.

———. Compiler. *Proceedings of the First Conference of Engineers of the Reclamation Service: With Accompanying Papers*. Washington, D.C.: Government Printing Office, 1904.

"'No Compromise' on Grand Canyon Dams: Sierra Club's Reply to Goldwater Plan." *U.S. News and World Report* 61 (December 12, 1966).

Noetzli, Fred A. "Arch Dam Temperature Changes and Deflection Measurements." *Engineering News-Record* 89, No. 22 November 30, 1922).

———. "Gravity and Arch Action in Curved Dams," with discussion. *Transactions of the American Society of Civil Engineers* 84 (1921), 1-135.

———. "Improved Type of Multiple-Arch Dam." *Transactions of the American Society of Civil Engineers* 87 (1924).

———. "Multiple-Arch Dams." In Edward Wegmann. *The Design and Construction of Dams*. 8th edition, New York: John Wiley & Sons Inc., 1927.

———. "The Relation Between Deflections and Stresses in Arch Dams." *Transactions of the American Society of Civil Engineers* 85 (1922), 284-307.

———. "With Mathematical Discussion and Description of Multiple Arch Dams." In Edward Wegmann. *The Design and Construction of Dams*. 8th edition, New York: John Wiley & Sons Inc., 1927.

Nygren, E. J. *Views and Visions: American Landscapes Before 1830*. Washington, D.C.: Corcoran Gallery of Art, 1986.

O'Brien, William Patrick, Mary Yeath Rathbun, Patrick O'Bannon, and Christine Whitacre. *Gateways to Commerce - The U.S. Army Corps of Engineers' 9-foot Channel Project on the Upper Mississippi River*. Denver: National Park Service, 1992.

O'Connell, Charles F., Jr. "The Corps of Engineers and the Rise of Modern Management, 1827-1856." In *Military Enterprise and Technological Change: Perspectives on the American Experience*. Merritt Roe Smith, editor. Cambridge, Massachusetts: MIT Press, 1985.

O'Connell, Harold. "Fort Peck Dam: Part 1 – The Project and Its Scope." *Compressed Air Magazine* (March 1935).

Olson, Cal. "Garrison Dam Fill Job Completed." *The Fargo Forum* (March 5, 1954).

Onuf, Peter S. *Statehood and Union: A History of the Northwest Ordinance.* Bloomington: Indiana University Press, 1987.

Opie, John. *The Law of the Land: Two hundred Years of American Farmland Policy.* Lincoln: University of Nebraska Press, 1987.

Outland, Charles F. *Man-Made Disaster: The Story of St. Francis Dam.* Glendale, California: Arthur H. Clark Company, 1963.

Oxx, Francis H. "The Ohio River Movable Dams." *Military Engineer* 27 (January-February 1935).

Pacific Constructors, Inc. *Shasta Dam and its Builders.* San Francisco: Schwabacher-Frey Company, 1945.

Packard, Fred M. "Echo Park Dam? Not By a Damsite!" *National Parks Magazine* 29 (July-September 1955).

Paehlke, Robert, editor. *Conservation and Environmentalism: An Encyclopedia.* New York: Garland Publishing, 1995.

Palmer, Tim. *The Columbia: Sustaining a Modern Resource.* Seattle: The Mountaineers, 1997.

———. *Endangered Rivers and the Conservation Movement.* Berkeley: University of California Press, 1986.

———. "Saving the Stanislaus: Must We Wear Chains to Keep Rivers Free?" *Sierra Club Bulletin* 64 (September/October 1979).

———. *The Snake River: Window to the West.* Washington, D.C.: Island Press, 1991.

Parkman, Aubrey. *History of the Waterways of the Atlantic Coast of the United States.* Washington, D.C.: Institute for Water Resources, January 1983.

Pattison, Kermit. "Why Did the Dam Burst?" *American Heritage of Invention & Technology* 14 (Summer 1998).

Paul, Charles H. "Core Studies in the Hydraulic-Fill Dam of the Miami Conservancy District." *Transactions of the American Society of Civil Engineers* 85 (1922), 1181-202.

———. "Flood Control in the Miami Valley, Ohio." *Transactions of the American Society of Civil Engineers* 85 (1922).

Paulson, John D. "Garrison Dam to Force Missouri to 'Detour' in '53." *The Fargo Forum* (November 1952).

———. "Great Amounts of Concrete, Steel Go into Dam." *The Fargo Forum* (December 2, 1951).

Pence, A. W. "The Fort Peck Dam Tunnels." *The Military Engineer* 29.

Penrose, Charles. "Utah Colonization Methods." *Proceedings of the Twelfth National Irrigation Congress.* Galveston, Texas, 1905.

Petersen, Keith C. "The Army Corps of Engineers and the Environment in the Pacific Northwest." (May 1982), Records of the Office of History, Headquarters, U.S. Army Corps of Engineers, Alexandria, Virginia.

———. "Battle for Ice Harbor Dam." *Pacific Northwest Quarterly* 86 (Fall 1995).

———. *River of Life, Channel of Death: Fish and Dams on the Lower Snake.* Lewiston, Idaho: Confluence Press, 1995.

Peterson, Elmer T. *Big Dam Foolishness: The Problem of Modern Flood Control and Water Storage.* New York: Devin-Adair Company, 1954.

Petroski, Henry. *Engineers of Dreams: Great Bridge Builders and the Spanning of America.* New York: Vintage Books, 1995.

Petulla, Joseph M. *American Environmental History.* San Francisco: Boyd and Fraser, 1977.

———. *Environmental Protection in the United States.* San Francisco: San Francisco Study Center, 1987.

"Pick Calls for Unity in Valley." *World Herald* (January 3, 1946).

"Pick Hears N.D. Tribes Plead for Lands Dam Would Flood." *Bismarck Tribune* (May 28, 1946).

"Pick-Sloan Disagree." *Bismarck Tribune* (June 1, 1949).

Pinchot, Gifford. *The Fight for Conservation*. Seattle: University of Washington Press, 1967; [originally, 1910].

Pinkett, Harold T. *Gifford Pinchot: Private and Public Forester*. Urbana: University of Illinois Press, 1970.

Pisani, Donald J. "Deep and Troubled Waters: A New Field of Western History." *New Mexico Historical Review* 63 (October 1988).

———. "Federal Water Policy and the Rural West." In *The Rural West Since World War II*, R. Douglas Hurt, editor. Lawrence: University Press of Kansas, 1998.

———. *From the Family Farm to Agribusiness: The Irrigation Crusade in California and the West, 1850-1931*. Berkeley: University of California Press, 1984.

———. "State vs. Nation: Federal Reclamation and Water Rights in the Progressive Era." *Pacific Historical Review* 51 (August 1982).

———. *To Reclaim a Divided West: Water, Law, and Public Policy 1848-1902*. Albuquerque: University of New Mexico Press, 1992.

———. *Water and American Government: The Reclamation Bureau, National Water Policy, and the West, 1902-1935*. Berkeley: University of California Press, 2002.

Pitzer, Paul C. *Grand Coulee: Harnessing A Dream*. Washington State University Press, 1994.

Plank, Ewart G. "The Town of Fort Peck." *The Military Engineer* 28, No. 161 (September–October 1936).

"Plans Changed–Foundation Going in for High Coulee Dam." *Wenatchee Daily World* (November 27, 1933).

Porter, Eliot. *Glen Canyon on the Colorado*, edited by David Brower. Salt Lake City: Peregrine Smith Books, 1988.

———. *The Place No One Knew: Glen Canyon on the Colorado*, edited by David Brower. San Francisco: The Sierra Club, 1963.

Potter, William E. *Engineering Memoirs*, oral history interview by Martin A. Reuss. Alexandria, Virginia: U.S. Army Corps of Engineers, July 1983.

———. "Tygart River Reservoir Dam." *The Military Engineer* 28, No. 161, (September-October 1936).

Powell, John Wesley. *Report on the Lands of the Arid Region in the United States*. Washington, D.C.: Government Printing Office, 1879.

———. *Tenth Annual Report of the United States Geological Survey, Part II - Irrigation*. Washington, D.C.: Government Printing Office, 1890.

"Power Line to Garrison Dam Advanced." *Fargo Forum* (January 8, 1946).

Powers, Richard P. "Future Appropriations Won't Have Restrictions Such As Indian Land Clause Which Hits Garrison Dam." *Minot Daily News* (January 29, 1947).

Powledge, Fred. *Water: The Nature, Uses, and Future of Our Most Precious and Abused Resource*. New York: Farrar Straus Giroux, 1982.

"Present Garrison Dam Cost Up to $268 Million." *The Bismarck Tribune* (June 7, 1951).

Price, Wesley. "What You Can Believe About MVA." *The Saturday Evening Post* (January 19, 1946).

"Program for Beginning Construction on the Central Valley Project." *Western Construction News* (January 1938).

" A Progress Report on the Stevenson Creek Test Dam." *Bulletin No. 2*, Engineering GFoundation, December 1, 1925, 8 pages.

"'Pup' Used at Garrison Dam Only One of its Kind." *The Minot Daily News* (September 27, 1949).

Purcell, Carroll W., Jr., editor. *Technology in America: A History of Individuals and Ideas*. 2nd edition. Cambridge, Massachusetts: MIT Press, 1991.

Purcell, Francis X., Jr. "Muskingum River Flood Control." *The Military Engineer* 28, No. 159 (May-June 1936).

Quivik, Frederick L. and Amy Slaton. "Owyhee Dam," Nyssa Vicinity, Malheur County, Oregon. Historic American Engineering Records (HAER No. OR-17), Prints and Photographs Division, Library of Congress

"Rampart Dam." *National Parks Magazine* 41 (August 1967).

"Rampart Dam—Alaska Study." *Izaak Walton Magazine* 31 (May 1966).

"Rampart Dam and the Perpetual Engineers." *Field and Stream* 71 (June 1966).

Randolph, Richard R. "Hydraulic Tests on the Spillway of the Madden Dam." *Transactions of the American Society of Civil Engineers* 103 (1938).

Rankine, W. J. M. "Report on the Design and Construction of Masonry Dams." *The Engineer* 33 (January 5, 1872).

Raushenbush, Stephen. "A Bridge Canyon Dam is Not Necessary." *National Parks Magazine* 38 (April 1964).

Reisner, Marc. *Cadillac Desert: The American West and Its Disappearing Water*. New York: Viking Press, 1986.

"Relocation Issue Vital to Project." *Grand Forks Herald* (January 26, 1947).

Report, 1927. *Proceedings of the American Society of Civil Engineers* (May 1928).

Reuss, Martin. "Andrew A. Humphreys and the Development of Hydraulic Engineering: Politics and Technology in the Army Corps of Engineers, 1850-1950." *Technology and Culture* 26, No. 1 (January 1985).

———. "The Army Corps of Engineers and Flood-Control Politics on the Lower Mississippi." *Louisiana History* 23, No. 2 (Spring 1982).

———. "Coping with Uncertainty: Social Scientists, Engineers, and Federal Water Resources Planning." *Natural Resources Journal* 32 (Winter 1992).

———. Oral history interview of William E. Potter. *In Engineering Memoirs*. U.S. Army Corps of Engineers, July 1983.

———. *Shaping Environmental Awareness: The United States Army Corps of Engineers Environmental Advisory Board, 1970-1980.* Alexandria, Virginia: Historical Division, U. S. Army Corps of Engineers, Office of Administrative Services, Office of the Chief of Engineers, 1980.

Reuss, Martin and Charles Hendricks. "The U.S. Army Corps of Engineers: A Brief History." Collections of the Office of History, U.S. Army Corps of Engineers, Fort Belvoir, Virginia.

Reuss, Martin, and Paul K. Walker. *Financing Water Resources Development: A Brief History.* Alexandria, Virginia: Historical Division, Office of Administrative Services, Office of the Chief of Engineers, July 1983.

Reynolds, Terry S. "The Engineer in 19th-Century America." In *The Engineer in America*, Terry S. Reynolds, editor. Chicago: University of Chicago Press, 1991.

Richardson, Elmo R. *Dams, Parks and Politics: Resource Development and Preservation in the Truman-Eisenhower Era.* Lexington: University Press of Kentucky, 1973.

———. *The Politics of Conservation: Crusades and Controversies, 1897-1913.* Berkeley: University of California Press, 1962.

Richardson, H. W. "Dredge Plant Characteristics and Hydraulic Fill Operation." Engineering News-Record (August 6, 1936).

———. "Tunnel Enlargement and Lining." *Engineering News-Record* (August 13, 1936).

Ridgeway, Marian E. *The Missouri Basin's Pick-Sloan Plan: A Case Study in Congressional Policy Determination.* Urbana: University of Illinois Press, 1955.

Rippon, C. S. "Methods of Handling and Placing Concrete at Shasta Dam." *Journal of the American Concrete Institute* 14 (September 1942).

Robins, Thomas M. "Improvement of the Columbia River." *Civil Engineering* 2, No. 9 (September 1932).

Robinson, Michael C. *History of Navigation in the Ohio River Basin.* Alexandria, Virginia: U.S. Army Corps of Engineers, Water Resources Support Center, 1983.

———. "The United States Army Corps of Engineers and the Conservation Community: A History to 1969" (draft manuscript, November 1982). Files, Office of History, Headquarters, U.S. Army Corps of Engineers, Alexandria, Virginia.

———. *Water for the West*. Chicago: Public Works Historical Society, 1979.

"The Garrison Story." In *The Story of Garrison Dam: Taming the Big Muddy*, Sheila C. Robinson,(compiler). Garrison, North Dakota: Dakota Trails Books, 1997.

Rocca, Al M. *America's Shasta Dam: A History of Construction, 1936-1945*. Redding, California: Redding Museum of History, 1995.

"Roller-Gate Dams for the Kanawa River." *Engineering News-Record* 111, No. 12 (September 21, 1933).

Roosevelt, Theodore. State Papers as Governor and President, 1899. In *The Works of Theodore Roosevelt, National Edition, 15*. New York: C. Scribner and Sons, 1926.

Rose, Gene. *San Joaquin: A River Betrayed*. Fresno: Linrose Publishing Company, 1992.

Rowley, William D. *Reclaiming the Arid West: The Career of Francis G. Newlands*. Bloomington: University Press of Indiana, 1996.

Ruddy, T. Michael. *Damming the Dam: The St Louis Corps of Engineers and the Controversy Over the Meramec Basin Project from Its Inception to Its Deauthorization*. St. Louis, Missouri: U.S. Army Corps of Engineers, 1992.

Russell, Edmund P., III. "Lost Among the Parts Per Billion: Ecological Protection at the United States Environmental Protection Agency, 1970-1993." *Environmental History* 2 (January 1997).

Sax, Joseph L. "Parks, Wilderness, and Recreation." In *Government and Environmental Politics*, Michael J. Lacey, editor. Washington, D.C.: Woodrow Wilson Center Press, 1989.

Scarpino, Philip V. *Great River: An Environmental History of the Upper Mississippi, 1890-1950*. Columbia: University of Missouri Press, 1985.

Schermerhorn, I. Y. "The Rise and Progress of River and Harbor Improvement in the United States." *Journal of the Franklin Institute* 139 (January-June 1895).

Schneiders, Robert Kelley. *Unruly River: Two Centuries of Change Along the Missouri.* Lawrence: University Press of Kansas, 1999.

Schnitter, Nicholas J. *A History of Dams: The Useful Pyramids.* Rotterdam, Netherlands: A. A. Balkema, 1994.

Schubert, Frank N. "From the Potomac to the Columbia: The Corps of Engineers and Anadromous Fisheries." Unpublished manuscript, December 1978, Office of History, Headquarters, U.S. Army Corps of Engineers, Ft. Belvoir, Virginia.

———, editor. *The Nation Builders: A Sesquicentennial History of the Corps of Topographical Engineers, 1838-1863.* Fort Belvoir, Virginia: Office of History, U.S. Army Corps of Engineers, 1988.

Schuyler, James D. "The Construction of the Sweetwater Dam." *Transactions of the American Society of Civil Engineers* 19 (1988).

———. "Recent Practice in Hydraulic-Fill Dam Construction." *Transactions of the American Society of Civil Engineers* 63 (June 1907).

———. *Reservoirs for Irrigation, Water-Power, and Domestic Water-Supply.* New York: John Wiley and Sons, 1902.

———. "Water Storage and Construction Of Dams." *USGS Eighteenth Annual Report.* Washington, D.C.: Government Printing Office, 1897.

Schuyler, Montgomery. *American Architecture and Other Writings.* Edited by William H. Jordy and Ralph Coe. New York: Atheneum, 1964.

Seelye, John. *Beautiful Machine: Rivers and the Republican Plan, 1175-1825.* New York: Oxford University Press, 1991.

"Sen. Murray to Try MVA Bill Again." *The Minot Daily News* (November 17, 1948).

"Senate Gives OK to O'Mahoney's Dam Amendment." *Bismarck Tribune* (March 15, 1946).

"Senator Murray Will Introduce TVA Bill." *North Dakota Union Farmer* (December 6, 1948).

Seybold, J. S. "Constructors Roll Nearly One Million Yards a Week Into Garrison Dam." *Civil Engineering* (October 1949).

Shallat, Todd. "Building Waterways, 1802-1861: Science and the United States Army in Early Public Works." *Technology and Culture* 31 (January 1990).

———. *Structures in the Stream: Water, Science, and the Rise of the U.S. Army Corps of Engineers*. Austin: University of Texas Press, 1994.

———. "Water and Bureaucracy: Origins of the Federal Responsibility for Water Resources, 1787-1838." *Natural Resources Journal* 32 (Winter 1992).

Shalowitz, Ann. "Endangered Darters 'Endanger' Tellico Dam." *Conservation News* 40 (March 15, 1975).

"The Shasta Dam on the Sacramento River, California." *Engineering* 157 (January 7, 1944).

"The Shasta Dam on the Sacramento River, California." *Engineering* 157 (January 21, 1944).

"The Shasta Dam on the Sacramento River, California." *Engineering* 157 (February 4, 1944).

"The Shasta Dam on the Sacramento River, California." *Engineering* 157 (February 25, 1944).

Shaw, Gaylord. "The Search for Dangerous Dams - A Program to Head off Disaster." *Smithsonian* 9 (April 1978).

Sherard, James L., Richard J. Woodward, Stanley F. Gizienski, and William A. Clevenger. *Earth and Earth-Rock Dams*. New York: John Wiley and Sons, Inc., 1963.

Sheriff, Carol. *The Artificial River: The Erie Canal and the Paradox of Progress, 1817-1862*. New York: Hill and Wang, 1996.

Sherill, Robert G. "Corps of Engineers: The Pork-Barrel Soldiers." *Nation* (February 14, 1966).

Sherow, James Earl. *Watering the Valley: Development Along the High Plains of the Arkansas River*. Topeka: University Press of Kansas, 1990.

"Silt, Cracks, Floods, and Other Dam Foolishness." *Audubon Magazine* 77 (September 1975).

Simmons, Ralph B. *Boulder Dam and the Great Southwest*. Los Angeles: The Pacific Publishers, 1936.

Skerrett, G. "Fish Over a Dam." *Scientific American* 159 (October 1938).

"Sloan Asks Solution to Souris Problems." *The Minot Daily News* (December 3, 1948).

Smith, Guy-Harold, editor. *Conservation of Natural Resources*. 4th Edition. New York: John Wiley & Sons, Inc., 1971.

Smith, Frank E., editor. *Conservation in the United States: A Documentary History*. Five volumes. New York: Chelsea House, 1971.

———. *The Politics of Conservation*. New York: Pantheon Books, 1966.

Smith, Karen. "The Campaign for Water in Central Arizona, 1890-1903." *Arizona and the West* 23 (Summer 1981).

———. *The Magnificent Experiment: Building the Salt River Project, 1870-1917*. Tucson: University of Arizona Press, 1986.

Smith, Merritt Roe, editor. *Military Enterprise and Technological Change: Perspectives on the American Experience*. Cambridge, Massachusetts: MIT Press, 1985.

Smith, Norman. *A History of Dams*. London: Peter Davies, 1971.

———. *Man and Water: A History of Hydro-Technology*. New York: Scribner's Sons, 1975.

Smith, V. Kerry, editor. *Environmental Resources and Applied Welfare Economics: Essays in Honor of John V. Krutilla*. Washington, D.C.: Resources for the Future, 1988.

Smythe, William E. *The Conquest of Arid America*. New York, privately printed: 1900.

Sontag, Raymond J. *A Broken World: 1919-1939*. New York: Harper & Row, 1971.

Sowles, Lawrence P. "Construction of Boulder Dam: Description of Methods of Pouring the Concrete." *Compressed Air Magazine* (April 1934).

Spurr, Stephen H. "Rampart Dam: A Costly Gamble." *Audubon Magazine* 68 (May-June 1966).

Stacy, Susan M. *When the River Rises: Flood Control on the Boise River, 1943-1985*. Boulder, Colorado: Institute of Behavioral Science, University of Colorado and College of Social Sciences and Public Affairs, Boise State University, 1993.

Stafford, Samuel. "Big Problem: Damming up the Flood of Antipathy." *Government Executive* 1 (December 1969).

"Stage Two Garrison Bids Let." *Mandan Pioneer* (September 22, 1948).

Stebbins, Theodore E., Jr., and Norman Keyes, Jr. *Charles Sheeler: The Photographs*. Boston: New York Graphic Society, 1987.

Stegner, Wallace. "Battle for the Wilderness." *New Republic* 130 (February 15, 1954).

———. *Beyond the 100th Meridian: John Wesley Powell and the Second Opening of the West*. Boston: Houghton Mifflin, 1954.

———. "Myths of the Western Dam." *Saturday Review* 48 (October 23, 1965).

Stein, Murray. "Problems and Programs in Water Pollution." *Natural Resources Journal* 2 (December 1962).

Steinberg, Theodore. *Nature Incorporated: Industrialization and the Waters of New England*. Amherst: University of Massachusetts Press, 1991.

———. *Slide Mountain, or, the Folly of Owning Nature*. Berkeley: University of California Press, 1995.

———. "'That World's Fair Feeling': Control of Water in 20th-Century America." *Technology and Culture* 34 (April 1993).

Steirer, William F., Jr. "Riparian Doctrine: A Short Case History for the Eastern United States." In *Historic U.S. Court Cases, 1690-1990: An Encyclopedia*. Edited by John W. Johnson. New York: Garland Publishing, 1992.

Sterling, Everett. "The Powell Irrigation Survey, 1888-1893." *Mississippi Valley Historical Review* 27 (1940).

Stevens, J. C. "Models Cut Costs and Speed Construction," *Civil Engineering* 6, No. 10 (October 1936), 674-7.

——— and R. B. Cochrane. "Pressure Heads on Bonneville Dam," *Transactions American Society of Civil Engineers* 109 (1944), 77-85.

Stevens, Joseph. *Hoover Dam: An American Adventure*. Norman: University of Oklahoma Press, 1988.

Stine, Jeffrey Kim. "Environmental Politics and Water Resources Development: The Case of the Army Corps of Engineers during the 1970s." Ph.D. dissertation, University of California, Santa Barbara, 1984.

Stites, Francis N. "A More Perfect Union: The Steamboat Case." In, *Historic U.S. Court Cases, 1690-1990: An Encyclopedia*. Edited by John W. Johnson. New York: Garland Publishing, 1992.

"Strawberry Valley Project," HAER UT-26, Prints and Photographs Division, Library of Congress.

Stren, Richard, Rodney White, and Joseph Whitney, editors. *Sustainable Cities: Urbanization and the Environment in International Perspective*. Boulder, Colorado: Westview Press, 1992.

"Strenuous Opposition to Relocation Program For Indians Develops." Mandan (North Dakota) *Daily Pioneer* (Thursday, December 5, 1946).

Strong, Douglas H. *The Conservationists*. Menlo Park, California: Addison-Wesley Publishing Company, 1971.

Strong, Paschal N. "The Great Flood of the Ohio." *The Military Engineer* (May-June 1937).

Sturgeon, Stephen Craig. *The Politics of Western Water: The Political Career of Wayne Aspinall*. Tucson: University of Arizona Press, 2002.

———. "Wayne Aspinall and the Politics of Western Water." Ph.D. dissertation, University of Colorado, Boulder, 1998.

"A Survey of Hydroelectric Developments." *Electrical Engineering* (July 1934).

Swain, Donald C. *Federal Conservation Policy, 1921-1933*. Berkeley: University of California Press, 1963.

Taylor, H., and O. C. Merrill. "Estimate of Cost of Examinations, etc. of Streams where Power Development Appears Feasible." House of Representatives, 69th Congress, 1st Session (December 7, 1925, to November 10, 1926), *Document No. 308*, Washington D.C.

Taylor, Joseph E., III. *Making Salmon: An Environmental History of the Northwest Fisheries Crisis*. Seattle: University of Washington Press, 1999.

"Ten River States Want Garrison Dam Built Fast." *Minot Daily News* (July 7, 1947).

"10,000 Montana Relief Workers Make Whoopee on Saturday Night." *Life* 1, No. 1 (November 23, 1936).

"Ten Year Plan for Flood Control." *St. Louis Globe Democrat* (July 17, 1947).

Tennessee Valley Authority. *Annual Report of the Tennessee Valley Authority for the fiscal year ended June 30, 1935*. Washington, 1936.

———. David E. Lilienthal. *Annual Report of the Tennessee Valley Authority for the Fiscal Year Ended June 30, 1935*. Washington, 1936.

———. *The Unified Development of the Tennessee River System*. Knoxville, 1936.

Terzaghi, Karl. *Erdbaumechanik*. Vienna: Franz Deuticke, 1925.

———. "Soil Mechanics–A New Chapter in Engineering Science." *Journal Institution of Civil Engineers*. London (June 1939).

———. *Theoretical Soil Mechanics*. New York: John Wiley and Sons, 1943.

Terzaghi, Karl, and Ralph B. Peck. *Soil Mechanics in Engineering Practice*. New York: John Wiley and Sons, 1948.

"Thaddeus Merriman (Memoir)." *Transactions of the American Society of Civil Engineers* 107 (1942).

"This Issue." *Engineering News-Record* 115, 9 (August 1935), 279.

Thomas, George. *The Development of Institutions under Irrigation with Special Reference to Early Utah Conditions*. New York: MacMillan Company, 1920.

Thomas, Harold A. and Emil P. Schuleen. "Cavitation in Outlet Conduits of High Dams." *Transactions of the American Society of Civil Engineers* 102 (1942).

Thorson, John E. *River of Promise, River of Peril: The Politics of Managing the Missouri River*. Lawrence: University Press of Kansas, 1994.

"Three Big Dam Operations Begun in the Northwest." *Engineering News-Record* (April 5, 1934).

Torres, Louis. *"To the Immortal Name and Memory of George Washington:" the United States Army Corps of Engineers and the Construction of the Washington Monument*. Washington D.C.: Office of the Chief of Engineers, 1985.

Treadwell, Charles. *The Cattle King*. Santa Cruz, California: Western Tanager Press, 1981.

"Truman Asks More Money for Garrison." *Minot News* (July 24, 1948).

"Truman Hope." *World Herald* (January 20, 1946).

"The Truth About the Army Engineers and Garrison Dam." *The Williston Daily Herald* (August 22, 1949).

Turner, Darrell J. "Dams and Ecology: Can They Be Made Compatible?" *Civil Engineering* 41 (September 1971).

Tweet, Roald D. *A History of the Rock Island District U.S. Army Corps of Engineers: 1866-1983*. Rock Island, Illinois: U.S. Army Engineer District, Rock Island, 1984.

———. *History of Transportation on the Upper Mississippi & Illinois Rivers*. Washington, D.C.: Government Printing Office, 1983.

"12 European Engineers to Visit Minot, Garrison Dam." *Minot Daily News* (September 5, 1952).

"$20,000,000 Approved for Garrison Project Includes Indian Funds." *Minot Daily News* (July 22, 1947).

"Tygart River Reservoir Dam." U.S. Army Corps of Engineers, Pittsburgh District, Revised November 1938.

"Undertakings Without Precedent." *Engineering News Record* (November 29, 1934).

USCOLD. *Lessons from Dam Incidents. USA-II*. New York: American Society of Civil Engineers, 1988.

U.S. Congress, House. Committee on Interior and Insular Affairs. *Central Valley Project Documents, Part One: Authorizing Documents,* Clair Engle, Chairman. 84th Congress, 1st session, 1955, House Document 416. Washington D.C.: Government Printing Office.

———. Committee on Interior and Insular Affairs. *Central Valley Project Documents, Part Two: Authorizing Documents,* Clair Engle, Chairman. , 84th Congress, 2nd session, 1956, House Document 416. Washington, D.C.: Government Printing Office

———. "Columbia River and Ninor Tributaries," March 29, 1932, House Document 103. Washington D.C.: Government Printing Office.

———. Committee on Rivers and Harbors. *The Rivers and Harbors Act of March 3, 1925*. 68th Congress, 1st session, March 31st and April 1st, 1924, 43 Stat., 1186, Section 3. Washington D.C.: Government Printing Office.

———. "Estimate of Cost of Examinations, etc. of Streams Where Power Development Appears Feasible," April 12, 1926, House Document 308. Washington D.C.: Government Printing Office.

———. "Estimate of Cost of Examinations, etc. of Streams Where Power Development Appears Feasible," prepared by H. Taylor and O. C. Merrill. 69th Congress, 1st session, December 7, 1925, to November 10, 1926, House Document 308. Washington D.C.: Government Printing Office.

———. "Message from the President of the United States," including a "Resolution of the Missouri River States Committee to Secure a Basin-Wide Development Plan," (sent to the President on August 21, 1944). 78th Congress, 2nd session, 1944, House Document 680. Washington D.C.: Government Printing Office.

———. *"Missouri River,"* 73rd Congress, 2nd session, February 5, 1934, House Document 238. Washington D.C.: Government Printing Office.

———. "Preliminary Examination of Reservoir Sites in Wyoming and Colorado," prepared by Hiram Martin Chittenden. 55th Congress, 2nd session, 1897, House Document 141, serial 3666. Washington D.C.: Government Printing Office.

———. "Preliminary Examination of Tennessee River Tributaries," prepared by Harold C. Fiske. 67th Congress, 2nd session, House Document 319. Washington D.C.: Government Printing Office.

———. "Report," prepared by Colonel Lewis A. Pick. 78th Congress, 2nd session, House Document 475. Washington D.C.: Government Printing Office.

———. "Report of the District Engineer," in *Ohio River*, October 31, 1933, prepared by W. A. Johnson. 74th Congress, 1st session, August 23, 1935, House Document 306. Washington D.C.: Government Printing Office.

———. *Tennessee River.* 71st Congress, 2nd session, March 24, 1930, House Document No. 328, part 2, plate 101. Washington D.C.: Government Printing Office.

———. *Tennessee River and Tributaries: North Carolina, Tennessee, Alabama, and Kentucky.* 67th Congress, 2nd session, House Document No. 319. Washington D.C.: Government Printing Office.

———. *Tennessee River and Tributaries: North Carolina, Tennessee, Alabama, and Kentucky,* 71st Congress, 2nd session, March 24, 1930, House Document No. 328. Washington D.C.: Government Printing Office.

———. *U.S. Board of Commissioners on the Irrigation of the San Joaquin, Tulare and Sacramento Valleys of the State of California,* 43rd Congress, 1st session, 1874, House Document 1, pt. 2. serial 1745. Washington D.C.: Government Printing Office.

———. 69th Congress, 1st session, April 13, 1926, House Document 308. Washington D.C.: Government Printing Office.

———, Senate. "Missouri River Basin: Conservation, Control, and Use of Water Resources of the Missouri River Basin in Montana, Wyoming, Colorado, North Dakota, South Dakota, Nebraska, Kansas, Iowa, and Missouri," 78th Congress, 2nd session, April 1944, Senate Document No. 191. Washington D.C.: Government Printing Office.

———, Senate. *Preliminary Report of the Inland Waterways Commission.* 60th Congress, 1st session, 1908, Senate Document 325. Washington D.C.: Government Printing Office.

———, Senate, 78th Congress, 2nd session, November 21, 1944, Senate Document 247. Washington D.C.: government Printing Office.

U.S. Department of Agriculture, *Report of Irrigation Investigations in California.* Washington D.C.: Government Printing Office, 1901.

U.S. Department of Commerce, Bureau of the Census, *Historical Statistics of the United States, Colonial Times to 1970,* Part 1. Washington, D.C.: Government Printing Office, 1975.

———. *Historical Statistics of the United States, Colonial Times to 1970,* Part 2. Washington, D.C.: Government Printing Office, 1975.

U.S. Department of Energy, Bonneville Power Administration; U.S. Army Corps of Engineers; and U.S. Department of the Interior, Bureau of Reclamation. *The Columbia River System: The Inside Story.* Portland, Oregon, 1991.

U.S. Department of the Interior. Bureau of Reclamation. *Boulder Canyon Project Final Reports: Part IV—Design and Construction.* Denver: Government Printing Office, 1941.

———. *Boulder Canyon Project Final Reports, Part V - Technical Investigations, Bulletin 1, Bureau of Reclamation, Trial Load Method of Analyzing Arch Dams.* Denver: Government Printing Office, 1938.

———. *The Colorado River: "A Natural Menace Becomes a Natural Resource:* Washington, D.C.: Government Printing Office, 1947.

———. *Dams and Control Works*. Washington, D.C.: Government Printing Office, 1938.

———. *The First Annual Report of the Reclamation Service.* Washington D.C.: Government Printing Office, 1903.

———. *Fourth Annual Report of the Reclamation Service.* Washington, D.C.: Government Printing Office, 1907.

———. *Ninth Annual Report of the Reclamation Service.* Washington, D.C.: Government Printing Office, 1911.

———. Frederick Haynes Newell, compiler. *Proceedings of the First Conference of Engineers of the Reclamation Service: With Accompanying Papers.* Washington, D.C.: Government Printing Office, 1904.

———. *Glen Canyon Dam and Powerplant*. Denver,: Government Printing Office 1970.

———. in coorperation with Iron Canyon Project Association. *Report on Iron Canyon Projects by the Office of the U.S. Reclamation Service at Portland, Oregon, October, 1914,* and Report Board of Review Appointed by the Secretary of the Interior, November, 1914.

———. *Sixth Annual Report of the Reclamation Service.* Washington, D.C.: Government Printing Office, 1908.

———. *The Story of Hoover Dam*. Washington, D.C.: Government Printing Office, 1953.

———. *Third Annual Report of the Reclamation Service, 1903-4*. Washington, D.C.: Government Printing Office, 1905.

———. Arthur Powell Davis. *Water Storage on Salt River, Arizona.* Washington D.C.: Government Printing Office, 1903.

U. S. Department of the Interior, National Park Service. *How to Prepare National Historic Landmark Nominations*. Washington, D.C.: Government Printing Office, 1999.

———. *National Register Bulletin 15: How to Apply the National Register Criteria for Evaluation.* Washington, D.C.: Government Printing Office, 1991.

———. Robie Lange. "National Historic Landmark Nomination: St. Clair Tunnel," February 1993.

U.S. Department of the Interior. Water and Power Resources Service. *Project Data.* Washington, D.C.: Government Printing Office, 1981.

Urquhart, Leonard Church, editor. *Civil Engineering Handbook.* 3rd edition. New York: McGraw Hill, 1950.

"Valley Plan Fight Shifts to Northwest." *World Herald* (January 27, 1946).

Van Faasen, Jerold B. *Making It Happen: A Sixty-Year Engineering Odyssey in the Northwest.* Seattle: Kip Productions, 1998.

Vivian, C. H. "Developing the Mighty Columbia." *Compressed Air Magazine* (September 1935).

Veltrop, Jan A. "Importance of Dams for Water Supply and Hydropower." In *Water for Sustainable Development in the Twenty-first Century*, Asit K. Biswas, Mohammed Jellali, and Glenn E. Stou, editors. Delhi: Oxford University Press. 1993.

Wagoner, L. and H. Vischer. "On the Strains in Curved Masonry Dams." *Proceedings of the Technical Society of the Pacific Coast* 6 (December 1889).

Walker, Paul K. "Developing Hydroelectric Power: The Role of the U.S. Army Corps of Engineers, 1900-1978." Washington, D.C.: Historical Division, Office of the Chief of Engineers, February 1979.

Walter, D. S. "Construction Plant and Methods—Friant Dam." *Civil Engineering* 12 (May 1942).

War Department, U.S. Army Corps of Engineers. "Report on the Slide of a Portion of the Upstream Face of the Fort Peck Dam." Washington, D.C.: U.S. Army Corps of Engineers, July 1939, 14 pages with 29 drawings and charts plus three appendices.

Warne, William A. *The Bureau of Reclamation*. New York: Praeger Press, 1973. Also a later edition: *The Bureau of Reclamation*. Boulder, Colorado: Westview Press, 1985.

"Water and Power for the Southwest." *National Parks Magazine* 39-40 (September 1965).

"Water Power Development on the Mississippi River at Keokuk, Iowa." *Engineering News* 66, No. 13 (September 28, 1911).

Water Resources Council. *The Nation's Water Resources*. Washington, D.C.: Water Resources Council, 1968.

Weatherford, Gary D. and F. Lee Brown, editors. *New Courses for the Colorado River: Major Issues for the Next Century*. Albuquerque: University of New Mexico Press, 1986.

Webb, Walter Prescott. *The Great Plains*. Boston: Ginn and Company, 1931.

Wegmann, Edward, Jr. *The Design and Construction of Dams*. 1st edition. New York: John Wiley & Sons, 1888.

———. *The Design and Construction of Masonry Dams: Giving the Method Employed in Determining the Profile of the Quaker Bridge Dam*. 2nd edition revised. New York: John Wiley and Sons, 1889.

———. *The Design and Construction of Dams*. 3rd edition. New York: John Wiley & Sons, 1900.

———. *The Design and Construction of Dams*. 6th edition. New York: John Wiley and Son, 1911.

———. *The Designs and Construction of Dams*. 8th edition. New York: John Wiley & Sons Inc., 1927.

Wheeler, William Bruce, and Michael J. McDonald. *TVA and the Tellico Dam, 1936-1979: A Bureaucratic Crisis in Post-Industrial America*. Knoxville: University of Tennessee Press, 1986.

Whipple, William, Jr. Autobiography, unpublished manuscript, 1978, copy in possession of David P. Billington.

———. "Comprehensive Plan for the Columbia Basin." *Transactions of the American Society of Civil Engineers*, 1950, paper No. 2473 (published in November 1950 as *Proceedings-Separate No. 45*).

White, Richard. *"It's Your Misfortune and None of My Own:" A History of the American West*. Norman: University of Oklahoma Press, 1991.

———. *The Organic Machine*. New York, Hill and Wang, 1995.

"Who Benefits by Boulder Dam?" *New Republic* 63 (July 30, 1930).

Wild, Peter. *Pioneer Conservationists of Western America*. Missoula, Montana: Mountain Press Publishing Company, 1979.

Williams, Albert N. *The Water and the Power: Development of the Five Great Rivers of the West*. New York: Duell, Sloan and Pearce, 1951.

Williams, Gardiner. Discussion on "Lake Cheesman Dam and Reservoir." *Transactions of the American Society of Civil Engineers* 53 (1904).

Williams, James C. *Energy and the Making of Modern California*. Akron: University of Akron Press, 1997.

Williams, R. B. "General Features of Friant Dam." *Civil Engineering* 12 (February 1942).

Willingham, William F. *Army Engineers and the Development of Oregon: A History of the Portland District, U.S. Army Corps of Engineers*. Washington, D.C.: U.S. Army Corps of Engineers, 1983.

———. *Water Power in the "Wilderness": The History of Bonneville Lock and Dam*. Portland, Oregon: U.S. Army Corps of Engineers, Portland District, 1988.

——— and Donald Jackson. "Bonneville Dam." *Historic American Engineering Record*, No. OR-11 (April 1989). Prints and Photographs Division, Library of Congress.

Wilson, James Q., editor. *The Politics of Regulation*. New York: Basic Books, Inc., 1980.

Wilson, Richard Guy. "Machine Age Iconography in the West." *Pacific Historical Review* 54 (1985).

Wiltshire, Richard L., P.E. "100 Years of Embankment Dam Design and Construction in the U.S. Bureau of Reclamation" (September 2002), paper prepared for Reclamation's centennial Symposium on the History of the Bureau of Reclamation at the University of Nevada - Las Vegas, June 18-19, 2002.

Wisner, George Y., and Edgar T. Wheeler. "Investigation of Stresses in High Masonry Dams of Short Spans," *Engineering News* 54, No. 60 (August 10, 1905).

Wolf, Frank. *Big Dams and Other Dreams: The Story of Six Companies Inc.* Norman: University of Oklahoma Press, 1996.

Wolfe, Henry C. "Concrete Structures for Spillway." *Engineering News-Record* (August 29, 1935).

———. "The Fort Peck Dam - The Project." *The Military Engineer* 27, No. 151 (January-February 1935).

Woodbury, David O. *The Colorado Conquest*. New York: Dodd, Mead & Company, 1941.

"Work Begun on Fort Peck Dam to Regulate Missouri River." *Engineering News-Record*, No. 9 (1933).

Worster, Donald. "Hoover Dam: A Study in Domination." In Worster's *Under Western Skies: Nature and History in the American West*. New York: Oxford University Press, 1992.

———. *Rivers of Empire: Water, Aridity and the Growth of the American West*. New York: Pantheon, 1985.

———. *A River Running West: The Life of John Wesley Powell*. Oxford: Oxford University Press, 2001.

"WPA Dam Fails at Kansas City." *Engineering News-Record* 119 (September 23, 1937).

Wuczkowski, R.. "Flussigkeitsbehälter." In *Handbuch für Eisenbetonbau*, Friedrich Ignaz Edler von Emperger, editor. Berlin: Wilhelm Ernst & Sohn, 1911 edition.

Young, J. M. "River Planning in the Missouri Basin." *The Military Engineer* 22 (March-April 1930).

Young, Walker R. "Boulder Dam Plays Its Part in Reclamation." *Reclamation Era* 27 (February 1937).

Zarbin, Earl A. "The Committee of Sixteen." *The Journal of Arizona History* 25 (Summer 1984).

———. *Roosevelt Dam: A History to 1911*. Phoenix: Salt River Project, 1984.

Zucker, Paul. *American Bridges and Dams*. New York: Greystone Press, 1941.

INDEX

308 reports 91-2, 121, 204, 237, 373
    Columbia River 191, 202-5, 207, 224, 359
    Corps expected 308 reports to lead to control
        over all major river basins 356
    Corps expected to control major river basins
        with navigation 356
    Cove Creek Dam 359
    Mississippi River flood 1927 111
    Missouri River 236, 241, 252, 270
    Most extensive and comprehensive engineering study
        of all times 236
    Opportunity to review navigable rivers in terms of
        flood control 356
    Rivers and Harbors Act (1927) 358
    Set the stage for multipurpose dams 121
    Tennessee River 358-61
    Twenty-four separate surveys 121
Abbot, Henry L. 111
    Corps levee policy 356
Acreage limitation
    Acreage limitation increased to 960 acres in 1982 347
Alamo Canal 140, 141
    Floods in 1905 opened uncontrolled flow 140
Alamo River 139
All-American Canal 140, 143, 154-6, 158, 171, 301
Allegheny River 353-4, 356, 370-1, 384, 466-8
Allegheny-Monongahela basin
    Nine-dam system of reservoirs 371
    Nine-reservoir flood control system 370
Ambursen Hydraulic Construction Company 79
Ambursen, Nils F. 79, 82, 366, 456
    Flat slab dams 79
American Planning and Civic Association
    Opposed Echo Park Dam 396
American River 303
American society
    Embraced the great transforming event of electricity 83

Analysis of design of dams
- Numerical values for stresses are always suspect 80
- U.S. Reclamation Service 65
- Uses for analysis 80

Anglen, Ralph 257
Arch dam 58, 64
- Intellectual rage of the 1920s was for concrete arches 79
- Analysis of design 64
- Arch dam theory 59
- Jones Falls Dam 59
- Relatively small amounts of material 59
- Thin arch dam 58
- Zola Dam 59

Arizona 10, 22, 26, 29, 74, 84, 95, 97-100, 103-4,
.... 109, 132, 135-7, 149, 155, 157, 159,
.... 169, 175, 181-2, 212, 219,
.... 385, 399, 405, 410, 425
- "Arizona Navy" 181
- Bartlett Dam on the Verde River 212
- Hydropower 386
- Lee's Ferry 136, 157
- Opposed Colorado River Compact 159, 181
- Parker Dam 181
- Prehistoric irrigation 21, 22

Arizona v. California 181
Arrowrock Dam 91, 107, 143, 146, 171, 334, 425, 430-1, 435, 458
ASCE awards
- Writings on concrete arch dams 79

Audubon Society 395, 411
Baker, Newton D.
- Secretary of War 113
- Sought to turn Muscle Shoals over to private companies 113

Ballinger, Richard
- Secretary of the Interior 37

Banks, Frank A. 214
- Calculating output 221

Barren River 16
Bartlett Dam 79, 212
Bashore, Harry W. 271, 274-5
Beach, Lansing 120
- Chief of Engineers 119

Bear Wallow Lake Dam
    Failed in 1976 406
Bechtel Company...171
Beggs, George
    Stevenson Creek Test Dam 78
Bensley, John
    Miller & Lux..307
    San Joaquin & Kings River Canal Company........................307
Bernard, Simon 15
Big Bend Dam 288
Big Creek, California
    Hydropower 80
Bixby, William H. 34
Black Canyon.. 134-136, 145, 153, 154, 166
    1940 there were ten different power lines out 179
    Allowed a less costly structure 153
    Attention shifts here from Boulder Canyon............................153
    Contains Boulder Dam...134
    Crowe, Frank 171
    Excavation began in 1932...176
    Explored for a dam 152, 153
    Forbidding environs..171
    Hydroelectric power delivered to Southern California...............178
    Initial generation of power in 1936-1937 179
    Metropolitan Water District of
        Southern California powerline...179
    Offered a small additional amount of hydropower...................152
    Selected because it provided larger reservoir 153
    Selected for Hoover Dam 152
    Selected for larger reservoir...153
    Work began before adequate housing was available 174
    Work relief 174
Board of Consultants For Tygart Dam
    Gerig, William 365
    Growden, James B. 365
    Harza, L. F...365
    Hill, Louis C. 365
    McAlpine, William 365
    Styer, Wilhelm D..365
Board of Engineers for Rivers and Harbors
    In 1922 on study of Tennessee River.....................................120

Board of Irrigation Commissioners
    Study central California 23
Bonneville Dam 91-2, 191-7, 199, 203, 205,
                                208, 210, 222-3, 225-6, 228, 239,
                                      253, 305, 363, 365, 385, 392
    Anadromous fish 226
    Anadromous fish research and laboratory 226
    Arguments against building 203
    Cement used 197
    Closed in January 1938 202
    Coffer dams 199
    Construction design issues 195
    Construction issues 195
    Contracts in 1934 196
    Design and construction proceeded on parallel tracks 196
    Displaced the Indian dip-net fishery at the Cascades 385
    Excess power in region 213, 215
    First model of hydraulics in a major,
        multipurpose, mainstream dam 196
    Fish hatcheries 226
    Fish ladder 226
    Fish ladders 202, 394
    Fishways 202
    Hydroelectricity 424
    Kaplan turbines 194
    Lower Columbia Fisheries Plan 226
    Power oversupply issue 202, 203
    Power used during World War II 203
    Public works project 196
    Puzzolan cement 197
    Salmon issue and the dam 202, 393, 394, 407, 408
    Second powerhouse 228
Bonneville Power Act
    Bonneville Power Administration 222
    Within the Interior Department 222
Bonneville Power Administration (BPA) 222, 223, 228
    Market federal power on the Columbia River 222
    Reclamation wanted to subsidize irrigation 222
    Wanted cheap power 222
Boulder Canyon 109, 134, 145, 147, 152-3, 156-9, 173
    Superior granite for foundation of dam 153

Boulder Canyon Project 75, 107, 122, 133, 144-5, 150, 153-4,
            .... 157-60, 162-5, 167,169, 171, 174, 180-1
    Agricultural and Los Angeles interests allied 160
    All-American Canal 171
    Alliance between Imperial Valley and Los Angeles 160
    Arizona could not prevent construction 181
    Budgeted at $177 million 171
    Cabot, Phillip 167
    Catalyst for the Colorado River Compact 182
    Colorado River Compact 182
    Construction approach 173
    Construction of Hoover 171
    Depended on political resolution of
        Colorado River water rights 159
    Financed by revenue from hydroelectric power 182
    Financial foundation is sale of power 110
    Generate hydroelectric power 171
    Guaranteed power sales 165
    How public and private power would share control 165
    Hydropower generation 179
    Hydropower incidental to use of water 166
    Industrial Workers of the World 174
    Irrigation 171
    Irrigation and flood control stimulated 160
    Irrigation stimulated 160
    Job seekers move to area 173
    Largest single federal contract ever let 171
    Municipal support carried 160
    Municipal water 171
    Opposed by private electric power 166
    Opposition by power companies 166
    Opposition to public power 165
    Politics of hydropower 165
    Private power published opposition pamphlets 167
    Promotion in Los Angeles 162
    Propaganda from the "Power Trust" 169
    Provide flood control 171
    Required settlement of water rights issues on Colorado 159
    Stock market crash of October 1929 174
    Symbolize political character and importance 179
    Tied to Imperial Valley 160

       Tied to Los Angeles municipal water supply needs....................160
       Very different from anything previously proposed....................165
Boulder Canyon Project Act 133-4, 169, 180
       Among the first large multipurpose water projects....................388
       Battle over passage 169
       Colorado River Compact 388
       Derives from more than the Depression 133
       Signed by Calvin Coolidge 133
Boulder Dam
       See "Hoover Dam"
Boulder Dam Association 163
Bourke-White, Margaret
       Fort Peck Dam 253, 255
       Photographer 253
       Sent to photograph Bonneville Dam..253
Brackenridge, W. A. 76
Brereton, Richard...307
Bridge Canyon Dam 409
Brower, David
       Sierra Club 398
Brown, Frank E.
       Bear Valley Dam..60
       California arch dam design 60
       Replaced by James D. Schuyler..61
       Sweetwater Dam 61
Brown, Lytle 361
       Missouri River 308 report..238
       Tennessee River 308 report..359, 360
Buffalo Bill Dam..................65, 91, 116, 146, 425, 431, 432, 445, 455, 459
Bureau of Reclamation 3, 131
       Arrowrock Dam 107
       Authorized 1902 92
       Bartlett Dam...79
       Boulder Canyon Project and Depression
             promote Reclamation..107
       Buttress dams - multiple arch 79
       Chose massive designs...82
       Davis, Arthur Powell, becomes director 107
       Davis, Arthur Powell, fired 107
       Decline in prestige 1910-20s 107
       Elephant Butte Dam...107
       Fact Finding Commission..109

    Financial difficulties 107
    Hydropower authorized 32
    Hydropower possibilities 32
    Issues of overlap with Corps..110
    Mead, Elwood, becomes Commissioner.................................109
    Mead, Elwood, most work was on already
        established projects..109
    Mead, Elwood, policies 108
    Mead, Elwood, unsullied by association with Reclamation.......109
    Mission in 1902 89
    Name changed to Bureau of Reclamation 107
    Proposed consolidation with Corps 281
    Reclamation's tenuous role in West
        beforeBoulder Canyon...107
    Salt River Project 107
    Stony Gorge Dam 79
    U.S. Reclamation Service 32
Burton Bill of 1906
    Brought government into Niagara River
        and Great Lakes water regulation 114
Butler, John S. 205-6, 208
    Columbia River 308 Report..204
Buttress dams 50, 58, 79-80, 118, 194, 318, 366, 429, 439, 456
    Flat-slab dams 58-9
    More complicated design and construction 59
    Multiple arch dams 79
    Relatively small amounts of material 59
    Require much less material..58
    Slab dams were usually relatively small scale..........................79
    Two types of buttress dams...79
Cabot, Phillip 167
    Opposes Hoover Dam..168
Cairo, Illinois 353
California 22-3, 27, 29, 109, 121, 129-30, 135-6,
        ....38-9, 141, 144, 145, 152, 154-8, 161-2, 164,
        ....168-9, 243, 301, 309-10, 312,
        ....314, 323-4, 344, 384, 410, 424
    See California Doctrine
    Aversion to 160 acre limit...314
    Bear Valley Dam 60, 429
    Big Creek 80
    California water rights law 8, 311

    Central Valley 302
    Central Valley Project 221, 305
    Clash of water doctrines 9
    Colorado River ambitions 109, 157-159, 163,174, 178, 179, 181
    Dam Safety Act passed (1929)..390
    Dual system of water rights 9
    Durham and Delhi colonies 109
    East Park Dam 433
    Evolution of water law..7
    Feared 160 acre limitation 314
    Feared the surrender of some autonomy...314
    Federal involvement in plan 328-330
    Federal restrictions...346
    Formation of water districts...164
    Friant Dam 221
    Grange 308
    Hydraulic mining 241, 313
    Hydropower 36, 80
    Imperial Valley..386, 405
    Klamath Project 315
    Lake Spaulding Dam 70
    Marshall Plan 320, 321
    Metropolitan Water District of Southern California 405
    Navigation..312
    Reclamation Service concentrated on
            projects at the periphery of 315
    Rice cultivation 319
    Riparian Doctrine...311
    Riparianism gained legal recognition 9
    Set a precedent in the application of prior appropriation................9
    Shasta and Friant Dams 304
    Spanish irrigation..21
    St. Francis Dam failure 389
    State plans for water 320-2, 324, 328, 331
    Water issues in the state 301
    Wright Act Districts 27
    Yuma Project..315
California Debris Commission 313-4, 317
    Formed in 1893..313
    Opposed reservoirs for flood control 314
    To protect navigability Sacramento River 313

California Development Company 139-42
    By early 1920s there was sentiment
        for All-American Canal 140
    Canal crossed Mexico 139
    Fast development in Imperial Valley 140
    Fights flooding of Imperial Valley 140
    Mexico reserved the right to draw half of the canal water 139
California Doctrine 8, 9
    Adopted along the Pacific Coast and in Great Plains 9
    Dual system of water rights 9
California Edison Company 76
    St. Francis Dam 78
    Stevenson Creek Test Dam 76
California National Bank 307
California state engineer report of 1927
    Basis of all subsequent debate on water plan 326
    Endorsed by federal/state commission 326
    Federal government should build Central Valley Project 326
California state engineer report of 1931 326
    Federal support more likely when Franklin D. Roosevelt became
        President 328
    Friant Dam 326
    How Central Valley water system would work 326
    Kennett Dam 326
    State law authorized 1933 326
    Voters authorize Central Valley bonds 328
California State Irrigation Association 321
California State Water Project 347
    Largely completed in 1970s 347
    Planning in 1950s 347
    Tehachapi Range 347
    Transfers water from Feather River to
        Los Angeles/San Diego 347
California water law
    Prior appropriation 9
    Riparian law 9
Campbell, Thomas 108
Canals
    Alamo Canal 139-40
    Alamo Canal flooding 140
    Alamo Canal under control 141
    All-American Canal 139, 142

    Bypassing obstructions 17
    California agriculture 301
    Central Valley of California 309
    Chesapeake to Ohio River 15
    Columbia Basin Project 206
    Contra Costa Canal 345
    Control of Colorado River 129
    Delta-Mendota Canal 344
    Early transportation efforts 5
    Financial risks 27
    Friant-Kern Canal 344
    From Friant Dam 326
    Hohokam Indian canals 21, 22, 99
    Hudson Reservoir and Canal Company 99
    Imperial Valley irrigation 139
    Irrigation canals 21
    Kingston Canal 59
    Lining to reduce salt loads 405
    Madera Canal 344
    Mexico receives water from 139
    Miller & Lux 310
    Mormon settlement 22
    Panama Canal 193
    Power canal Salt River 102
    Salt River Project 21
    San Joaquin & Kings River Canal
          and Irrigation Company 23
    San Joaquin & Kings River Canal Company 307
    San Joaquin River canal 323
    Studies of foreign canals 309
    Swilling Irrigation Canal Company 22
    Twin Falls Canal 26
    Upper Sacramento River 320
Canyon Lake Dam
    Failed in 1972 406
Carey Act 26, 28-9
Carey, Joseph M. 26
Carnegie Institute of Technology
    Contract to model spillway 364
Carquinez Straits 303
Carson River Project 451
Casagrande, Arthur 244, 258, 261

      Embankment designer 240
      Seepage through dams 240
Cascades, Columbia River 385
Cavitation 90
Celilo Falls 385
Central Arizona Project 181-2, 405, 409-10
      Authorized in 1968 410
      Required federal financing 182
Central California
      Agriculture 22
      Irrigation colonies 23
Central Utah Project 396
Central Valley
      California's most important agricultural region 302
      Early studies of the state engineer 322, 323
      Iron Canyon Dam 316
      Iron Canyon Project 316
      Kings River 310
      Miller & Lux 310, 311
      Riparian water rights 311
      State Engineer report 1927 326
Central Valley Project 301, 347
      Acreage limitation 345-6
      Acreage limitation increased to 960 acres in 1982 347
      American River 303
      Bensley, John 307
      California, most economically important
          state in the West 301
      California, "buy-back" Central Valley Project in 1950s 347
      Central Valley agriculture 301
      Central Valley topography 301
      Design of Friant Dam 330
      Design of Shasta Dam 330
      Downey, Sheridan 346
      Feather River 303
      First phase completed 1945 344
      Friant Dam 304, 305
      Functioned with remarkable efficiency 345
      Historical background 305, 320
      Kaweah River 303
      Kennett Dam 305
      Kern River 303

    Kings River 303
    Kings River excluded from Reclamation project 330
    Large scale project 305
    Marshall Plan 305
    Marshall, Robert, of the U.S. Geological Survey 305
    Merced River 303
    Miller & Lux 303, 306-7, 346
    Mining dams pioneered large-scale water storage 301
    "Move the Rain" 304
    Pine Flat Dam excluded from project 330
    Pit River 303
    Rice culture 319
    Rice gave impetus to project 319
    Rivers and Harbors Act (1937) 329
    Sacramento/San Joaquin River Delta 303
    San Joaquin Valley 303
    Shasta Dam 304-5, 330
    State seeks money from Federal Emergency
        Administration of Public Works 329
    Topography 302
    Tulare Lake 303
    Tulare Valley 303
    Tuolumne River 303
    Water swap 345
    Yuba River 303
Chaffey, George
    Imperial Valley development 139
Chandler, Harry
    Land owner in Mexico 139
Channel
    Improvements 196
    St. Louis to St. Paul 18
Chanoine wicket 18, 372
Chanoine, Jacques 18
Chapman, Oscar
    Approved Echo Park Dam 396
    Reverses decision on Echo Park Dam 397
Chattanooga District 91
Cheat River 361
Cheesman Dam, Colorado 68
Chelan County Public Utilities District 192

Chippewa Indians...20
    Wild rice fields..20
Chippewa River 19
Chittenden Survey of 1897 26
      Recommends five reservoirs..26
Chittenden, Hiram...27
Chorpening, C. H. 283
Citizens Committee on Natural Resources
      Opposed Echo Park Dam...397
Colorado..10, 26-7, 29, 92, 132, 135-7, 157, 395, 396
    Colorado Doctrine...10
    Hiram Chittenden survey..26
    North Platte River 157
Colorado Agricultural College..364
Colorado Desert
    See Imperial Valley
Colorado Development Company...135
Colorado River 1, 129-31, 133-8, 140-5, 155-9, 166, 186, 263
    90 percent of flow from the Rocky Mountains 135
    Alamo River...140
    Alamo River canal 140
    All-American Canal...143, 154, 301
    Arizona concerns 181
    "Arizona Navy"..181
    Arizona tried to oppose Hoover on
        constitutional grounds...181
    Arizona v. California..181, 410
    Average annual flow about 14 million acre-feet....................129
    Basin river system..135
    Boulder Canyon Project...159
    Bridge Canyon Dam 409
    California aspirations for 109, 157, 159, 161, 163
    California eyes the resource...130
    Central Arizona Project..181, 409
    Central Utah Project...396
    Colorado River Aqueduct 161-2, 164, 169, 171,
        ....179, 181, 386
    Colorado River Aqueduct Association..................................162
    Colorado River Aqueduct required electricity
        for pumping 164
    Colorado River Basin Salinity Control Act 405
    Colorado River Board...150

Colorado River Compact 154-5, 157-9, 180-1
Colorado River Storage Project 182, 395
Construction of Hoover Dam 149
Delta 138, 387-8
Design of Hoover Dam 146
Drainage basin 135
Early 1920s most flow unused 161
Fall/Davis Report 142
Federal Power Commission rejected private
    power applications 180
Flood control 109, 167, 171
Flood into Imperial Valley 140
Flows through seven western states 129
Gila River 405
Glen Canyon Dam 183, 399, 404
Grand Canyon dams 185, 410
Grand Canyon National Park 410
Grand River 24
Green River 24
Hoover Dam 143-4, 147, 386
Imperial Valley 135, 138, 160, 301, 388
John Wesley Powell 24, 137
Kincaid Act 143
Los Angeles 130, 161
Los Angeles desires partners 163
Los Angeles filed claim to 1500 cubic feet per second 161
Los Angeles sought water for municipal development 161
Los Angeles threatened future water rights
    claims of other state 161-2
Lower Basin 157
Marble Canyon Dam 409
Metropolitan Water District of Southern Califorrnia 169
Mexico 405
Most flow unused and unclaimed in early 1920s 157
Official ratification of Colorado River Compact 159
Palo Verde Valley 138
Parker Dam 181
Phil Swing 154, 156
Private power opposition to Hoover Dam 144
Recent reassessments of Hoover Dam 185
Salinity 405
Salton Sea 138

    Short-lived steamboat use 137
    Siting of Hoover Dam 145, 152
    Southern Pacific Railroad 140
    Strict application of the appropriation doctrine 157
    Studies of 142-3
    Swing-Johnson Act 169
    Upper Basin 157
    Wellton-Mohawk Irrigation District 405
    Yampa and Green Rivers 182
Colorado River Aqueduct 162, 164, 169, 171, 179, 181, 386
    Large amount of electricity required 164
Colorado River Aqueduct Association
    Phil Swing 162
Colorado River Basin Salinity Control Act 405
Colorado River Board ....
    Approved design of Hoover 150
Colorado River Commission 157
Colorado River Compact 154, 158-9, 182, 203, 204
    Arizona bitterly opposed 159
    Arizona concerns 159
    Arizona refused to ratify 159
    Arizona still feared California's claims 159
    Arizona v. California 159
    Final ratification in 1929 159
    Motives of various states 158
    Negotiated in 1922 in Santa Fe 158
    Ratified March of 1929 159
    Ratified without Arizona 159
    Terms of 158
    Why California supported 158
Colorado River Storage Project 182, 398
    Approved 398
    Authorized without Echo Park 398
    Echo Park Dam 398
    Financed by revenue from hydroelectric power 182
Colorado River System
    Colorado River Storage Project 182
    Dinosaur National Monument 396
    Echo Park Dam 182, 395, 396
    Green River 182, 395, 396
    National Park Service 182
    Sierra Club 182

    Wilderness Society 182
    Yampa River 182, 395
Columbia Basin Commission 208, 211
Columbia Basin Project 217
    Dam and pump plan 206
    Gravity Plan 206
Columbia River 129, 191, 193-4, 202, 205-6, 209, 223, 235, 253
    308 Report 121, 191, 202-4, 359
    Almost completely untamed in 1932 193
    Anadromous fish 393
    Bonneville Dam 194
    Bonneville Power Administration 222
    Bureau of Reclamation 192, 207
    Bureau of Reclamation study 223
    Chelan County Public Utilities District 192
    Columbia River 308 report 121
    Columbia Basin Commission 211
    Columbia Basin Project 206, 213, 222, 226, 228
    Columbia Valley Authority 217, 222, 228, 278
    Comprehensive Plan (1948) 224
    Dalles Dam eliminated Indian fishing at Celilo Falls 385
    Excess power in area 213
    Fisheries conflicts 392
    Fisheries issues 407
    Fisheries issues 392, 393
    Fishways 202
    Flooding 223
    Grand Coulee Dam 192, 206, 210, 213,
    .... 218, 219, 221, 387, 463
    Grand Coulee Third Powerhouse 228
    Hatcheries 226, 394
    Huge flow required fresh thinking 194
    Indian dip-net fishery 385
    Largest hydroelectric producer in the world 191
    Lower Columbia Fisheries Plan 226
    Main Control Plan 1948 224
    NcNary Dam 227
    Operation for power and flood control 225
    Power network 226
    Restructuring 223
    Rock Island Dam 192
    Rock Island Rapids 192

Salmon	202
Salmon issues	202, 392, 393, 395, 407
U.S. Army Corps of Engineers	207

Columbia River Comprehensive Plan
- Main Control Plan 224

Columbia River Control Plan 228

Columbia River flood (1948)
- Bonneville Dam 223
- Vanport 223

Columbia Valley Authority 217, 222, 228
- Compromise was the Bonneville Power Act 222
- Opposed by Corps 228
- Opposed by Reclamation 228
- Pushed by Henry "Scoop" Jackson 278
- Truman, Harry S. 228
- U.S. Army Corps of Engineers opposed 217
- U.S. Bureau of Reclamation opposed 217

Committee on Flood Control 270

Conconully Dam
- Hydraulic fill construction 243

Concrete
- A mere substitute for stone masonry 83
- Easier simply to put the new cast material into old familiar for 83
- Made the building of integrated structures possible 83
- Performance of equations replaced the performance of structures 83
- Stimulated the search for new forms 83

Concrete dams
- Chief Engineer Reclamation Service 93
- Design and analysis 49
- Gravity dams 50
- Intellectual rage of the 1920s was for concrete arches 79
- Massive tradition 49, 50
- Structural tradition 49, 58
- Trial-load method 49

Connecticut River 129

Conservation
- Management of natural resources and their efficient use 31
- Progressive Era 31

Consolidated Builders Incorporated (CBI) 219

Constant-angle arch dams 61

Cooke, Morris L. ... 364
    Public Works Administration ... 364
Coolidge, Calvin ... 114, 133
    How public and private power would share control ... 165
    Signs Boulder Canyon Project Act ... 133
    Supported leasing of Muscle Shoals ... 114
Cooper, Hugh L. ... 111, 114-9
    Electrical Development Company of Ontario ... 114
    McCall Ferry Hydroelectric powerplant ... 114
Corps of Engineers
    See U.S. Army Corps of Engineers
Coulomb, Charles A.
    Early earthworks designer ... 239
Cove Creek Dam ... 359, 361-2
    See Norris Dam
    Designer C. M. Hackett ... 360
    Power capacity ... 360
    Precursor to Tygart Dam ... 360
Cove Creek Dam designs
    Used to design Tygart Dam ... 362
Coyne, André ... 212
Cret, Paul ... 363-365
Cross, Hardy ... 78
    Critical of complex dam analysis ... 78
Crowe, Frank
    At Shasta Dam ... 334
    Built Parker Dam for Six Companies Inc. ... 334
Cummins, W. F. ... 263
Cylinder formula ... 59-61, 64, 68, 70, 72, 80-1
    Zola Dam ... 59
Dalles Dam 194
    Eliminated fishing site of Celilo Falls ... 385
Dam
    Wall or barrier that resists the pressure of water stored in reservoir 64
Dam design approaches
    Sliding circle stability analysis ... 240
    Trial-load method ... 149, 150
Dam design techniques
    1920s the ideal was for more rigor ... 81
    Cylinder formula ... 59-61, 64, 68, 70-72, 80, 81
    Spillway cross section ... 90

  Trial-load analysis 73, 430-432
  Trial-load method 73
Dam design theory at the beginning of the twentieth century 61
Dam safety 389
Dam Safety Assurance Program 407
Dams
  Conflict between flood control, power generation, and navigation 91
  Environmental effects 383
  Environmentalists worried about alkali buildup 404
  Environmentalists worried about
    herbicides and pesticides 404
  Environmentalists worried about intense irrigation 404
  Environmentalists worried about riverine ecology 404
  Environmentalists worried about salinity 404
  Environmentalists worried about silting 404
  Every dam site unique 392
  Focus on structural solutions 388
  Headwaters dams 18-20, 353
  Improve navigation 16
  Modern opposition 383
  Naturally competing uses 91
  Nineteenth century federal government dams only for navigation 89
  Nineteenth century gravity dam technology 89
  Palisades Dam 405
  Questioning the inherent value of dams
    post World War II 395
  Social effects 383
  The least reversible form of river alteration 404
  Uniqueness of each major river basin 92
Dams, criticisms 385
  Assumptions about flood control 387
  Block fish migration 383
  Change natural flow 383
  Changes in 1960s 399
  Destroy or decommission 383
  Displace farmers 383
  Displacement of people 384
  Drew capital away from sounder technology 383
  Drew, Elizabeth B. 400
  Earth Day 400
  Emphasis shifted from value of dams to value of scenic
    and wild rivers 385

    Environmental movement 400
    Environmental preservation 400
    Flood natural landscape 383
    Flooding 387
    Free the rivers 383
    Recreation 400
    Reduce water quality 383
    Silting 387
Dams, uses of
    Harnessing rivers into service to humankind 383
    Hydropower generation 383
    Irrigation 383
    Maximize 'efficient use' of a vital resource 383
    Navigation 383
    Soil conservation 383
    Symbols of American achievement 384
    Transforming the hinterland 384
    Urban use 384
    Water storage 383
Darcy, Henri-Philibert-Gaspard
    Early earthworks designer 240
Darland, A. F.
    Columbia Basin Commission 211
Davis, Arthur Powell 25, 65, 76, 92, 94-5, 97-8, 100-4, 143
    Assistant chief engineer 94, 142
    Chief Engineer 94
    Director of USRS 107
    Fall/Davis Report 144
    Fired as director USRS 107, 160
    Proposed hydroelectric power be the
        key to Hoover financing 144
    Use of electricity revenues to finance Hoover 144
    Work on Colorado River 142-7, 151-3
Davis Dam 199
Davis Island
    Concrete instead of masonry 18
    Lock development 18
Davis Island Dam
    Concrete instead of masonry 18
Davis, Raymond
    Stevenson Creek Test Dam 78

de Laplace, Pierre S.
    Early earthworks designer 240
de Sazilly, J. Augustin Tortene 52-4, 60, 453
    Design of gravity dams 52
Decisions about designs based on a variety of factors 62
Delano Heights, Montana 255
Delaware River 129, 411
    Flood of 1955 411
    Tocks Island 411
    Tocks Island Dam 411
Delocre, F. Emile 53-4, 60, 453
    Alicante Dam 54
    Almansa Dam 54
    Furens Dam 54
    Profile of equal resistance 53
Delta-Mendota Canal 344
Derleth, Charles Jr. 76
DeVoto, Bernard
    Opposed Echo Park Dam 397
Dill, Clarence 210, 212
    Grand Coulee Dam 208
Dinosaur National Monument 182, 396
    Enlarged 1938 396
    Established 1915 396
Downey, Sheridan
    Opposes acreage limitation 346
Drury, Newton
    Director of National Park Service 397
    Opposed Echo Park Dam 397
Dual system of water rights
    California Doctrine 9
Dworshak Dam 199
Early studies of the California State Engineer
    Concrete gravity design for Shasta Dam 326
    Hydropower seen as financing project 324
    Kennett Dam 324, 325
    Need for federal assistance 324
    Report in 1923 322
    Shasta Dam 324, 326
    Tulare Lake 323

Eastwood, John S. 61
    Big Creek 80
    First multiple arch dam 80
    Hume Lake in the Sierra Nevada 80
    In Fresno, California 79
    Lake Hodges Dam 80
    Mountain Dell Dam 82
    Multiple arch dams 79-83, 431
    Palmdale Dam 82
    Several dams he made the arches three hinged 81
Echo Park Dam 182-4, 395-8
    Approved 398
    California aspirations for water 397
    Colorado River Storage Project 398
    Dinosaur National Monument 396
    Fight to save parks from inundation and exploitation 395
    Glen Canyon Dam substituted for defeated Echo Park 398
    Interconnection with Glen Canyon 398
    Opposed by preservationists 396
    Project approved without Echo Park Dam 398
    Split Mountain Reservoir 397
    Threats to fight entire Colorado River Storage Project 398
Edison, Thomas
    Supports Henry Ford bid to buy Muscle Shoals 114
Effects of hydraulic mining techniques
    California Debris Commission 313
    Caused flooding 313
    Clogged Sacramento River 313
Electric power
    See hydropower
Electrical Development Company of Ontario 114
Elephant Butte Dam 91, 107, 143, 146, 425-6, 445, 461
Ellet, Charles, Jr.
    Diversified flood control program 356
    Suggests levees and reservoirs for flood control in 1852 356
Embankment dams
    Massive embankments 50
Engineering Foundation 76, 78, 389
Environmental effects
    Increased salinity 389
    Silt reduction 388
Fact Finder's Commission 107

Fall, Albert
    Fall/Davis Report 144
    Secretary of the Interior 144
Fall/Davis Report 142, 154
Farm Bureau
    Supports Henry Ford bid to buy Muscle Shoals 114
Feather River 303
Federal appropriations for river and harbor projects 17
Federal Emergency Management Agency
    Responsibility for coordinating dam safety 407
Federal government
    Ability to plan and complete big construction jobs 131
    Major contributions to the art of hydraulic engineering.............131
Federal Guidelines for Dam Safety published (1979) 407
Federal land distribution 3
    Railroads 4
    Rapid transfer of public lands 4
    Stimulate economic development..4
Federal Power Commission 133
    Established by Federal Water Power Act of 1920....................113
    Oversight private power companies 133
Federal Trade Commission
    Study in 1928 about private power
        influencing public opinion 168
Federal Water Power Act of 1920 113
Fellenius, Wolmar
    Sliding circle stability analysis 240
Finley, William L. 385
Fish and Wildlife Coordination Act of 1934
    Ensure migration of fish past dams...393
Fisher, Walter L.
    Secretary of the Interior 37
Fiske, Harold C. 91-2, 119-21, 203, 358
    308 reports 91
    Father of 308 reports 121
Flat slab dam 79
Flinn, Alfred
    Stevenson Creek Test Dam 78
Flood Control 3, 26, 34, 38, 89, 91-2, 113, 322, 384
    308 reports 356
    Assumption rivers be predictable and
        not threaten construction..387

Bypass channels on the Sacramento 314
Colorado River 132
Columbia River 226-228
Conflicts with irrigation and navigation 317
Controversy on Lower Colorado 168, 169
Corps uncertainties about 223
Debate over the ability of large dams to control floods 283
Delaware River flood of 1955 411
Federal role expanded 34
Flood Control Act of 1917 .... 314, 358
Flood Control Act of 1936 388
Flood Control Act of 1970 404
Flood control new to Corps in 1930s 354
Flood plain management 388
Floods at Hoover in February 1932 175
Focus on structural solutions 388
Fort Peck Dam 281, 282
Fort Randall Dam 292
Garrison Dam 281
Great Missouri River flood of 1993 292
History of flooding in the Ohio River Basin 356
Hoover Dam 110, 386, 388
House Committee on Flood Control 110, 269
Kings River 346
Kinzua Dam 371, 372
Levees on the Sacramento River 314
Lower Colorado River 166-8
Mainstem Missouri dams affected
    sedimentation patterns 292
Miami Conservancy District 357, 371
Mississippi River 273
Missouri River 269, 272-4, 291, 292
Missouri River floods (1943) 270
Muskingum Watershed Conservancy District 371
Nonstructural alternatives 403
Ohio River 308 Report 356
Ohio River flood of 1907 356
Ohio River flood of 1913 356
Ohio River flood of 1936 370
Ohio River flood of 1937 353
Ohio River floods of 1935 and 1936 373
Ohio River floods of 1939 371

    Pick, Lewis A. 269
    Pittsburgh Flood Commission..356
    Rivers and Harbors Act of 1920 119
    Rivers and Harbors Act of 1927 358
    Sacramento River...329, 335
    Tygart Dam 353, 354, 363, 367, 371
    U.S. Army Corps of Engineers 14, 15
Flood Control Act of 1917..110, 314, 358
    Lower Mississippi River...358
    Sacramento River...358
Flood Control Act of 1936
    Established flood control for navigable waters..........................388
Flood Control Act of 1970
    Multi-objective planning...404
Flood Control Bill of 1944..276
Flood control studies
    Had to include comprehensive assessment of watershed 110
Flooding 20
    Fort Berthold Reservation partially flooded by
        Garrison Reservoir..279
    Missouri River floods 1947..279, 281
    Standing Rock Sioux Reservation lost land................................385
Floods
    1948 flood pushed Corps toward
        flood control on Columbia 224
    308 reports on Missouri 237, 252
    308 reports to consider flood control..122
    Alamo Canal...140
    California Development Company 141
    Caused by mining debris in California 313
    Colorado River...109
    Columbia River in 1948...223
    Congress first allocated funds for flood control.............................110
    Control of lower Colorado..143
    Corps plans to minimize flood control on the Columbia............223
    Designing embankment dams to handle 287
    Early twentieth-century floods
        commanded increased attention......................................110
    Flood Control Act of 1917...110
    Flood Control Act of 1936...354
    Flood control dam on Colorado River 142
    Gave new direction to the Corps...110

    Glen Canyon Dam 288
    Grand Coulee flood in 1935..215
    Hoover Dam and floods 288
    Hydropower as a side benefit of Hoover Dam............................144
    Imperial Irrigation District..142
    Imperial Valley...135, 140, 154, 160
    Imperial Valley flood of 1905 ....140, 141
    Imperial Valley flood stopped in 1907 141
    Imperial Valley irrigation..135
    Irrigation water from Hoover..142
    Kings River 303, 307
    Lake Powell and floods..288
    Managing reservoirs for flood control and hydropower 225
    Maximum probable flood on the Missouri 252
    Miami Valley region floods...110
    Mississippi River flood 1927 ....111, 122
    Missouri River 235
    Missouri River flood of 1881 288
    Missouri River floods 1951..283
    Missouri River floods 1952..283
    Muscle Shoals 113
    Natural irrigation by flooding 307
    New direction for the twentieth-century Corps 110
    Ohio floods of 1913 110
    Ohio River flood of 1937 353
    Ohio River floods of 1935 and 1936.....................................353, 373
    Passing floods at Hoover..148
    Pittsburgh flood of 1907..110
    Raising political pressure 110
    Sacramento River..303
    Salt River floods in 1905 102
    San Joaquin River 303
    Storage of seasonal floods...136, 324, 346
    Storage of water 24, 27, 100, 320
    Storage of water in California...301
    Tulare River 307
    Vanport...223
Forchheimer, Philipp
    Early earthworks designer 240
Ford, Henry
    Bids to buy Muscle Shoals...114
Fort McDowell...22

Fort Peck Dam 245, 363
    Board of Consultants report..259
    Board of Consultants to study the slide 258
    Bourke-White, Margaret...253
    Casagrande, Arthur 266
    Casagrande, Arthur, added to Board of Engineers
        after slide 258, 261
    Completed in 1940..265
    Conflict between Upper Missouri Basin states
        and Lower Missouri 272
    Construction methodology...246
    Corps responds to concerns of President Roosevelt...............261
    Cummins, W. F..263
    Debate between academics and practitioners 266
    Design 252
    Dredges 246
    Earth..239
    Gerig, William 264
        On hydraulic fill...259
    Gilboy, Glennon...261, 266
    Hydraulic fill construction 241, 244
    Job creation a high priority..238
    Justin, Joel, added to Board of Engineers after slide.................258
    Kittrell, Clark 255, 257, 259
    Larkin, T. B..255
    Mead, Warren J. 263, 264
    Merriman, Thaddeus...261, 263, 264, 267
    Middlebrooks, Thomas 259, 265-267
    Moore, Douglas 256
    Provided many jobs 245
    Public-Works Project Number 30 238
    Ray Kendall 256
    Rebuilding after the slide 258
    Reminiscence of Fort Peck Dam slide 1938............................256
    Schley, Julian 264
    Slide at Fort Peck Dam 256
    Sturdevant, C. L. 259
    Terzaghi, Karl 265-267
    Tourlotte, Eugene..257
    Tyler, M. C. 261
    Upstream slope flattened in rebuilding slide.............................259
    Van Faasen, Jerold 256

    Van Stone, Nelson 256
    Westergaard, H. M. 263
    Workers' shanty towns 255
Fort Peck, Montana 255
Fort Randall Dam 277, 288
    Completed mid-1956 290
    Construction begins 1948 290
    Designed by Omaha District 290
    Flooded Sioux land 291
    Lake Case 291
    Pickstown 290
    President Eisenhower dedicated first power unit 1954 291
Frederick Snare Corporation 365
French Engineering Tradition 11, 12
Friant Dam 326
    Authorization 1935 342
    Completed in 1942 344
    Concrete gravity dam 330
    Construction 342
    Construction begins 1939 342
    Construction methodology 342
    Friant-Kern Canal 344
    Griffith Company and Bent Brothers consortium
        receives contract 342
    Millerton Lake 329
    Straight concrete gravity type 331
Friant-Kern Canal 344
Garfield, James R. 35, 108
Garrison Dam 277
    Bonner, John 282
    Completed in 1955 287
    Construction begins in 1948 281
    Corps and Reclamation differ on capacity
        of Garrison Dam 282
    Dedicated 1956 287
    Design and construction 280
    Design controversies 278
    Diversion of Missouri River 284, 285
    Fort Berthold Indian Tribal Business Council
        signed a contract 281
    Issues with Mandan, Arikara, and Hidatsa 280
    Missouri River flood of 1947 279

    Missouri River States Committee 281
    Missouri-Souris irrigation project..284
    Missouri-Souris Project never was funded 287
    Morrison-Knudsen contractor..283
    Peter Kiewit Sons Company was the contractor..........................283
    Pick, Lewis A...278
    President Eisenhower visits for formal
        closure ceremonies...285
    Reservoir would flood parts of
        Fort Berthold Reservation..279
    Riverdale, North Dakota 278
    Rolled fill embankment dam..280
    Smith, Forrest...282
    Spillway 287
    Spillway dimensioned to carry a huge flood................................287
Gatun Dam
    Hydraulic fill construction 244
Gavins Point Dam..288
General Dam Act of 1906 38
    Amendment in 1910..38
    Corps reconcile power and navigability 38
    Federal role in navigability...38
    Required private dams to build navigation facilities 38
General Survey Act 1824...355
Gerig, William..259, 264, 266, 365
Gibbon v. Ogden 13-4, 132
    Chief Justice John Marshall..14
Gibson Dam 75, 82, 109, 431, 445, 462
Gila River...405
Gilboy. Glennon..261
    Casagrande, Arthur 244
    Hydraulic fill construction 244
Glen Canyon Dam...................... 145, 152, 182-4, 383, 384, 398, 399, 463
    Arch design insufficient as a gravity dam...................................184
    Brower, David..399
    Ecological effects...404
    Environmental effects 404, 425
    Interconnected with Echo Park Dam 398
    Reservoir to protect Upper Colorado River Basin
        water rights 399
Goldmark, Henry 80
    Multiple arch dam in Utah..80

Gould, E. Sherman
    High Masonry Dams 54
Grand Canyon dams 409
    Coal fired plants suggested 410
    Not approved 410
    Nuclear plants suggested 410
    Opposed by David Brower 410
    Part of Reclamation's Central Arizona Project 410
Grand Coulee
    Third Powerplant 228
Grand Coulee Dam 383, 387
    Banks, Frank A. 214
    Bid accepted for low dam 212
    Bids for low dam opened 212
    Centerpiece of largest reclamation project undertaken 387
    Civil engineering wonder of the United States 384
    Consolidated Builders Incorporated (CBI) 219
    Construction complete 1941 221
    Construction methodology 216, 221
    Design issues 210
    Excess power in area 213
    Final plans for high dam 218
    First large turbine completed 221
    Ickes, Harold, approves high dam 216
    Largest concrete structure in the world 217
    Low dam v. high dam 211
    Mead, Elwood, asks to build high dam in 1934 214
    Moving from low to high dam 214
    Multiple arch dam studied 211
    MWAK 213
    Power market too small 214
    Primary construction issue 220
    Progress of construction 218
    Reclamation report supports high dam 213, 214
    Rivers and Harbors Act of 1935 217
    Site not economical for low dam 213
    Water over spillway 1942 221
Grandval Dam 212
Grange 18
Granite Reef Diversion Dam 103
    Inflation 104

Grant, U. S, III
    Opposed Echo Park Dam 396
Gravity dam 64
    Alicante 51
    Almans 50
    Don Pedro Bernardo Villa de Berry 51
    Elche arch dam 51
    Gravity dam theory 51
    Rellue arch dam 51
Green River 16, 135, 396
Growden, James
    Consultant on Tygart Dam 369
Gulf of California 1, 136, 138
Gulf of Mexico 33, 121, 449, 451
Gull Lake 19
Gunnison River 135
Hackett, C. M.
    Design of Cove Creek Dam 360
Hadley, Ebenezer
    Imperial Valley development 138
Haggin, Ali 311
Harper, Sinclair O. 331
Harts, William 90-1, 120
    Eight criticisms of the high dams and storage reservoirs 90
    Spillways and other dam cross sections 90
Harza, L. F.
    Tygart Dam 364
    Tygart Dam consulting 366
Hayden, Carl 155
    Concerned about water rights on the Colorado River 155
Hazen, Allen
    Hydraulic fill construction 243
Heliopolis 15
Hetch Hetchy 34, 35
    Hydroelectric power 35
    O'Shaughnessy Dam on the Tuolumne completed (1925) 35
    Transfer site to San Francisco (1913) 35
Hill, Louis C.
    Tygart Dam consulting 365, 368-70
Hohokam Indians 99
Hollow dams 58
    More complicated design and construction 79

Holyoke Dam across the Connecticut River ... 118
Hoover Commission
    Opposed valley authorities like TVA ... 281
Hoover Dam ... 91-2, 107, 110, 122, 129, 131, 133, 135, 143, 145, 149, 151-4, 159-60, 162, 169, 171, 179-82, 185, 193, 199, 211, 213, 217-20, 253, 334, 362, 383, 386-8, 427, 430-3

    1940 - Ten different power lines out of Black Canyon ... 179
    America's largest civil engineering contractors ... 171
    Arizona could not prevent construction ... 181
    Arizona tried to oppose Hoover on constitutional grounds ... 181
    Bids for construction ... 212
    Boulder Dam Association ... 167
    Chosen from among many possible sites ... 145
    Civil engineering wonder of the United States ... 384
    Cleaning foundation ... 176
    Colorado River Aqueduct ... 386
    Concrete curing heat ... 178
    Concrete placement in 1933 ... 176
    Construction ... 171
    Dam and powerhouses declared complete in 1936 ... 178
    Dedication of dam ... 179, 180
    Designated "Hoover Dam" in 1930 ... 135
    Diversion of river ... 175
    Diversion tunnels ... 175
    Diverting the Colorado River ... 175
    Drawing off excess heat from concrete ... 178
    Excavation began in 1932 ... 176
    Expectation of material progress ... 386
    Federal Power Commission rejected private power applications on Colorado River ... 180
    First multipurpose dam ... 135
    Flood control ... 109, 386
    Flood control dam on the lower Colorado River ... 142, 144
    Floods in February 1932 ... 175
    Focus shifted to construction in 1931 ... 169
    Great monument to American engineering ... 122
    Hydroelectric power delivered to Southern California ... 178
    Hydropower as a side benefit of Hoover Dam ... 144
    Hydropower generation ... 166, 169, 386
    Imperial Valley ... 109, 130
    Initial generation of power in 1936-1937 ... 179

    Irrigation water 386
    Kaiser, Henry 171
    Lake Mead 218, 237, 288
    Largest single federal contract ever let to that time....................171
    Los Angeles..130
    Metropolitan Water District of Southern California
        powerline 179
    Morrison-Knudsen..171
    Name change...135, 179, 219, 220
    Name changed back to Hoover Dam in 1947 135
    Name changed to Boulder Dam in 1933......................................135
    Named 135, 171
    New dimensions to Reclamation program...................................180
    Opposition to dam..169
    Originally known as Boulder Dam 135
    Political character of dam-building..131
    Power operations leased to private power 169
    Precursors..91
    Public Works Administration funding..179
    Recent reassessments...185
    Renamed Boulder Dam 1933 135
    Renamed Hoover in 1947 135
    Revolutionized federal participation in water projects................135
    Selection of the Black Canyon site...152
    Silt capture 388
    Six Companies Inc. 171, 174
    Southern California champions 130
    "Swallows the floods"..388
    The "public power" issue..144
    Threat to private power...144
    Use of electricity revenues to finance Hoover 144
    Utah Construction Company 171
    Winning bid less than $49 million...171
    Work began before adequate housing was available 174
    World War II 386
    World War II contributions 386
Hoover, Herbert 111, 133, 154, 158, 203, 326
    Closely associated with the Boulder Canyon Project................154
    Secretary of Commerce 111
    Supports Henry Ford bid to buy Muscle Shoals.........................114
Horse Mesa Dam..107
Houk, Ivan..78

House Committee on Flood Control 110
House Document No. 308
    Provided in 1926 111
    Resulted in 308 reports and studies 111, 121
    Set the stage for multipurpose dams 121
House Rivers and Harbors Committee
    Requests cost estimates for comprehensive
        survey of rivers 111
Howell, C. H.
    Refines trial-load analysis 73-4, 462
    Trial-load method 73
Hudson Reservoir and Canal Company 101
Hudson River 129
Hugh L. Cooper & Co.
    Muscle Shoals 118
Hume-Bennett Lumber Company 80
Humphreys, Andrew A. 111
    Corps levee policy 356
Hungry Horse Dam 199
Hurricane Agnes in 1972 .... 354, 372
Hurricane Hazel in 1954 371
Hydraulic fill dams 241
    Failures 243
Hydroelectric power
    Bonneville Dam 394
    Corps supports private development 359
    Echo Park Dam 396, 397
    Environmental opposition 410
    Grand Canyon dams 409, 410
    Grand Coulee Dam 387
    Hoover Dam 386
    Multiple-purpose movement 32
    Salmon and fisheries issues 408
    Teton Dam 391
    Third great river issue 110
    Urban growth 386-7
    Works projects develop 384
Hydropower 36, 62, 165, 237, 276
    See "mill power"
    Appleton, Wisconsin 36
    As repayment for irrigation 107
    At Niagara Falls 36

Bureau of Reclamation 131-2
Calculation of power..221
California 80
Central Valley of California................................... 321-2, 324, 325
Central Valley Project 341, 344
Colorado River..136, 144
Colorado River Storage Project 182
Columbia River...204
Comprehensive surveys of all navigable streams 120
Cooper, Hugh Lincoln... 114-6
Corps approval of projects...38
Corps of Engineers issue... 110, 111
Could be transmitted to user...84
Dams could now become multipurpose structures 83
Developed in Europe 36
Development in U.S...36
Eastwood, John S...79
Effects on navigation 38, 238
Federal Power Commission 113, 133
Federal Water Power Act 113
Glen Canyon Dam 183
Government regulation of...37
Grand Canyon dams..185
Grand Coulee Dam 208
Hetch Hetchy 35
Hoover Dam................................. 110, 145, 148, 152, 165-6, 178
Hoover Dam highly controversial...165
Hoover Dam power revenues 180
Issue of public power...111
Jorgensen, Lars 61
Keokuk...353
"Marshall Plan"...321
Metropolitan Water District's Colorado Aqueduct 171
Muscle Shoals 111, 117, 119, 133
Near Juneau, Alaska..61
Pre-World War I 38
Private facilities 37
Reclamation issue 110
Reclamation Project Act of 1939 269
Requiring dam owners to maintain and
    operate navigation facilities...38
Revenues at Hoover...144, 157

    Roosevelt Dam 99
    Rulings of Federal Power Commission 180
    Sale of power could now justify the cost of high dams 83
    Salmon Creek Dam 61
    Salt River Project 99, 102, 107
    Society embraced the great transforming
        event of electricity 83
    Tonto Dam 99
    Waterway developments 38

Hydropower
    Transforming the hinterland 384

Ickes, Harold 210

Idaho 29
    Anderson Ranch Dam 458
    Arrowrock Dam 107, 143, 171, 458
    Carey Act 26
    Colorado Doctrine 10
    Morrison-Knudsen 171, 334
    Owyhee Project 109
    Teton Dam 391

Imperial Irrigation District 142-3
    Desire for flood protection 142
    Desire for protection from Mexican interference 142
    Flood control could store water for irrigation 142
    Sought a flood control and storage dam 142
    Sought All-American Canal 142

Imperial Valley 130, 135, 138-44, 154-7, 160-1, 171, 180, 301, 386, 388, 405
    Diverted over 3 million acre-feet per year
        from Colorado River 161
    Flooding of the valley 135
    Floods affected development 135
    Huge tract of desert 135
    Required canal lining 405
    Required drainage system 405
    Salinity issues 405
    Salt-tolerant crops 405

Imperial Valley floods
    Affected storage on Colorado River 135

Indian dip-net fishery 385

Industrial Workers of the World 174

Industrialization of U.S. 2

Inland Waterways Commission 33-4
Iron Canyon Project 317-9
    Iron Canyon Dam 316
    Significant in planning preceding Central Valley Project 319
    Vicinity of Redding, California 316
Iron Canyon Project Association 316
Irrigation Survey 92
Izaak Walton League 393, 395
Jackson Report
    Flood Control Act of 1917 314
    Jackson, Thomas H. 314
Jackson, Thomas H. 314
Jakobsen, B. F. 72, 74
Jaquith, A. C.
    Refines trial-load analysis 73, 74, 462
    Johnson, Hiram 156
    Closely associated with the Boulder Canyon Project 154
    National Irrigation Congress (1893) 169
    Supported Boulder Canyon Project 154
    Swing-Johnson Acts 154
Johnstown flood 390
Jones Falls Dam 59
    Jorgensen, Lars 68, 70, 72, 146, 147
    Constant angle arch dam 61
    Lake Spaulding Dam 70
    Salmon Creek Dam 61
Justin, Joel D. 258, 269
    Embankment designer 241
    Hydraulic fill construction 244
Kaiser, Henry 171, 218
    Kanawha River 16, 374
    Huntington, West Virginia 374
    Roller gate dams 374
Kansas 29
    California Doctrine 9
    Kansas City 270, 292
    Kansas River 276
    Pick Plan of the Corps 269
    Pick-Sloan Missouri Basin Program 292
Kansas City 270
Kansas City District 237
Kansas River 273

Kaufmann, Gordon
　　Surface design/treatment of Hoover 151
Kaweah River 303
Kendall, Ray 256
Kennett Dam 319, 326
　　See Shasta Dam
Kentucky River 16
Keokuk and Hamilton Water Power Company
　　Hydroelectric dam 115
Keokuk Dam 91, 114-8, 194, 353
　　Cooper, Hugh 116
Kern River 303, 307, 311, 323
　　Divert water to Los Angeles 321
Kettner, William 155-6
Keuntz, Oscar O.
　　Columbia River 308 Report 204
Kincaid Act 144, 155
　　Resulted in Fall/Davis Report of 1922 144
　　Study storage dam and All-American Canal 143
Kings River 307, 310, 323, 330
　　Acreage limitation 346
　　Flood control 100 percent funded by federal government 346
　　Lake Buena Vista 303
　　Reclamation and Corps struggle over construction 346
　　U.S. Army Corps of Engineers 346
Kinzua Creek 371
Kinzua Dam 371-2
　　Nine thousand acres of Seneca land was taken 384
　　Seneca Indians lose land 371
Kittrell, Clark 255, 257, 259
Klamath Project 315
Kutz, C. W. 120
Laguna Diversion Dam 143
Lahontan Dam
　　Spillway repair 406
Lake Case
　　Flooded Sioux land 291
Lake Hodges Dam 80
Lake Itasca
　　Source of Mississippi 19
Lake Lynn 361
Lake Winnibigoshish 19, 20

Lane, E. W. 388
Lane, Franklin
  Secretary of the Interior 106
Larkin, T. B. 255
Lee's Ferry 152
Leighton, Marshall O. 34
Lock and Dam Number 15
  Completed 1934 374
Lodore Canyon...396
Long, Stephen H. 16
Los Angeles 110, 130, 160-5, 169, 171, 178, 301,
         ....315, 321, 386, 390
  Claims 550,000 acre feet of Colorado River water 161
  Colorado River Aqueduct Association..162
  Colorado River as water source 161
  Colorado River-to-Los Angeles aqueduct......................................161
  Filed claims to Colorado River 162
  Imperial Valley...161
  Los Angeles Department of Water and Power...............................162
  Los Angeles River...160
  Mulholland, William 160
  Owens River 161, 315
  Palo Verde 161
  St. Francis Dam..389
  St. Francis Dam failure 389
  State Water Project..347
  Threatened future water rights claims of other states 161
Los Angeles Department of Water and Power 162
Los Angeles River...160
Lower Columbia Fisheries Plan 226
Luce, Henry
  Fort Peck Dam 253
Lukesh, Gustave..204
Lux v. Haggin
  Kern River..311
  Upheld riparian water rights 311
  Wright Irrigation District Act...311
Madden Dam...364, 368
  Panama Canal 364
Magee, Clair..283
Magnusson, Carl E.
  Supported high Grand Coulee..215

567

Mahan, Dennis Hart
    Engineering textbook 12
Maillart, Robert 74, 83
Main Control Plan 224, 226
    Columbia River Comprehensive Plan 224
Mainstream dams
    Common in the twentieth century 89
    Low built across wide stretches of river 90
Marble Canyon Dam 409
Markham, Edward M.
    Chief of Engineers during Tygart Dam design 363
Marshall Plan
    Irrigation, navigation, power 321
    Kern River 321
    Large dam on upper Sacramento 320
    Moved toward large Central Valley Project 320
    Opposed by Pacific Gas and Electric 321
    Opposed by private power companies 321
    Stanislaus River 321
    Supplanted by other ideas using similar ideas 321
    Supported by California State Irrigation Association 321
    Water to Los Angeles and San Francisco 321
Marshall, Robert
    Marshall Plan 320
Martin, Charles 209-10
Martin, Clarence
    Grand Coulee construction 217
Mason City 218
Massive tradition of dam building
    Amount of material 58
    Analysis of the relationship between
        "gravity" and "arch" 59
    Built along a curved axis like an arch dam 58
    Dam design 49
    Distinction between massive and structural 58
    Large amounts of construction material 59
    Nineteenth-century gravity dam technology 59
    Often expensive 59
Matthes, Gerard 119
Maxwell, George 28
    Irrigation advocate 26
    National Irrigation Association 26

McAlpine, William
    Board for Tygart Dam construction 365-6
McCall Ferry Dam 91, 114-6, 118
    Construction method 115
    Cooper, Hugh 115
    Low gravity dam made of unreinforced concrete 115
    Structural type that became characteristic of
        mainstem dams 115
McCall Ferry Hydroelectric powerplant
    Susquehanna River 114
McCall Ferry Power Company 114
McKay, Douglas
    Approves Echo Park Dam 398
    Echo Park Dam issue 397
    Inherits Echo Park Dam issue 397
McNary, Charles 209-10, 217, 228
    Running mate with Wendell Wilkie 227
Mead, Elwood 108, 211, 217
    Approved work with small scale models 77
    At Grand Coulee site 213
    Columbia Basin Project 208
    Death in 1936 218
    Multiple arch dam studied 211
Mead, Warren J. 263-4
    Consults on Fort Peck Dam 363
    Fort Peck Dam 363
    Tygart Dam 363
Mechanization of agriculture 2
Meigs, Montgomery 115
Meramec Basin Project
    National Environmental Policy Act 411
Meramec Park Dam 411
Meramec River 411
Merced River 303
Merrill, William E. 17, 18
Merriman, Thaddeus 261, 263-4
    Consultant on Colorado River Aqueduct 263
    Consultant to Metropolitan Water District of Southern California 263
    Consultant to Tennessee Valley Authority on Norris Dam 263
Metropolitan Water District of
        Southern California 160, 163, 165, 169, 263
    Colorado River ambitions 179, 181

    Colorado River Aqueduct 181
    Could impose property taxes 165
    Established in 1927 using state legislation 164
    Imperial Valley...405
    Included many small cities 163
    Legality of...165
    Organized to build the Colorado Aqueduct 169
    Parker Dam 181
    Ruled constitutionally valid 165
    To build Colorado River Aqueduct 164
Mexico 1, 114, 135, 136, 139, 158, 425
    Colorado River salinity 405
    Salinity issues 405
    Treaty with United States (1944)..405
    Wellton-Mohawk Irrigation District 405
Miami Conservancy District 371
    Begins construction in 1918 357
    Dayton area 357
    Established in 1915 357
    Five large reservoirs...357
    Flood of 1922...57
    Floods of 1959 357
Middlebrooks, Thomas 259, 265, 268
Milk River Project..98
Mill power...83
    Dependent upon the vagaries of river flow....................................83
    Local 83
    Low dams..84
    Required mechanical transmission 83
Miller & Lux 307, 310, 311
    Ali Haggin..311
    Lux v. Haggin 311
    Riparian water rights...311
Milner Dam..26
Minneapolis..18-20, 115, 353, 374
Mississippi River 13, 19
    Basin 1
    Basin floods...356
    Comprehensive planning 470
    Cooper, Hugh L. 115, 116
    Lakes-to-the-Gulf..451
    Mississippi River Commission 18, 203, 238

  Missouri River 235
  Navigation 15, 17, 313
  Powell, J. Wesley, explores 24
  Private dams on 37
  Reservoirs recommended 19
  Survey of reservoir sites 26
  Wisner, George Y., work on 97

Mississippi River
  Flood control 388

Mississippi River flood of 1927 111
Mississippi River Power Company of Boston 116
Missouri River flood of 1881 288
Missouri Basin Interagency Committee 282
  Pick, Lewis A. 282
  Sloan, William 282
Missouri River 228, 235, 237, 245-6, 249, 269, 273, 275-7,
          .... 279-80, 292, 293, 411
  308 Report 121, 204, 236, 237
  Amount of water not remarkable 235
  Big, unique, and temperamental 235
  Effects of Pick-Sloan Missouri Basin Program on 292
  Flooding 26, 27, 269-70, 279, 283, 292
  Fort Peck Dam 245, 249, 251, 263, 268, 272, 461
  Garrison Dam 284, 288, 425, 462
  Indians lose lands to dams 385
  Lewis and Clark expedition 236
  Mapping of the river 236
  Missouri River Basin Compact 284
  Missouri River Division 283
  Missouri River States Committee 275, 279, 284
  Missouri Valley Authority 228, 273, 275-6, 278-80, 283
  Mouth near St. Louis 235
  Navigable below Yankton, South Dakota 235
  Oahe Dam 385, 425
  Pick, Lewis A. 269
  Pick Plan of the Corps 269, 270, 278, 288
  Pick-Sloan Missouri Basin Program 292, 424
  Sixteenth greatest watershed in the world 235
  Sloan, William Glenn 269
Missouri River 308 report
  Board of Engineers for Rivers and Harbors 238
  Chief of Engineers 238

  Construction of a huge dam at Fort Peck, Montana 237
  House Committee on Rivers and Harbors 238
  Mississippi River Commission 238
  Separate studies of each of
    the twenty-three major tributaries.................237
Missouri River Authority
  Supported by President Truman....................278
Missouri River Basin Compact................................284
Missouri River Division..283
Missouri River flood 1951 283
Missouri River States Committee 275, 281, 284
  Missouri River Basin Compact....................284
  Missouri River States Committee 279
  Recommends Missouri River Basin Compact 284
Missouri Valley Authority
  Hoover Commission opposition 281
  Magee, Clair supports..........................283
  Murray, James Edward 275
  Not to be jeopardized by Pick-Sloan program..........276
  Proponents hostile to Lewis A. Pick 283
  Rankin, John Elliot 283
  Supported by James Murray 281
Missouri-Souris reclamation project............................281
  Fort Peck reservoir water........................281
  Never funded................................287
  Project dropped 283
Monongahela River................16, 353-5, 362, 370, 371, 466-9
  Drought of 1930..............................361
Montana 29, 226
  Colorado Doctrine..............................10
  Delano Heights 255
  Early hydroelectric power........................36
  Fort Peck...................................255
  Fort Peck Dam 237, 239, 245, 253, 255-6, 424, 461
  Fort Peck Dam slide...........................256
  Fort Randall Dam 292
  Franklin D. Roosevelt visits Fort Peck Dam site........256
  Gallatin River................................235
  Garrison Dam................................282
  Gibson Dam..............................74, 462
  Jefferson River...............................235
  Lower Yellowstone Project......................451

    Madison River...235
    Milk River Project 98
    Missouri River flood control..292
    Missouri Valley Authority..275
    New Deal 255
    Origin of the Missouri River..235
    Park Grove 255
    Pick, Lewis A...269
    Pick Plan of the Corps 269, 270
    Square Deal..255
    Sun River 74
    Sun River Project..74
    Three Forks 235
    Wheeler (town of)..255
    Wilson (town of) 255
Moore, Douglas..256
Moore, George Holmes
    Gravity dam design..56
Morgan, Arthur E.
    Chairman of TVA board..362
Mormon Flat Dam..107
Mormon irrigation
    City Creek in Salt Lake City...22
    Precedent for later pioneers 22
    Small projects 22
Morris, Robert M. 371
Morrison-Knudsen 171
Movable dams..17
    Replacement..373
Muir, John 35, 36, 191
    Columbia as a wild river...191
    Hetch Hetchy 35
    Views on conservation 31
Mulholland, William..390
    Los Angeles/Owens Valley aqueduct...161
    Owens River 161
Multiple arch dams 79
    John S. Eastwood...79
    Least expensive designs...81
    Still in service 79

Multiple-purpose dams
- Early twentieth century 89
- Mission of Corps 1900 89

Multiple-purpose river development 32
Multipurpose debate 91
- Broaden historic missions 91

Mundt, Karl E. 385
Murray, James Edward 275, 276, 281
Muscle Shoals 37, 111, 113-4, 117-9, 133
- Burgess, Harry 117
- Cooper, Hugh L., assigned to design by Corps 118
- Corps hydroelectric plant a completely new venture 117
- Ford, Henry, bids to buy 114
- Ford withdraws bid offer 114
- Hugh L. Cooper & Co. 118
- Nitrate plant 117
- Raised issue of public power 111
- World War I caused national controversy 111

Muscle Shoals Dam
See Wilson Dam
- Abundant, inexpensive electric power 111
- Early name of Wilson Dam 37
- Nitrate production 111

Muscle Shoals Inquiry Commission 114
Muskingum River 16, 371
Muskingum Watershed Conservancy District
- Total of 14 flood control dams 371

MWAK
- Atkinson-Kier Company 213
- Consolidated Builders Incorporated (CBI) 219
- Mason, Walsh Construction Company 213

Nashville District 91
National Conservation Convention 132
National Dam Inspection Act (1972) 407
National Dam Inspection Program 407
National Environmental Policy Act (NEPA) (1969) 401, 411
- Citizen participation 401
- Council on Environmental Quality (CEQ) 401
- Environmental impact statements 401
- Environmental Protection Agency (EPA) 402
- NEPA process 401
- Tool to slow down or impede construction 401

National Industrial Recovery Act
    Bonneville Dam 194
National Irrigation Association 26, 28
National Park Service 183, 395, 396
    Opposed Echo Park Dam 396
National Parks Association 398
    Opposed Echo Park Dam 398
National Waterways Commission 34
National Wildlife Federation...393, 395
Native American
    Chippewa Indians affected..20
    Damages paid..20
    Fisheries...20
    Wild rice fields..20
Native Americans
    Concerned about Columbia River salmon 393
Navajo Dam
    Seepage 406
Navigation
    See River navigation 110
Navigation improvements
    General Survey Act 355
Nebraska 29, 269
    California Doctrine 9
    North Platte Project..98, 451
    Only state wholly in the Missouri Basin..276
    Pick Plan of the Corps 270
    Wyoming Doctrine..10
Nevada 26, 29, 157-8, 175
    Carson River Project...451
    Colorado River...136
    Hoover Dam.................... 109, 133, 148-9, 169, 177, 219, 386, 425
    Lahontan Dam...406
    Newlands Project..97, 451
    Truckee River Project 451
New Deal 133, 165, 180, 217, 253, 269, 306, 387, 424, 430, 453
New Deal, Montana 255
New Mexico...10, 26, 29, 135, 137
    Colorado River Compact 158
    Colorado River Storage Project 395
    Colorado River System 157
    Elephant Butte Dam...107, 143, 425, 461

    Navajo Dam 406
    Rio Grande River 451
    San Juan River 135
Newell, Frederick Haynes 25, 27-8, 30, 65, 92-107
    Association with Francis G. Newlands 28
    Becomes director 94
    Dismissed by Franklin Lane 106
    Drawn to the construction of huge structures 106
    Financial problems of Reclamation 106
    First Chief Engineer 30
    Huge structures as symbols of safety 106
    Huge structures as symbols of Reclamation's ability 106
    Irrigation in the United States 28
    Lingered on until 1914 106
    Original hopes and ambitions for Reclamation lost 106
    Original repayment terms should be enforced 105
    Political character of dam building 131
    Resignation 107-8, 206
    USGS, Chief Hydrographer 28
Newlands Act 29, 450
    See Reclamation Act
Newlands, Francis G. 26, 28-9, 33-4, 98, 358, 450, 452-3
    Waterways Commission 453
Newlands Project 97
Newlands waterways bill 452
Noetzli, Fred 72, 74, 76, 78-81
    Additional rigor and safety 80
    Analysis of Eastwood work 80
    Arch compression stresses 80
    Curved dam has greater resistance to overturning than
        straight dam 72
    Curved Dams as Cylinder Hoofs 71
    Death 211
    Education at the Federal Technical Institute in Zurich 78
    Mass did not mean safety but form,
        properly conceived, did 72
    Mountain Dell Dam studies 81
    On arch dams 70-2, 76-8
    Safety in gravity dams 71
    Stevenson Creek Test Dam 77
    Three types of curved dams 72
    Trial-load analysis 211, 366

Norris Dam...362
    Reclamation creates final plans for TVA 362
Norris, George 91, 203
    Opposes sale of Muscle Shoals..114
    TVA..91, 276, 358
North Dakota..10, 29
    California Doctrine 10
    Elbowoods 279
    Garrison Dam...278-9, 283-4, 287
    Missouri River flood control...292
    Missouri-Souris reclamation project.....................281, 284, 287
    Pick, Lewis A
        And Indians...279
North Platte Project...98, 425
Oahe Dam...277, 288
    Loss of Indian land 385
    Standing Rock Sioux Reservation lost land...........................385
Ohio River...14
    308 Report...121, 356
    Corps of Engineers..313
    Dams mostly for navigation...353
    Flooding history..356
    Floods..353
    Floods of 1935 and 1936 353
    Gallipolis locks and dam..374
    Historic dams in basin..466
    Long, Stephen H. 16
    Miami River flood (1913)...357
    Movable dams on...18
    Navigation...17, 313, 355
    Navigation after 1865 17
    Navigation modernized by 373
    Nine foot channel by 1929..372
    Nine foot channel complete 372
    Ohio River Commission 18
    Opposition to dams 17
    Shallow-draft steamboats..355
    Slackwater locks and dams 16
    Snags...16
    Statistics on the river...353
    Traffic increase 1955 to 1977 373
    Transportation on 353

    Tygart Dam 353
    William E. Merrill proposals 18
    Wing dams 16
Ohio River flood of 1907 356
Oklahoma 10, 29
    California Doctrine 9
O'Mahoney, Joseph C.
    Pick-Sloan Missouri Basin Program
        introduced to Senate 273-5, 279, 281
Oregon 29, 205, 208, 209
    Bonneville Dam 203, 210
    California Doctrine 10
    Corps created an outdoor hydraulic
        laboratory in Portland 196
    First hydroelectric dam was completed at Oregon City 458
    Hydroelectric power generation 36
    Hydropower network 226
    Klamath Project 315, 451
    Owyhee Dam 75, 109, 178, 425, 430, 463
    Umatilla Project 451
Orland Project 315
    East Park Dam 315
    Stony Creek 315
    Stony Gorge Dam 316
O'Sullivan, James
    Columbia Basin Commission 208
    Grand Coulee 208
Owens River 161, 315
Owyhee Dam 75, 82, 91, 109, 178, 199, 212, 425, 430-2, 463-4
    Heavy arch design 82
    Tallest dam in the world 430
Pacific Constructors Inc.
    Hire Frank Crowe to supervise work at Shasta Dam 334
Pacific Gas and Electric 36, 321
Packard, Fred M.
    National Parks Association 398
Page, John 331, 342
Page, R. J. B. 283
Palo Verde Valley 138, 157, 161
Panama Canal
    Gatun Dam 193
    Madden Dam 364

Park Grove, Montana 255
Parker Dam 181
Parker, Theodore
    Had rights to tainter gates 375
Pathfinder Dam 65, 67, 70, 91, 97, 116, 146,
                                                            184, 425, 431-2, 445, 455
Pelletreau, Albert
    Constant-angle arch dams 61
Phelan, James D. 35
Phoenix 21-2, 99-102, 181, 387
Physical testing of dams..76
Pick, Lewis A.
    Retired from Corps in 1953 287
    Returns to Missouri Basin..278
Pick Plan of the Corps
    Pick's plan was in place by 1964 291
Pick-Sloan Missouri Basin Program
    Corps and Reclamation differ on capacity
        of Garrison Dam 282
    Flood Control Bill of 1944..276
    Not to jeopardize Missouri Valley Authority 276
    Reybold, E. 271
    Young, Milton R. 279
Pinchot, Gifford 32, 35, 106
Pine Flat Dam 330
    Acreage limitation..346
    Built by Corps..346
    Reclamation and Corps struggle over construction 346
Pine River..19
Pit River 303
Pittsburgh 16, 18, 110, 353-4, 356, 362, 370-2
    Flood control..371
    Flood of 1907 ....110, 356, 358
    Flood of 1954...371
    Flood protection 362, 370
    Floods of 1935 and 1936 353
    Pittsburgh Flood Commission...356
    Protection of nine, high, flood control dams...................................354
Pittsburgh District 361, 363-4
Plank, Ewart G.
    Town manager at Fort Peck 256
Pokegama Falls 19

Political character of dam-building 131
Portland Hydraulic Laboratory of the Corps
    Construction design issues at Bonneville Dam 196
    Did some 100 model studies 196
    Function transferred to Vicksburg in 1982 196
    Kaplan turbine tests 195
Potter, William E.
    Integrates Missouri River dams through computers 286
Powell, John Wesley 23, 24, 30, 92
    1868 on Grand and Green 24
    1869 in Grand Canyon 24
    1869 on Colorado 24
    1871 in Grand Canyon 24
    1893 National Irrigation Congress 25
    Irrigation Survey 25
    Report on the Lands of the Arid Region of
        the United States 24
    Resigned from USGS 25
    Spokesman for development of the West's water resources 24
    Union Army 24
Powell, R. G.
    Meeting about Tygart delay 369
Profile of equal resistance 52-4, 56-7, 464
    Gileppe Dam 54
    Vyrnwy Dam 54
Progressive Era 31
    Conservation 31
Public Works Administration
    Cooke, Morris L. 364
    Grand Coulee allotment 216
Puls, Louis 184
Puzzolan cement 197
    Reduced heat of hydration 199
Pyle, Ernie
    Fort Peck Dam 253
Quaker Bridge Dam 89
Ralston, William 23, 307-8, 310
Rampart Dam 410
    Block salmon migration 410
    Cancelled by Lyndon B. Johnson Administration 410
    Electricity for six million people 410
    Inundate 11,000 square miles 410

　　　　Inundate waterfowl-producing habitat................................410
　　　　Yukon River..410
Rankin, John Elliot..283
Rankine, W. J. M.
　　　　Confirms work of de Sazilly's and Delocre...............................54
　　　　"Middle third"...54
Raritan River 308 report..121
Reclamation Act...26, 29, 32, 105, 180, 450
　　　　Depression caused rethinking of government projects...............132
　　　　June 17, 1902..29
　　　　Newlands Act..29
　　　　No provision for hydropower 32
　　　　Repayment of costs..29
　　　　Residents opposed increased construction costs.......................105
　　　　State governments would play no role in the program's
　　　　　　implementation 29
Reclamation Project Act of 1939..269
Red River Gorge
　　　　Dam opposed there 411
　　　　Red River Gorge Legal Defense Fund....................................411
Red River Gorge Legal Defense Fund..411
Refuse Act..20
Reilly, Thomas L...371
Reinforced concrete 83
　　　　Reactions to by designers 83
　　　　Slow design change 83
Report on the Physics and Hydraulics of the Mississippi River................111
Reybold, E... 272-275
　　　　Chief of Engineers 271
　　　　Pick-Sloan Missouri Basin Program......................................271
Rise of the steamboat..13
Ritter, J. R. 388
River and harbor projects
　　　　Corps flood control work 14
　　　　CVP authorized 1935..329
　　　　Effect of *Gibbons v. Ogden* case..14
　　　　Federal projects increase..17
　　　　No benefits to West..26
　　　　Value 1802-1823...13
River basins
　　　　Economic development plans..32

River navigation 6, 32-3, 38, 91, 113, 117, 119, 383-4, 387-8
    308 Report Tennessee 119-0
    308 reports 111, 120-1, 237, 356
    Attention shifts to other uses 400
    Bureau of Reclamation 269
    California controls water 317
    California Water Plan 324
    Colorado River 137, 181
    Columbia River 191-2, 196, 202, 204, 207, 209, 223
    Conflicts with water mills 6
    Corps assigned flood control 34
    Corps control navigation 317
    Corps focus in early Republic 13
    Corps reconcile power and navigability 38
    Corps responsibility in Nineteenth Century 110
    Costs for 308 reports 111
    Early 1800s Corps mostly removed snags and
        deepened channels 15, 16
    Early Corps work 12
    East coast rivers 5
    Federal nineteenth-century dams for navigation only 89
    Federal Origins 3, 4
    Federal policy 5
    Federal supervision of hydropower 113
    Federal Water Power Act of 1920 113
    Flood Control Act (1917) 314, 358
    Flood control conflicts 354
    Flood control responsibility added in twentieth century 110
    Gibbons v. Ogden 13
    Inland Waterways Commission 33, 34
    Interference of hydropower 37
    Interstate commerce 14
    Keokuk Dam 91, 353
    Mississippi River 13, 15, 19-20, 313, 353, 355
    Missouri River 235, 237-8, 269, 272, 274-5, 282, 291-2
    Modernization of the Ohio 372-3
    Monongahela River 361
    Navigation 371
    Ohio River 13-5, 17-8, 313, 353, 355, 372
    Ohio River 308 report 356
    Ohio River after the Civil War 17
    Ohio River slackwater system 1830s 16

    Pick's plan was in place by 1964 291
    Reclamation Project Act of 1939 269
    Refuse dumping in rivers 20
    Riparian rights and 6
    Rivers and Harbors Act of 1826 14
    Rivers and Harbors Act of 1852 16
    Rivers and Harbors Act of 1899 20
    Rivers and Harbors Act of 1909 34
    Rivers and Harbors Acts of 1890 and 1899 37
    Sacramento River 304, 312-4, 316-7, 320, 324, 329
    Snagboats 355
    Steamboat case 13
    Survey of rivers by Corps 91
    Tennessee River 359, 360
    TVA threat to Corps 356
    Tygart Dam 353, 363, 367
    Tygart Dam drought in 1939 371
    Tygart Dam floods in 1939 371
    Upper Mississippi River 373, 374, 376
    U.S. Army Corps of Engineers 89
    Wilson Dam 91
Rivers and Harbors Act
    First 14
Rivers and Harbors Act (1826)
    Model for Corps legislation 15
Rivers and Harbors Act (1866) 17
Rivers and Harbors Acts of 1890 and 1899
    Congress permits dams 37
    General Dam Act of 1906 38
    Rivers and Harbors Acts of 1890 and 1899 37
Rivers and Harbors Act (1899)
    Refuse Act 20
    Regulate dam construction 37
Rivers and Harbors Act (1920)
    Tennessee flood protection 119
Rivers and Harbors Act (1925)
    Corps and FPC to estimate cost of comprehensive
        surveys of river 120
Rivers and Harbors Act (1927)
    Authorized 308 reports 358
Rivers and Harbors Act (1935)
    Authorized Grand Coulee 217

Rivers and Harbors Committee..33
Roberts, William Milnor 17
    Navigation on the Ohio 17
Rock Island 374
Rock Island Dam...192
Rock Island Rapids 192
Rockwood, Charles
    Imperial Valley development 138
Roller gates 374
    American style submersible roller gate 374
    Description..374
Roosevelt Dam........................65, 84, 89-92, 98-107, 115, 146, 193, 365
    40 miles from the railroad..101
    Apache Trail.. 101-2
    Cement mill..102
    Completed in 1911 103
    Completely paid off in 1956 106
    Construction methods 102
    Contract out work 102
    Cost-overruns...104
    Costs exceeded original estimates 104
    Diversion tunnel...102
    Electric power system..104
    Heavy flooding...103
    Hydroelectric power component..132
    Hydropower plant to generate electricity for
        construction equipment..102
    Inflation..104
    John M. O'Rourke & Company..102
    Low bid of $1,147,000...103
    Masonry curved gravity design 100
    Mimic major structures of the East..100
    One of the West's best reservoir sites 106
    Power canal, Roosevelt Dam 102
        Contracting 102
        System 19-miles long 102
    Powerplant 107
    Prelude to the big dam era 84
    Reclamation builds own cement mill.......................................103
    Remote location 101
    Residents expressed dissatisfaction with Reclamation.............105
    Residents opposed increased construction costs......................105

    Residents tried to reduce repayment requirements 105
    Unrealistic expectations and inexperience 104
Roosevelt, Franklin D. 92, 119, 133, 135, 179, 180, 193-6, 202,
                .... 208-10, 214, 216, 222, 228, 238, 253, 259
                .... 264, 272-8, 306, 328-9, 361-2, 364, 373, 396
    At Grand Coulee site 213
    Central planning 193
    Columbia as economic vehicle 191
    Dedicates Bonneville Dam 191
    Dedicates Bonneville Dam in 1937 201
    Dedicates Hoover Dam 179
    Expresses concern about safety of Fort Peck after slide 259
    Muscle Shoals 119
    Supports Bonneville Dam 210
Roosevelt, Theodore 27-9, 31-3, 90-1, 100, 105,
                                  .... 132, 140-1, 204, 358
    Articulated multipurpose nature of rivers 90
    Broaden historic missions 91
    Conservation message to Congress (1907) 31
    Interest water conservation 29
    Largest steamboat parade in history 115
    Muir, John 35, 37
    National Conservation Convention at the White House 132
    Refuses to fight flooding of Imperial Valley 140
Ruckelshaus, William
    Director of EPA 402
Sacramento River 110, 302-5, 310, 313
    Dabney, T. G. ....
    Effects of hydraulic mining techniques 313
    Flood Control Act of 1917 358
    Flood control plan 313
    Navigable stream 313
    Navigation 313
    Navigation concerns 313-7, 319-21, 323-5,
                                  .... 329-0, 335, 338, 345
    Thomas H. Jackson 314
Sacramento River Valley 22, 302, 310, 313, 316-7, 319, 321
    Wheat culture 22-3
Sacramento/San Joaquin River Delta 303
Salmon Creek Dam 70
    Juneau, Alaska 61

Salt River 21-2, 89, 95, 99-100, 102, 107
    Watershed 107
Salt River Project 74, 97-100, 104-7
    Almost all land in private hands in 1903 100
    Completion of Roosevelt Dam not timely or economical 100
    Horse Mesa Dam 107
    Mormon Flat Dam 107
    One of Reclamation's first major projects 100
    One of the Reclamation Service's first five
        authorized projects 99
    Stewart Mountain Dam 107
    Struggle with the Water Users' Association
        over repayment 99
Salt River Valley 100, 103, 105
Salt River Valley Water Users' Association 99, 105, 106
Salton Sea 141-2
San Francisco
    1906 earthquake and fire 35
San Francisquito Valley
    St. Francis Dam 390
San Joaquin & Kings River Canal and Irrigation
    Company (SJ&KRC&IC) 307
    Brereton, Richard 307, 308
    Miller & Lux 307
    San Joaquin & Kings River Canal Company 307
San Joaquin & Kings River Canal Company
    Bensley, John 307
    Miller & Lux 307
San Joaquin River 1, 23, 76, 303-11, 319, 321,
                              323-4, 326, 342-4, 347
San Juan River 135
Sandy Lake 19
    Navigation lock 19
Savage, John (Jack) L. 75-6, 78, 146, 149-51, 184, 211,
                                  213, 219-20, 364
    Consulted for Tygart Dam design 364
    Design engineer 146
    Revised Weymouth's design of Hoover 149
Schley, Julian
    Chief of Engineers 264

Schuyler, James D. 99
    Hydraulic fill construction 241
    Sweetwater Dam 61
Scientific forestry practices..31
Shasta Dam 384
    Aggregate supplied under separate contract 333
    Concrete gravity dam..330
    Construction begins 1938 ....333, 341
    Construction methodology...335
    Construction schedule...341
    Conveyor belt system to deliver sand and gravel 333
    Embankment design considered 331
    Hydropower 341
    Keener, Kenneth on design...331
    Last bucket of concrete 1945 341
    Pacific Constructors Inc. receive contract................................334
    Powerhouse 326
    Primary construction contract in 1938.....................................333
    State seeks money from Federal Emergency Administration
        of Public Works..329
    Water storage begins 1944 341
Shingler, Donald G. 283
Shoshone Dam
    See Buffalo Bill Dam
Shoshone Project..98, 425
Shreve, Henry M. 15
Sierra Club 182, 395, 398, 411
Sierra Nevada...22, 35, 80, 156, 302, 303, 315
Sioux City 270
Six Companies Inc. 171
Slackwater
    Navigation.. 16-8, 372
Slater, Willis A. 81
    Stevenson Creek Test Dam 78, 81
Sloan, William Glenn...273-6, 281
    Missouri-Souris reclamation project..281
    Studies Missouri Basin for Reclamation 269, 273
Smoot, Reed
    Utah Power and Light Company 166
Snag removal 15
Snag removal and channel deepening
    Early Nineteenth Century 15

Snagboats 16
    Heliopolis 15
Snake River
    Fish kill due to low flows..405
South Dakota 10, 29
    Belle Fourche Dam 146
    California Doctrine 9
    Canyon Lake Dam 406
    Garrison Dam..462
    Missouri River 235, 237
    Missouri River flood control..292
Southern California Edison Company 76, 169
Southern Pacific Railroad 141
    Fights flooding of Imperial Valley 140
Spillway design...287
    Hathaway, Gail..287
Split Mountain Reservoir
    See Echo Park Dam..397
Spokane..206
Spring Valley Water Company..35
Square Deal, Montana...255
St. Anthony, Minnesota...19
St. Francis Dam.................................. 78, 370, 389-90, 406, 435, 437
    Failure 389
    William Mulholland...437
Stanislaus River 321
State Water Commission
    Created in California 1914...320
"Steamboat Case" 13
    Gibbons v. Ogden..13
Steiwer, Frederick..209
Stevenson Creek Test Dam 75-82, 389
    Conclusions reached by the committee in late 1927.....77
    Extensive bibliography ends the report 78
Stewart Mountain Dam..107
Stewart, Senator William
    Opposed Irrigation Survey..25
Stone and Webster of Boston
    Consulting electrical power engineers..........................116
Stony Gorge Dam...79, 316
    Large-scale, flat-slab, buttress dam................................316

Storage dams
    Usually marked by height 89
Storage of seasonal waters
    Kings River 346
    Pine Flat Dam 346
Straus, Michael
    Commissioner of Bureau of Reclamation 397
    Supported Echo Park Dam 397
Strawberry Valley Project 146
Structural dams 50
Structural engineering
    Begins with iron bridges 62
Structural tradition
    Analysis of the relationship between
        "gravity" and "arch" 59
    Depends upon its shape 58
    Distinction between massive and structural 58
    Flat-slab and multiple-arch buttress dams 58
    Multiple-arch dams 58
Sturdevant, C. L. 259
Styer, Wilhelm D. 365
    At Cove Creek Dam 362
    Board of consultants For Tygart Dam 365
    Pittsburgh District Engineer 362
    Tygart Dam 362-4
        Approved 363
        Design 365, 367
Suisun Bay 303
Sullivan, John L. 15
Sun River 74
Susquehanna River 114, 129
Swilling, "Jack"
    Canal 22
    Salt River Valley 22
Swing, Phil 166
    Championed Colorado River storage dam 156
    Colorado River Aqueduct Association 162
    Imperial Valley 154
    Most important and persistent political proponent of
        Hoover Dam 154
    National Irrigation Congress (1893) 169
    Pushed All-American Canal 154

- Pushed Boulder Canyon Project 154
- Swing-Johnson bills 154

Swing-Johnson Act 156
- The first version 156

Swing-Johnson Act of 1928 .... 156, 162
- Fear that California desired to control Colorado River 157
- Passed 169
- Propaganda from the "Power Trust" 169
- Strong financial commitment from Los Angeles 160

Swing-Johnson bills 154
Swing-Johnson bill 158
Taft, William Howard
- Appoints Richard Ballinger 37

Tainter gates
- Description 375
- Makes roller gates obsolete 375
- Originated in the United States 375
- Proved reliable in the upper Mississippi 376
- Tainter, Jeremiah Burnham, patented a design 375

Tainter, Jeremiah Burnham
- Patented Tainter gate design 375

Tedesko, Anton 83
Telford, Thomas
- Early earthworks designer 239

Tellico Dam
- Bell, Griffin 409
- Dam was built 409
- Endangered Species Act 409
- Exempted from Endangered Species Act 409
- Little Tennessee River 408
- Snail darter, an endangered fish 409

TVA 409
- U.S. Fish and Wildlife Service 409

Tennessee River 110-1, 114, 117, 119, 193, 277, 353, 356
- 308 report 121, 204, 358-61
- Corps studies on the river 92
- Cove Creek Dam 360
- Excluded from FPC reviews 113
- Muscle Shoals 37
- Norris Dam 425
- Snail darter 409
- Taken away from Corps 358

Tellico Project 408
Wilson Dam 89, 360, 363
Tennessee Valley Authority 3, 91-2, 119, 133, 203, 217, 222, 228, 263,
 ..... 275-7, 281, 291, 356, 359, 361-2, 408-9, 423
    Built 26 dams 383
    Father of TVA 91
    Fiske, Harold C. 91
Tennessee Valley Authority Act 362
Terzaghi, Karl 265, 268
    Embankment designer 240
    Hydraulic fill construction 244
Teton Dam
    Failure 391
    National Dam Inspection Act in 1972 407
Texas 243
    California Doctrine 9
    Denison Dam 470
Thayer, Sylvanus 12
The Design and Construction of Dams
    State of federal dams 1900 89
Thin arch dam 58
Tocks Island
    National Wild and Scenic Rivers System 411
Tocks Island Dam 411
Tonto Dam
    See Roosevelt Dam 99
Tonto Dam site 99-101
    Hydroelectric power component 99
    Remote location 101
Topock Dam 167
Topock, California
    Damsite 167
Totten, Joseph G. 15
Tourlotte, Eugene 257
Trial-load method of dam analysis 73, 184, 430-2
    Acquires name publicly 73
    Cat Creek Dam 73
    Considering a series of cantilevers 74
    Deadwood Dam 73
    Gibson Dam 73, 74
    Hoover Dam 73
    Horse Mesa Dam 74

>   Houk, Ivan 75
>   Immensely complex procedure 74
>   Maillart, Robert..74
>   Mulholland, William 79, 520
>   Owyhee Dam 73
>   Parker Dam 73
>   Savage, John L..75
>   Seminoe Dam..73
>   Stucky, Alfred 74

Truckee Project 451
>   *See* Newlands Project

Truckee-Carson Project.. 97, 451
>   *See* Newlands Project

Truman, Harry S. 228
Tucson 181
Tulare Lake 323
Tulare River 307, 323
Tule (or Tulare) River 303
Tuolumne River 35, 303
TVA
>   *See* Tennessee Valley Authority

Twin Falls Canal 26
Two types of buttress dams
>   Flat slab..79
>   Multiple arch 79

Tygart Dam 353-4, 356, 359, 361, 363-4, 369-71, 454, 466-7
>   Architectural treatment by Paul Cret 363
>   Built for flood control and navigation..363
>   Completed 1938 370
>   Contract let..365
>   Contractor was Snare Company 365, 367, 369
>   Cooke, Morris L..364
>   Costing out benefits 371
>   Costs of project 370
>   Cove Creek Dam was the precursor 360
>   Designs 362
>   Design 362-3, 365-6, 368
>   Design dimensions...367
>   Flood of 1912...367
>   Floods in 1939 371
>   Foundation and spillway design 363
>   Growden, James, concerns..369

    Hydraulic studies were pioneering 368
    In West Virginia on Tygart River 362
    Largest Corps dam east of the Mississippi 363
    Mixing of design and construction got out of hand 369
    Monongahela River..371
    Morris, Robert M. 371
    Outlets had steel liners 368
    Powell, R. G., meeting about delays in project.............................369
    Project approved 363
    Public Works Administration 363
    Reclamation creates final plans for TVA 362
    Reilly, Thomas L..371
    Reservoir increased in size 363
    Spillway tests at the Carnegie Institute of Technology................365
    Supports navigation in 1939 371
    Tygart River 362
    Used Norris Dam designs 362
    Wellons, Charles 364
    Work begins in 1934 362
Tygart River 362
Tyler, M. C. 261, 264, 269
U.S. Army Corps of Engineers 3, 8, 11-13, 15, 16, 21, 26, 30, 33-4,
                           ....37-8, 49, 63, 84, 89, 118-21, 191, 193-6,
                             ......205-6, 208, 217, 223, 228, 271-2, 282-3,
                                        ....356, 360, 365, 421

    308 Report Tennessee River 359
    308 Reports...204, 237, 241, 356
    Anadromous fish 394, 407, 408
    Authorized on Tennessee 119
    Board of Engineers for Internal Improvements 15
    Bonneville Dam 194, 196, 197, 199-202, 204, 210, 222
    Build locks on the Ohio 355
    Builds Pine Flat Dam..346
    California hydraulic mining...313-4
    Central Valley of California 306, 308, 328
    Central Valley Project 329
    Chittenden, Hiram...26
    Civil works tradition 13
    Columbia River focused on navigation and power.....................223
    Combination of roller and tainter gates 374
    Cooper, Hugh L. 118
    Cove Creek Dam..359-362

Criticism of 408
Dam No. 4 374
Dam No. 24 375
Dam Safety Assurance Program 407
Decidedly American style submersible roller gate 374
Design issues nineteenth and early twentieth century 90
Designed larger tainter gates 375
DeVoto, Bernard 397
Elliptical tainter gates 374
Engineering department of the federal government 14
Environmental Advisory Board (EAB) 403
Environmental community 403
Environmental Resources Branch 403
Establishment 11
Evolves dam gate design 374
Favors private power 359
First model of hydraulics in a major, multipurpose,
    mainstream dam 196
First used tainter gates (1889) 375
Fish 226
Fisheries-Engineering Research Laboratory 1955 407
Flood control 15, 371
Flood control 100 percent funded by federal government .. 346
Flood Control Act of 1917 .... 314, 358
Flood control and non-structural alternatives 403
Flood control dams to protect Pittsburgh 354
Floods of 1935 and 1936 373
Fort Peck Dam 237-8, 241, 244, 249, 250,
                                         .... 252, 255-6, 263-4, 268
Fort Peck Dam slide 258-67
Fort Randall Dam 290
Frederick Snare Corporation 365
French engineering tradition 11, 12
Friant Dam 329
Garrison Dam 278-281, 287, 288
Garrison Dam and Indians 279-281
Gas supersaturation 407
Gatun Dam 244
General Dam Act of 1906 38
General Survey Act 14
Genesis of the Pick-Sloan Plan 272-6
Grand Coulee Dam 208

Grand Coulee region 192
Grant, U. S., III 396
Hydropower and floods bring
      Corps into big dam business 111
Inland Waterways Commission 33-4
Irrigation survey 26
"Jackson Report" 314
Kansas City District 237
Keeping anadromous fish in their natural habitat 408
Kennett Dam 328, 329
Kentucky's Red River Gorge 411
Landowners in the King River watershed 346
Largest dam east of the Mississippi 363
Leader in embankment dams 49
Levees 89-92, 110-1, 114-5, 117, 356
Locks on Ohio approved 1879 355
Loses control of Tennessee River 358
Madden Dam 364, 368
Maintaining the navigability of major streams 313
Mead, Warren 363
Meramec Park Dam 411
Miami Conservancy District 371
Mississippi and Ohio Rivers 313
Mississippi flood control 356
Missouri River 237
Missouri River Division 283, 285
Muscle Shoals 117-8
Muskingum Watershed Conservancy District 371
Nashville District 91
National Environmental Policy Act 401
National Marine Fisheries Service 408
Navigation 13, 15, 372
Navigation dams 236
Navigation projects 14
New flood control mission 354
Nine foot channel 272
Nitrogen Task Force 408
Norris Dam 359
Obstructions to navigation 38
Ohio Basin 354
Ohio River 372-374

Ohio River converted to nine-foot navigation channel by 1929	355
Ohio River navigation	355
Opposes multiple-purposes	34
Opposition arguments	395, 396, 400
Panama Canal	364
Pick, Lewis A., Chief of Engineers	281
Pick-Sloan Missouri Basin Program	271, 276-278, 280, 283, 290, 292
Pickstown	290
Pine Flat Dam	346
Pittsburgh District	361-2
Projects controversial	15
Projects stopped	401
Proposed consolidation with Reclamation	281
Rampart Dam	410
Red River Gorge Legal Defense Fund	411
Reestablishment	11
Reybold, E.	271
Rising environmental criticism	402, 403
Rivalry between Corps and Reclamation over Grand Coulee	208
Riverdale	278
Rivers and Harbors Act (1826)	14
Rivers and Harbors Act (1852)	16
Roller gates	374
Run-of-the-river dams	89
Sacramento River	313-4
Snag and channel work	15
Store water for navigation	19
Supports federal construction of Shasta Dam in 1933	328
Survey of the Miami River	359
Surveys of navigable stream	120
Tainter gates	375, 376
Tennessee Valley studies	92
Traditional mission of river navigation	354
TVA	362
Tygart Dam	353, 354, 362, 363, 365-371
Upper Mississippi River	18-20, 373-375, 377
Waterways engineering	15
Woods, Rufus	208

U.S. Bureau of Reclamation 3, 8, 21, 25-6, 29, 30, 32, 34, 49,
................64-5, 73, 76-9, 82, 84, 90-2, 95,
................97-8, 211, 228, 314-5, 395, 400-3
   Acquires hydropower responsibilities 98
   Acreage limitation............314, 346-7
   Arrowrock Dam 171
   Authorized as bureau in 1902 92
   Banks, Frank A. 214
   Bartlett Dam............212
   Bashore, Harry............271
   Begins to fill Hoover Dam 178
   Bonneville Power Act 222
   Bridge Canyon Dam 185, 410
   Brower, David............398, 410
   Buy-back the Central Valley Project by California............347
   Central Arizona Project............410
   Central Valley Project 304-5, 312, 329-30, 347
   Colorado River Board 150
   Colorado River flood control 132
   Colorado River planning............143-5, 182
   Colorado River Storage Project 395
   Colorado River studies (1902)............142
   Columbia Basin............206, 208
   Columbia Basin Commission 211
   Columbia River focus on irrigation and dams............223
   Columbia Valley Authority 217, 228
   Conconully Dam............243
   Construction engineers 95
   Construction of Hoover Dam............177-179
   Consulting engineers............95
   Consulting "engineering boards"............95
   Cost-overruns............98
   Creates final plans for Norris Dam for TVA............362
   Crowe, Frank 171, 174
      At Shasta Dam 334
   Dam safety programs 392
   Darland, A. F., consulting engineer............211
   Davis Dam 199
   Depression caused rethinking of government projects............132
   Design of Hoover Dam............148-151, 171, 177
   DeVoto, Bernard 397
   Different authorities than those of Corps of Engineers 272

Dinosaur National Monument	182
District engineers	95
Dominy, Floyd E.	410
Downey, Sheridan	346
Early built massive embankment dams	146
East Park Dam	315
Echo Park Dam	182, 396
Engineering aides	95
Engineering assistants	95
Fall/Davis Report	144
First two decades electric power production ancillary to irrigation	132
Focus on Boulder Canyon	145, 146
Fresno	315
Friant Dam	331, 342
Garrison Dam	282, 284, 287
Glen Canyon Dam	183-4, 398
Grand Canyon	185
Grand Canyon dams	185
Grand Coulee Dam	212-4, 218-9, 221-2, 387
Grand Coulee region	208
Hoover Dam construction	211-2
Hoover Dam hydropower	166
Hoover Dam, Industrial Workers of the World	174
Hoover Dam planned as gravity masonry dam	146
Hoover Dam provides municipal water	180
Hoover Dam thin arch design considered	146
Hungry Horse Dam	199
Ideal of promoting the family farm	131
Interests gradually expanded	132
Interests of large-scale farmers	160
Iron Canyon Dam	317
Iron Canyon Project	318, 319
Iron Canyon Project Association	316-7
Issue of lower Colorado River development	143
Kaufmann, Gordon designs Hoover Dam's appearance	151
Kennett Dam	330
Kincaid Act	144
Laguna Diversion Dam	143
Lahontan Dam	406
Large-scale farmers in the Imperial Valley	160
Lengthy delays in project completion	98

Looks at Columbia Basin..192
Marble Canyon Dam...185, 410
Mid-1920s began to foster a range of benefits.........................132
Mission differs from Corps..89
Mission evolved into planning river basins..............................132
Missouri River States Committee...275
Missouri-Souris Project never was funded...............................287
Missouri-Souris reclamation project...281
Missouri Valley Authority..273, 277
Multiple arch dam at Grand Coulee...212
Multiple arch design at Shasta rejected.....................................331
Multiple purpose planning...269
Multipurpose dams did not derive from some master plan........132
Multipurpose projects..132
Navajo Dam...406
Okanogan Project...243
Opposes state safety laws during
        Hoover Dam construction..175
Original focus on irrigation...98
Orland Project..315
Owyhee Dam..212
Parker Dam...181
Pick-Sloan Missouri Basin Program..............................275-6, 282
Pine Flat Dam...346
Planning engineers..95
Plans for Missouri River..271-3
Plans Norris Dam for TVA...362
Pre-1920s hydropower...165
Primary construction contract..333
Promoting the family farm...160
Pueblo Dam..199
Reclamation Act...358
Reclamation and Corps needed to cooperate
        on the Missouri..273
"Reclamation law"..347
Reclamation Project Act of 1939...269
Reclamation Service held a conference of
        engineers at Ogden..65
Reclamation supported Six Companies, Inc..............................174
Regulation of lower Colorado River..388
Repayment issues...160

  Rivalry between Corps and Reclamation
    over Grand Coulee 208
  Rivers and Harbors Act (1937) 329
  Safety Evaluation of Existing Dams 407
  Salmon issues 394
  Savage, John L. (Jack) 211
  Shasta 331
  Shasta bids 333
  Shasta Dam 330, 333
    Aggregate 333
    Primary construction contract 333
  Shift of energies to goals beyond agriculture 132
  Siting of Hoover Dam 152
  Six Companies 174
  Sloan Report 273-4
  Sloan, William Glenn 269
  Slowly evolved 132
  Stony Creek, California 315
  Stony Gorge Dam 316
  Studied low multiple-arch dam for Grand Coulee 211
  Subsidize Columbia Basin Project with
    Grand Coulee electricity 222
  Supervisory engineers 95
  Tensions between municipal and agricultural water 160
  They Would Rule the Valley 346
  Trial-load method 49, 74
  Trial-Load method of design at Hoover 149
  TVA 362
  Walter, Raymond F. 330
  Wedded to massive forms 211
  Weymouth Report 147
  Young, Walker 171
U.S. Fish and Wildlife Service
  Listed the snail darter as an endangered species 409
U.S. Geological Survey 92
  Created 24
  Irrigation Survey 25, 92
  Irrigation Survey closed 25
  To look at western irrigation 23
U.S. Reclamation Service
  See Bureau of Reclamation 82
  Conflict between form and mass 64

       Detailed analytic study of dams from
the structural perspective 64
       Renamed the Bureau of Reclamation 21
       Storage dams 90
       Structures were for storage 89

Uplift 57
       Austin, Pennsylvania 57
       Safety of gravity dams 57

Upper Mississippi Basin Commission
       Coordinated planning 376

Upper Mississippi River
       Dam No. 24 375
       Environmental management 376
       Gigantic laboratory for hydraulics,
structures, mechanisms 376
       Headwater reservoirs 19
       Keokuk Dam 353
       Lock and Dam Number 15 374
       Minneapolis-St. Paul benefits 19
       Navigation improvements 18
       Nine foot channel approved in 1929 373
       Shipping increases 1939-1976 376
       Twenty-seven dams by 1940 374

Utah 26, 29
       Central Utah Project 396
       Colorado Doctrine 10
       Colorado River 135, 137, 157
       Colorado River Storage Project 395
       Echo Park Dam 182, 396, 426
       Glen Canyon Dam 463
       Glen Canyon Dam and reservoir 425
       Green River 135, 182
       Mormon colonization 22
       Reclamation Service held a conference of
engineers at Ogden 65
       Rudimentary Mormon irrigation 22
       Strawberry Valley Project 146, 451
       Utah Construction Company 171
       Yampa River 182

Utah Construction Company 171

Van Faasen, Jerold
       Reminiscence of Fort Peck Dam slide 1938 256

Van Stone, Nelson ... 256
Vanport ... 223
Variable-radius arch dams ... 61
Verde River ... 212
Walcott, Charles D.
    Director of the U.S. Geological Survey ... 95
    Director of the U.S. Reclamation Service, 1902-1907 ... 95
Walker, John S.
    Spillways and other dam cross sections ... 90
Walla Walla District ... 227
Walter Bouldin Dam
    Failed in 1975 ... 406
Walter, Raymond F. ... 211, 213, 219, 330
Warren, Francis E. ... 26-7
Washburn, William D. ... 19
Washington ... 29, 36, 205, 215, 226
    California Doctrine ... 9
    Columbia Valley Authority ... 278
    Ephrata ... 208
    Grand Coulee Dam ... 203, 463
    Hanford Reservation ... 387
    Okanogan Project ... 243
    Richland ... 387
    Spokane ... 206
    Wenatchee ... 206
Water law
    California Doctrine ... 9
    Colorado Doctrine ... 10
    Evolution of ... 5
    Evolution of in the West ... 7
    Prior appropriation ... 10
    Wyoming Doctrine ... 10
Water Power Act (1920) ... 110, 358
Water rights legal systems
    California Doctrine ... 9
    Colorado Doctrine ... 10
    Wyoming Doctrine ... 10
Watkins, Lewis H.
    Tennessee River 308 report ... 358-9
Weaver, Theron ... 224

    Northwest Division of Corps 223
Weeks, John W.
    Announced government would sell Muscle Shoals 114
    Secretary of War 113
Wegmann, Edward 54, 72, 89-90, 122, 166, 244, 268
    Cavitation 90
    Considered only masonry gravity dams 89
    He did not treat earth embankments 89
    New Croton Dam 54
    Quaker Bridge Dam 54
    Spillways and other dam cross sections 90
    State of federal dams 1900 89
    The Design and Construction of Dams 54
    The Design and Construction of Dams (1888) 54, 89
    Uplift pressure 90
Wellton-Mohawk Irrigation District
    Saline drainage diverted 405
Wenatchee 206
West Penn Power Company 361
Westergaard, H. M. 263
    Earthquake effects on dams 390
    Stevenson Creek Test Dam 78
Weymouth Report 147-9, 151, 153
Weymouth, Frank E. 76, 146-9, 1513
    Chief Engineer 146
    Proposed construction method for Hoover Dam 149
Wheeler, E. T. 65-67, 70, 74
Wheeler, Montana 255
Wheeler, Raymond A. 223, 224
Whipple, William, Jr. 223, 226-228
    Fisheries issues 226
    Flood of 1948 223
    Hydropower network 226
    Main Control Plan 224
    Portland District of Corps 223
    Walla Walla District 227
White River 135
Widstoe, John 108
Wilbur, Ray Lyman 133, 169, 171, 179
    Hoover Dam 171
    Names Hoover Dam 133, 179
    Power leases for Hoover 169

Wild and Scenic Rivers Act of 1968
- River preservation ... 406
- Wilderness Society ... 182, 395
- Wiley, A. J. ... 72
- Williams, Gardiner ... 61
- Wilson Dam ... 84, 89-92, 111-2, 114-5, 117-8, 122, 133, 193-4, 203, 239, 255, 277, 359-60, 363
  - Bid from Henry Ford to buy ... 114
  - Cavitation ... 90
  - Its crucial problems ... 90
  - Locks opened ... 122
  - Prelude to the big dam era ... 84
  - Pure gravity structure ... 90
  - Raised many questions ... 113
  - Taken away from Corps ... 119
  - Two nitrate plants ... 112
  - Work begins in 1918 ... 112
  - Work completed in 1925 ... 112
- Wilson, Montana ... 255
- Wilson, Woodrow ... 38, 91, 106, 111, 117, 396, 453
  - Muscle Shoals ... 111, 117
- Wing dams ... 115, 355
  - Improve channels ... 16
  - Navigation ... 16, 18
- Wisconsin River ... 19
- Wisner, George Y. ... 65, 68, 70, 92, 96-7, 464
- Woods, Rufus ... 206, 211
  - Grand Coulee ... 208
- Wozencroft, Oliver
  - Imperial Valley development ... 138
- Wright Act Districts
  - California ... 27
- Wright, C. C.
  - Irrigation advocate ... 27
- Wright Irrigation District Act
  - Authorized irrigation districts in California ... 311
- Wyman, Theodore ... 237
  - Fort Peck Dam ... 237, 252
  - Missouri River 308 report ... 237-8, 252, 270
- Wyoming ... 10, 26, 29, 135-7, 146, 157, 395-6
  - Buffalo Bill Dam ... 65, 425, 431
  - Buffalo Bill Dam ... 459

    First state engineer 108
    Green River 135
    Hiram Chittenden survey 26, 27
    North Platte Project 98, 451
    North Platte River 157
    Pathfinder Dam 65, 146, 425, 431, 464
    Shoshone Dam 146
    Shoshone Project 98, 451
    Wyoming Doctrine 10
Wyoming Doctrine 10
Yampa and Green Rivers
    Echo Park Dam 182
Yampa Canyon 396
Yampa River 135
Yosemite National Park 34-5
Young, C. C. 326
Young, Milton R. 279
Young, Walker (Brig) R.
    Project Engineer at Hoover 146, 171, 388
Yuba River 303
Yuma, Arizona 143
Yuma Project 315
Zola Dam 59, 60
Zola, Francois 59

www.ingramcontent.com/pod-product-compliance
Lightning Source LLC
Chambersburg PA
CBHW080226180526
45167CB00006B/2231